SB
945
.F8
F78
1989

Fruit flies.

$188.75

FRUIT FLIES
THEIR BIOLOGY, NATURAL ENEMIES AND CONTROL

World Crop Pests

Editor-in-Chief
W. Helle
University of Amsterdam
Laboratory of Experimental Entomology
Kruislaan 302
1098 SM Amsterdam
The Netherlands

Volumes in the Series

1. Spider Mites. Their Biology, Natural Enemies and Control

   Edited by W. Helle and M.W. Sabelis
   - A.   1985   xviii + 405 pp.   ISBN 0-444-42372-9
   - B.   1985   xviii + 458 pp.   ISBN 0-444-42374-5

2. Aphids. Their Biology, Natural Enemies and Control
   Edited by A.K. Minks and P. Harrewijn
   - A.   1987   xx + 450 pp.   ISBN 0-444-42630-2
   - B.   1988   xix + 364 pp.   ISBN 0-444-42798-8
   - C.   1989   ISBN 0-444-42799-5

3. Fruit Flies. Their Biology, Natural Enemies and Control
   Edited by A.S. Robinson and G. Hooper
   - A.   ISBN 0-444-42763-5
   - B.   ISBN 0-444-42750-3

4. Armoured Scale Insects. Their Biology, Natural Enemies and Control
   Edited by D. Rosen
   - A.   ISBN 0-444-42854-2
   - B.   ISBN 0-444-42902-6

5. Tortricoid Pests
   Edited by L.P.S. van der Geest and H.H. Evenhuis

World Crop Pests, 3A

FRUIT FLIES THEIR BIOLOGY, NATURAL ENEMIES AND CONTROL

Volume 3A

Edited by

A.S. ROBINSON

Research Institute ITAL, Wageningen, The Netherlands

G. HOOPER

Australian Plague Locust Commission, Department of Primary Industries, Barton, Canberra ACT 2600, Australia

ELSEVIER
Amsterdam – Oxford – New York – Tokyo   1989

ELSEVIER SCIENCE PUBLISHERS B.V.
Sara Burgerhartstraat 25
P.O. Box 211, 1000 AE Amsterdam, The Netherlands

Distributors for the United States and Canada:

ELSEVIER SCIENCE PUBLISHING COMPANY INC.
655, Avenue of the Americas
New York, NY 10010, U.S.A.

ISBN 0-444-42763-5 (Vol. 3A)

Printed in The Netherlands

Preface

The past two decades have seen a remarkable upsurge in the volume of research being done on fruit flies. In the mid-1960s there were approximately 30 scientists engaged in independent research on these insects. Today there are well over 200, representing more than 50 countries.

One of the first formal international meetings of fruit fly specialists took place in Rome in 1968. It marked the formation of the Working Group on Fruit Flies established under the auspices of the International Biological Programme. It was particularly significant because, in addition to assembling representatives from different countries, it brought together, effectively for the first time, workers on the two major species groups of the Family Tephritidae, the temperate and tropical forms. The working group meetings and associated I.B.P. activities helped to establish a tradition of collaboration among fruit fly workers which persists to this day, and which has surely facilitated the compilation of this impressive compendium on fruit fly research.

In recent years, contact and collaboration between fruit fly workers have been sustained and nurtured by frequent international meetings, but an additional and most effective cohesive force has been the news bulletins which have been produced by Dr Ernst Boller, of the Swiss Federal Research Station, and sent regularly to fruit fly workers throughout the world. I believe that our gratitude to Dr Boller for this valuable service should be put on record.

The current burgeoning of fruit fly research takes place during a period when demand for fruit is increasing and when fruit production industries, as a result are growing substantially in importance in many parts of the world. In western countries, dietary patterns are changing progressively towards unprocessed or 'natural' foodstuffs, particularly those which, like fruit and vegetables, have a high natural fibre content. Many developing countries, in response to this increasing demand are moving towards the expansion of their fruit industries. Unfortunately, it is almost universally the case that these countries have indigenous populations of serious pest tephritids which consistently frustrate both production and marketing of their fruit.

The solution of the fruit fly problems of third world countries is perhaps the major challenge currently facing the world's fruit fly specialists, and although some impressive initial programmes have been commenced, for example the campaign against medfly in Mexico and Central America, an immense amount remains to be done. Eradication technologies, such as the sterile insect technique and, to a lesser extent, male annihilation, have received heavy emphasis in recent years. Far less work has been done on procedures which involve co-existing with fruit flies, such as integrated management within individual orchards, or 'area control' methods which seek to suppress breeding populations over extensive areas and to keep them at levels below economic pest status. Yet either of these procedures may be a more appropriate strategy than eradication in many geographic or socio-economic situations.

Throughout most of the world, the front line of our attack on fruit flies is still the harrassed orchardist with his cans of insecticide. It is a sobering thought that despite the huge volume of research that has been done, and the immense amount of information that has been elicited, the weapons we are able to place in his hands have changed remarkably little in the past two decades.

The time is certainly ripe to take stock of this abundance of information which continues inexorably to accumulate. These volumes, which comprise a most comprehensive series of summaries and reviews of the current state of our knowledge, prepared by experts in the various fields of research, serve that purpose admirably. They will undoubtedly become an authoritative reference work for the student and the professional practical entomologist as well as for the research scientist.

M.A. BATEMAN

Contents

Contributors to this Volume

A. AVERILL
New York State Agricultural, Experiment Station, New York 14456, U.S.A.
B. BARTLETT
c/o Graeme Ramsay, D.S.I.R., Private Bag, Auckland, New Zealand
M.A. BATEMAN
6, Prince Alfred Parade, Newport, New South Wales 2106, Australia
T. BURK
Creighton University, Department of Biology, Omaha, Nebraska 68178, U.S.A.
E. BUSCH-PETERSEN
IAEA Laboratories, A-2444 Seibersdorf, Austria
R.T. CUNNINGHAM
USDA/ARS, Tropical Fruit and Vegetable Research Laboratory, P.O. Box 4459,
    Hila, Hawaii 96720, U.S.A.
R.A.I. DREW
Entomology Branch, Department of Primary Industries, Meiers Road, In-
    dooroopilly, Queensland 4068, Australia
D. ENKERLIN
Inst. Tech. de Estudias Supenoves de Monterrey, Sucursal de Correos J., 64849
    Monterrey, N.L., Mexico
P. FIMIANI
Cattedra di Entomologia Agraria, Universita degli Studi de Basilicata, via N.
    Sauro 85, 85100 Potenza, Italy
P. FISCHER-COLBRIE
Bundesanstalt für Pflanzenschutz, Trunnerstrasse 5, A-1021 Wien, Austria
B.S. FLETCHER
CSIRO, Division of Entomology, Entomology Research Station, 55 Hastings
    Road, Warrawee, New South Wales 2074, Australia
R.H. FOOTE
Box 166, Lake of the Woods, Locust Grove, Virginia 22508, U.S.A.
L. GARCIA R.
Inst. Tech. de Estudias Supenoves de Monterrey, Sucursal de Correos J., 64849
    Monterrey, N.L., Mexico
P.D. GREANY
USDA/ARS, Insect Attractants, Behavior & Biology Research Laboratory,
    P.O. Box 14565, Gainesville, Florida 32604, U.S.A.
D.L. HANCOCK
5 Northampton Crescent, Hillcrest, Bulawayo, Zimbabwe
E.J. HARRIS
USDA/ARS, Tropical Fruit and Vegetable Research Laboratory, P.O. Box 2280,
    Honolulu, Hawaii 96804, U.S.A.

G.H.S. HOOPER
Austrialian Plague Locust Commission, Dept. Primary Industries, Barton,
   Canberra, Australian Capital Territory 2600, Australia
D.J. HOWARD
Biology Department, Museum of Northern Arizona, P.O. Box 720, Flagstaff,
   Arizona 86001, U.S.A.
O.T. JONES
Biological Control Systems Ltd., Treforest Industrial Estate, Treforest, Mid
   Glamorgan CF37 5SU, Great Britain
V.C. KAPOOR
Department of Zoology, Punjab Agricultural University, Ludhiana 141004,
   India
B.I. KATSOYANNOS
University of Thessaloniki, Lab. Applied Zoology and Parasitology, GR-540 06
   Thessaloniki, Greece
J. KOYAMA
Shikoku National Agricultural Station, Zentsuji, Kagawa 765, Japan
A.C. LLOYD
Entomology Branch, Department of Primary Industries, Meiers Road, In-
   dooroopilly, Queensland 4068, Australia
F. LOPEZ M.
Inst. Tech. de Estudias Supenoves de Monterrey, Sucursal de Correos J., 64849
   Monterrey, N.L., Mexico
P. MADDISON
D.S.I.R., Entomology Division, Private Bag, Auckland, New Zealand
B.E. MAZOMENOS
Democritos, Nuclear Research Centre, Aghia Paraskevi, P.O. Box 60228, Att-
   iki, Athens, Greece
A. MEATS
School of Biological Sciences, Zoology Building, University of Sydney, Sydney,
   New South Wales 2006, Australia
J.L. NATION
Department of Entomology, University of Florida, 202 Newell Hall, Gaines-
   ville, Florida 32611, U.S.A.
A.L. NORRBOM
Systematic Entomology Laboratory, USDA, c/o U.S. National Museum, NHB
   168, Washington, DC 20560, U.S.A.
R.J. PROKOPY
Department of Entomology, University of Massachusetts, Amherst, Massach-
   usetts 01003, U.S.A.
M.J. RICE
Department of Entomology, University of Queensland, St. Lucia, Queensland
   4067, Australia
B.D. ROITBERG
Center for Pest Management, Department of Biological Sciences, Simon Fraser
   University, Burnaby, British Columbia V5A 1S6, Canada
J. SIVINSKI
USDA/ARS, Insect Attractants, Behaviour and Basic Biology Res. Lab., P.O.
   Box 14565, Gainesville, Florida 32604, U.S.A.
P.H. SMITH
CSIRO, Division of Entomology, P.O. Box 1700, Canberra City, Australian
   Capital Territory 2601, Australia
J.A. TSITSIPIS
Institute of Biology, "Democritos" National Research Center, P.O. Box 60228,
   GR-15310 Aghia Parskevi, Greece
D.L. WILLIAMSON
USDA/ARS, Subtropical Crop Insect Research Laboratory, 509 West Fourth
   Street, Weslaco, Texas 78596, U.S.A.

PART 1

TAXONOMY AND ZOOGEOGRAPHY

Chapter 1.1 Taxonomic Characters used in Identifying Tephritidae

R.A.I. DREW

1 INTRODUCTION

The Tephritidae, particularly the subfamily Dacinae, are one of the most difficult groups of insects to study taxonomically. It has been hard to find sound taxonomic characters and this has led to major problems. The most common errors are synonyms, homonyms, misidentifications and establishment of supra-specific groups based on questionable characters. Prolific speciation within the Tephritidae has resulted in large numbers of species (Hardy (1977) reports some 3700 species are known worldwide) and almost certainly many sibling species. Only now can some of these be recognised with the aid of recently developed taxonomic techniques.

It is not possible to discuss all characters that have been used, consequently this chapter will be confined to a discussion of those that are thought to be most important. The characters used in the subfamily Dacinae have been illustrated by Drew (1982). There will always be some difference of opinion in the definitions of supra-specific groups due to the varying degrees of importance that workers place on the different characters.

2 FAMILY LEVEL CHARACTERS

The characters most commonly used to define the Tephritidae are as follows:

Head.   Ptilinal fissure present; incurved *i.or.* bristles present; vibrissae absent.
Thorax.   Costal vein broken at two positions bordering second costal cell; subcosta complete and separate from R_1; R_1 setulose above; R_{4+5} unbranched; CuA short and joining 1A well back from wing margin, if anal cell long and acute then CuA is angulate or indented; cell M closed; mesonotal suture broadly interrupted medially; scutellum present; scutellar bristles not at apices of long spines; no setulae on lower margin of metathoracic spiracle; legs not elongate, tibiae without preapical dorsal bristles.
Abdomen. Abdomen not elongate; females with a well developed ovipositor; males with well developed surstyli and claspers. Hennig (1973) used (a) abdominal segment VII of female with tergum and sternum fused into a tube; (b) male with a small asymmetrical post-abdomen retracted beneath segment V; (c) male with a tubular shaped aedeagus spiralled and lying beneath the post-abdomen.

Chapter 1.1 references, p. 7

3 SUBFAMILY LEVEL CHARACTERS

3.1 Subfamily Dacinae

The characters most commonly used to define this subfamily are:

Head. Antennal segment 3 elongate, three or more times longer than wide; lacking ocellar and post-ocellar bristles.

Thorax. Reduction in bristles, lacking dorsocentral, presutural, sterno-pleural and usually humeral; cell M in wing two times wider than cell Cu, cell M approximately two times longer than wide, cell Cu with an elongate apical lobe, dense pattern of microtrichia in wing of male at distal end of anal lobe; wings generally hyaline except for a dark costal margin and anal streak.

Abdomen. Abdomen often short and broad; mostly with a pair of shining spots on tergum V of male above tergal glands; usually with a pecten of cilia on posterolateral margins of tergum III of male; female with two spermathecae composed of tightly set coils.

3.2 Subfamily Tephritinae

The characters most commonly used to define this subfamily are:

Head. Antennal segment 3 short, less than three times longer than wide; chaetotaxy not reduced; occipital row consisting of setae that are stubby, thick, flat and scale-like, usually white or yellow-white.

Thorax. Chaetotaxy not reduced; dorsocentral bristles before or near the supra-alar bristles; cell M narrow equal in width to cell Cu and over 4 times longer than wide, apical lobe of cell Cu short, no dense pattern of microtrichia in wing of male between anal cell and wing margin; wings usually with a spotted colour pattern; mesonotum tomentose and covered with flat recumbent scale-like white setae; vertical suture of mesopleuron absent or rudimentary.

Abdomen. Generally more slender and straight sided than in the Dacinae; no shining spots on tergum V and no pecten of cilia on tergum III; abdominal tergum VI in female equal to or longer than V; females with 2 to 3 spermathecae but never formed of tight set coils.

3.3 Subfamily Trypetinae

The main diagnostic characters of this subfamily are similar to those for the Tephritinae except as follows:

Head. Occipital row setae thin, pointed and dark coloured; in the Adramini antennal segment 3 is similar to that in the Dacinae.

Thorax. Dorsocentral bristles usually behind the supra-alar bristles; mesonotum rarely tomentose and never with scale-like white setae; wings rarely spotted and usually covered with elaborate colour patterns; vertical suture of mesopleuron well developed.

Abdomen. Tergum VI of female shorter than tergum V. In the Adramini the chaetotaxy of the head and thorax is reduced and similar to that in the Dacinae.

3.4 Subfamily Schistopterinae

This subfamily is represented by one genus (*Rhabdochaeta* de Meijere). It is similar to the Tephritinae and Trypetinae except that it possesses a deep cleft in the costa at the extremity of the subcosta, two developed bristles on the anterior margin of this cleft and distinctive wing colour patterns consisting of a large dark formation over centre of wing with radiating arms to wing margin.

4 GENERIC/SUBGENERIC LEVEL CHARACTERS

4.1 Subfamilies Tephritinae and Trypetinae

These subfamilies are generally divided into distinct genera without significant disagreement between workers. Subgenera are not used. The main characters used to define genera are:

Head. Shape of proboscis; ratio of width to height of head; number of *i.or.* and *s.or.* bristles; antennal segment 3 pointed or rounded at apex, size of arista and whether or not the arista is pubescent, plumose or bare.

Thorax. Position of dorsocentral bristles in relation to mesonotal suture; number of scutellar bristles; presence or absence of prescutellar bristles; whether or not the humeral, dorsocentral, mesopleural or sternopleural bristles are absent, present, weak or well developed; general wing shape and some wing colour patterns; the shape of some wing veins; ratio of length of subcostal cell to costal cell; whether or not R_1, R_{4+5}, CuA_1 and CuA_2 are bare or setose; cell Cu pointed or truncate at distal end; shape of apex of female ovipositor; occasionally some scutellum and mesonotum colour patterns; presence or absence of spines or bristles on ventral surfaces of femora; presence or absence of 1 or 2 spurs on apex of mid-tibiae.

4.2 Subfamily Dacinae

It has been extremely difficult to establish enough reliable characters at the generic/subgeneric level for this subfamily. As a result there has been a great deal of conjecture over whether most supra-specific groups should be given generic or subgeneric status. Also there has been considerable shifting of species between supra-specific groups. The early taxonomists such as Fabricius (1805) and Walker (1857, 1865) used one genus (*Dacus* Fabricius). Tryon (1927) also used genera (without subgenera) while Malloch included species in four subgenera of *Dacus*. Perkins (1937, 1939) and May (1963) insisted on using genera but Hardy (1951, 1955) and Drew (1972) used two genera and a series of subgenera of *Dacus*. Perkins and May did not present sound taxonomic arguments for rejecting the subgeneric concept. The only reason given by Perkins (1937) was that 'the Dacinae contained so many species, is so widely distributed and is so important that it should be divided into clearly defined genera'. The argument of May (1963) was equally vague, stating that the subfamily was naturally divided into two groups on the basis of fusion or non-fusion of abdominal terga. Hardy (1951) presented reasons of more taxonomic significance stating that it was not sound practice to establish genera on some of the available characters such as single chaetotaxic characters which vary within species and secondary sexual characters (e.g. pecten of cilia on abdominal tergum III of male).

Chapter 1.1 references, p. 7

Munro (1984) has added considerable confusion to the group by dividing a comparatively small African fauna of 175 species into 45 genera many of which are based on characters that have little or no significance at the species level e.g. face colour and facial spots, presence or absence of anal stripe in wing, width of costal band in wing, presence or absence of post-sutural yellow vittae, presence or absence of yellow on hypopleural calli, presence or absence of microtrichia in cell M and leg colour patterns. This classification of Munro cannot be applied to the much larger South-East Asian and South Pacific fauna. It is clear that a great deal of study remains to be done if a worldwide classification is going to be achieved.

The characters most commonly used at the generic/subgeneric level are as follows:

Head.   The sum of lengths of antennal segments compared with vertical length of head.

Thorax.   Presence or absence of scutellar, prescutellar, supra-alar and humeral bristles.

Abdomen. Abdominal terga fused or free; abdominal shape i.e. strongly petiolate with length of tergum I equal to or greater than width, to elongate oval and oval with the length of tergum I less than width; presence or absence of a pecten of cilia on posterolateral corners of abdominal tergum III; length of posterior lobe of surstylus on male genital segment; size of concavity on posterior margin of abdominal sternum V of male.

5 SPECIES LEVEL CHARACTERS

In all subgenera the characters most commonly used to define species are colour patterns. In the Tephritinae, Trypetinae and Schistopterinae most species have distinctive colour patterns across the wing. These remarkable patterns are significant in courtship and mating behaviour and are the primary characters used to identify species. Other characters used are colour patterns on the face, antennae, mesonotum, legs and abdominal terga. In the Dacinae where most species do not possess elaborate wing colour patterns the following characters are used:

Head.   Face colour including presence or absence of facial spots.

Thorax.   Width of coloured costal band on wing, presence or absence of other dark colour patterns on wing, colour of costal cells and presence or absence of dense microtrichia in these cells; colour of mesonotum; presence or absence, shape and size of the yellow longitudinal vittae on the mesonotum; size of the mesopleural stripe; presence or absence of a yellow band connecting the humeral and notopleural calli; colour of the humeral and notopleural calli; colour of the scutellum; colour of legs.

Abdomen. Colour patterns on abdomen particularly terga III to V; shape of the tip of the apical segment of the ovipositor and shape of spicules (scales) at the distal end of the second segment of the ovipositor, as seen under the scanning electron microscope; pattern of teeth and ridges on the male surstyli and claspers under the scanning electron microscope.

6 DISCUSSION

The taxonomic characters presently used in the Tephritidae are a mixture of reliable and less reliable ones. Further investigation is required into the generic/subgeneric character states in the Dacinae with the aim of developing an acceptable classification of the entire world fauna. It is becoming increasingly evident that we are now faced with a very large number of species and many sibling (or cryptic) species. The definition of these cryptic species is going to require intensive research into the genetic characters e.g. cytology and tissue enzyme (isozyme) patterns, biological characters e.g. mate recognition and sterility experiments, gathering biological information such as host records, the chemistry of male pheromones particularly in the Dacinae, scanning electron microscopy of significant body structures such as the female ovipositors and male surstyli and claspers. This approach was recently applied to the study of two cryptic species of the *Dacus tryoni* complex in Australia (Drew and Lambert, 1986).

7 REFERENCES

Drew, R.A.I., 1972. The generic and subgeneric classification of Dacini (Diptera: Tephritidae) from the South Pacific area. Journal of the Australian Entomological Society, 11: 1–22.

Drew, R.A.I., 1982. Taxonomy. In: R.A.I. Drew, G.H.S. Hooper and M.A. Bateman (Editors), Economic Fruit Flies of the South Pacific Region. Queensland Department of Primary Industries, Brisbane, 139 pp.

Drew, R.A.I. and Lambert, D.M., 1986. On the specific status of *Dacus (Bactrocera) aquilonis* and *Dacus (Bactrocera) tryoni* (Diptera: Tephritidae). Annals of the Entomological Society of America, 79: 870–878.

Fabricius, J.C., 1805. Systema Antliatorum secundum ordines, genera, species adiectis synonymis, locis, observationibus, descriptionibus. 373 pp. (Brunswick).

Hardy, D.E., 1951. The Krauss collection of Australian fruit flies (Tephritidae-Diptera). Pacific Science, 5: 115–189.

Hardy, D.E., 1955. A reclassification of the Dacini (Tephritidae-Diptera). Annals of the Entomological Society of America, 48: 425–437.

Hardy, D.E., 1977. Family Tephritidae. In: M.D. Delfinado and D.E. Hardy (Editors), A Catalog of the Diptera of the Oriental Region, Vol. III. The University Press of Hawaii, Honolulu, 854 pp.

Hennig, W., 1973. Diptera (Zweifluger). In: Handbuch der Zoologie, Vol. 4, pt. 2, Section 31. Walter de Gruyter, Berlin, 337 pp.

May, A.W.S., 1963. An investigation of fruit flies (Fam. Trypetidae) in Queensland 1. Introduction, species, pest status and distribution. Queensland Journal of Agricultural Science, 20: 1–82.

Munro, H.K., 1984. A taxonomic treatise of the Dacidae (Tephritoidea, Diptera) of Africa. Entomology Memoir Department of Agriculture and Water Supply Republic of South Africa, No. 61, 313 pp.

Perkins, F.A., 1937. Studies in Australian and Oriental Trypaneidae. Part 1. New Genera of Dacinae. Proceedings of the Royal Society of Queensland, 48: 51–60.

Perkins, F.A., 1939. Studies in Oriental and Australian Trypetidae. Part 3: Adraminae and Dacinae from New Guinea, Celebes, Aru Is., and Pacific Islands. University of Queensland Papers Department of Biology, 1: 1–35.

Tryon, H., 1927. Queensland fruit flies (Trypetidae), Series 1. Proceedings of the Royal Society of Queensland, 38: 176–224.

Walker, F., 1857. Catalogue of the Dipterous Insects collected at Singapore and Malacca by Mr. A.R. Wallace, with Descriptions of New Species. Proceedings of the Linnean Society of London, 1: 28–35.

Walker, F., 1865. Description of new species of the Dipterous insects of New Guinea. Proceedings of the Linnean Society of London, 8: 102–130.

Chapter 1.2 The Taxonomy and Distribution of Tropical and Subtropical Dacinae (Diptera: Tephritidae)

R.A.I. DREW

1 INTRODUCTION

The Dacinae are distributed throughout the tropical and subtropical regions of the world and, except for the African and American continents, they are endemic to the tropical and subtropical rainforests. These forests contain a very diverse flora comprised of thousands of species. Webb and Tracey (1981) listed 545 plant genera in Australian rainforests and 750 in Papua New Guinea lowland and lower montane rainforests, 400 of which are not recorded in Australia. It is in association with this very diverse habitat that the Dacinae have speciated, utilising the soft fleshy fruits for larval food.

Considering the known fauna and taxonomic work presently being undertaken the Dacinae probably contains at least 800 species (Africa 200, Asia/South East Asia 300, South Pacific region 300). Papua New Guinea with some 200 species probably has more than any other single land mass and appears to be in the centre of origin of the group.

The first species of Dacinae to be described were *Musca ferruginea* Fabricius (1784), *Musca oleae* Gmelin (1790) and *Dacus armatus* Fabricius (1805). From these early studies a large number of entomologists have been involved in describing species and developing supraspecific groupings. Francis Walker (1857, 1859, 1860, 1861, 1862, 1864, 1865) described species from material collected on the voyages of Alfred Russel Wallace but he had confused generic and family concepts (Hardy, 1959).

In recent times large sections of the fauna have been described and revised by Hardy (1973, 1974, 1977) for South East Asia, Drew (1971, 1972a, 1972b, 1973, 1974) for the South Pacific Region and Munro (1984) for the African continent.

In this chapter the present status of our knowledge of the taxonomy of Dacinae will be reviewed.

2 SUBFAMILY CLASSIFICATION

The definition of the subfamily is based primarily on antennal segment 3 elongate, reduced chaetotaxy on the head and thorax, an elongate apical lobe extension on cell Cu, dense microtrichia at end of this lobe in males, a pair of shining spots on tergum V, usually with a pecten of cilia on posterolateral margins of tergum III in males and females with two coiled spermathecae (Hardy, 1973, 1974). This definition has been widely accepted by most dipterists and, except for the occasional borderline species, it delineates a distinct group and is clearly workable. In contrast, Munro (1984) elevated the group to family status and redefined it as having five terga in the dorsum of the abdomen, tergum and sternum VI in the female forming a separate collar around the ovipositor, a pair of shining spots on tergum V, a pair of glands on anterior corners of sternum II, wing vein R_{2+3} close to costa, cell M wide and hyaline, anal cell extended into a lobe, subcosta ending bluntly with a short darkened streak directed towards costa, females with two spermathecae. Munro also gave supplementary characters as genal bristle present, 1 *s.or.* and 2 *i.or.* bristles present, reduced chaetotaxy of head and thorax, antennae 3-segmented and with an arista, yellow markings on thorax well developed, anterolateral knobs on tergum I each with a gland, aedeagus characteristic. This is a departure from the standard family level characters in Diptera which are mainly major differences in wing venation. Munro also does not make clear how this new family definition relates to those for the other close dipteran families. The major problem in raising the Dacinae to family level is that it weakens the generic definitions by reducing the generic level characters to those that are even questionable at species level. It is doubtful if the classification of Munro will be widely accepted particularly as family and subfamily definitions are generally required to cater for the worldwide fauna and not for that of one continent.

3 GENERIC/SUBGENERIC CLASSIFICATION

It is at this level that taxonomists have had greatest difficulty in developing a widely accepted classification that would encompass most of the worldwide fauna. A very large number of genera and subgenera have been defined but the basic problem of an inadequate number of valuable characters still persists.

Although the very first two dacine species described were placed in genus *Musca*, the commonly used and longest standing generic name has been *Dacus* Fabricius. At the beginning this genus was used to pool a conglomeration of species now belonging to various families e.g. Tephritidae, Otitidae, Platystomatidae. As more and more species were described it was necessary to define more supraspecific groups. Some workers insisted on making all such groups genera (e.g. Perkins, 1937; May, 1963), while others used two genera with most supraspecific groups being subgenera of *Dacus* (Hardy, 1951, 1955; Drew, 1972a). The classification most widely used at present is a division of the subfamily into two genera viz. *Callantra* Walker and *Dacus* and then the subsequent division of *Dacus* into subgenera. The finding of valuable characters on the male genitalia and abdominal sternum V (Drew 1972a) has assisted in defining more stable subgenera but this has still not overcome the problems raised by Hardy (1951, 1955) that we use characters based on one sex and also single chaetotaxic characters. The latter are a problem especially the supra-alar bristles which can be present or absent in the one species. This particular chaetotaxic character is used to separate subgenus *Afrodacus* Bezzi (*sa.* bristle absent) from subgenus *Bactrocera* Macquart (*sa.* present). It is most likely that some species presently placed in *Afrodacus* are just aberrant *Bactrocera*.

In taking an overview of the Dacinae, some interesting facts are now coming to hand that are leading to a new classification. An examination of the distribution of numbers of species indicates that the centre of origin and prolific speciation is in the Papua New Guinea region. The South East Asian and South Pacific fauna (some 450 known species) is unique in having free abdominal terga (except for approximately 50 species) while the African fauna is equally unique in having 175 known species, all of which have fused abdominal terga (except for 5 species). The known endemic plant hosts of the African species belong primarily to two families viz., Asclepiadaceae and Cucurbitaceae. The known endemic plant hosts of the South East Asian and South Pacific species with fused abdominal terga are Asclepiadaceae and Cucurbitaceae which generally grow away from the rainforests which contain most of the hosts of the South Pacific Dacinae. Consequently the initial division of the subfamily based on fusion and non-fusion of abdominal terga appears to be much more significant geographically and biologically than that presently used to divide the group into *Callantra* and *Dacus*. This will also overcome another problem in that we now have enough 'Callantra' like species with so much variation in abdominal shape and antennal length that it is difficult to maintain this as a well defined group warranting generic status. Consequently in a monograph presently being prepared on the South Pacific Dacinae in which almost 300 species are being treated (Drew unpub. data) a new classification is used based on the division of the subfamily into two genera:

(a) *Dacus* — species with fused abdominal terga, mostly African, primarily infesting endemic Asclepiadaceae and Cucurbitaceae.

(b) *Bactrocera* — species with free abdominal terga, mostly South East Asian and South Pacific, primarily infesting endemic tropical and subtropical rainforest fruits.

In this classification, the difficult 'Callantra' types will be placed in subgenera of *Dacus* based on characters not befitting of generic status. *Callantra* will be one subgenus. This new classification will mean that our major Asian and Pacific pest species will now be placed in genus *Bactrocera* and not *Dacus* which will be a problem in our economic entomology 'world'. However we cannot ignore the mounting taxonomic, geographic and biological evidence that comes through the collection and knowledge of increasing numbers of species. This is an advantage that earlier taxonomists did not have.

The new generic classification of Munro (1984) is not relevant to the South East Asian and South Pacific fauna which contains the bulk of the species. 6 of the 12 characters upon which it is based are only applicable at the species level and another two are of no value at that level in the Asian and Pacific fauna. If indeed the African species can be classified on the characters proposed by Munro then it proves that that fauna is very unique and distinct from the rest of the world fauna.

From the information explained above, it is most likely that the Dacinae originated in the Papua New Guinea area and speciated prolifically. An ancestral form with fused abdominal terga from the South East Asian/Pacific region was probably the precursor of the African fauna.

4 SPECIES LEVEL STUDIES

Good progress has been made at this level. Following on the long line of workers and species descriptions started by Fabricius in 1784, Hardy (1951, 1954, 1955, 1970, 1973, 1974, 1977) and Hardy and Adachi (1954) have described many species and produced an excellent catalogue of the South East Asian fauna. Drew (1968a, 1968b, 1971, 1972a, 1972b, 1979), Drew and Hancock (1981) and Drew and Hardy (1981) have described a large number of South Pacific species. The known South Pacific fauna will be increased by more than 100 species when the current monograph presently being prepared is published (Drew, unpublished data).

The major difficulties presently being encountered at this level are in the identification of sibling (or cryptic) species. It is now evident that there are large numbers of such species not separable on the standard external morphological characters. We are now encountering, for example, species very close or identical on external morphological characters but which occupy different host plants and/or different geographic regions. We must now address this problem and make definite progress towards defining these species. Examples of these are *Dacus opiliae* Drew and Hardy (a sibling species of the dorsalis complex), *Dacus tryoni* (Froggatt) and *Dacus aquilonis* (May) (two very closely related species in the tryoni complex). The latter two species have been the subject of an intensive study using genetic and scanning electron microscope techniques (Drew and Lambert, 1986). Such studies in the Dacinae are now essential in order to further define the species and relate them taxonomically to the different biological characteristics now being observed.

The value of polytene chromosomes to taxonomy has been well illustrated in the Drosophilidae. The first successful attempt at preparing these chromosomes in Tephritidae was in *Ceratitis capitata* (Bedo, 1986). In this work, polytene chromosome preparations were made from trichogen cells at the bases of the lower pair of superior orbital bristles in males and from thoracic trichogen cells in both sexes. Further development of these techniques will be of great value in taxonomic studies on sibling species.

The value of biological and ecological studies to taxonomy at the species level cannot be underestimated. Two areas of investigation will eventually make great contributions to our understanding of the dacine species, viz., larval and adult feeding. As we research and understand more the host fruit records, the bacterial diets of the larvae and adults, the complex chemical attractants produced by the bacteria in the host plants and the influence of these odours on mating and oviposition behaviour, then we will develop a clearer understanding of the speciation processes that continue within the populations and a better understanding of the species themselves. Conversely it is clear that a sound knowledge of the taxonomy of the group will assist in ecological research and in understanding the ecology of the species.

In contrast to the temperate subfamilies, the Dacinae are restricted to an area of the world which includes a large number of developing countries. There is major crop loss in some areas and restriction on trade in others. The socioeconomic impact of fruit flies on these countries is enormous and demands that increasing resources be expended in fruit fly research. There is still a major need for more taxonomic research to define the species that are major pests and in most countries there is a complete lack of reference collections for identification purposes. Without such research and facilities, Plant Quarantine and fruit fly control programmes will be severely hindered.

For such research efforts, people well trained in taxonomy are required. In the history of dacine taxonomy it is very clear that because fruit flies caused major economic impacts on society, there was a rush to identify the species involved by the economic entomologists of the day. These worked tirelessly but left a legacy of taxonomic problems and misidentifications. Large amounts of funding and applied research were based on much of this work. This problem is particularly evident in Australia where Froggatt, Tryon,Perkins and May were confronted with evolving and increasing fruit fly problems in developing horticulture industries. If we learn from this historical review we must ensure that trained taxonomists are made available to undertake research in developing countries.

5 DISCUSSION

Good progress is being made on the basic description of species in some countries. In others within South East Asia and the South Pacific region further work is essential together with the establishment of good reference collections. Extensive research in applying new genetic and scanning electron microscope techniques is long overdue in order to distinguish the complexes of sibling species.

The generic/subgeneric classification is also badly in need of revision and this will demand further investigation.

Biological and ecological research in specific areas such as larval and adult feeding will also make major contributions to our understanding of species.

6 REFERENCES

Bedo, D.G., 1986. Polytene and mitotic chromosome analysis in *Ceratitis capitata* (Diptera; Tephritidae). Canadian Journal of Genetics and Cytology, 28: 180–188.

Drew, R.A.I., 1968a. Two new species of Dacinae (Diptera: Trypetidae) from New Britain. Journal of the Australian Entomological Society, 7: 21–24.

Drew, R.A.I., 1968b. Two new species of Dacinae (Diptera: Trypetidae) from New Guinea. Journal of the Australian Entomological Society, 7: 77–79.

Drew, R.A.I., 1971. New species of Dacinae (Diptera: Trypetidae) from the South Pacific area. Queensland Journal of Agricultural Science, 28: 29–103.

Drew, R.A.I., 1972a. The generic and subgeneric classification of Dacini (Diptera: Tephritidae) from the South Pacific area. Journal of the Australian Entomological Society, 11: 1–22.

Drew, R.A.I., 1972b. Additions to the species of Dacini (Diptera: Tephritidae) from the South Pacific area with keys to species. Journal of the Australian Entomological Society, 11: 185–231.

Drew, R.A.I., 1973. Revised descriptions of species of Dacini (Diptera: Tephritidae) from the South Pacific area. I. Genus *Callantra* and the *Dacus* Group of subgenera of genus *Dacus*. Queensland Department of Primary Industries, Division of Plant Industry Bulletin No. 652, 39 pp.

Drew, R.A.I., 1974. Revised descriptions of species of Dacini (Diptera: Tephritidae) from the South Pacific area. II. The *Strumeta* group of subgenera of genus *Dacus*. Queensland Department of Primary Industries, Division of Plant Industry Bulletin No. 653, 101 pp.

Drew, R.A.I., 1979. The genus *Dacus* Fabricius (Diptera: Tephritidae) — Two new species from Northern Australia and a discussion of some subgenera. Journal of the Australian Entomological Society, 18: 71–80.

Drew, R.A.I. and Hardy, D.E., 1981. *Dacus (Bactrocera) opiliae*, a new sibling species of the dorsalis complex of fruit flies from Northern Australia (Diptera: Tephritidae). Journal of the Australian Entomological Society, 20: 131–137.

Drew, R.A.I. and Lambert, D.M., 1986. On the specific status of *Dacus (Bactrocera) aquilonis* and *Dacus (Bactrocera) tryoni* (Diptera: Tephritidae). Annals of the Entomological Society of America, 79: 870–878.

Drew, R.A.I., Hancock, D.L. and Romig, M.C., 1981. Australian Dacinae (Diptera: Tephritidae) — New species from Cape York Peninsula, a discussion of species complexes and key to species. Australian Journal of Zoology, 29: 49–91.

Fabricius, J.C., 1784. Entomologia Systematica emendata et aucta. Vol. 4, Hafniae, 472 pp.

Fabricius, J.C., 1805. Systema Antliatorum secundum ordines, genera, species adiectis synonymis, locis, observationibus, descriptionibus. 373 pp. (Brunswick).

Gmelin, J.F., 1790. Editio decima tertia aucta reformata cura, Systema Naturae, Vol 1, part 5, Insecta, p. 2844; Lipsiae.

Hardy, D.E., 1951. The Krauss collection of Australian fruit flies (Tephritidae — Diptera). Pacific Science, 5: 115–189.

Hardy, D.E., 1954. The *Dacus* subgenera *Neodacus* and *Gymnodacus* of the world. Proceedings of the Entomological Society of Washington, 56: 5–23.

Hardy, D.E., 1955. The *Dacus (Afrodacus)* Bezzi of the world (Tephritidae, Diptera). Journal of the Kansas Entomological Society, 28: 3–15.

Hardy, D.E., 1959. The Walker types of fruit flies (Tephritidae — Diptera) in the British Museum Collection. Bulletin of the British Museum of Natural History, B. Entomology, 8: 159–242.

Hardy, D.E., 1970. Tephritidae (Diptera) collected by the Noona Dan Expedition in the Philippine and Bismarck Islands. Entomologiske Meddelelser, 38: 71–136.

Hardy, D.E., 1973. The fruit flies (Tephritidae-Diptera) of Thailand and bordering countries. Pacific Insects Monograph, 31: 1–353.

Hardy, D.E., 1974. The fruit flies of the Philippines (Diptera: Tephritidae). Pacific Insects Monograph, 32: 1–266.

Hardy, D.E., 1977. Family Tephritidae. In: M.D. Delfinado and D.E. Hardy (Editors), A catalog of the Diptera of the Oriental region, Vol. III. The University Press of Hawaii, Honolulu, 854 pp.

Hardy, D.E. and Adachi, Marian S., 1954. Studies in the fruit flies of the Philippine Islands, Indonesia, and Malaya Part 1. Dacini (Tephritidae-Diptera). Pacific Science, 8: 147–204.

May, A.W.S., 1963. An investigation of fruit flies (Fam. Trypetidae) in Queensland 1. Introduction, species, pest status and distribution. Queensland Journal of Agricultural Science, 20: 1–82.

Munro, H.K., 1984. A taxonomic treatise of the Dacidae (Tephritoidea, Diptera) of Africa. Entomology Memoir Department of Agriculture and Water Supply Republic of South Africa, No. 61, 313 pp.

Perkins, F.A., 1937. Studies in Australian and Oriental Trypaneidae. Part 1. New Genera of Dacinae. Proceedings of the Royal Society of Queensland, 48: 51–60.

Walker, F., 1857. Catalogue of the Dipterous Insects collected at Singapore and Malacca by Mr. A.R. Wallace, with Descriptions of New Species. Proceedings of the Linnean Society of London, 1: 28–35.

Walker, F., 1859. Catalogue of the Dipterous Insects collected in the Aru Islands by Mr. A.R. Wallace, with Descriptions of New Species. Proceedings of the Linnean Society of London, 3: 77–131.

Walker, F., 1860. Catalogue of the Dipterous Insects collected at Makassar in Celebes by Mr. A.R. Wallace, with Descriptions of New Species. Proceedings of the Linnean Society of London, 4: 149–160.

Walker, F., 1861. Catalogue of the Dipterous Insects collected in Batchian, Kaisoa and Makian, and at Tidon in Celebes, by Mr. A.R. Wallace, with Descriptions of New Species. Proceedings of the Linnean Society of London, 5: 270–303.

Walker, F., 1862. Catalogue of the Dipterous Insects collected at Gilolo, Ternate, and Ceram, by Mr. A.R. Wallace, with Descriptions of New Species. Proceedings of the Linnean Society of London, 6: 4–23.

Walker, F., 1864. Catalogue of the Dipterous Insects collected in Waigiou, Mysol, and North Ceram by Mr. A.R. Wallace, with Descriptions of New Species. Proceedings of the Linnean Society of London, 7: 202–238.

Walker, F., 1865. Descriptions of New Species of Dipterous Insects of New Guinea. Proceedings of the Linnean Society of London, 8: 102–130.

Webb, L.J. and Tracey, J.G., 1981. Australian rainforests: patterns and change. In: A. Keast (Editor), Ecological Biogeography of Australia, Vol. 1. D. Junk, The Hague pp. 605–694.

Chapter 1.3 The Taxonomy and Zoogeography of the Genus *Anastrepha* (Diptera: Tephritidae)

ALLEN L. NORRBOM and RICHARD H. FOOTE

1 INTRODUCTION

Anastrepha Schiner is the largest New World genus of Tephritidae with over 190 known species (Norrbom, 1985). This genus and *Rhagoletis* Loew include most of the fruit flies of major economic importance native to the Western Hemisphere; only the introduced Mediterranean fruit fly and several species of *Dacus* cause comparable losses. In Table 1.3.1 we have listed the most destructive species, but many others are potential pests. The most notorious species, *Anastrepha ludens* (Loew), *Anastrepha fraterculus* (Wiedemann), *Anastrepha obliqua* (Macquart), and *Anastrepha serpentina* (Wiedemann), have adopted numerous exotic commercial plants as hosts and attack such important crops as mango, citrus, guava, sapodilla, and *Spondias, Eugenia,* and *Syzygium* species. Other *Anastrepha* attack cassava, *Inga* or *Passiflora* species, cucurbits, or other plants of minor agricultural importance. *Anastrepha* is endemic to the American tropics and subtropics, although many of the pest species have a wide distribution within this range. Adults of most species are easily recognized as belonging to *Anastrepha*, but many are difficult to identify to the species level.

TABLE 1.3.1

Pest species of *Anastrepha*, their common names, and the most important crops they commonly attack

Species	Common name	Crops attacked
antunesi Lima	–	*Spondias* spp.
bistrigata Bezzi	–	guava
distincta Greene	Inga fruit fly	*Inga* spp.
fraterculus Wiedemann	South American fruit fly	Citrus, guava, *Eugenia, Prunus, Syzygium* spp.
grandis Macquart	–	cucurbits
ludens Loew	Mexican fruit fly	citrus, mango, peach
manihoti Lima	–	cassava
montei Lima	–	cassava
obliqua Macquart	West Indian fruit fly	mango, *Spondias* spp.
pickeli Lima	–	cassava
serpentina Wiedemann	Serpentine fruit fly	sapodilla, star apple, citrus, *Pouteria* spp.
sororcula Zucchi	–	guava, *Eugenia*
striata Schiner	Guava fruit fly	guava
suspensa Loew	Caribbean fruit fly	guava, *Eugenia, Syzygium, Annona* spp., tropical almond

Chapter 1.3 references, p. 24

In this chapter we discuss the major taxonomic and zoogeographic features of *Anastrepha*, including its recognition, past and current taxonomic status, distribution, immature stages, and host plants. We also discuss the needs as we perceive them for the most effective development of knowledge about the genus.

2 RECOGNITION AND RELATIONSHIPS

Anastrepha belongs to the tephritid subfamily Trypetinae which, together with the subfamily Dacinae, includes most of the fruit fly species that breed in fleshy type fruits. The Trypetinae are characterized by the lack of derived features found in other tephritid subfamilies, thus the group may be paraphyletic. Most species are relatively large and have banded wings; they may be recognized by the color, shape, and position of certain body bristles. Foote (1980) presents a description of these and other characters and provides identification keys for distinguishing the New World tephritid subfamilies, tribes, and genera, including *Anastrepha*.

Anastrepha is most closely related to the genus *Toxotrypana* Gerstaecker, which formerly has been placed in the Dacinae because of superficial resemblance in wing pattern and convergent reduction of bristles. Both *Anastrepha* and *Toxotrypana* possess derived features of the male and female terminalia, such as a T-shaped apical sclerite in the distiphallus (glans) and a basally expanded rasper with elongate dorsal scales, which clearly indicate their

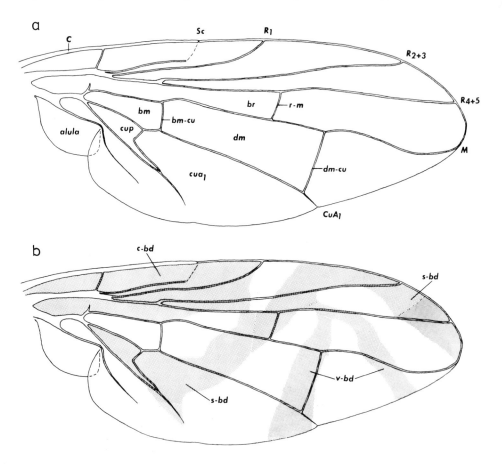

Fig. 1.3.1. Wing of *Anastrepha nigrifascia* Stone showing (A) venation and (B) color pattern; M = medial vein, c-bd = costal band, s-bd = s-band, v-bd = v-band.

relationship (Norrbom, 1985). The immunological data of Kitto (1983) also support this hypothesis.

Traditionally, *Anastrepha* has been diagnosed from all other tephritids by the apical curvature of the medial wing vein (Fig. 1.3.1). It is strongly curved forward in most species, meeting the costa without a visible angle, but in a few it is only weakly curved as in *Toxotrypana*. These species have well developed thoracic bristles and all but one of them have wing patterns with crossbands, unlike *Toxotrypana* which have weak or absent thoracic bristles and only a broad costal band. In addition to the above-mentioned genitalic characters, *Anastrepha* can be distinguished from other New World Trypetinae except *Toxotrypana* by the following combination of characters: dorsocentral bristles closer to a line through acrostichals than to a line through post-sutural supra-alars; scutellum with two pairs of marginal bristles; ocellar bristles much weaker than orbitals (except in *Anastrepha tripunctata* Wulp); face with median carina; and katepisternal bristle often weak or absent. The adult can be further characterized as a medium to large, yellow brown fly, sometimes with white or dark brown, usually ill-defined, body markings. The legs are usually entirely yellow brown and the wing pattern (Fig. 1.3.1) consists of yellow to dark brown bands. Typically a short costal, an S-shaped and a V-shaped band are present (Fig. 1.3.1), although parts of these are missing, expanded, or fused in many species. A few species have quite different patterns. Recognition of *Anastrepha* larvae and eggs will be discussed in the section of this chapter on immature stages.

3 TAXONOMIC HISTORY AND CURRENT TAXONOMIC STATUS

The first species of *Anastrepha* to be discovered in taxonomic studies of the Tephritidae were described in the genera *Dacus* and *Trypeta* by Wiedemann (1830) and by Macquart and Walker in publications appearing as early as 1835 to 1858. Schiner (1868) proposed the genus *Anastrepha* with *Dacus serpentinus* Wiedemann as the type species. Loew (1873), who first used the apical curve of the medial vein as a diagnostic character, very shortly later recognized the same concept under his name *Acrotoxa* (type species *Dacus fraterculus* Wiedemann), which is now considered to be a synonym of *Anastrepha* (Bezzi, 1909). Bezzi (1919a, b) and Hendel (1914) presented early reviews of the genus, including 19 and 33 species, respectively, but most of the significant alpha-level taxonomy of *Anastrepha* was published in a period of eight years by Greene (1934), Lima (1934, 1937–1938), and Stone (1939a,b, 1942a,b), who together described about three-fifths of the currently recognized species. Stone's papers, which included 142 species, comprise the last comprehensive treatment of the group. Forty six nominal species have since been described (Steyskal, 1977b; Caraballo, 1985; Zucchi, 1979a,c, 1982, 1984) and regional reviews have been published by Fernandez (1953) and Caraballo (1981) for Venezuela, Korytkowski and Ojeda (1968) for north-western Peru, Blanchard (1961) for Argentina, and Zucchi (1978) for Brazil. Stone's monographs, supplemented by Steyskal's (1977b) updated version of his keys, remain the most useful identification tools for the genus, although since 1977, 26 nominal species have been newly described, synonymized, or otherwise changed in status (Zucchi 1979a,b,c, 1981a,b, 1982, 1984: Caraballo, 1985). The senior author is currently revising *Anastrepha* on a comprehensive basis. Norrbom (1985) includes a preliminary infrageneric classification, revisions of four species groups, and a further updated key. By our current count, which includes new synonymies and undescribed species (Norrbom, 1985), there are 194 species.

Chapter 1.3 references, p. 24

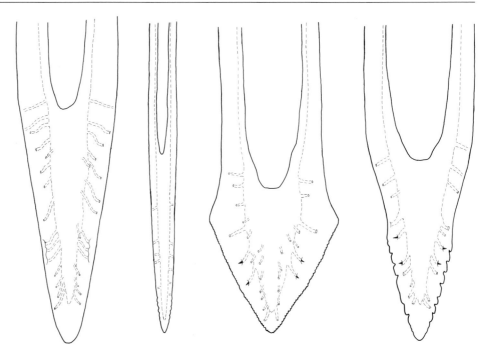

Fig. 1.3.2. Ovipositor tips of representative *Anastrepha* species; from left, *parallela* Wiedemann, *pallens* Coquillett, *distans* Hendel, and *suspensa* Loew.

In addition to *Acrotoxa*, the taxa *Phobema* Aldrich (1925) (type species *P. atrox* Aldrich), *Pseudodacus* Hendel (1914) (type species *A. daciformis* Bezzi), and *Lucumaphila* Stone (1939a) (type species *L. sagittata* Stone) are also considered synonyms of *Anastrepha* (Steyskal, 1977a). These taxa were proposed for species with the apically curved medial vein, and with additional distinctive derived characters (autapomorphies); their recognition as genera would require further splitting of *Anastrepha*, however, and the similar recognition of other equivalent species groups (Norrbom, 1985). Because these groups are established mainly on the basis of male genitalic characters, we prefer to maintain *Anastrepha* as a single, easily recognizable taxon with the names mentioned above as synonyms.

Species level taxonomy in *Anastrepha* has been based largely upon characters of the female ovipositor (Fig. 1.3.2). Except in species with distinctive wing patterns or some other external characters, it has therefore been difficult if not impossible to identify males. For example, in Steyskal's (1977b) key, only 25 species can be identified in the absence of female genitalic characters. Although characters of the male genitalia have been largely ignored, they are taxonomically useful. The shape of the surstyli (Fig. 1.3.3), length of the aedeagus, and sclerotization of the proctiger and distiphallus (glans) permit the identification of males to at least the species group level; in many species they allow complete determination (Norrbom, 1985).

Intraspecific variation in *Anastrepha* is not well understood. Many of the widely distributed species such as *A. fraterculus* and *Anastrepha hamata* (Loew) are geographically variable and may be complexes of localized sibling species. Baker et al. (1944) observed differences in host preference and wing pattern between Mexican and Brazilian populations of *A. fraterculus* and Mendes (1958) and Bush (1962) reported karyotype variation. Morgante et al. (1980) also found differences in isozymes among geographic populations within Brazil. Stone (1942a), who studied *A. fraterculus* throughout its entire range, found no characters to consistently subdivide it, however. Thorough biosyste-

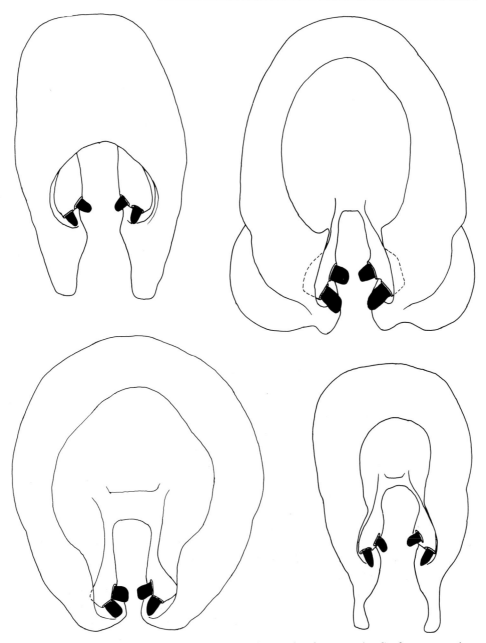

Fig. 1.3.3. Male genitalia in posterior view (proctiger and aedeagus omitted) of representative *Anastrepha* species; clockwise from upper left, *zuelaniae* Stone, *galbina* Stone, *quararibeae* Lima, undescribed species near *stonei* Steyskal.

matic studies involving populations from various localities and hosts are needed to resolve the status of *A. fraterculus* and other polymorphic *Anastrepha* species.

4 IMMATURE STAGES

The immature stages of the Tephritidae are poorly understood taxonomically, as are those of most Muscomorpha (Teskey, 1981b). Considerable advances have been made recently in standardizing the morphological terminology used for larvae (Teskey, 1981a), but the fact that the immatures of only a small percentage of taxa have been studied remains as a major problem. Within

Chapter 1.3 references, p. 24

Anastrepha, for example, descriptions are available for the third-stage larvae of only 13 species and for the eggs of only 11. Further, most of these descriptions are inadequate and are based on limited material; therefore it has been impossible to analyze character variation properly. Study material is difficult to obtain because of the time required to locate infested host plants, and to rear the immatures and carefully associate them with identifiable adults (Anon, 1982). Considering the high economic importance of fruit flies, however, it is surprising that more resources have not been allocated for the solution of this problem. The immature stages, especially the larvae, are usually the first to be encountered in economic situations, and the lack of identification capabilities for them is a severe handicap to workers in regulatory and pest management programs. Gary Steck, Lynn Carroll (Texas A & M University), and Stan Jones (Pennsylvania State Unversity) have initiated comparative taxonomic studies on immature tephritids in attempts to meet these needs, and the senior author is incorporating taxonomic characters of immatures where available into his revisionary work.

It is not possible to recognize *Anastrepha* immatures with certainty because those of few related taxa have been described. The third-stage larvae of the *Anastrepha* that have been studied possess only small cuticular tubercles below the posterior spiracles on the eighth abdominal (caudal) segment. These tubercles are larger in known larvae of the Dacinae and other Trypetinae, except *Toxotrypana* in which they are absent (Berg, 1979). Third-instar *Anastrepha* resemble other fruit-infesting tephritids in most other respects. They are active, white to yellow-brown, subcylindrical maggots, about five times as long as wide, tapering anteriorly and truncate posteriorly. The anterior spiracles are bicornuate with 8–34 openings. The posterior spiracles are sessile with the peritreme and spiracular plate weakly sclerotized. The three elongate-oval rimae of the spiracular openings are subparallel.

Berg (1979) produced the most comprehensive key to *Anastrepha* larvae, including six species. A more thorough analysis of variation in these and other species is needed to evaluate the characters he used. Although the characters intergrade in some species, they appear to be potentially useful and together with host plant and geographic data may allow at least an educated guess at species identification. They even clearly distinguish certain species from the others now known. For example, *Anastrepha grandis* (Macquart) have a high number of openings on the anterior spiracle (Fischer, 1932) and *Anastrepha sagittata* (Stone) have a very high ratio of the length to width of the rimae of the hind spiracles (Baker et al., 1984). Other characters also show taxonomic potential. *A. sagittata, Anastrepha pallens* Coquillett, and *Anastrepha interrupta* Stone have short hind spiracular hairs (Baker et al., 1944; Norrbom, 1985) and *Anastrepha limae* Stone have a different cuticular spination pattern than four other species studied by Norrbom (1985); the rows of spines on the dorsum of the metathorax are more numerous in *A. limae* than in the other species. The significance of all these data is unclear, however, without more comprehensive studies on many more species and larger sample sizes.

The pupae of *Anastrepha* have not been studied taxonomically; the puparium possesses only remnants of the larval characters. The eggs (Fig. 1.3.4), however, show considerable taxonomic potential. They are whitish and elongate-ovoid, more strongly tapering to varying degrees at the end opposite the micropyle. Although no diagnostic characters at the genus level have been found to date, great variation in general shape is evident, and several species have distinctive lobes. *A. obliqua* have a short lobe on the micropyle end (Emmart, 1933, as *A. fraterculus*), whereas *Anastrepha pittieri* Caraballo and *Anastrepha nigrifascia* Stone, the only species of the *robusta* species group for which eggs have been examined, have elongate, basally constricted lobes on the micropyle end (Norrbom, 1985).

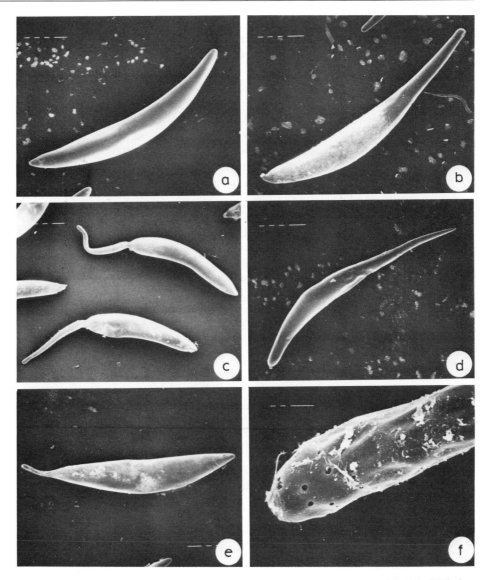

Fig. 1.3.4. Eggs of *Anastrepha* species, micropylar end at left, (A) *leptozona* Hendel, (B) *ludens* Loew, (C) *pittieri* Caraballo, (D) *cordata* Aldrich, (E) *obliqua* Macquart, (F) *obliqua*, apex of lobe at micropylar end.

5 ZOOGEOGRAPHY

The distribution of *Anastrepha* extends from the extreme southern United States (Rio Grande valley and southern Florida) to northern Argentina, approximately between 27°N and 35°S latitudes, with the greatest diversity occurring in the tropical parts of that range (Stone, 1942a). Collecting for fruit flies within the Neotropics has been spotty, so that it is not possible at this time to conduct a detailed analysis of the historical biogeography of *Anastrepha*, but a few observations can be made.

Many species of *Anastrepha* are widespread, especially the economically important ones. *A. obliqua*, commonly known as the West Indian fruit fly, is not restricted to the Antilles at all; it is probably the most widely distributed species, occurring throughout almost the entire range of *Anastrepha*. Other species, including *A. fraterculus, Anastrepha striata* Schiner, and *A. serpentina*, occur from Mexico or Texas to Argentina. Zoogeographic analysis of these pest species is difficult because they may have been spread by man. It is doubtful, for instance, whether *A. fraterculus* is native to the Galapagos Islands (Foote, 1982). *A. ludens*, probably originally restricted to northern Mexico (Baker et al., 1944), now occurs at high elevations into Central America and perhaps Peru. *Anastrepha suspensa* (Loew), probably better termed the Greater Antillean fruit fly than the Caribbean fruit fly because of its limited original distribution (Whervin, 1974), has been similarly introduced into Florida (Weems, 1965).

Many other *Anastrepha* species are known from limited areas or single localities. Much of this represents true endemism (e.g., in the case of *A. suspensa*), but many other apparently limited distributions are probably the result of limited collecting. For example, *Anastrepha fenestrata* Lutz and Lima is known only from Panama and Amazonia, Brazil (Stone, 1942a), although it probably occurs in much of the intermediate area. Many other species are known only from Panama, particularly the Canal Zone, which is probably the best known limited tropical area of the range, thanks to exhaustive trapping and rearing collections made by James Zetek and staff of the former Division of Fruit Fly Investigations, USDA, during the 1930's and 1940's. Similar efforts are needed in other areas and will undoubtedly extend our knowledge of the range of many species. For example, the survey being conducted by La Programa Moscamed, Sanidad Vegetal, Mexico, in the state of Chiapas, has already discovered a number of *Anastrepha* species previously known only from Panama (P. Liedo, M. Aluja, pers. comm.).

The Antilles (excluding Trinidad) and Florida have a distinctive *Anastrepha* fauna; except for *A. obliqua* and *Anastrepha dissimilis* Stone, the species occurring there are endemic. There are also species restricted to the higher latitudes of the range that might be regarded as subtropical, for example, *A. ludens, Anastrepha dentata* (Stone), and *Anastrepha bicolor* (Stone) in the north, and *Anastrepha punctata* Hendel and *Anastrepha elegans* Blanchard in the south. There may also be distinctive tropical Mesoamerican and Amazonian faunal elements and perhaps Andean endemics (e.g., *Anastrepha atrox* (Aldrich)), but further collecting is needed to confirm the limited distributions of the species involved.

It does appear that there is little correlation between phylogenetic relationships and endemism in *Anastrepha*. There are a number of small monophyletic groups which are endemic, for example, *Anastrepha stonei* Steyskal and a related new species in the Caribbean, but most of the larger species groups are widely distributed. This suggests a high potential for biogeographical analysis at low taxonomic levels, that is, by comparing distributions within different species groups, once the ranges and relationships of the species are better known.

6 HOSTS

Anastrepha species breed almost exclusively in the pulp or seeds of mainly fleshy-type fruits; *Anastrepha pickeli* Lima sometimes attacks tubers of cassava (Peña and Bellotti, 1977), but they also breed in the seed pods. An extremely broad range of hosts is attacked by *Anastrepha*; the published records compiled by Norrbom (1985) include 254 plant species in 41 families. This impressive diversity is misleading in several respects, however. Some records are undoubtedly based on misidentifications, and many reported hosts are probably rarely, if ever, attacked in the field. These include incidental hosts attacked only in the vicinity of normal hosts with heavy fly populations (Fischer, 1934; Swanson and Baranowski, 1972), and others known only from laboratory rearings (e.g.,

Baker et al., 1944; McAlister, 1936). Seven generalist species also account for over two-thirds of the records and a third of the hosts are exotic. Although shifts to these plants, which include citrus, mango, and other commercial crops, are significant, they could not have played a role in *Anastrepha* evolution until very recently. On the other hand, no hosts are known for over half of the *Anastrepha* species and there must be many unknown native hosts remaining to be discovered.

The degree of host specificity in *Anastrepha* is not well understood; it probably varies considerably within the genus. Many species have been reared from single or only a few closely related hosts, although this may represent an artifact of limited collecting rather than true specificity in many cases. However, a species such as *Anastrepha pallens* Coquillett, known only from *Bumelia celastrina* H.B.K. (McPhail and Berry, 1936) is probably narrowly oligophagous if not monophagous. Other species, including the major pests discussed in the introduction to this chapter, are broadly polyphagous; for example, *A. suspensa* has 96 reported hosts. Even these species have distinct preferences, however; for example, in Jamaica, *A. suspensa* is the main species attacking Myrtaceae and Sapotaceae, whereas *A. obliqua* is found mostly on Anacardiaceae (Whervin, 1974). Both species are rarely reared from the other's normal hosts, however; thus secondary hosts may play an important role in maintaining a fly population when preferred hosts are unavailable (Swanson and Baranowski, 1972). Zwölfer (1983) discussed tephritid life strategies and classified most of the pest species of *Anastrepha* in the Type 1 category, whose members exhibit a range of behaviors connected with broad polyphagy, such as often mating on non-hosts, lek formation by males, and non-synchronization with host phenology. If some *Anastrepha* are more host specific, they may lack these behavioral traits, as Landolt and Hendrichs (1983) and Landolt (1984) found in *Toxotrypana curvicauda* Gerstaecker, which breeds mainly in papayas. The seed feeding and pulp feeding *Anastrepha* species may also exhibit behavioral differences; for example, in nutritional requirements and the length of the premating period, and they should likewise have different impacts on their hosts' reproductive capabilities.

There is limited congruence between phylogenetic relationships within *Anastrepha* and host plant associations. Excluding exotic hosts and laboratory rearings, almost all of the hosts known for the species groups revised by Norrbom (1985) belong to the Sapotaceae. Other probably monophyletic groups within the genus appear to prefer, or to be restricted to, *Passiflora* or Bombacaceae. Host shifts have obviously been frequent in *Anastrepha*, however, as there is considerably greater host variability in other species groups. A better understanding of this congruence and of the importance of host shifts in the evolution of *Anastrepha* will require further analysis of *Anastrepha* phylogenetic relationships and much more intensive collecting to discover more of the native host plants.

7 FUTURE NEEDS

In view of its rising importance to New World agriculture, the genus *Anastrepha* requires a great deal of further biosystematic study. Earlier in this chapter we indicated specific areas where research is needed, but we re-emphasize here those we consider to be most crucial.

1. Comprehensive revisionary studies of fruit fly taxa such as those being conducted by the senior author on *Anastrepha*, should be emphasized. Long-term goals of this research should include (a) development of a phylogenetically based, predictive classification as a basis for analyzing all types of bio-

systematic data from an evolutionary perspective; (b) further development of a computerized biosystematic data base, including host plant and distribution data; and (c) production of comprehensive keys for males and immatures as well as females.

2. Survey efforts, which in the past have consisted mainly of trapping adults in commercial orchards, should be drastically modified. Much more useful information could be gathered by including more native hosts in surveys and by rearing specimens from fruit. Such rearing programs would provide (a) host information badly needed to formulate and carry out quarantine regulations, and (b) immature specimens, properly associated with adults, that could be made available for taxonomic studies.

3. Special effort should be made to encourage continued taxonomic studies of fruit fly immatures and to foster cooperation between taxonomists and economic workers to collect material. Permanent, well curated collections of tephritid immatures should also be developed (Anon., 1982), and in this regard the senior author will provide instructions for proper rearing and preservation techniques and will accept specimens for deposit in the collection of the U.S. National Museum of Natural History, where they will be made available to all workers.

8 REFERENCES

Aldrich, J.M., 1925. New Diptera or two-winged flies in the United States National Museum. Proceedings of the United States National Museum 66 (Art. 18): 1–36.

Anon., 1982. Report of *Anastrepha* larvae workshop. United States Department of Agriculture, Brownsville, Texas, September 22–23, 1982. Unpublished.

Baker, A.C., Stone, W.E., Plummer, C.C. and McPhail, M., 1944. A review of studies on the Mexican fruit fly and related Mexican species. United States Department of Agriculture Miscellaneous Publication, 531: 1–155.

Berg, G.H., 1979. Pictorial key to fruit fly larvae of the family Tephritidae. O.I.R.S.A., San Salvador, 36 pp.

Bezzi, M., 1909. Le specie dei generi *Ceratitis, Anastrepha, y Dacus*. Bolletino del Laboratoria di Zoologia generale e Agraria delle R. Scuola Superiore d'Agricultura in Portici, 3: 273–313.

Bezzi, M., 1919a. Una nuova especie brasiliana dei genere *Anastrepha*. Loc. cit., 13: 3–14.

Bezzi, M., 1919b. Descoberta de uma nova mosca das fructas no Brasil. Chacares y Quintaes, 19: 373–374.

Blanchard, E.E., 1961. Especies argentinas del género *Anastrepha* Schiner (sens. lat.) (Diptera: Tephritidae). Revista de Investigationes Agricolas 15(2): 281–342.

Bush, G.L., 1962. The cytotaxonomy of the larvae of some Mexican fruit flies in the genus *Anastrepha*. Psyche, 69: 87–101.

Caraballo, J., 1981. Las moscas de frutas del género *Anastrepha* Schiner, 1868 (Diptera: Tephritidae) de Venezuela. M.S. Thesis, Universidad Central de Venezuela, Maracay.

Caraballo, J., 1985. Nuevas especies del género *Anastrepha* Schiner, 1968 (Diptera: Tephritidae) de Venezuela. Boletin Entomologia Venezuelana, 4: 25–32.

Emmart, E.W., 1933. The eggs of four species of fruit flies of the genus *Anastrepha*. Proceedings of the Entomological Society of Washington, 35: 184–191.

Fernandez Yepez, F., 1953. Contribucion al estudio de las moscas de las frutas del género *Anastrepha* Schiner (Diptera: Tephritidae) de Venezuela. II. Congreso Ciencas Nat. Afin. (Caracas), No. 7: 5–24.

Fischer, C.R., 1932. Nota taxonomica e biologica sobre *Anastrepha grandis* Macq. (Diptera: Trypetidae). Revista de Entomologia, 2: 302–310.

Fischer, C.R., 1934. Variacão das cerdas frontaes e otras notas sobre duas especies de *Anastrepha* (Diptera: Trypetidae). Loc. cit., 4: 17–22.

Foote, R.H., 1980. Fruit fly genera south of the United States (Diptera: Tephritidae). United States Department of Agriculture Technical Bulletin 1600, 79 pp.

Foote, R.H., 1982. The Tephritidae (Diptera) of the Galapagos Archipelago. Memoirs of the Entomological Society of Washington, 10: 48–55.

Greene, C.T., 1934. A revision of the genus *Anastrepha* based on a study of the wings and on the length of the ovipositor sheath. Proceedings of the Entomological Society of Washington, 36: 127–179.

Hendel, F., 1914. Die Bohrfliegen Südamerikas. Königliche Zoologischen und Anthropologisch-Ethnographischen Museums zu Dresden, Abhandlungen und Berichte Berlin, (1912) 14 (3): 1–84.

Kitto, G.B., 1983. An immunological approach to the phylogeny of the Tephritidae. In: R. Cavalloro (Editor), Fruit Flies of Economic Importance. A.A. Balkema, Rotterdam, pp. 203–211.

Korytkowski, C. and Ojeda Peña, D., 1968. Especies del género *Anastrepha* Schiner, 1868 en el nor-oeste Peruano. Revista Peruano Entomologie (Lima), 11: 32–70.

Landolt, P.J., 1984. Behavior of the papaya fruit fly, *Toxotrypana curvicauda* Gerstaecker (Diptera: Tephritidae), in relation to its host plant, *Carica papaya* L. Folia Entomologia Mexicana, 61: 215–224.

Landolt, P.J. and Hendrichs, J., 1983. Reproductive behavior of the papaya fruit fly, *Toxotrypana curvicauda* Gerstaecker (Diptera: Tephritidae). Annals of the Entomological Society of America, 76: 413–417.

Lima, A. da Costa, 1934. Moscas de frutas do gênero *Anastrepha* Schiner, 1868. Memorias Instituto Oswaldo Cruz, 28: 487–575.

Lima, A. da Costa, 1937–1938. Novas moscas de frutas do gênero *Anastrepha*. O Campo, 8:34–38 (June), 60–64 (October), 9: 61–64.

Loew, H., 1873. Monographs of the Diptera of North America. Part III. Smithsonian Miscellaneous Collections, 11(3) (publication 256): 1–351.

McAlister, L.C., Jr., 1936. Observations on the West Indian fruit fly in Key West in 1932–33. Journal of Economic Entomology, 29: 440–445.

McPhail, M. and Berry, N.O., 1936. Observations on *Anastrepha pallens* reared from wild fruits in the lower Rio Grande Valley of Texas during the spring of 1932. Journal of Economic Entomology, 29: 405–410.

Mendes, L.O.T., 1958. Observacoes citologicas em "Moscas dos frutas". Bragantia, 17: 29–39.

Morgante, J.S., Malavasi, A. and Bush, G.L., 1980. Biochemical systematics and evolutionary relationships of Neotropical *Anastrepha*. Annals of the Entomological Society of America, 73: 622–630.

Norrbom, A.L., 1985. Phylogenetic analysis and taxonomy of the *cryptostrepha, daciformis, robusta,* and *schausi* species groups of *Anastrepha* Schiner. PhD. Dissertation, Pennsylvania State University, University Park.

Peña, J.E. and Bellotti, A.C., 1977. Estudios sobre los moscas del tallo y fruto de yuca: *Anastrepha pickeli* y *Anastrepha manihoti*. Revista Columbiana Entomologia, 3: 79–86.

Schiner, I.R., 1868. Diptera (Article 1), In: Reise der Österreichische Fregatte Novara um die Erde, Zoologie 2 (Arbeit 1, Section B), Wien [Vienna], 388 pp.

Steyskal, G.C., 1977a. Two new neotropical fruit flies of the genus *Anastrepha*, with notes on generic synonymy. Proceedings of the Entomological Society of Washington, 79: 75–81.

Steyskal, G.C., 1977b. Pictorial key to species of the genus *Anastrepha*. Entomological Society of Washington, Washington, D.C., 25 pp.

Stone, A., 1939a. A revision of the genus *Pseudodacus* Hendel. Revista de Entomologia, 10: 282–289.

Stone, A., 1939b. A new genus of Trypetinae near *Anastrepha*. Journal of the Washington Academy of Sciences, 29: 340–350.

Stone, A., 1942a. The fruit flies of the genus *Anastrepha*. United States Department of Agriculture Miscellaneous Publication 439, 112 pp.

Stone, A., 1942b. New species of *Anastrepha* and notes on others. Journal of the Washington Academy of Sciences, 32: 298–304.

Swanson, R.W. and Baranowski, R.M., 1972. Host range and infestation by the Caribbean fruit fly, *Anastrepha suspensa* (Diptera: Tephritidae), in south Florida. Proceedings of the Florida State Horticultural Society, 85: 271–274.

Teskey, H.J., 1981a. Morphology and terminology – larvae, pp. 65–88. In: J.F. McAlpine, et al. [coordinators], Manual of Nearctic Diptera, Vol. 1. Monograph 27, Research Branch, Agriculture Canada, Ottawa, 674 pp.

Teskey, H.J., 1981b. Key to families – larvae. Loc. cit., pp. 125–147.

Weems, H.V., Jr., 1965. *Anastrepha suspensa* Loew (Diptera: Tephritidae). Florida Department of Agriculture, Bureau of Entomology, Circular No. 38: 1–4.

Whervin, L.W., van, 1974. Some fruit flies (Tephritidae) in Jamaica. PANS 20: 11–19.

Wiedemann, C.R.W., 1830. Aussereuropäische zweiflügelige Insekten. 2: xii 684 pp. Hamm.

Zucchi, R.A., 1978. Taxonomie das especies de *Anastrepha* Schiner, 1868 (Diptera: Tephritidae) assinaladas no Brasil. PhD. Dissertation, Escola Superior de Agricultura "Luis de Queiros", Universidad de São Paulo, Piricicaba.

Zucchi, R.A., 1979a. Novas especies de *Anastrepha* Schiner, 1868 (Diptera: Tephritidae). Revista Brasileira Entomologia, 23: 35–41.

Zucchi, R.A., 1979b. Sobre os tipos de *Anastrepha parallela* (Wiedemann, 1830), de *A. striata* Schiner, 1868e de *A. zernyi* Lima, 1934 (Diptera: Tephritidae). Loc. cit., 23: 263–266.

Zucchi, R.A., 1979c. Duas novas especies de *Anastrepha* Schiner, 1868 (Diptera: Tephritidae). Loc. cit., 23: 115–118.

Zucchi, R.A., 1981a. Notas taxonomicas sobre *Anastrepha consobrina* Loew, *A. zikani* Lima, e *A. amnis* Stone (Diptera: Tephritidae). Loc. cit. 25: 5–8.

Zucchi, R.A., 1981b. *Anastrepha* Schiner, 1868 (Diptera: Tephritidae): Novas sinonimies, Loc. cit., 25: 289–294.

Zucchi, R.A., 1982. A new species of fruit fly of the genus *Anastrepha* Schiner, 1868 (Diptera: Tephritidae) from Brazil. Anais da Sociedade Entomológica do Brasil, 11: 251–254.

Zucchi, R.A., 1984. A new species of *Anastrepha* (Diptera: Tephritidae) from the Amazon Region. Loc. cit., 13: 279–280.

Zwölfer, H., 1983. Life systems and strategies of resource exploitation in tephritids. In: R. Cavalloro (Editor), Fruit Flies of Economic Importance. A.A. Balkema, Rotterdam, pp. 16–30.

Chapter 1.4 A Contribution towards the Zoogeography of the Tephritidae

PETER A. MADDISON and BRENDAN J. BARTLETT

1 INTRODUCTION

The study of the zoogeography of any group of animals is governed by the information available on their systematics and distribution. The well-tried maxim 'absence of evidence is not evidence of absence' rings true in this context.

For the Tephritidae, it seems that the state of the art is that for some groups which are economically important, such as the subfamilies Dacinae and Trypetinae, the systematic and distributional evidence is well advanced, when compared with that for the seed flies in the subfamily Tephritinae. As an example of the state of taxonomic knowledge of this family, the situation of the Hawaiian fauna can be cited. Before 1980, 15 species had been described; of these 15, 11 were introduced or of cosmopolitan occurrence. After the work of Hardy and Delfinado (1980), the number of species of Tephritidae known from Hawaii has risen to 36 (Table 1.4.1).

Several large-scale eradication schemes against fruit-flies have been conducted; since these have involved fruit rearing and attractant trapping, extensive collections have been made. Because specimens of non-economic species have often been trapped (in low numbers) at the lures or bred from native host plants during the course of such work, they have been available to taxonomists for description and for systematic study.

Some groups of Tephritinae (seed flies) have also been well studied because of the potential of some of these seed-feeding species for biological control of weeds, particularly those in the family Asteraceae (Compositae). However it is probable that the fauna of non-economic Tephritidae awaiting description is still large and this, if the Hawaiian experience is repeated, will be particularly true for the subfamily Tephritinae. It goes almost without saying that the distribution data is also only as good as the collections that have been made; though it is important to note that this is only partly true – if the collected material is held in such a way as not to be accessible to the taxonomists, then so also is the information associated with such specimens.

2 SOURCES OF INFORMATION

In preparing this account of zoogeography the authors have been somewhat constrained by the available information on distribution. The published lists of Diptera of the Afrotropical Region (Crosskey, 1980), the Oriental Region (Delfinado and Hardy, 1977), the Northern American Region (Stone et al., 1965) and the South American Region (Foote, 1967) are primary sources. These have been

Chapter 1.4 references, p. 34

TABLE 1.4.1

The Tephritidae of Hawaii (Data from Hardy and Delfinado, 1980)

Subfamily	No. of species	No. described before 1980	No. of endemic sp.
Dacinae	2	2	0
Oedaspinae	2	2	0
Tephritinae	31	10	25
Trypetinae	1	1	0

supplemented with several papers by Drew (1972–82) on Australasia and the Pacific Region; by Hardy and Delfinado (1980) on Hawaii, by Hardy (1974) on the Philippines, by Malloch (1933) on Patagonia, by Curran (1934) on British Guiana, and by Tan and Lee (1982) on Malaysia. Also consulted were several maps on insect distribution produced by the Commonwealth Institute of Entomology which however deal with only a few major species of fruit fly. The distributional information available from these and other references cited has been collated, so that information on the distribution of about 2500 species was available.

An additional problem is that of the 'state of the art' of Tephritid systematics. Because of the morphological similarity of some species, there have been a number of misidentifications in the past. These records, unless rectified in a major publication, can and do persist in the literature. Examples include the records of *Dacus facialis* Coquillett from New Caledonia and of *Dacus psidii* (Froggatt) from French Oceania and Western Samoa (Dumbleton, 1934). Though these records have not been substantiated by the revisions of Drew (1974) and Drew et al. (1982) of the Pacific Dacinae, the records persist in the literature and are raised periodically in relation to inter-island quarantine in the South Pacific. A similar case persists for *Dacus dorsalis* Hendel from Northern Australia. The recent description (Drew and Hardy, 1981) of the sibling species *Dacus opiliae* Drew and Hardy for this Northern Australian population has not been noticed by some agriculturalists, who continue to record *D. dorsalis* from Australia.

One further difficulty in this context is the tendency of some authors to accept records from quarantine sources as gospel. However, anyone who has worked with quarantine services, will realise that a tourist who has visited several countries will be often vague about where actually he/she bought some particular produce. The same doubt can occur over commercial shipments where trans-shipment occurs. One example of what can happen is the record of *Dacus melanotus* Coquillett in peaches from the Cook Islands. Though the locality is right, there was to our knowledge only one peach tree on the Cook Islands at this period and its fruit never escaped the attention of the local children!

3 BEHAVIOUR OF TEPHRITIDAE

Several authors indicate that fruit flies are capable of wide dispersal, as for example the statement by Hely et al. (1982):- 'The adults of [*Dacus tryoni*] are strong fliers capable of travelling some miles in a lifetime. In other countries related flies have been shown to travel up to 64 kilometres (wind-assisted) over water.' Further studies have shown a dispersal of at least 100 km by adult flies. These flights usually occur either between emergence from the soil and the onset of sexual maturity or after disappearance of a fruit species used for oviposition or to and from overwintering areas (Zwölfer, 1982). Miyahara and Kawai (1979) indicate dispersal up to 200 km. Whether these are true 'migration' or merely dispersal movements is debatable.

However the 'normal' behaviour of these flies in the presence of abundant fruit trees is that they show low vagility (using this term in the sense of 'intrinsic ability of a species to spread' (Udvardy, 1969)). Typical observations of the behaviour of Dacinae is that the flies show little willingness to move out of an area – if disturbed a fly may move a few metres to a neighbouring fruit or leaf or may return to the same site from which it was distributed. Territorial behaviour has been recognised in *Urophora* and is probable in other Tephritinae. Fruit marking by ovipositing fruit flies (Dacinae) has also been observed (Zwölfer, pers. comm.).

So in one sense their behaviour could be termed 'territorial' (though they don't vigorously defend a territory) or 'philopatric' (in the sense of species that are inclined to remain in their 'home range'). In ecological terms, the majority of Tephritidae are stenotropic ('able to inhabit only special habitat types, because their tolerance limits are narrow or their requirements are specific' (Udvardy, 1969)). A few species, on the contrary, are eurytropic ('able to adapt to many different kinds of habitat because their tolerance limits are wide and/or their requirements are small' (Udvardy, 1969)). Amongst the latter are Dacinae like *D. dorsalis, Dacus tryoni* (Froggatt) and *Dacus cucurbitae* Coquillett, and Tephritinae like *Ceratitis capitata* (Wiedemann). However it should be noted at this point that some of the success of this latter group of species is due to assisted distribution by man. There is ample evidence that the widely distributed fruit flies can travel as larvae inside fruit shipments. The question as to why only a few species have become major pests may lie in their polyphagy as to fruit colonised and their eurytropic nature. Using the terminology of Southwood (1981) the Tephritidae belong to the 'r-strategists' since they occupy niches which are of a temporary nature. However most species do not show the 'boom and bust' strategy of extreme r-strategists and the lack of pronounced migratory tendencies put them at the low end of the r-spectrum.

4 HOST PLANTS

As indicated above, the larval feeding habits of the Tephritidae range from strict monophagy through oligophagy to polyphagy. As with any group of insects, the feeding habits of many species, particularly those of the Tephritinae, are unknown. However, it is probable that the majority of these are restricted to the fruits or seeds of one or a few hosts.

For a monophagous species, it is self-evident that the distribution of the species will never be greater than that of its food plants, though other ecological factors may well restrict its range inside that available from its hosts' distribution. There may be temporal factors (e.g. fruiting seasons) which affect the species ability to colonise other areas. It should be noted at this point that this monophagy has great value for biological control of weeds e.g. *Eutreta xanthochaeta* Aldrich has been widely introduced for the control of *Lantana camara* Linnaeus (Verbenaceae), and *Tetraeuarestra obscuriventris* Loew for the control of *Elephantopus* spp. (Asteraceae).

Oligophagous and polyphagous species are often of widespread distribution, particularly if the fruiting/seeding seasons of the host plants overlap. However the distribution of these species may be restricted by other environmental factors – physical or climatic barriers (deserts, mountain ranges) or biological factors (vegetational type, temporal fruiting/seeding patterns of potential

Chapter 1.4 references, p. 34

hosts, etc.). Mankind has had major effects on the distribution of Tephritidae in 3 ways: (1) in increasing the distributional range of weed hosts (particularly genera of Asteraceae (Compositae) such as *Bidens* and *Elephantopus* which are hosts of the Tephritinae; (2) in the wide cultivation of fruit trees associated with human habitation increasing the host availability in an area, and (3) creating 'corridors' by which fruit-flies may have dispersed far beyond their natural home range, a product of both shifting and plantation agriculture.

5 DISTRIBUTION OF SELECTED GENERA

Figures 1.4.1–1.4.5 indicate the distribution of described species of 6 genera of Tephritidae.

The large Neotropical genus *Anastrepha* (Figs. 1.4.1 A and B) has 152 species distributed from Central America and the West Indies to Argentina and Chile.

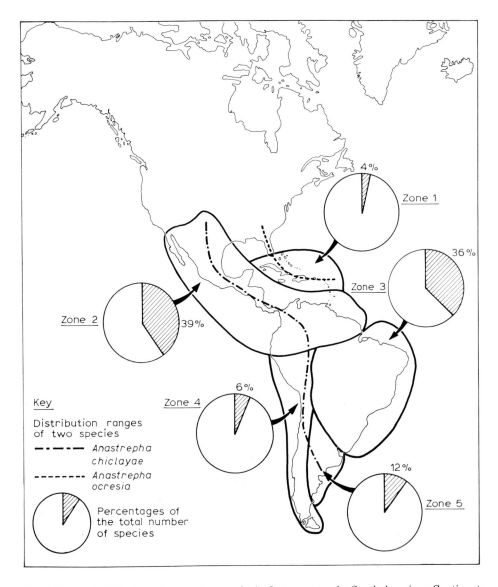

Fig. 1.4.1. (A). Distribution of genus *Anastrepha* in five zones on the South American Continent; 152 species.

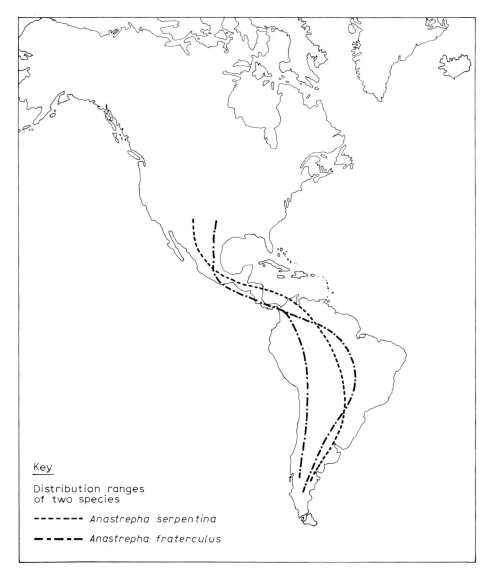

Fig. 1.4.1. (B). Distribution of two species of the genus *Anastrepha*.

Most species are of limited distribution within the tropical and subtropical forests – this limitation of range could be due in part to physical, climatic and gross vegetational factors, but is more likely to be a result of specificity on fruits of particular host plants. However so little information is available on the host range of these species that it is difficult to comment further. The few wide-ranging species of this genus (e.g. *A. fraterculus*) seem either to be poly-phagous or to have a host plant that is widespread, either naturally or as a result of spread by man.

The four species of *Ceratitis* include one widespread, polyphagous species, the medfly (*C. capitata*) and three very localised species restricted to East Africa, Madagascar and nearby Islands (Fig. 1.4.2). Historically some of the spread of the medfly has been documented. The origin of the medfly could lie either in eastern/southern Africa and Madagascar, or as one of those species of animals and plants that are believed to have evolved over the tract between the Mediterranean region and southern Africa (Hagan et al., 1981). Since this species seems to have travelled well with man it seems reasonable to suggest that its origins lie with its congeners in Southern Africa/Madagascar and that

Chapter 1.4 references, p. 34

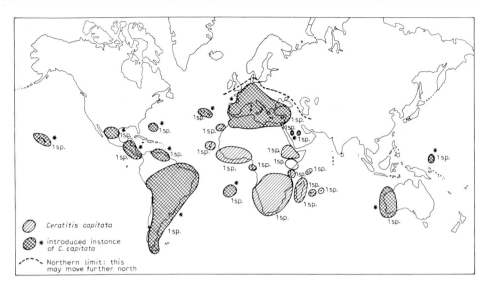

Fig. 1.4.2. Distribution map of genus *Ceratitis*; 4 species.

it travelled along the east coast of Africa to the Mediterranean with the Arab coastal trade.

The distribution of the widespread genus *Tephritis* (Fig. 1.4.3) seems to show several centres of speciation – in Australasia, India, North America and western and southern South America with single species in north-east Africa and southern Africa. One species is Holarctic occurring in Scandinavia, Alaska and north-west Canada and U.S.A. The larval hosts of this genus are seeds of plants of the family Asteraceae – one of the most widespread plant families. This would seems to be an 'old' genus, with elements which show a Gondwana-like, and others a Tethyan distribution. Until studies of the inter-relationships of the species from different areas are made, it is difficult to comment further on the zoogeography of this group.

The genus *Myoleja* has two, or possibly three, speciation centres (Fig. 1.4.4). There are 27 species in the Indo-Pacific region, one in western U.S.A. and two in eastern U.S.A. The larvae of species of this genus feed on plants in the family Asteraceae (Compositae). This trans-Pacific distribution is shown in other

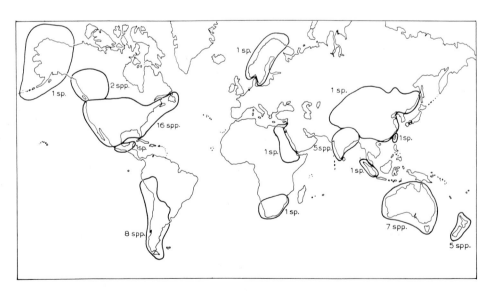

Fig. 1.4.3. Distribution map of genus *Tephritis*; 46 species.

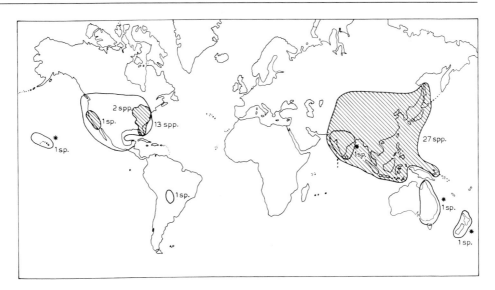

Fig. 1.4.4. Distribution map of the genera: (1) ○ *Procecidochares*, total 13 species; (2) ○ *Myoleja*, total 30 species: ○*: introduced.

groups such as the Cucullinae (Noctuidae), Hepialidae and Sabatincoid Micropterigidae in the Lepidoptera (Dugdale, pers. comm.) and in various plant groups (Van Steenis, 1962). Further comment should await a study of the taxonomic relationships within this genus.

Procecidochares has a native distribution within the New World (Fig. 1.4.4). There are 13 species of localised distribution in North America, and a single species in South America (Fig. 1.4.4). The larval hosts of this genus are plants of the family Asteraceae (Compositae) and are restricted to *Ageratina riparia* and related genera. The restricted distribution of species in the genus may be due to their host specificity, a factor recognised in the utilization of *P. utilis* for biological control of *Eupatorium adenophorum* in Australasia, Hawaii and India.

Species of the genus *Dioxyna* (Fig. 1.4.5) include one widespread species (which is however absent from the Americas) and three other species groups – one in central and southern North America and the West Indies (2 species), another in central and southern South America (3 species), and a third in the

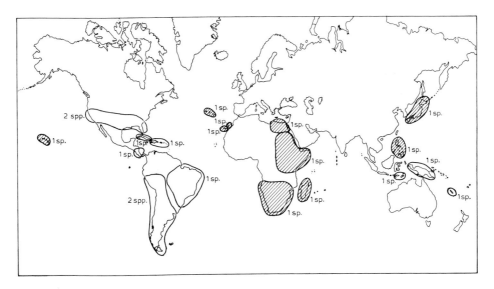

Fig. 1.4.5. Distribution map of genus *Dioxyna*; total 8 species; ⊘ *Dioxyna sororcula*.

Chapter 1.4 references, p. 34

Melanesian/Philippines area (2 species). The larval hosts of this genus are plants of the family Compositae. The distribution suggests again that this might be an 'old' genus, with trans-Pacific links as in *Myoleja*. The origin of the widespread species, *D. sororcula* is uncertain but it is possibly east Asian, becoming more wide-spread because of the man-induced spread of its host plants, (genera such as *Bidens, Coreopsis, Lactuca*).

6 CONCLUSION

What is known of the distribution and biology of the members of the Tephritidae shows evidence that mankind has played an important part in altering the distribution of some of the more polyphagous species as well as of certain oligophagus species where the hosts have been spread by man. The frequency with which fruit flies are intercepted in trade (106 interceptions in New Zealand between 1979 and 1982) indicates that the movement of fruits is important in the spread of these species.

Similarly, since the hosts of many of the Tephritinae are Asteraceae, their range has been capable of extension if the host has become a weed, as indeed have many of the plants in this family. The oligophagy of these flies has enabled some species to be utilized for the biological control of these weed species, (e.g. *Procecidochares utilis, Tetraeuaresta obscuriventris*).

However, remarks on the zoogeography of many genera within this family must await further information on the host associations as well as global taxonomic studies on the inter-relationships of species in those genera with species described over a broad distributional range.

The Tephritidae in Hawaii seem to have speciated in the same way though not to the same extent as the Drosophilidae (25 endemic species as opposed to 800 + for the latter). Some recent studies (Beverly and Wilson, 1985) have indicated that, based on protein molecular analysis, the evolution of the Hawaiian Drosophilids started a little more than 40 million years ago, seven times longer than the present islands existed. It would be interesting to find out if the Hawaiian Tephritidae had a similar origin.

7 ACKNOWLEDGEMENTS

We would like to record thanks to the colleagues who commented and gave their assistance in the preparation of this review. These include Dr R.C. Craw and Mr J.S. Dugdale, Entomology Division, D.S.I.R. Auckland, Mr P.S. Dale, Ministry of Agriculture and Fisheries, Lynfield, Auckland, and Dr H. Zwölfer, Lehrstuhl Tierökologie, Universität Bayreuth. We are also grateful to Dr R.A.I. Drew, Department of Primary Industry, Indooropilly, Queensland, for additional comments on the manuscript.

8 REFERENCES

Beverley, S.M., Wilson, A.C., 1985. Ancient Origin for Hawaiian Drosophilinae inferred from protein comparisons. Proceedings of the National Academy of Science, U.S.A., 82 (14): 4753–4757.

Crosskey, R.W., 1980. Catalogue of the Diptera of the Afrotropical Region British Museum (Natural History) London, pp. 518–554.

Curran, C.H., 1934. The Diptera of Kartabo, Bartica District British Guiana. Bulletin of the American Museum of Natural History, 66 (3): 432–436.

Delfinado, M.D. and Hardy, D.E., 1977. A catalogue of the Diptera of the Oriental Region. Vol III. Suborder Cyclorrhapha (excluding Division Aschiza). The University Press of Hawaii, Honolulu, pp. 44–134.

Drew, R.A., 1972. Additions to the species of Dacini (Diptera: Tephritidae) from the South Pacific area with Keys to Species. Journal of the Australian Entomological Society, 11: 185–231.

Drew, R.A., 1972. The Generic and Subgeneric Classification of Dacini (Diptera: Tephritidae) from the South Pacific Area. Journal of the Australian Entomological Society, 11: 1–22.

Drew, R.A., 1973. Revised Descriptions of Species of Dacini (Diptera: Tephritidae) from the South Pacific area I. Genus *Callantra* and the *Dacus* Group of Subgenera of Genus *Dacus*. Queensland Dept. of Primary Industries, Division of Plant Industry Bulletin No. 652, pp. 1–39.

Drew, R.A., 1974. Revised Descriptions of Species of Dacini (Diptera: Tephritidae) from the South Pacific area. II. The *Strumeta* Group of Subgenera of Genus *Dacus*. Queensland Dept of Primary Industries, Division of Plant Industry Bulletin No. 653, pp. 1–101.

Drew, R.A., 1975. Zoogeography of Dacini (Diptera: Tephritidae) in the South Pacific Area. Pacific Insects, 16 (4): 441–454.

Drew, R.A. and Hardy, D.E., 1981. *Dacus* (Bactrocera) *opiliae*, a new sibling species of the *dorsalis* Complex of Fruit Flies from Northern Australia (Diptera: Tephritidae). Journal of the Australian Entomological Society, 20: 131–137.

Drew, R.A., Hooper, G.H.S. and Bateman, M.A., 1982. Economic Fruit Flies of the South Pacific Region. 2nd edition. Watson Ferguson and Co., Brisbane, pp. 1–139.

Dumbleton, L.J., 1934. A list of insect pests recorded in South Pacific territories. South Pacific Commission Technical Paper: 79. 202 pp.

Foote, R.H., 1967. A Catalogue of the Diptera of the Americas South of the United States 57. Family Tephritidae (Trypetidae, Trupaneidae). Departmento de Zoologia, Secretaria da Agricultura, Sao Paulo, pp. 1–91.

Hagan, K.S., Allen, W.W. and Tassan, R.L., 1981. Mediterranean Fruit Fly: the worst may be yet to come. California Agriculture, 35 (3/4): 5–7.

Hardy, D.E., 1974. The Fruit Flies of the Philippines (Diptera: Tephritidae). Pacific Insects Monograph 32. Entomology Department, Bernice, P. Bishop Museum, Honolulu, Hawaii, U.S.A., pp. 1–266 + 6 plates.

Hardy, D.E. and Delfinado, M.D., 1980. Insects of Hawaii Volume 13 Diptera: Cyclorrhapha. The University Press of Hawaii, Honolulu, 451 pp.

Hely, P.C., Pasfield, G. and Gellatley, J.G., 1982. Insect Pests of Fruit and Vegetables in New South Wales. Inkata Press, Melbourne, 312 pp.

Malloch, J.R., 1929–48. In: 'Diptera of Patagonia and South Chile. Based mainly on material in the British Museum (Natural History), Part VI. Brachycera (Cyclorrhapha) (Aschiza and Acalyptrata). pp. 263–296. Printed by Order of the Trustees of the British Museum.

Miyahara, Y. and Kawai, A., 1979. Movement of Sterilized Melon Fly from Kume Is. to the Amani Islands. Applied Entomology and Zoology, 14 (4): 496–497.

Southwood, T.R.E., 1981. Bionomic strategies and population parameters. pp. 30–52. In: R.M. May (Editor), Theoretical Ecology. Principles and Applications 2nd Edition. Blackwell Scientific Publications, Oxford, pp. 1–489.

Stone, A., Sabrosky, C.W., Wirth, W.W., Foote, R.H. and Coulson, J.R., 1965. A catalog of the Diptera of America North of Mexico. pp. 658–678. Agricultural Research Service, United States Department of Agriculture Washington, D.C.

Tan, K. and Lee, S., 1982. Species Diversity and abundance of *Dacus* (Diptera: Tephritidae) in five ecosystems of Penang West Malaysia. Bulletin of Entomological Research, 72: 709–716.

Udvardy, M.D.F., 1969. Dynamic Zoogeography with special reference to land animals. Van Nostrand Reinhold Company, New York, 445 pp.

Van Steenis, C.G.G.J., 1962. The land-bridge theory in botany. Blumea, 11: 235–372.

Zwölfer, H., 1982. Life Systems and Strategies of Resource Exploitation in Tephritids. In: Fruit flies of economic importance. Proceedings of the CEC/IOBC International Symposium, Athens, 16–19 November 1982, edited by R. Cavalloro, pp. 16–30.

PART 2

PEST STATUS

Chapter 2.1 Mediterranean Region

P. FIMIANI

1 INTRODUCTION

The most important fruit flies in the Mediterranean region are the Olive fruit fly: *Dacus oleae* (Gmelin), the European Cherry fruit fly: *Rhagoletis cerasi* (Linnaeus) and the Mediterranean fruit fly: *Ceratitis capitata* (Wiedemann). *D. oleae* causes heavy damage to the olive crop and is the key-pest in olive groves. *R. cerasi*, which occurs in central and mediterranean Europe, is an important pest in some areas, while in other areas it causes only negligible damage. However, some quarantine regulations by certain countries mean that even a small number of infested fruit can cause economic problems.

C. capitata is the most serious pest because it is widely distributed in the Mediterranean region, it has the highest number of generations per year, and it attacks a large variety of fruit.

2 *DACUS OLEAE* (GMELIN)

One of the most noxious insect pests to olive fruit, *D. oleae*, annually causes severe economic losses. These effects are more pronounced along the Mediterranean coasts where the favourable climatic conditions allow a high number of generations per year. It pupates to spend winter in the soil, except in those areas where the presence of fruits on trees enables larval and adult activity to continue. The olive culture, though sometimes underestimated, is an extremely important resource in the Mediterranean. Considering the high heterogeneity of production, variability of infestation, and the difficulty of evaluating the losses caused by *D. oleae*, make a precise assessment of its pest status difficult to give. Destroyed pulp, fruit fall which is often caused by the summer generation, and reduced oil quality are the most frequent damage caused by this pest.

Grison (1962) reported the following losses in the Mediterranean area: 5% in Spain, 25% in Italy, 30–35% in Greece, 20–40% in Yugoslavia, 20–60% in Israel, and 15–20% in Cyprus, but there are wide annual fluctuations. For example in Liban (Mechelany, 1982) in 1980 infestation was 27% in Chouf, 46% in Koura and 72% in Coté whereas in the following year the values were 7, 12 and 14% respectively. Losses are also appreciable in Mediterranean Africa. In general larvicides are used when infestation is in excess of 15% for olives destined for oil production, and 5% for table olives. Bait sprays are based on trap catches.

The technical and scientific aspects of this insect have been reviewed comprehensively by Martelli (1967) and Delrio (1979); in addition there are those by Young (1977) and Zwolfer (1985).

Chapter 2.1 references, p. 47

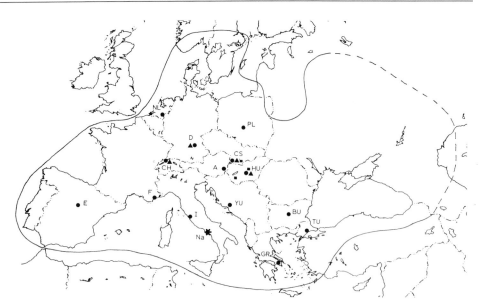

Fig. 2.1.1. Distribution of *Rhagoletis cerasi* (by Boller and Bush, 1974) and location of some samples from *Prunus* (●), *Lonicera* (▲) and *Berberis* (■).

3 *RHAGOLETIS CERASI* (L.)

This almost monophagous fly is a severe pest in many Mediterranean areas, and is known as the "European Cherry Fruit Fly" in order to distinguish it from its American *Rhagoletis* counterpart.

It occurs in almost the entire cherry producing area (Bezzi, 1927) of Europe, the southern limits of its distribution are the Mediterranean peninsulae and the islands of Baleares, Sardinia and Sicily (Fig. 2.1.1). It has recently been found in Crete (Neuenschwander et al., 1983), while Cyprus is still free from this pest (Orphanides, personal communication). Recent studies have shown that there are two different strains of *R. cerasi* in Europe (Boller and Bush, 1974; Boller et al., 1976; see Chapter 5.2, Vol. 3B) because of partial reproductive incompatibility.

The insect also can attack sour cherries or the fruit of some species of *Lonicera*. Infestation usually occurs just before the cherry ripens so that the extent of damage at harvest is variable (Boller et al., 1970). In some cases the adult population is very abundant and consequently fruit infestation can exceed 90%. However, the same level of infestation also can be reached with a lower adult population. In Italy infestation can vary greatly from one area to another (Fimiani et al., 1981). In some cases a correlation between adult population level was evident, in others high infestation was observed with a relatively low number of adults and vice versa. In some areas around Naples (Italy) more than one larva per fruit was often observed (Fimiani, 1984). The pest is also prevalent in Portugal, Spain, Yugoslavia, Greece and Turkey (Perko, 1983; Fimiani, 1986; Zumreoglu et al., 1981). It is impossible to separate the infested cherries from the crop and this often results in the loss of the entire crop because in many countries, an infestation tolerance limit of 2% makes export difficult. A large bibliography and synopsis on this pest has been published (Haisch et al., 1978).

4 *CERATITIS CAPITATA* (WIEDEMANN)

This important pest infests more than 250 types of fruit grown commercially in the Mediterranean area and is thus of great importance.

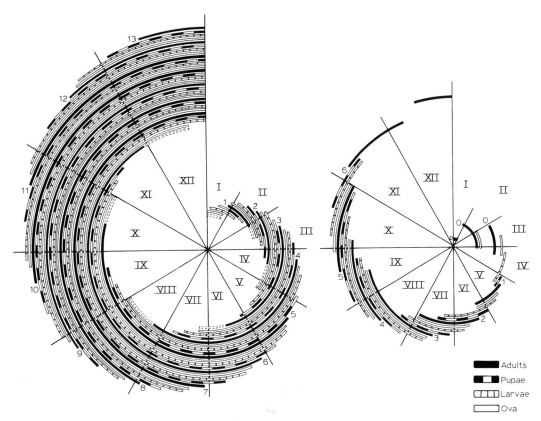

Fig. 2.1.2. The annual development of medfly under laboratory conditions (left) showing 13 generations compared with open air cages, near Naples (right) showing 6 generations.

In several parts of this region as well as in other subtropical and tropical areas, it exploits all or part of its biological potential and thus causes severe damage. In addition to the favourable climate, these areas offer an increased number of host fruits owing to the introduction of new species or cultivars. Outside these optimal or favourable areas i.e. in Central Europe, it is regarded as a secondary pest (Fimiani and Tranfaglia, 1972), (Fig. 2.1.2).

Furthermore the quarantine regulations of some importing countries and restrictions on the use of fumigants impose limits on the marketing of some species of fruit.

Of the various polyphagous fruit flies present in various parts of the world viz. *Anastrepha* spp., *Dacus* spp., and others, only *C. capitata* occurs in the Mediterranean area. In the past it was known as peach fly or orange fly but is now commonly called the medfly.

4.1 European Mediterranean area

4.1.1 Iberian peninsula

Medfly, then called *Ceratitis hispanica*, was first seen near Malaga in Spain and is the earliest record of this pest in Europe (De Breme, 1842). Later, in 1894, more medfly were found in Sevilla (Arroyo, in A.N.I.A., 1970) and a few years later still more were recorded in Portugal (Coutinho, 1898). In the first half of the 20th century the pest caused some problems in Spain with infestation of fruits, occurring in the months between the end of June and the beginning of August. Some export difficulties were experienced because of infested grapes (La Gasca, 1925). Medfly are also a key-pest in the Canary Islands (Mellado et al., 1966). Because of the different climatic conditions the pest is not evenly distributed in Spain, but damage caused to fruit production both in quality and quantity, led to mandatory control measures being implemented by the Agricultural Department (Traver in A.N.I.A., 1970).

Recently, especially in peach orchards where the fly causes large problems, it is under "Supervised control" (Cabezuelo and Sampayo, 1975). Depending on temperature and host availability, the number of generations can reach five in favourable areas, such as on the southern coast of Andalusia (Muñiz and Gil, 1984) where the insect is present all year.

4.1.2 France

High fruit production, as reported by French writers of long ago, was considerably reduced by a fly then called *Mouche a dard* (Ghesquiere, 1949) and medfly was reported on the French mediterranean coast in 1885, and occasionally in the area of Paris (Giard, 1900). The insect was again found in Paris by Lesne (1915) and in the Argenteuil area on peaches, apricots and late pears (Lesne, 1921).

The Côte d'Azur and nearly all of Provence appear to be the most favourable areas for medfly, and four generations per year are possible. The first generation develops on apricots and prickly pears the second and third generations develop on peaches in July to August, and the fourth in September on apples, pears or persimmons. However occasionally infestations can occur at more northern latitudes e.g. Lyon and Bordeaux (Diezeude, 1929; Balachowsky, 1932 in E.P.P.O., 1957).

The development of an autumnal generation is attributed to the presence of figs, persimmons and, as in Antibes, to the exotic plant *Brumelia lycioides* (Poutiers, 1938). However even in these areas, climatic conditions can reduce the fly population (Milaire, 1964). The question of whether the fly is endemic to France, and so the importance of quarantine was underlined to prevent introduction (Trouillon, 1976).

4.1.3 Italy

Medfly were first recorded in Calabria in 1863 and in the same area peaches were 100% infested (Martelli, 1910) and, subsequently became widely distributed (Costantino, 1930). In the Po Valley and around Bologna 2–3 generations per year are reported. The infestations are mainly on apricots and peaches in July, and on late peaches and pears in August (Ferrari, 1966).

In the Rome coastal area extremely variable infestations are recorded on peaches depending on zones and varieties but in September damage levels of

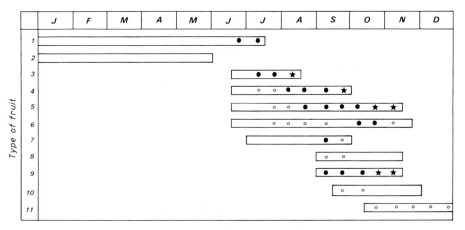

Fig. 2.1.3. Medfly situation on various types of fruit, near Naples during one year. In the center: the presence of various types of fruit and the different intensity of larval infestation on 1. sour oranges; 2. medlars; 3. apricots; 4. peaches; 5. pears; 6. figs; 7. plums; 8. Indian figs; 9. persimmons; 10. apples; 11. oranges. ○: light infestation: ●: medium infestations;★: heavy infestation.

70–90% can be reached and even total crop destruction can result (Cirio and Capparella, 1972). In the Naples coastal area heavy infestations on apricots, peaches, pears and persimmons are reported (Fig. 2.1.3) from June till October (Fimiani, 1972; Fimiani and Pandolfo, 1973, 1975) with a similar situation on the island of Ischia, (Fimiani and Sollino, 1987). As the many different kinds of fruit grown on Procida Island are all susceptible to infestation, conditions there are extremely favourable for the medfly. On peaches infestation can rise from 5–25% up to 90–92% in August and September (Cirio et al., 1972). In Sicily medfly is also a serious pest. In a peach orchard an average of 60 larvae per fruit was found, with a maximum of 80 (Monastero, 1957). Infestations are also now reported for the areas where early ripening varieties are cultivated. The same author reports that during the years 1953–55 even green fruits had been heavily infested by the beginning of June. In Sardinia medfly is a pest of apricots and peaches (Contini et al., 1978).

4.1.4 Balkan Peninsula

Although probably already present in various Yugoslavian areas, medfly was reported for the first time in 1947 near Split (Kovacevic, 1965). Studies on the Yugoslavian coast were carried out (Tominich, 1951) and a severe outbreak occurred in 1958, causing damage to 90% of the crop. Migration and movement on the Adriatic coast was studied and heavy infestations were observed on peaches (Kovacevic, 1965).

In Greece the medfly is also an important pest and was first reported in 1915 (Papageorgiou, 1915) and described as mandarin fruit fly. It was found to cause extensive damage to citrus orchards in Attiki and on the island of Aegina. Subsequently it has been reported both on the mainland and on some other islands (Raftopoulos and Kourmoussis, 1937; Mourikis, 1965). The fly can have up to 6 generations per year with apricot, pear, peach, local varieties of orange, and mandarin being the most favourable host. The Navel orange appeared unsuitable for the larva, but was used as oviposition host. The adult population is at a maximum in July and in September, and at minimum during winter (December–March). At that time the population is found primarily as pupae in the soil, and sometimes as larvae in Bitter orange (Mourikis, 1965).

The pest is also present on several Greek islands and extensive damage has been reported on Chios, mainly on mandarins. Citrus fruits are important there as they are for export and so very low injury level is required (Mourikis in E.P.P.O., 1963). Recent studies on the same island show that fig trees bearing ripe fruits are important both as food source for the adults and place for larval development for *C. capitata* and should be taken into consideration in fruit fly management programmes.

In Patras the first oviposition punctures were noted early in May on *Eryobotria japonica* (Raftopoulos and Kourmoussis, 1937) but in the Leonidion, Corint areas and elsewhere oviposition in apricots begins by mid-May (Mourikis, 1965). The fly is present in Northern Greece (Tzimos, 1961) and has been found recently in the Thessaloniki region where in autumn fruits such as pears and peaches are attacked (M.E. Tzanakakis and B.I. Katsoyannos, personal communication).

4.1.5 Turkey

Medfly were reported here for the first time in 1904 (Vieira, 1952) and have subsequently become widely spread. Damage to lemons and oranges were considered to be of no great importance (E.P.P.O., 1954).

It appears that the western and southern parts of Turkey are the two main citrus growing districts where medfly are found, while its presence in other areas is rare. Favourable hosts are citrus, peaches, apricots, plums, pears, pomegranates, figs and quinces. In 1961 damage to about 10% of the crop was reported in Ismir (W. Turkey) where medfly has 4–5 generations per year. The population density peaks in late August and September. After the peach harvest, flies disperse to other fruits. The pest can have 8–9 generations per year in southern Turkey. The first flies appear on apricots, then spread to citrus and other hosts. On citrus the population peak is reached at the end of October (Giray, 1966; Tuncyurek, 1972).

Research in the Aegean region have shown that *C. capitata* is an important pest of citrus since tolerance by export markets of infested fruit is practically zero. Medfly occurs in almost the entire Aegean region, except for Edremit, Hairan and Sultanhizar. Warm winters provide very favourable conditions for the pest (Akman and Zumreoglu, 1973).

4.2 Oriental Mediterranean areas

Although medfly was already present in this area at the end of last century and some larvae were found in the market of Jaffa, it was only in the first quarter of this century that the insect became an important nation-wide pest in the area now known as Israel (Rivnay, 1968). In the spring of 1915, the entire crop of an orange grove dropped owing to attack by the fly. The climate and the availability of host plants are given as the principal reasons for the successful and rapid establishment of the fly in that country. With the exception of three winter months the fly is present all year round and the climate of the coastal plan enables the fly to produce 8–9 generations per year. The host fruits are guava, clementine, tangerine and grape-fruits in autumn followed by Jaffa oranges in winter and Valencia oranges in spring. The summer hosts are apricots, peaches, figs, cactus, etc. (Rivnay, 1968).

Infestation of citrus groves was investigated many years ago (Bodenheimer, 1928) and for some years severe losses were reported (Ballard, 1936). In Israel three periods of infestation are described. The autumnal one is economically very important because of the dropping of fruit and interference with harvesting and shipping of the fruit. The fruit attacked during this period can be easily recognized by the oviposition spot, the so-called *samra* (Rivnay, 1936; Rivnay et al., 1938). The winter generation causes less damage owing to the unfavourable climatic conditions. The spring generation causes severe damage because of favourable weather conditions. Extreme weather conditions i.e. frost, severe *Khamsin*, inadequate rainfall etc. exert a regulating influence on medfly population. Moderate weather, on the other hand, promotes rapid development

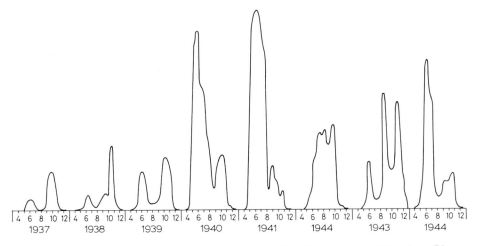

Fig. 2.1.4. Variation in medfly populations at Rehovot (Israel) over 8 years (data from Rivnay, 1941–1951).

and reproduction of the pest (Avidov, in E.P.P.O. 1957). A good outline of the situation in Israel was given by Rivnay (1951, 1956) and it was stressed again that the amount of damage in any particular area depends on climatic conditions and horticultural practices (Fig. 2.1.4).

The Shamanti orange can escape injury as its ripening season in winter coincides with a quiescent period of the fly; on the other hand the early ripening varieties such as clementine, tangerine and grape-fruit suffer the autumn attack while the late varieties such as Valencia orange suffer damage in the spring. The situation for summer fruits is worse; apricots and peaches cannot be cultivated in the plains because of heavy attack. In the hills only fruit which ripen in quiescent period of the fly can be grown. Where the fly population is dense, varieties of plums, apples and pears are heavily infested. The situation has become still more serious in recent years as a result of the introduction of many exotic subtropical species and varieties of fruits. Of these guava is most favoured and the entire crop can become infested unless collected before ripening. Also mango and avocado are favourable hosts.

The pest is also well established in Lebanon (Jalloud, 1968), especially on citrus trees.

Cyprus has large numbers of fruit trees both cultivated and wild. Medfly are a serious pest of citrus, peaches, apricots, plums, figs, pomegranate etc. Present at the beginning of this century as a pest of many fruits, especially citrus and peaches (Gennadios, 1914) where it remained at low level until the twenties (Anonymous, 1922). However high damage was subsequently reported (Wilkinson, 1925) on oranges, tangerines, figs and apricots. A more recent survey conducted during 1969–71 showed that the fly population was very low in winter and spring, increased during summer, to peak in fall (Serghiou, 1975). High catches were observed on figs, a very susceptible host.

4.3 African Mediterranean areas

Medfly is the insect that probably causes the most damage to the fruit crops of North Africa (Martin, in E.P.P.O., 1957). However the nature and intensity of attack vary according to the geographical situation of orchard. It is on the coast and in the regularly irrigated areas that its development is more abundant.

Recognized for a long time in Egypt (Adair, 1920) medfly is reported among the 38 Egyptian Trypaneids as very harmful (Efflatoun, 1924) and a list of host fruits was subsequently compiled (El-Ghawaby in E.P.P.O., 1957). After this period a decrease in the peach cultivation primarily due to medfly attack was registered. The infestation was so severe that nearly every fruit had been attacked (Hanna in E.P.P.O., 1957). In this country damage increases annually and the losses inflicted upon fruit crops are extremely high (Hafez et al., 1973). The peak for medfly adults generally occurs in November. The guava is the most favourable host for larvae, followed by mandarin and Navel orange, while sour orange seems to be unfavourable for oviposition (Hafez et al., 1973). November offers moderate temperature and rainfall and thus affords the most favourable climatic conditions for medfly activity.

In several parts of Libya medfly is a pest of peaches, apricots, citrus etc. and in some areas also pepper *(Capsicum)* is infested (Di Cairano Vitale, 1933). The period of most flight seems to coincide with the end of the year. Observation carried out in the principal citrus areas showed a long period of fruit susceptibility with larval attack occurring from the end of November till the beginning of February (Martin, in E.P.P.O., 1957). The damage pattern is varied; the most important areas being Sorman, Zavin and Tagiura, while in the interior orchards are very limited. The grape-fruits are heavily infested both in number

of stings and larval infestation. Among the oranges, the Navel and Demia are
the most susceptible. More recently expansion of fruit cultivation has increased medfly infestation.

In 1855 *C. capitata* was reported for the first time in Tunisia (La Gasca, 1925)
and many larvae were subsequently found in July–August on peaches, apricots
and persimmons; in August, September and October on pears and apples and in
January on orange (Guillochon, 1916). Peaches are excellent hosts (Soria and
Jana, 1962). In the same period the secondary host-plants in Tunis were investigated (Soria, 1962) while the fruits of sour orange were identified as being a
dangerous "reserve" for flies (Delanoue and Soria, 1962).

Damage by medfly in Algeria was recorded in the 1850s in the surrounds of
Algiers (Vieira, 1952). Damage to oranges was subsequently recorded by Laboulbene (1871), still using the name *Ceratitis hispanica*. Fruits attacked by medfly
can be found in all seasons in the region of Sahel and Mitidja. However variety,
situation, and the season, cause large fluctuations.

Economically citrus is the main host, especially the early varieties such as
Thomson and Washington Navel but other fruits can suffer infestation in
different periods. Medlars collected in June can show more than 80% infestation. The late varieties of apricots, although not very extensively cultivated
can show infestations of 34–73%. Damage is heavy on late apricots, peaches,
pears, pomegranate and early oranges. In the coastal belt medfly can be trapped
almost all year round. There are however two seasons of major flight, the first
in June–July, the second from August to November Periods of "Sirocco" or low

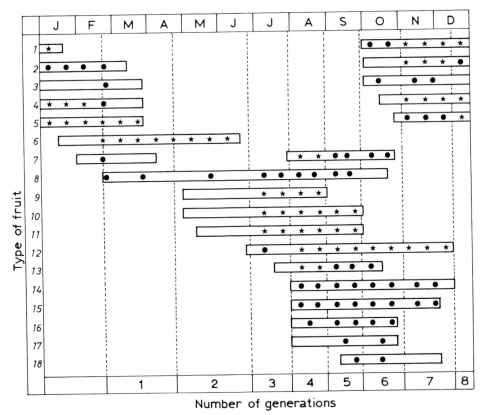

Fig. 2.1.5. Medfly situation on various types of fruit in Madeira during one year as derived from
the data of Vieira (1952). Below: the number of fly generations. In the center: the presence of
various types of fruit and the different intensity of larval infestation on 1. persimmons; 2. guavas;
3. perseas; 4. citrus 5. anonas; 6. medlars; 7. pitangas; 8. papayas; 9. apricots; 10. figs; 11. peaches;
12. pears; 13. fejioas; 14. apple macas; 15. apple peros; 16. pomegranates; 17. quinces. ●: medium
infestation; ★: heavy infestation.

temperature considerably reduce the flight activity (Martin in E.P.P.O., 1957).

In Morocco medfly is also one of the most noxious fruit pests (Bouhlier et al., 1935). Citrus trees are highly infested and many spray applications are required during the season. Here also the wild host plant *Argania spinosa* is present and constitutes a dangerous focus (Sacantanis, 1955).

4.4 Extramediterranean islands

Although not far from the Mediterranean region, some groups of Atlantic islands have quite different climatic conditions and vegetation, e.g. Canary Islands, where SIT applications for control of medfly were carried out. The island of Madeira is also infested with medfly.

About five years after Wiedemann's description (1824) of medfly as *Trypeta capitata*, it was reported for the first time in Madeira and in a short time it became a major pest of citrus on the island, and subsequently has shown its full potential of expansion and polyphagy (Vieira, 1952).

Placed in the Atlantic, at about 35° of latitude, corresponding to Morocco, the island of Madeira has a unique climate and vegetation, including tropical, subtropical and Mediterranean plants, all of which are very favourable for the fly (Natividade, in Vieira 1952). Figure 2.1.5 shows the medfly situation in Madeira on an annual basis. The most susceptible fruits are peaches, apricots, figs, oranges, grapefruits, pears, medlars, pitangas (*Eugenia* spp.). Other hosts, although not represented in high number, are heavily infested e.g. coffee, araças (*Psidium cattleianum*), persimmons, fejoia.

Medlars, being very susceptible, are thus a very good host for the development of the insect and have contributed significantly to its dispersal. This fruit is heavy cultivated in the island, up to 450 m and with its fruit already ripening in January it is utilized by the first generation of the year.

Except in case of very cold winters the infestation is always of great economic importance till May, reaching and sometimes exceeding 80%. On the island the insect has 8 generation per year, the most important being the three in the period August–October and the 8th from mid-December to mid-February.

5 REFERENCES

Adair, E.W., 1920. Note on Fruit flies occurring in or which might be introduced into Egypt. Agricultural Journal of Egypt, 10: 18–20.

Akman, K. and Zümreoglu, A., 1973. A survey of the Mediterranean Fruit Fly (*Ceratitis capitata* Wied.). Plant protection Research Annual, 141–142.

A.N.I.A., 1970. Conferencia mixta OEPP/OILB sobre la *Ceratitis capitata*. Boletin de la Asociacion Nacional de Ingenieros Agronomos, 209: 451–507.

Anonymous, 1922. Mediterranean Fruit Fly. Citrus Agricultural Journal, 17: 41–42.

Ballard, E., 1936. Report of the Government Entomologist. Reports of Department Agriculture and Forestry of Palestine. Jerusalem, 1936, pp. 178–180.

Bezzi, M., 1927. Sulla distribuzione geografica della Mosca delle ciliege (*Rhagoletis cerasi*, Diptera). Bollettino del Laboratorio di Zoologia Generale ed Agraria, Portici, 20: 7–16.

Bodenheimer, F., 1928. Contribution towards the knowledge of the Citrus Insects in Palestine. I. Preliminary Report on the work of the Palestine Breeding Laboratory at Petah-Tikuwa (1924–27). Palestine Citograph, Tel Aviv I(5–6) pp. 16.

Boller, E.F. and Bush, G.L., 1974. Evidence for genetic variation in population of the European Cherry Fruit Fly, *Rhagoletis cerasi* (Diptera:Tephritidae) based on physiological parameters and hybridation experiments. Entomologia Experimentalis et Applicata, 17: 279–293.

Boller, E., Haisch, A., Russ, K. and Vallo, V., 1970. Economic importance of *Rhagoletis cerasi* L., the Feasibility of Genetic control and resulting research problems. Entomophaga, 15: 305–313.

Boller, E., Russ, K., Vallo, V. and Bush, G., 1976. Incompatible Races of European Cherry Fruit fly, *Rhagoletis cerasi* L. (Diptera: Tephritidae). Their origin and potential use in Biological Control. Entomologia Experimentalis et Applicata, 20: 237–247.

Bouhlier, R., De Francolini, J. and Perret, J., 1935. Essais attractifs pour la destruction de *Ceratitis capitata* Wied. Revue de Zoologie Agricole et Appliquee, 34: 149–152.

Cabezuelo, P. and Sampayo, M., 1975. Primeros resultados de un programma de lutta dirigida en melocotonero. Boletin del Servicio de Plagas Forestales, 1: 107–138.

Cirio, U. and Capparella, M., 1972. Reperti sulla presenza della *Ceratitis capitata* Wied. (Dipt. Tryp.) lungo la costa laziale. Primo anno di osservazioni. Atti IX Congresso Nazionale di Entomologia, Siena, pp. 65–81.

Cirio, U., De Murtas, I., Gozzo, S. and Enkerlin, D., 1972. Preliminary ecological observation of *Ceratitis capitata* Wied. on the island of Procida with an attempt to control the species using the sterile male technique. Bollettino Laboratorio Entomologia Agraria Portici, 30: 175–188.

Contini, C., Delrio, G., Luciano, P. and Ortu, S., 1978. Variazioni delle popolazioni di *Ceratitis capitata* Wied. e programmazione della lotta nella frutticoltura sarda. Atti giornate fitopatologiche 1978, CLUE Bologna: 223–230.

Costantino, G., 1930. Contributo alla conoscenza della Mosca della Frutta (*Ceratitis capitata*) (Dipt. Tryp.). Bollettino Laboratorio Zoologia Generale ed Agraria Portici, 23: 238–322.

Coutinho, A., 1898. A Mosca da laranja e do Pêsego. Agricultura Contemporanea, 9(8).

De Breme, F., 1842. Note sur le genre *Ceratitis* de M. Mac Leay (Diptera). Annales de la Societé Entomologique de France, 11: 183–190.

Delanoue, P. and Soria, F., 1962. Les fruits de l'oranger amer (*Citrus bigaradia* Risso) reserve dangereuse en Tunisie de Mouches des fruits (*Ceratitis capitata* Wied.). Annales de l'Institut National de la Recherche Agronomique, 35: 185–204.

Delrio, G., 1979. *Dacus oleae* Gmel., Bibliografia 1966–1979. CNR Progetto finalizzato "Fitofarmaci/ Fitoregolatori", Firenze pp. 46.

Di Cairano Vitale. 1933. La lotta contro la Mosca della Frutta. Bollettino del Regio Ufficio Servizi Agrari, Tripol, 2(7–8): 3–5.

Dieuzeude, R., 1929. Sur la présence de la Mouche des fruits (*Ceratitis capitata* Wied.) en Gironde. Revue de Zoologie Agricole et Appliquee, 28: 183–186.

Efflatoun, H.C., 1924. A de la Monograph of Egyptian Diptera (Part II, Fam. Trypaneidae). Memoires Societe Royale d'Entomologie d'Egypte, 2(2), pp. 132.

E.P.P.O., 1954. *Ceratitis capitata* Wied. Report of the international conference on Mediterranean Fruit Fly. E.P.P.O publications, Series A, n.11, pp. 27.

E.P.P.O., 1957. *Ceratitis capitata* Wied. Report of the Second international Conference on Mediterranean Fruit Fly. E.P.P.O Publications Series A, n.18, pp 67.

E.P.P.O., 1963. *Ceratitis capitata* Wied. Report of the International Conference in Wien. E.P.P.O. Publications Series A, n.34, pp. 104.

Ferrari, R., 1966. Presenza e attivitá della Mosca della Frutta (*Ceratitis capitata* Wied.) nel bolognese. Bollettino Osservatorio per le malattie delle Piante, Bologna I(1): 65–82.

Fimiani, P., 1972. Osservazioni biologiche sulla Mosca della frutta (*Ceratitis capitata* Wied.) effettuate nella zona di Monte di Procida (Napoli) negli anni 1968–69. Bollettino Laboratorio Entomologia Agraria, Portici, 30: 71–87.

Fimiani, P., 1984. Ricerche sulla Mosca delle ciliege (*Rhagoletis cerasi* L.) in Campania. II. Distribuzione ed intensità delle infestazioni larvali. Bollettino Laboratorio Entomologia agraria "F. Silvestri", 41: 11–22.

Fimiani, P., 1986. Phenological observation on European Cherry Fruit Fly (*Rhagoletis cerasi* L.) in the South areas. Proceedings 'CEC.IOBC ad hoc meeting Fruit flies of Economic importance, Hamburg'. Balkema Publisher, Rotterdam, pp. 169–173.

Fimiani, P. and Pandolfo, F., 1973. Fluttuazioni delle popolazioni di adulti di Mosca della frutta nel litorale flegreo. Informatore Fitopatologico, 23: 13–16.

Fimiani, P. and Pandolfo, F., 1975. Rilievi bioecologici del 1973 sulla Mosca della frutta in biotopi della terraferma prospiciente l'isola di Procida. Atti giornate di Studio CNEN, Roma II: 35–46.

Fimiani, P. and Sollino, G., 1987. Observations on Fruit Flies of the Island of Ischia. International Symposium on Fruit Flies of Economic importance, Roma. Proceedings, Balkema Publisher, Rotterdam, pp. 7.

Fimiani, P. and Tranfaglia, A., 1972. Influenza delle condizioni climatiche sull'attività moltiplicativa della *Ceratitis capitata* Wied. Annali Facoltà Scienze agrarie Portici, 6: 190–199.

Fimiani, P., Frilli, F., Inserra S., Monaco, R. and Sabatino, A., 1981. Ricerche coordinate su aspetti bioecologici della *Rhagoletis cerasi* L. in Italia. Bollettino Laboratorio Entomologia agraria "F. Silvestri", 38: 159–211.

Gennadios, P.G., 1914. Dictionary of Phytology, Athens.

Ghesquiere, J., 1949. Le probléme de la Mouche des fruits en France. L'agriculture pratique.

Giard, A., 1900. Sur l'existence de *Ceratitis capitata* Wied. var. hispanica Breme, aux environ de Paris. Compte Rendu de l'Academie des Sciences de Paris, 131: 436–439.

Giray, H., 1966. Ege bôlgesinde Kültur bitkilerine ariz olan Trypetidae (meyve sinekklery) familyasi turleri ve: konukculari üzerinde arastirmalar. Ege Universitesi, Ziraat Fakultesi Yayinlari, 126: 1–161.

Grison, P., 1962. Dévelopment et perspective de la lutte biologique. Entomophaga, 7: 325–335.

Guillochon, L., 1916. Concerning the Fruit Fly (*Ceratitis capitata*) in Tunis. International Review Science and Practice Agricolture. Monthly Bullettin Agricolture and Plant disease Rome, 6: 1562–1563.

Hafez, M., Abdel Malek, A., Wakid, A. and Shoukry, A., 1973. Studies on some ecological factors affecting the control of the Mediterranean Fruit Fly, *Ceratitis capitata* in Egypt by the use of sterile male technique. Zeitschrift für Angewandte Entomologie, 73: 230–238.

Haisch, A., Boller, E., Russ, K., Vallo, V. and Fimiani, P., 1978. *Rhagoletis cerasi* L.. Synopsis and Bibliography. WPRS Bullettin 1978 I/3 pp. 43.

Jalloud, A. 1968. Observations sûr divers traitments effectuées au Liban en 1966 contre *Ceratitis capitata*. Fruits, 23(8).

Kovacevic, Z., 1965. Bemerkungen über die Populations-Bewegungen der Mittelmeerfruchtfliege (*Ceratitis capitata* Wied.) an der Jugoslavischen Adriaküste. Anzeiger für Schädlingskunde, 38: 151–153.

Laboulbene, A., 1871. Note sûr les dommages causés par la *Ceratitis hispanica* aux fruits des orangers dans nos possesion d'Algerie. Annales de la Societé Entomologique de France, 5/I: 439–443.

La Gasca, F., 1925. La Mosca mediterrànea y la Uva de Almeria. Conferencia Accademia de San Ignacio, Almeria, pp. 44.

Lesne, P., 1915. La Mouche des fruits aux environ de Paris. Bullettin de la Societé Etude et Vulgarization Zoologie Agricole, 14: 91–93.

Lesne, P., 1921. Un foyer de multiplication de la Mouche des fruits (*Ceratitis capitata* Wied.) aux environ de Paris. Comptes Rendus Hebdomadaires Academie des Seances de Paris, 172: 490–491.

Martelli, G., 1910. Alcune note intorno ai costumi e ai danni della Mosca delle arance (*Ceratitis capitata* Wied.) Bollettino Laboratorio di Zoologia Agraria, Portici, 4: 120–127.

Martelli, G.M., 1967. Contributo alla Bibliografia della Mosca delle olive, *Dacus oleae* Gmel. Annali Facoltà di Agraria dell'Università di Bari, 21: pp. 106.

Mechelani, E., 1982. Evaluation de l'attaque de *Prays oleae* et *Dacus oleae* durant l'année 1981 au Liban. Working Paper 23. III Sess. Res. FAO Subnetwork on Olive protection. Jaen (Spain).

Mellado, L., Caballero, F., Arroyo, M. and Jmenez, A., 1966. Ensayo sobre erradicacion de *Ceratitis capitata* Wied. por el método de los "machos estériles" en la isla de Tenerife. Boletino de Patologia Vegetal y Entomologia, 29: 89–117.

Milaire, H., 1964. Observations sur la mouche méditerranéenne des fruits en 1963. Phytoma, 16: 26–27.

Monastero, S., 1957. Gli insetti più nocivi agli agrumi e i metodi per combatterli. Bollettino dell'Istituto di Entomologia Agraria di Palermo, 14: 131–165.

Mourikis, P.A., 1965. Data concerning the development of the immature stages of the Mediterranean Fruit Fly (*Ceratitis capitata* Wied. Dipt. Tryp.) on different host fruits and on artificial media under laboratory conditions. Annales de l'Institute phytopatologique Benaki N.S., 7: 59–105.

Muñiz, M. and Gil, A., 1984. Desarollo y reproducion de *Ceratitis capitata* Wied. en condiciones artificiales. Boletin Servicio de Defesa contra plagas y Ispecion fitopatologica. Fuera de serie, n.2. pp. 140.

Neuenschwander, P., Russ, K., Höbans, E. and Mikelakis, S. 1983. Ecological Studies on *Rhagoletis cerasi* L. in Crete for the use of the incompatible insect technique. In: R. Cavalloro (Editor), Fruit Flies of Economic Importance, Balkema, Rotterdam, pp. 41–51.

Papageorgiou, P., 1915. The mandarin fly. Agricultural Bulletin of the Royal Agricultural Society, 12: 258–260.

Perko, S., 1983. Vrijednosti vizuelnih mamaka u dijagnosticirnju pojave muhe Trešnjrice (*Rhagoletis cerasi* L.) na maraski. Agriculturae Conspectus Scientificus, 61: 231–240.

Poutiers, R., 1938. Notes biologiques sur la Mouche des fruits (*Ceratitis capitata* Wied.) Revue de Pathologie Végétale, 25: 208–217.

Raftopoulos, I. and Kourmoussis, A., 1937. Some conclusion on efforts for the experimental control of the Mediterranean Fruit Flies. Agricultural Bulletin of the Greek Agricultural Society, 30: 163–171 and 223–232.

Rivnay, E., 1936. Infestation of oranges by the Mediterranean Fruit Fly during the Autumn in Palestine. Hadar, 9: 134–137.

Rivnay, E., 1951. Mediterranean Fruit Fly population and its activity in the citrus grove in Israel. IV Congres international d'Agriculture mediterranéenne.

Rivnay, E., 1956. Studies on the control of the fruit fly in Valencia grovas in Israel. Ktavim, 6: 101–109.

Rivnay, E., 1968. Biological control of pests in Israel. Diptera; The Mediterranean Fruit Fly, *Ceratitis capitata* Wied. Israel Journal of Entomology, 3: 104–113.

Rivnay, E., Nadel, M. and Littauer, F., 1938. The reaction of the orange fruit to the autumn attack of the Mediterranean Fruit Fly and its economic status. Hadar, 11: 16.

Sacantanis, K., 1955. La forêt d'Arganier, le plus grand foyer naturel de la mouche des fruits

existant au Maroc. Comptes Rendus de Seances Mensuelles de la Societe des Sciences Naturelles et Physiques du Maroc, 2: 51–52.

Serghiou, C., 1975. The sterile male technique for control of the mediterranean fruit fly, *Ceratitis
capitata* Wied, in the Mediterranean Basin. Proceeding of Symposium on "Sterility Principle for
insect control" IAEA, 11–28.

Soria, F., 1962. Plantes-hotes secondaires de *Ceratitis capitata* Wied. en Tunisie. Annales de l'Istitut
National de Recherche Agronomique, Tunis, 32: 95–108.

Soria, F. and Jana, A., 1962. Contribution a l'étude de l'epoque de receptivité des péches aux
attaques de la Ceratite en Tunisie. Annales de l'Institut National de Recherche Agronomique,
Tunis, 32: 109–123.

Tominich, A., 1951. La Mouche des Fruits, *Ceratitis capitata* Wied., sur le litoral yugoslave. Biljna
zastita, 3. Zagreb.

Trouillon, L.L., 1976. Un succès du contrôle phytosanitaire, la disparition. en France de la Mouche
des fruits. Phytoma, 28: 27–29.

Tuncyurek, M., 1972. Recent advances in the Bio-ecology of *Ceratitis capitata* Wied. in Turkey.
OEPP Bullettin 6: 77–81.

Tzimos, K., 1961. Mouche mediterraneenne in North Greece. Agr Takidromos, 172: 141–142.

Vieira, R.M.S., 1952. A Mosca da fruta (*Ceratitis capitata* Wied.) na Ilha da Madeira. Gremio dos
Exportadores de Frutas da Ilha da Madeira, pp. 219.

Wilkinson, D.S., 1925. Entomological Notes. Cyprus Agricultural Journal, 20: 9–10.

Young, E.C., 1977. Annotated Bibliography on Economically important Fruit Flies. Diptera:
Tephritidae (= Trypetidae) for the years 1955–1976. FAO GRE/69/525 Technical Report n.1
(UNDPW L0661) pp. 63.

Zumreoglu, A., Akaman, K., San S. and Ulu, O., 1981. Ege bölgesinde kiraz zehirli yem dal ilaclama
yontemini uygulama olanaklri uzerinde arastirmalar. Bitky koruma Bülteni (1980), 21: 218–230.

Zwolfer, H., 1985. Bibliography of Fruit Fly Literature, A review of recent publications (1977–1983)
LOBC/WRPD Bullettin 1985/VIII/2 pp. 35.

Chapter 2.2 Southern Africa

D.L. HANCOCK

1 INTRODUCTION

Although fruit flies are important pests of horticultural crops in Southern Africa, little work has been done on them, particularly with regards to their biology. Part of the reason for this may lie in the relative ease of their control by bait and cover sprays, especially in large orchards or plantings. It is in smallholdings, gardens and village communities that control is more difficult. Repeated and improper use of insecticides in cover sprays is not only costly but also environmentally hazardous and unintentional effects on parasites and natural enemies of other pests may create further problems.

An understanding of the flies' biology, such as habitat preferences, host preferences (both wild and cultivated), behaviour, mating strategies etc., and the development of efficient population monitoring techniques through the use of attractants, may prove of great benefit in reducing insecticide use in integrated pest management systems. Most of the biological information available is due to the work of Munro (1924, 1925, 1926, 1929, 1933, 1935, 1939, 1947, 1952, 1953, 1967, 1985).

The current state of our knowledge of the biology of Southern African fruit flies, although still meagre, is reviewed below. Notes on the identification of economic species have been provided by Munro (1964) and Hancock (1981).

2 SUBFAMILY TRYPETINAE

The taxonomic limits of this subfamily in Africa are not well understood. Cogan and Munro (1980) recognised Adraminae, Euphrantinae, Trypetinae and Acanthonevrinae as separate subfamilies but these grade into one another and are best regarded as tribes. Cogan and Munro (1980) referred to the Trypetinae (= Trypetini) only *Rivelliomima* Bezzi and an undescribed genus. This latter belongs to the Tephritinae (it lacks scapular bristles and a mesopleural suture and breeds in the flower-heads of Compositae) and even *Rivelliomima* is of doubtful placement in this tribe; it and *Xanthanomoea* Bezzi appear to be related closely to the Oriental genus *Cycasia* Malloch. These three genera were placed in the tribe Rivelliomimini by Hancock (1986).

Biologies in this subfamily are rather varied. The only Southern African member of the Adramini, *Munromyia nudiseta* Bezzi, infests the seeds of wild olives (Oleaceae) but has not been reported from cultivated varieties (Munro, 1924). *Sosiopsila* Bezzi, often referred to the Adramini, lacks the long hairs on the pleuroterga and the apical projection of the anal cell and appears to belong to the Phytalmiini. In the Euphrantini, species of *Coelopacidia* Enderlein

Chapter 2.2 references, p. 57

tunnel in the stems of *Senecio* (Compositae) and *Polemannia* (Umbelliferae), whilst species of *Coelotrypes* Bezzi (which includes the African *"Rhacochlaena"* spp.) infest the buds of *Ipomoea* (Convolvulaceae). Hosts of the other Southern African genera *Trypanophion* Bezzi, *Conradtina* Enderlein and *Celidodacus* Hendel are unknown. In the Acanthonevrini, *Themarictera laticeps* Loew breeds in the fruit of *Boscia* (Capparidaceae) whilst species of *Afrocneros* Bezzi and *Ocnerioxa* Speiser infest the stems of *Cussonia* (Araliaceae). The latter two genera, together with *Ptiloniola* Hendel, were placed in the Euphrantini by Cogan and Munro (1980) but they lack the long hairs on the pleuroterga characteristic of that tribe and the Adramini and, with a tactile aculeus, appear to be better placed in the Acanthonevrini. The Trypetini contains an assortment of genera with varied hosts. *Hoplandromyia* Bezzi mines the leaves of *Canthium* (Rubiaceae), *Taomyia* Bezzi has been bred from *Sansevieria* (Agavaceae), *Notomma* Bezzi forms twig galls on *Dicrostachys* and *Acacia* (Leguminosae) and *Scleropithus* Munro breeds in fruits of *Strychnos* (Strychnaceae). The only trypetine reported as an economic pest, *Zacerata asparagi* Coquillett, was placed in the tribe Zaceratini by Hancock (1986).

2.1 Pest species

Zacerata asparagi (the asparagus fly) is a small black species with dark wing bands that breeds in native species of *Asparagus* (Liliaceae) with thick shoots. The larvae hollow out the shoots and pupate within them, many per shoot. It occasionally infests cultivated *Asparagus* in South Africa (Munro, 1925, 1964; Annecke and Moran, 1982).

3 SUBFAMILY DACINAE

Approximately 182 species of this subfamily occur in Africa. Amongst the species with free abdominal tergites (*Bactrocera* group of subgenera) only *Dacus oleae* Gmelin is of economic importance in Southern Africa. *Dacus cucurbitae* Coquillett is an introduced pest of cucurbits in East Africa, whilst *Dacus biguttulus* Bezzi occurs in wild olives (Oleaceae) but has not been recorded from cultivated varieties (Munro, 1924).

Species of the *Dacus* group of subgenera, with fused abdominal tergites, placed in various subgenera, infest the fruits of Cucurbitaceae and Passifloraceae or the pods of Asclepiadaceae, Apocynaceae and Periplocaceae, whilst a few species breed in the flower buds of Passifloraceae, the flowers, leaves or stems of Asclepiadaceae, or the stamens of male flowers of Cucurbitaceae. Occasionally, other plants may be attacked.

Subgenera *Dacus* and *Didacus* contain species of major economic importance. Species belonging to subgenus *Dacus* show a marked preference for forest and moist woodland habitats in Southern Africa, appearing only occasionally in drier woodland types, whilst those belonging to subgenus *Didacus* prefer drier woodland types and do not occur in forests.

3.1 Pest species

D. oleae (the olive fruit fly) is known in Africa from the Western Cape Province of South Africa northwards to Ethiopia but it does not appear to be as important a pest as it is in Mediterranean countries. In South Africa it attacks cultivated olives from around February. It is frequently parasitized by braconid and chalcidoid wasps, one of which, *Bracon celer*, is reported to attain up to 87% parasitism of *D. oleae* in cultivated olive groves in the Cape (Annecke and Moran, 1982). The adults apparently overwinter and the female lays 200 or more eggs, one per fruit, which hatch in 2–3 days. There are several generations per year, and in warm weather a generation may be completed in about 5 weeks (Annecke and Moran, 1982). Apart from cultivated olives, the wild olive *Olea verrucosa* is also attacked in South Africa (Munro, 1924, 1925).

Dacus bivittatus Bigot (the greater or two-spotted pumpkin fly) is a major pest of cucurbits over most of Africa. In Southern Africa it is of economic importance only in the moister northern and eastern regions and occurs throughout the year. It is a common species in forests. In drier areas, such as the Transvaal (Bot, 1965) and southwestern Zimbabwe, it is of sporadic occurrence during the wet summer months and is presumably a migrant. This species attacks a wide range of cucurbits, especially pumpkins and marrows and has recently been reared from oyster-nuts in Zimbabwe. It occasionally infests tomatoes, pawpaws and granadillas. In Sao Tomé it has been reported infesting coffee berries (Schmidt, 1967).

Dacus punctatifrons Karsch has a similar distribution to that of *D. bivittatus*, being a scarce summer visitor to the central Transvaal (Bot, 1965) and only once recorded from southwestern Zimbabwe, whereas it is common in the moister northern and eastern areas. It also infests cucurbits but to a much lesser extent than *D. bivittatus*.

Dacus telfaireae Bezzi (the oyster-nut fly) occurs in Tanzania, Malawi and Zimbabwe and is associated with forests or rich riverine vegetation. In Southern Africa it is known only from the eastern districts of Zimbabwe and was common near oyster-nuts. This species has been recorded as a pest of oyster-nuts *(Telfairea pedata)* in East Africa.

Dacus lounsburyi Coquillett and *Dacus pallidilatus* Munro, two species with large apical wing spots, are minor pests of cucurbits in Southern Africa, seldom, if ever, reaching economic proportions. *D. lounsburyi* has been recorded from scattered localities throughout South Africa and Zimbabwe, whilst *D. pallidilatus* has been recorded only from northern and eastern Zimbabwe.

Dacus ciliatus Loew (the lesser pumpkin or cucurbit fly) is a major pest of cucurbits throughout Africa, showing a strong preference for small fruits and squashes measuring less than 50 mm in diameter and with a soft skin. Groups of 3–9 eggs are laid beneath the rind; these hatch within 1 or 2 days and, in summer, the larvae feed in the pulp for 5–6 days. Pupation is in the soil beneath the fruit, adults emerging some 2–4 weeks later (Annecke and Moran, 1982). Adults occur throughout the year but are more numerous in summer. Apart from cucurbits, tomatoes, green beans and even cotton bolls may be attacked.

Dacus frontalis Becker is another widespread species similar in appearance to *D. ciliatus*. It also attacks cucurbits but to a much lesser extent than *D. ciliatus*.

Dacus vertebratus Bezzi (the jointed pumpkin fly or melon fly) is another widespread species infesting cucurbits, with a strong preference for water-melons. Like *D. bivittatus* and *D. ciliatus*, it is a major pest in Southern Africa. Adults occur throughout the year, being most numerous in summer.

3.2 Male attractants

Three compounds have been found attractive to male Dacinae in Southern Africa, cue-lure (4-(*p*-acetoxyphenyl)-2-butanone) and vert-lure (methyl-*p*-hydroxybenzoate) and propyl-*p*-hydroxybenzoate (Hancock, 1985b).

Cue-lure attracts the following economic species: *D. bivittatus, D. punctatifrons, D. telfaireae, D. pallidilatus* and *D. frontalis.* Vert-lure attracts *D. vertebratus* and appears to be specific for this species. Propyl-*p*-hydroxyben-

Chapter 2.2 references, p. 57

zoate also attracts *D. vertebratus. D. ciliatus* does not respond to any of these lures (nor to methyl eugenol) and the response of *D. lounsburyi* is unknown.

Vert-lure, determined only recently, appears to have a similar action to cue-lure. Only males are attracted with neither teneral nor old and worn specimens appearing in traps, despite the close proximity, at one stage, of a roosting swarm of mixed sexes and ages.

3.3 Communal roosts

Munro (1925, 1985) noted that most, if not all species of Southern African Dacinae have a habit of frequenting definite roosting places, usually large leaves. The flies, often two or more species, congregate on the undersides, returning soon afterwards if disturbed. This behaviour was recorded for *D. bivittatus, D. ciliatus, Dacus brevis* Coquillett, *Dacus binotatus* Loew, *Dacus plagiatus* Collart, *Dacus viator* Munro, *Dacus inopinus* Munro, *Dacus ficicola* Bezzi and *D. vertebratus*. One roost contained all but the first two of these species.

Four such roosts have been observed in Zimbabwe and contained males and females of *D. vertebratus, D. brevis, D. plagiatus, D. binotatus, Dacus serratus* Munro, *Dacus amphoratus* Munro, *Dacus botianus* Munro and *Dacus umbeluzinus* Munro, in combinations of up to six species clustered together on leaves of avocado, fig, mulberry or citrus.

In a swarm that persisted for a month (Cooper, pers. comm.), adults arrived each evening and dispersed in the morning except during cold or wet weather, when they remained at the roost. All rested head uppermost beneath a small group of leaves and were tightly packed together. Defaecation occurred on the leaves but there was no evidence of feeding, nor of courtship. As well as mature examples of both sexes, the roost contained teneral specimens. *D. amphoratus, D. botianus, D. brevis, D. plagiatus* and *D. umbeluzinus* were unaffected by attractant traps hung in close proximity, whilst mature males of *D. vertebratus* responded to vert-lure.

It appears likely that these swarms appear as a result of a nearby source of adult food and move on when the food supply is exhausted, though why they roost communally in such a small area is unknown. The same 2 or 3 leaves are used throughout the period of occupancy, which is often several months.

4 SUBFAMILY CERATITINAE

The Ceratitinae are a closely-knit group of fruit and bud infestors, to which the Oriental bamboo-shoot breeders (Gastrozonini) appear to be allied. They differ from the Trypetinae in having only 2 spermathecae in the female (3, rarely 2 in Trypetinae) and in the shape of the anal cell extension, and appear to be more closely related to the Dacinae than to the Trypetinae. Several species referred here to *Ceratitis* MacLeay are often placed in the genera *Pterandrus* Bezzi or *Pardalaspis* Bezzi; these were reduced to subgenera and redefined by Hancock (1984). This group contains some of the most important pests of deciduous fruits, *Ceratitis capitata* (Wiedemann) and *Ceratitis rosa* Karsch being especially injurious to a wide range of fruit crops. Several species are pests of coffee berries, but records of *Trirhithrum occipitale* Bezzi from this host (Munro, 1929; Annecke and Moran, 1982) actually refer to *T. nigerrimum* Bezzi (Munro, 1934). In Southern Africa, genera such as *Ceratitis, Trirhithrum* Bezzi and *Trirhithromyia* Hendel attack various fruits and berries, whilst *Carpophthoromyia* Austen breeds in *Drypetes* (Euphorbiaceae) and *Perilampsis* Bezzi breeds in the berries of mistletoe (Loranthaceae). Hosts for other genera,

such as *Nippia* Munro, *Clinotaenia* Bezzi and *Leucotaeniella* Bezzi, are unknown, although the latter two genera appear to belong to the *Gastrozonini*
(Hancock, 1985c).

4.1 Pest species

C. capitata (the Mediterranean fruit fly) has a wide range of native and cultivated hosts and is a major pest throughout Africa. Studies in the Western
Cape Province of South Africa have shown that populations vary from one area
to another, depending upon availability of hosts, and that migration occurs
between these areas. Numbers were found to increase in spring in orchards
with early ripening fruit such as apricots, then move to peach orchards in late
summer and early autumn, then to vineyards and finally, in late autumn and
winter, to citrus orchards (Myburgh, 1964; Annecke and Moran, 1982). Subtropical and deciduous fruits such as coffee, granadillas, guavas, litchis, mangoes, apples, apricots, figs, grapes, peaches, pears, plums and quinces are
readily attacked; citrus is stung but the larvae normally do not develop.
Mulberries and youngberries are also attacked.

Ceratitis rosa (the Natal fruit fly) is another major pest with a wide host range.
In Southern Africa it is primarily a species of the northern and eastern parts,
extending as far as the Eastern Cape. In Zimbabwe it has not been recorded
west of the central watershed, although it is very common in both Bulawayo
and Harare. A larger species than *C. capitata*, where it occurs it tends to
displace the latter from several hosts. In Mauritius, where it was introduced
accidentally, *C. rosa* displaced *C. capitata* as the major pest within 4 years
(Hancock, 1984). This species attacks avocados, guavas, litchis, mangoes, pawpaws, apples, apricots, figs, grapes, peaches, pears, plums, quinces and citrus.
Occasionally tomatoes are also attacked. Apart from native fruits, *Solanum*
(Solanaceae) serves as a host in some areas. Its life history is similar to that of
C. capitata and has been described by Carnegie (1962).

Ceratitis rubivora Coquillett (the blackberry fruit fly) infests the berries of
Rubus spp., such as blackberries, raspberries, youngberries and loganberries.
It occurs in the eastern parts of Southern Africa, from the Cape to northeastern
Zimbabwe, and further north to East Africa. One, rarely two, larvae occur in
a berry. Although wild blackberries are often heavily infested, this species
appears to be of minor importance in cultivated berries.

Ceratitis pedestris (Bezzi) (the Strychnos fruit fly) is an occasional pest of
tomatoes in Southern Africa but its importance must be regarded as very
minor. Native hosts are various species of *Strychnos* (Strychnaceae).

Ceratitis cosyra (Walker) (the Marula fruit fly) normally breeds in *Sclerocar-
ya caffra* (marula: Anacardiaceae) and other wild fruits. Occasionally it infests
early peaches in the summer rainfall areas, and subtropical fruits such as
mangoes, avocados, guavas and custard apples. It is a common bushveld species.

Ceratitis quinaria (Bezzi) (the Five-spotted fruit fly) is another widespread
bushveld species that occasionally attacks peaches, apricots, guavas and figs,
especially in summer rainfall areas.

Ceratitis discussa (Munro) has been reported as a minor pest of citrus in
eastern Zimbabwe (Carnegie, 1962). It is similar in appearance to *C. cosyra* but
the black thoracic patches are less distinct. It also breeds in wild *Annona* fruits
(Annonaceae) and could become a pest of custard apples.

Trirhithrum nigerrimum Bezzi is a dark species with the sexes dimorphic in
wing pattern. Widespread in Africa, it has been reared from coffee berries in
Natal and elsewhere.

Chapter 2.2 references, p. 57

4.2 Male attractants

In Africa, male Ceratitinae have been found to respond to a number of attractants, principally methyl eugenol, trimedlure and terpinyl acetate (Ripley and Hepburn, 1935; Georgala, 1965; Hancock, 1985a). Methyleugenol attracts non-economic species of *Perilampsis, Nippia* and *Ceratitis* subgenus *Pardalaspis* Bezzi, whilst the other two lures attracts several economic species of *Ceratitis*.

Trimedlure has been found to attract *C. capitata, C. rosa, C. rubivora* and *C. pedestris*. Terpinyl acetate attracts all species of *Ceratitis* so far tested, including *C. capitata, C. rosa, C. rubivora, C. pedestris, C. cosyra* and *C. quinaria*. No data are available for *C. discussa* and species of *Trirhithrum* show no response to any of the lures.

4.3 Mating habits

Little work has been done on this topic but studies on *C. capitata* and *C. rosa* (Myburgh, 1962; Prokopy and Hendrichs, 1979) show that light intensity plays an important part in the timing of mating. *C. capitata* mated during the day, on fruit in the early morning and late afternoon, beneath leaves from mid-morning to early afternoon. Mating in this species was inhibited by light intensities below 200 foot candles. *C. rosa* however mated in the evening, the optimum light intensity being between 20 and 0 foot candles, with copulation broken off on the following day. As is usual in Tephritidae, mating took place on the host plant, males calling in females with the aid of a pheromone.

5 OTHER SUBFAMILIES

The remaining subfamilies Myopitinae, Terelliinae, Aciurinae and Tephritinae (including Platensinini, Schistopterinae, etc.) contain species which form galls on, or infest flower-heads of, Compositae (Myopitinae, Tephritinae), Labiatae and Verbenaceae (Aciurinae) or Acanthaceae (Aciurinae, Platensinini). Except for *Craspedoxantha marginalis* (Wiedemann), which infests flower-heads of garden plants such as *Dahlia* and *Zinnia* (Compositae), none are of economic importance, although a few species destroy the seeds of weeds and have some beneficial use. *Craspedoxantha* Bezzi does not readily fit in any of the above subfamilies but appears to belong to the Terelliinae. It differs from the Tephritinae in the shape of the anal cell and presence of a distinct mesopleural suture.

6 QUARANTINE THREATS

Several species of economic importance elsewhere in the Afrotropical Region are a potential threat to Southern Africa, requiring continual quarantine vigil.

6.1 From elsewhere in Africa

Dacus cucurbitae Coquillett is an important pest of cucurbits in Southeast Asia that has been introduced to East Africa and the Mascarenes but has not yet been recorded in Southern Africa. This species responds to cue-lure. *Trirhithrum coffeae* Bezzi, as its name suggests a pest of coffee berries, is widespread in western and eastern Africa. It is similar in appearance to *T. nigerrimum*.

6.2 From Madagascar and the Mascarenes

A number of species occur in these countries which could become pests in Southern Africa should they be introduced. These are *Dacus demmerezi* Bezzi and *D. cucurbitae*, both pests of cucurbits, *Ceratitis catoirii* Guérin-Méneville and *Ceratitis malgassa* Munro, pests of fleshy fruits, *Trirhithromyia cyanescens* Bezzi, a pest of tomatoes and *Trirhithrum manganum* Munro, a further coffee pest related to *T. nigerrimum* (Orian and Moutia, 1960; Hancock, 1984). *D. demmerezi* and *T. cyanescens* occur both in Madagascar and the Mascarenes, *C. malgassa* and *T. manganum* in Madagascar, and *D. cucurbitae* and *C. catoirii* in the Mascarenes. The Madagascan *Ceratitis tananarivana* Hancock appears to be related to *C. rosa* and, although its hosts are unknown, is also a potential pest species.

7 REFERENCES

Annecke, D.P. and Moran, V.C., 1982. Insects and Mites of Cultivated Plants in South Africa. Butterworths, Durban & Pretoria, 383 pp.

Bot, J., 1965. Census of a Dacine population (Diptera: Tripetidae). Journal of the Entomological Society of Southern Africa, 28: 166–178.

Carnegie, A.J.M., 1962. Problem of fruit flies and melon flies (family Trypetidae) in Southern Rhodesia. Rhodesia Agricultural Journal, 59: 229–235.

Cogan, B.H. and Munro, H.K., 1980. Family Tephritidae. In: R.W. Crosskey (Editor), Catalogue of the Diptera of the Afrotropical Region. British Museum (Natural History) London, pp. 518–554.

Georgala, M.B., 1964. The response of the males of *Pterandrus rosa* (Karsch) and *Ceratitis capitata* (Wied.) to synthetic lures (Diptera: Trypetidae). Journal of the Entomological Society of Southern Africa, 27: 67–73.

Hancock, D.L., 1981. Some economic Zimbabwean fruit flies (Diptera: Tephritidae). Hortus (Zimbabwe), 27 (1980): 11–15.

Hancock, D.L., 1984. Ceratitinae (Diptera: Tephritidae) from the Malagasy subregion. Journal of the Entomological Society of Southern Africa, 47: 277–301.

Hancock, D.L., 1985a. Two new species of African Ceratitinae (Diptera: Tephritidae). Arnoldia Zimbabwe, 9 (21): 291–297.

Hancock, D.L., 1985b. New species and records of African Dacinae (Diptera: Tephritidae). Arnoldia Zimbabwe, 9 (22): 299–314.

Hancock, D.L., 1985c. A key to *Clinotaenia* Bezzi and related genera (Diptera: Tephritidae) with description of a new species. Transactions of the Zimbabwe Scientific Association, 62 (9): 56–65.

Hancock, D.L., 1986. Classification of the Trypetinae (Diptera: Tephritidae), with a discussion of the afrotropical fauna. Journal of the Entomological Society of Southern Africa, 49: 275–305.

Munro, H.K., 1924. Fruitflies of wild olives. Entomology Memoirs, Department of Agriculture, Union of South Africa, 2: 5–17.

Munro, H.K., 1925. Biological notes on South African Trypaneidae (fruitflies) I. Entomology Memoirs, Department of Agriculture, Union of South Africa, 3: 39–67.

Munro, H.K., 1926. Biological notes on South African Trypaneidae (Trypetidae) (Fruitflies) II. Entomology Memoirs, Department of Agriculture, Union of South Africa, 5: 17–40.

Munro, H.K., 1929. Biological notes on the South African Trypetidae (Fruitflies: Diptera). III. Entomology Memoirs, Department of Agriculture, Union of South Africa, 6: 9–17.

Munro, H.K., 1933. Records of South African fruitflies (Trypetidae, Dipt.) with descriptions of new species. Entomology Memoirs, Department of Agriculture, Union of South Africa, 8: 25–45.

Munro, H.K., 1934. A review of the species of the subgenus *Trirhithrum* Bezzi (Trypetidae, Diptera). Bulletin of Entomological Research, 25: 473–489.

Munro, H.K., 1935. Biological and systematic notes and records of South African Trypetidae (Fruitflies: Diptera) with descriptions of new species. Entomology Memoirs, Department of Agriculture, Union of South Africa, 9: 18–59.

Munro, H.K., 1939. Studies in African Trypetidae with descriptions of new species. Journal of the Entomological Society of Southern Africa, 1: 26–46.

Munro, H.K., 1947. African Trypetidae (Diptera): A review of the transition genera between Tephritinae and Trypetinae, with a preliminary study of the male terminalia. Memoirs of the Entomological Society of Southern Africa, 1: 1–284.

Munro, H.K., 1952. A remarkable new gall-forming Trypetid (Diptera) from Southern Africa, and its allies. Entomology Memoirs, Department of Agriculture, Union of South Africa, 2 (10): 329–341.

Munro, H.K., 1953. Records of some Trypetidae (Diptera) collected on the Bernard Carp Expedition to Barotseland, 1952, with a new species from Kenya. Journal of the Entomological Society of Southern Africa, 16: 217–226.

Munro, H.K., 1964. Some fruitflies of economic importance in South Africa. Department of Agricultural Technical Services, Pretoria, 18 pp.

Munro, H.K., 1967. Fruitflies allied to species of *Afrocneros* and *Ocnerioxa* that infest *Cussonia*, the umbrella tree or kiepersol (Araliaceae). (Diptera: Trypetidae). Annals of the Natal Museum, 18: 571–594.

Munro, H.K., 1985. A taxonomic treatise on the Dacidae (Tephritoidea, Diptera) of Africa. Entomology Memoirs, Department of Agriculture, Republic of South Africa, 61: i–xi, 1–313.

Myburgh, A.C., 1962. Mating habits of the fruit flies *Ceratitis capitata* (Wied.) and *Pterandrus rosa* (Ksh.). South African Journal of Agricultural Science, 5: 457–464.

Myburgh, A.C., 1964. Orchard populations of the fruit fly, *Ceratitis capitata* (Wied.), in the Western Cape Province. Journal of the Entomological Society of Southern Africa, 26: 380–389.

Orian, A.E. and Moutia, L.A., 1960. Fruit flies (Trypetidae) of economic importance in Mauritius. Revue Agricole et Sucriere de L'ile Maurice, 39: 142–150.

Prokopy, R.J. and Hendrichs, J., 1979. Mating behaviour of *Ceratitis capitata* on a field-caged host tree. Annals of the Entomological Society of America, 72: 642–648.

Ripley, L.B. and Hepburn, G.A., 1935. Olfactory attractants for male fruit-flies. Entomology Memoirs, Department of Agriculture, Union of South Africa, 9: 3–17.

Schmidt, C.T., 1967. A fruit-fly attacking coffee cherries in Sao Tomé.Garcia de Orta (Lisboa), 15: 329–331.

Chapter 2.3 Indian Sub-Continent

V.C. KAPOOR

1 INTRODUCTION

Fruit flies are responsible for most of the damage to fruits and vegetables in this sub-continent and many ornamental and oilseed Compositae plants are also attacked. Despite severe damage to fruits and vegetables little work on taxonomy, ecology, behaviour and control has been done in this region. The last decade a few papers have appeared which have started to fill this gap (Kapoor, 1970; Kapoor and Malla, 1978; Kapoor et al., 1980).

Fruit flies are gradually increasing their host range (Kapoor and Agarwal, 1983), but little is known of the host range of many of them. Of the total of about 310 species known from this sub-continent, hosts are known for only about 40 species. The most economically important species are *Dacus cucurbitae* Coquillett, *Dacus zonatus* (Saunders) and *Dacus dorsalis* Hendel. The occurrence of *Dacus caudatus* Fabricius is doubtful as the earlier records were misidentifications of *D. cucurbitae*. The fruit fly *Dacus correctus* (Bezzi) is known from Sri Lanka, India and Thailand and may possibly be present in Pakistan, Burma, Nepal and Bangladesh. Infestation in mango, peaches, bael, jujube, jamun and orange has been low and it frequently occurs with other major fruit flies such as *D. zonatus* and *D. dorsalis*.

Dacus ciliatus Loew is an Ethiopian species and was known in the Indian literature as *Dacus brevistylus* Bezzi. It is fairly well represented in this sub-continent and infests a large number of cucurbits but squash melon is preferred. It sometimes is referred to as the 'Ethiopian melon fly'. The reason for its low pest profile is attributed to competition by its most dominant economic homologue, *D. cucurbitae*. *Dacus diversus* Coquillett is another fruit fly which is showing increased activity. It was originally reported from India and Sri Lanka but now has invaded Burma, Pakistan and Thailand. It attacks a variety of fruits and vegetables e.g. guava, mango, banana, citrus, jamum and various cucurbits. *Dacus tau* (Walker), an Oriental species, is also increasing its pest status. It was hitherto known as *Dacus hageni* or *D. nubilus* in the Indian literature and sometimes confused with *D. caudatus* and *D. cucurbitae*. It is fairly well represented in this sub-continent attacking various fruits and vegetables such as mango, sapodilla, citrus, cucurbits and tomatoes. *Dacus longistylus* Wiedemann is an African species and fairly well distributed in this region on the non-economic Ak plant, *Calotropis procera* (Ait.), which is also of African origin. However, shortage or absence of this host plant may induce the Ak fruit fly to attack other cucurbitaceous plants (Batra, 1955). There was a similar occurrence with *Dacus brevistylus* Bezzi (now *D. ciliatus*) in upper Egypt; this species was once a pest of only Ak plants but later became a serious pest of cucurbitaceous plants wherever the Ak plants were either absent or in

Chapter 2.3 references, p. 61

short supply (El Zoheiry, 1951). Fletcher (1919) reported the olive fruit fly, *Dacus oleae* Gmelin in Cherat, North West Frontier Province of Pakistan and Pruthi and Batra (1938) recorded its presence on wild and cultivated olives in north-west India. However, since then it has not been reported from this region. *Callantra eumenoides* Bezzi was once recorded attacking Ivy gourd, *Coccinia grandis* Linnaeus in Burma (Shroff, 1919)

Amongst the trypetine fruit fly pests, the ber fly (*Carpomyia vesuviana* Costa) and Baluchistan melon fly (*Myiopardalis pardalina* Bigot) are of some economic importance. Unlike dacine species which are active during the hot season, the ber fly is active during winter from November to April. Its infestation peaks from February to March when fruits of all late varieties of ber ripen. The fruits are actually attacked in November when they are small and tender and sometimes no fruit remain unattacked. *M. pardalina* attacks have been reported although they are of limited economic importance. It was first reported from Baluchistan attacking melons (Fletcher, 1917, 1919, 1921). The fly is currently present throughout Pakistan except Sind. It has been reported from wild cucurbit, *Cucumis trigonus* Roxburg and melons in India (Fletcher, 1919, 1921; Misra, 1919; Clausen et al., 1965), but not in recent years.

The Mediterranean fruit fly, *Ceratitis capitata* (Wiedemann) has been wrongly reported from this sub-continent. Wiedemann (1824) mentioned its distribution as 'Ostindien' which was interpreted as East India in subsequent papers. Munro (1938) reported five specimens of this species from Pusa, Bihar in India from peaches but the source of these fruits was not known. Since then the fly has not been recorded from any part of this region.

The *Acanthiophilus helianthi* Rossi is the only tephritine fruit fly which occasionally causes serious damage to safflowers in this region. It is fairly well represented in northern India and Pakistan, and undoubtedly occurs in other parts of the sub-continent. It is a potential economic pest of safflower and corn and may (Bhatia and Singh, 1939; Menon et al., 1968; Agarwal and Kapoor, 1982) achieve serious pest status once safflower is well established in this region. The fly is very active from March to May.

2 SPECIES OF MAJOR IMPORTANCE

2.1 *Dacus cucurbitae* Coquillett

In India this insect destroys more than half of all vegetables including cucurbits (melons and gourds), tomato, chillies, papaya, peach, date, citrus and grapes. It is active throughout the year except for a short period during the cold months of January and February. While it prefers to oviposit in young green and tender fruits sometimes eggs are laid in the corolla and other floral parts. The larvae will feed on the flowers and sometimes even on the stems of the cucurbit vines forming characteristic galls (Narayanan, 1953). The population peaks during July and August.

2.2 *Dacus zonatus* (Saunders)

This species was first reported from Bengal and it is widespread in India, Pakistan, Nepal, Bangladesh, Sri Lanka, Burma and evidently over South east Asia. While mango, guava, peaches are the most preferred hosts, citrus, jujube, sapodella, pear, apples, pomegranate and vegetables such as gourds, melons, brinjals are also attacked.

This fly is active throughout the year except during the cold winter months of January and February. Sometimes both this species and *D. cucurbitae* are reared from the same hosts. In Spring ber is commonly attacked (Grewal, 1981). In May and June, the fly does serious damage to peaches. Due to its preference for this fruit, it is known as the peach fruit fly. Later in the season cucurbits are attacked. During the monsoon and post-monsoon periods infestation occurs in mango and guava with attack on the latter being very severe, especially in northern plains.

In some regions this species occurs with *D. dorsalis* in mango. There is a possibility that this species may displace *D. dorsalis*. There has always been competition between these two species but earlier reports indicated that *D. dorsalis* usually outnumbered *D. zonatus* in mixed infestations.

Since this species is gradually increasing its host range and causing significant damage, it may become more important than the current major pest species, *D. cucurbitae*.

2.3 *Dacus dorsalis* Hendel

The Oriental fruit fly was reported from Koshun in Taiwan and is now widely distributed over the Oriental region, Micronesia and the Hawaiian Islands. In the Orient it is regarded as one of the most important species attacking almost all types of fleshy fruits but especially mango, guava, carambola, jamun, papaya, etc. It has been known as *Dacus ferrugineus* Fab. in the Indian literature.

The fly is active form early April to late November and in March, guava, which is plentiful in the plains, is an important host. The damage to guava peaks from July to the end of September. Attacks continue throughout the season in areas having equable climatic conditions i.e. subtropical south. From April to May, fruits like loquat, apricot and plum are attacked. Later in July, the attack of the fly is very conspicuous on mango, peach and pear. On peaches the attack of this species is usually mixed with *D. zonatus* which has always been more dominant.

3 REFERENCES

Agarwal, M.L. and Kapoor, V.C., 1982. *Acanthiophilus helianthi* (Rossi) and *Chetostoma completum* Kapoor (Diptera: Tephritidae) as serious pests of *Centaurea cyanus* L. (Compositae) in India. Journal of Entomology Research, Delhi, 6 (1): 102–104.

Batra, H.N., 1955. *Dacus longistylus* Wied. becoming a pest of cucurbitaceous plants in India. Indian Journal of Entomology, 17: 278–279.

Bhatia, H.L. and Singh, M., 1939. *Acanthiophilus helianthi* Rossi, a new pest of safflower in Delhi. Indian Journal of Entomology, 1: 110.

Clausen, C.P., Clancy, D.W. and Chock, Q.C., 1965. Biological control of the Oriental fruit fly (*Dacus dorsalis* Hendel) and other fruit flies in Hawaii. USDA Technical Bulletin No. 1322, pp. 102.

El Zoheiry, M.S., 1951. Review of Applied Entomology, Ser. A., p. 301.

Fletcher, T.B., 1917. Proceedings of second Entomological Meeting, Pusa, India, II: 306; 1919. Annotated list of Indian crop pests, III: 41; 1921. Additions and corrections to the list of Indian crop pests. IV: 14–15.

Grewal, J.S., 1981. Relative incidence of infestation by two species of fruit flies in Ludhiana, Punjab, India. Indian Journal of Ecology, 8: 123–125.

Kapoor, V.C., 1970. Indian Tephritidae with their recorded hosts. Oriental Insects, 4 (2): 207–251.

Kapoor, V.C. and Malla, Y.K., 1978. Tephritids of Nepal and India (A taxonomic review). Journal of Natural History Museum, Nepal, 2 (3): 117–125.

Kapoor, V.C. and Agarwal, M.L., 1983. Fruit flies and their increasing host plants in India. In: R. Cavalloro (Editor), Fruit flies of Economic Importance. Balkema Publishers, Rotterdam, The Netherlands, pp. 252–257.

Kapoor, V.C., Hardy, D.E., Agarwal, M.L. and Grewal, J.S., 1980. Fruit fly systematics of the Indian sub-continent. Export India Publications, Jalandhar, India, 113 pp.

Menon, M.G.R., Mahto, Y. and Kapoor V.C., 1968. *Centaurea americana* as new host plant record for the fruit files, *A. helianthi* Rossi and *Craspedoxantha octopunctata* Bezzi in Delhi, India. Indian Journal of Entomology, 30 (4): 316.

Misra, C.S., 1919. Index to Indian fruit pests. Proceedings of Third Entomological Meeting, Pusa, India, 2: 564–595.

Munro, H.K., 1938. Studies on Indian Trypetidae (Diptera). Record Indian Museum, 40: 21–37.

Narayanan, E.S., 1953. Fruit flies of orchards and kitchen gardens. Indian Farming, 3 (4): 8–11, 29–31.

Pruthi, H.S. and Batra, H.N., 1938. Some important fruit pests of North-West India. Misc. Bull. I.C.A.R., New Delhi, No. 19, pp. 1–113.

Shroff, K.D., 1919. A list of the pests of vegetables in Burma. Proceedings of Third Entomological Meeting, Pusa, India, 1: 351.

Wiedemann, C.R.W., 1824. Munus rectoris in ex museo Acedemia Christian Albertina. Analecta entomologica ex museo regio Hafniae maxinae congesta, 60 pp.

Chapter 2.4 South-East Asia and Japan

J. KOYAMA

1 INTRODUCTION

The fruit flies of economic importance in the area of south-east Asia and Japan are the melon fly, *Dacus cucurbitae* Coquillett, and the oriental fruit fly, *Dacus dorsalis* Hendel (Commonwealth Institute of Entomology, 1956, 1960). The northern limit of the distribution of both species was in Taiwan and the Mariana Islands until around 1920, but then the distribution expanded northward up to the Amami and Ogasawara (Bonin) Islands of Japan (Fig. 2.4.1). To exclude this invasion of both fruit flies, some eradication programs have been conducted by the Japanese and its prefectural governments (Table 2.4.2). As a result, the northern distribution limit of the oriental fruit fly had been pushed down to the south of Yaeyama and Ogasawara Islands by 1986. The melon fly in Kume Island was eradicated successfully in 1978. The original description is not incorrect, but the distribution pattern, by 1986, had been changed significantly by an eradication program (cf. Table 2.4.2).

2 DAMAGE

The melon fly infests all kinds of cucurbit fruits and several other host fruits from 16 families, 55 species and 9 varieties. The oriental fruit fly attacks citrus, mango, guava and many kind of subtropical and tropical fruits from 40 families, 169 species and 6 varieties (Umeya and Sekiguchi, 1967a, b). Table 2.4.1 shows main host fruits infested by the oriental fruit fly in several localities in Japan and South-East Asia.

In most cases, adult females oviposit eggs under the surface of host fruit and larvae develop in the fruit. The fruit infested by the larvae either rots or is unmarketable. However, more important is the quarantine problem in the areas where the fruit is grown for export. For example, Japan is a highly industrialized country with a considerable population, so needs to import subtropical and tropical fruits from South-East Asia. However, the Japanese government strictly restricts importation of these host fruits from areas in which *D. dorsalis* and *D. cucurbitae* occur regardless of the actual infestation. The reason for this restriction is that mainland Japan which has no fruit fly problem would be threatened with the establishment of these fruit flies by the import of possibly infested host fruit.

To permit importation of fruit from these areas, some fumigation procedures specified and supervised by Japanese government, or eradication of the species from appropriate areas are necessary. These fumigation or eradication procedures impose considerable expense.

Chapter 2.4 references, p. 66

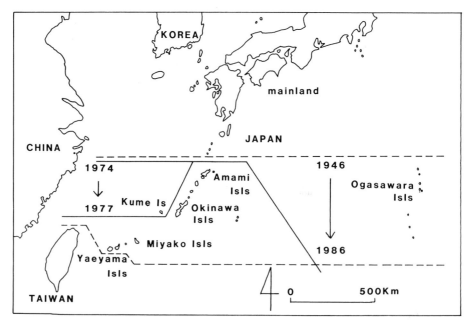

Fig. 2.4.1. Northern limit of distribution of the melon fly *Dacus cucurbitae* (——) and the oriental fruit fly *Dacus dorsalis* (–––) in the area around Japan. Numerals and arrows indicate years and pushing down of the limit by the eradication programs, respectively.

3 CONTROL

There are two approaches to the fruit fly problem. One is to apply insecticidal control and the other involves eradication of pest species.

TABLE 2.4.1

Main host fruits of the oriental fruit fly in South-East Asia and Japan[1]

Common name	Scientific name	Japan	Taiwan[2]	Philippines[3]	Malaysia[4]	Indonesia[5]
Avocado	*Persia americana*			×		
Bachang	*Mangifera foetida*				×	
Banana	*Musa paradisiaca*	×			×	×
Chilli	*Capsicum annuum*	×			×	×
Guava	*Psidium guajava*	×	×	×		
Jackfruit	*Artocarpus heterophyllus*			×		
Kwini	*Mangifera odorata*				×	
Mango	*Mangifera indica*		×	×	×	×
Orange	*Citrus spp.*	×	×			×
Papaya	*Carica papaya*	×		×	×	
Peach	*Prunus persica*	×				
Plum	*Prunus domestica*	×				
Sentol	*Sandoricum koetjape*			×		
Sineguelas	*Spondias purpurea*			×		
Soursop	*Annona muricata*				×	
Starfruit	*Averrhoa carambola*		×		×	×
Sugar apple	*Annona squamosa*		×			
Tomato	*Lycopersicon esculentum*	×				
Wax apple	*Eugenia javanica*		×		×	
Yellow sentol	*Sandoricum indicum*				×	

[1] × marks indicate main host fruits reported in each country.
[2] Yao and Lee (1978), Chu, personal communication (1984).
[3] Golez (1981).
[4] Tan and Lee (1982).
[5] Tjiptono, personal communication (1985).

TABLE 2.4.2

Current state of the Japanese and Taiwanese programs for the control and/or eradication of fruit flies[1]

Program	Region	Species	Method[2]	Period	Eradication[3]
Kagoshima	Amami Is. 1238 km²	Oriental fruit fly	MA	1968–1980	+
		Melon fly	SIT	1979–	±[4]
Tokyo	Ogasawara Is. 73 km²	Oriental fruit fly	MA	1975–1976	−
		Oriental fruit fly	SIT	1976–1985	+
Okinawa	Okinawa Is. 1434 km²	Oriental fruit fly	MA	1977–1982	+
		Melon fly	SIT	1972–	±[5]
	Mikayo Is. 227 km²	Oriental fruit fly	MA	1982–1984	+
		Melon fly	SIT	1984–	−
	Yaeyama Is. 584 km²	Oriental fruit fly	MA	1982–1986	+
Taiwan	Taiwan 35961 km²	Oriental fruit fly	MA, SIT	1975–	−

[1] Prepared from Habu et al. (1980), Iwahashi (1977), Koyama (1980, 1982), Koyama et al. (1984), Lee and Chang (1980), Tanaka (1980), Ushio et al. (1982) and personal communications.
[2] MA: Male annihilation method with poisoned methyleugenol, SIT: Sterile insect technique.
[3] Eradication was achieved (+), partially achieved (±) or not achieved (−) by 1986.
[4] Eradicated from Kikai Is. in 1985.
[5] Eradicated from Kume Is. in 1978.

To reduce damage by the oriental fruit fly, insecticide spraying with or without protein hydrolysate was tested in the Philippines on mango fruit and was of value when applied when the fruit were ripening (Golez, 1981). Residues on the fruit are a potential problem. However, in general insecticidal control of fruit flies is not very prevalent in South-East Asia (K.H. Tan, Pudjo Tjiptono, personal communication, 1985), primarily because control measures such as protein bait spraying and poisoned attractant baits are not very effective and are not cost-effective.

Male annihilation (MA) with poisoned attractant and/or sterile insect technique (SIT) have been applied more extensively for the purpose of eradication. Table 2.4.2 summarizes the current state of the eradications in programs of Japan and Taiwan to eradicate the melon fly and the oriental fruit fly. Detail of Japanese programs will be given in Vol. 3B (Chapter 9.5.2). In Taiwan, control relied on the use of the male annihilation technique from 1984 (Chu, personal communication, 1984).

To remove the ban on exportation to mainland Japan, fumigation with methyl bromide or ethylene dibromide is applied for some kinds of fruit in the following countries. In the Philippines, mango is fumigated. In Taiwan, ponkan orange, tankan orange, liutin variety of sweet orange, mango, papaya and litchi are fumigated prior to exportation. In south-western islands of Japan where the melon fly and the oriental fruit fly have not been eradicated, fumigation is applied for melon, string bean, citrus, tomato, papaya, guava and passion fruits destined for mainland Japan.

4 SURVEILLANCE SYSTEM

In Japan, more than 1,300 monitor traps baited with methyl eugenol and cue-lure are maintained to detect possible invasions of the oriental fruit fly and the melon fly respectively. If any trap catches flies, control measures are

Chapter 2.4 references, p. 66

applied as fast as possible to prevent establishment and possible expansion of the infestation. This surveillance system may protect mainland Japan from invasion of these fruit flies.

5 REFERENCES

Commonwealth Institute of Entomology (1956). Distribution maps of pests. *Dacus (Strumeta) cucurbitae* Coq. (Melon Fruit-fly) Series A. (Agricultural) Map No. 64.

Commonwealth Institute of Entomology (1960). Distribution maps of pests. *Dacus dorsalis* Hend. (Dipt., Trypetidae) (Oriental Fruit-fly) Series A (Agricultural) Map No. 109.

Golez, H.G., 1981. Population studies and control of oriental fruit fly *Dacus dorsalis* Hendel (Diptera: Tephritidae) in Guimaras Island. Thesis for Master of Science (Entomology) submitted to the Faculty of the Graduate School, University of the Philippines at Los Baños.

Habu, N., Iga, M. and Numazawa, K., 1980. Progress of the eradication program of the oriental fruit fly, *Dacus dorsalis* Hendel, on the Ogasawara (Bonin) Islands. Proceedings of a Symposium on Fruit Fly Problems, Kyoto and Naha, 1980. National Institute of Agricultural Sciences, Yatabe, Ibaraki 305, Japan, pp. 123–142.

Iwahashi, O., 1977. Eradication of the melon fly, *Dacus cucurbitae*, from Kume Is., Okinawa with the sterile insect release method. Researches on Population Ecology, 19: 87–98.

Koyama, J., 1980. The Okinawa Project of eradicating fruit flies. Proceedings of a Symposium on Fruit Fly Problems, Kyoto and Naha, 1980. National Institute of Agricultural Sciences, Yatabe, Ibaraki 305, Japan, pp. 99–106.

Koyama, J., 1982. The Japan and Taiwan Projects on the control and/or eradication of fruit flies. Sterile Insect Technique and Radiation in Insect Control, International Atomic Energy Agency, Vienna, 1982, pp. 39–51.

Koyama, J., Teruya, T. and Tanaka, K., 1984. Eradication of the oriental fruit fly (Diptera: Tephritidae) from the Okinawa Islands by a male annihilation method. Journal of Economic Entomology, 77: 468–472.

Lee, L.W.Y. and Chang, K.K., 1980. The Taiwan project. Proceedings of a Symposium on Fruit Fly Problems, Kyoto and Naha, 1980, National Institute of Agricultural Sciences, Yatabe, Ibaraki 305, Japan, pp. 85–97.

Tan, K.H. and Lee, S.L., 1982. Species diversity and abundance of *Dacus* (Diptera: Tephritidae) in five ecosystems of Penang, West Malaysia. Bulletin of Entomological Research, 72: 709–716.

Tanaka, A., 1980. Present status of fruit fly control in Kagoshima Prefecture. Ibid., pp. 107–121.

Umeya, K. and Sekiguchi, Y., 1967a. Explanation of the important pests for plant quarantine 2. *Dacus dorsalis*. Yokohama Shokubutsu-bôeki News, No. 332: 2–5. (in Japanese).

Umeya, K. and Sekiguchi, Y., 1967b. Explanation of the important pests for plant quarantine 3. *Dacus cucurbitae*. Ibid., No. 333: 2–4 (in Japanese).

Ushio, S., Yoshioka, K., Nakasu, K. and Waki, K., 1982. Eradication of the oriental fruit fly from Amami Islands by male annihilation (Diptera: Tephritidae). Japanese Journal of Applied Entomology and Zoology, 26: 1–9 (in Japanese with English summary).

Yao, A.L. and Lee, W.Y., 1978. A population study of the oriental fruit fly, *Dacus dorsalis* Hendel (Diptera: Tephritidae), in guava, citrus fruits, and wax apple fruit in northern Taiwan. Bulletin of the Institute of Zoology, Academia Sinica, 17: 103–108.

Chapter 2.5 Australia and South Pacific Islands

G.H.S. HOOPER and R.A.I. DREW

1 INTRODUCTION

There is a large endemic tephritid fauna in the area defined by this chapter, i.e. Australia, Papua New Guinea and the islands of the South Pacific. The pest species belong to the subfamily Dacinae except for the introduced Mediterranean fruit fly, *Ceratitis capitata* (Wiedemann) in Australia. Within Australia 78 species have been described and further new species are known. Within the remainder of the area approximately 200 species are known and since collecting has not been as extensive as in Australia, an increase in species might be anticipated.

Apart from the Mediterranean fruit fly, which was introduced into Australia from Europe around 1897 (but now only occurs in the western coastal fringe of the continent), and more recently the establishment in northern Queensland in 1972 of *Dacus frauenfeldi* Schiner from Papua New Guinea, the pest species of Australia comprise endemic species which proved able to shift from endemic rain forest fruits to introduced crop plants. Similarly, in the islands of the South Pacific the species of concern are endemic with the exception of the Queensland fruit fly, *Dacus tryoni* (Froggatt) which has extended its range, as a consequence of man's activities, as far east as Easter Island from which it was eradicated twice in the early 1970's. However, in Papua New Guinea the introduced melon fly, *Dacus cucurbitae* Coquillett is well established and occasional specimens of *D. tryoni* have been recorded.

There are several aspects to the fruit fly problem in the area under consideration. Firstly, a number of species cause damage to crops and the loss plus control measures are a direct cost to agricultural production. Secondly, there are both intra and international quarantine problems resulting in either denial of potential markets, or the added cost of appropriate disinfestation procedures to farmers in fruit fly areas. Thirdly, just as the major pest species (other than *C. capitata*) in Australia are endemic and have demonstrated an ability to shift from endemic to introduced hosts, it is likely that many developing South Pacific countries will encounter a similar problem with their endemic tephritid species. In many of these countries agriculture and horticulture are currently practised at a subsistence level only. When there is a departure from this to large monocultures it must be expected that more fruit fly species will achieve pest status. Fourthly, the possible establishment within the area of exotic species is of concern.

Table 2.5.1 indicates to which of the male lures, methyleugenol, cue-lure and trimedlure, the species considered in the text respond. Further information on this aspect can be found in Drew (1974, 1982) and Drew and Hooper (1981). An approximate indication of the distribution of the species considered can be

TABLE 2.5.1

Male lures, cue-lure (CL), methyleugenol (ME), trimedlure (TML), to which the species discussed respond, and the distribution of each species within the areas designated in Fig. 2.5.1

Species	Responding to				Distribution in areas designated in Fig. 2.5.1
	CL	ME	TML	None	
Dacus aquilonis (May)	*				F
Dacus atrisetosus Perkins				*	A
Dacus bryoniae (Tryon)	*				A, B, F
Dacus cucumis French				*	B, C, F
Dacus cucurbitae Coquillett	*				A
Dacus curvipennis Froggatt	*				G
Dacus decipiens Drew				*	A (New Britain)
Dacus dorsalis Hendel		*			A
Dacus facialis Coquillett	*				H (Tonga)
Dacus frauenfeldi Schiner	*				A, B
Dacus jarvisi (Tryon)				*	B, C, F
Dacus kirki Froggatt	*				H, J (plus Niue Is)
Dacus melanotus Coquillett[1]	*				I
Dacus musae (Tryon)		*			A, B
Dacus neohumeralis Hardy	*				A, B, C
Dacus passiflorae (Froggatt)	*				H (plus Niue Is)
Dacus psidii (Froggatt)	*				G
Dacus trivialis Drew	*				A
Dacus tryoni (Froggatt)	*				A, B, C, D, G, J
Dacus umbrosus (Fabricius)		*			A
Dacus xanthodes Broun		*			H, I
Ceratitus capitata (Wiedemann)			*		D, E

[1] Drew (1974, 1982) incorrectly recorded this species as responding to ME; recent work (M.A. Bateman, pers. comm.) confirms the above response.

gained by using the information in Table 2.5.1 in conjunction with Fig. 2.5.1. For presentation in this chapter the South Pacific region is divided into 10 areas.

2 PEST SPECIES AND THEIR DISTRIBUTION

2.1 Papua New Guinea and the Bismarck Archipelago (Area A, Fig. 2.5.1)

A number of serious pest species occur in this area. *Dacus musae* (Tryon) is a major pest of bananas and will oviposit in immature green bananas as well as mature fruit. *Dacus bryoniae* (Tryon) has been recorded from banana and capsicum in Papua New Guinea. *D. cucurbitae* is a major pest of vegetable crops. It occurs in lowland areas of Papua New Guinea and islands to the east as far as Bougainville Island. It was recently introduced into the Shortland Islands but was subsequently eradicated. *Dacus atrisetosus* Perkins occurs at higher altitudes in mainland Papua New Guinea and infests vegetable crops (particularly tomatoes and cucurbits) which are attacked by *D. cucurbitae* at lower altitudes.

Dacus decipiens Drew occurs only on New Britain and is a serious pest of pumpkins. *D. frauenfeldi* occurs in large populations throughout area A and is a serious pest of mango. *Dacus umbrosus* (F.) is abundant in lowland areas and infests citrus and breadfruit throughout the area. *Dacus trivialis* Drew has been bred in large numbers from grapefruit, peach and guava in mainland Papua New Guinea.

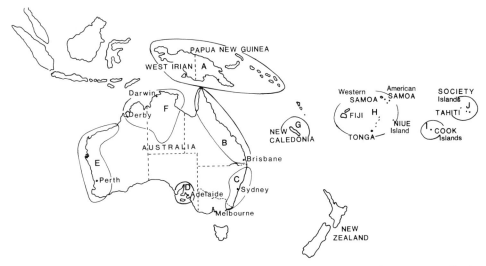

Fig. 2.5.1. Map of Australia and South Pacific Islands designating areas of approximate distribution of fruit fly species referred to in the text and Table 2.5.1.

Dacus neohumeralis Hardy appears to be endemic in mainland Papua New Guinea but to date has not been reared from fruit. Only the occasional specimens of *D. tryoni* have been collected in Papua New Guinea and considerable doubt exists as to whether this species is established there.

The fruit fly fauna of Irian Jaya is virtually unknown. A few specimens were recently received from the western extremity and some attracted to methyl eugenol were *Dacus dorsalis* Hendel.

2.2 Australia (Areas B, C, D, E, F, Fig. 2.5.1)

As might be expected from the size of the Australian continent, and the consequent range of climatic zones, the tephritid problem is rather complex. The most severe problems occur in the wet tropical and sub-tropical areas of eastern Queensland (Area B), in the northernmost third of the Northern Territory (Area F), and the northern tip of Western Australia (Area E). However, because of the extent of horticultural crops in Queensland the problem there is most severe.

In eastern Australia (Areas B, C) the major pest is the Queensland fruit fly, *D. tryoni*. In addition to attacking a wide range of fruit and vegetable crops, *D. tryoni* breeds in numerous native fruits, particularly in areas of rainforest. Another closely related species, *D. neohumeralis* has a more restricted range (Area B), but can be as damaging as *D. tryoni*; in some commercial fruits both species occcur in approximately equal numbers. In Queensland and eastern New South Wales, these tephritids must be controlled on a regular basis. A further species attacking commercial fruit is *Dacus jarvisi* (Tryon). In Queensland it has been recorded from deciduous fruits, persimmons and guava.

D. musae is the major pest of banana, attacking both immature and ripe fruit, but occurs only in northern Queensland. *Dacus cucumis* French is a serious pest of all cucurbits, tomatoes and papaya in north-eastern Australia (Area B).

Other species have been recorded as minor or potential pests in Australia (Drew, 1982) based on incidental occurrences in fruit. It is possible that these could become economic pests with changing patterns of fruit production.

Although efficient control measures are available and implemented to limit direct loss to commercial crops in Queensland and coastal New South Wales,

Chapter 2.5 references, p. 72

movement of vegetables and tropical fruits to southern areas of Australia is subject to quarantine restrictions. During the major period of fruit fly activity (September–May) fruit and vegetables entering Victoria must either be certified as originating in a fruit fly free area or be certified as having received an approved disinfestation treatment. Even more stringent limitations occur with other states.

In the inland fruit growing areas of the Goulburn, Murray and Murrumbidgee River valleys *D. tryoni* is of major economic importance. This is not because of direct damage, but because of loss of markets, or increased production costs arising from disinfestation procedures for fruit destined for export to other Australian states or overseas. In these normally fruit fly free areas the detection of larvae in fruit, or more than 2 flies in a trap (cue-lure) results in post-harvest treatment of all fruit within 3–80 km of the "outbreak" depending on the potential market.

In South Australia (Area D, Fig. 2.5.1) *D. tryoni* and *C. capitata* are the species of concern. Despite inspection at roadblocks and a requirement that fruit from other states be certified as originating from a fruit fly free area, or as having been subjected to a disinfestation treatment, outbreaks of both species have occurred regularly in Adelaide since 1947. To date these outbreaks have not occurred in the commercial fruit areas of the state. The presence of fruit fly in the citrus orchards of the Murray River area, and any consequent eradication program, would present special problems for the existing biological control of California red scale. Each fruit fly outbreak in Adelaide is eradicated by protein hydrolysate baiting and limited cover spraying.

The major tephritid pest species in Western Australia (Area E, Fig. 2.5.1) is *C. capitata*. While it has long been established in the south-western area of the state its range extends up the western coast and recently it has been detected at Derby. There is some concern that it could spread further north. As elsewhere in the world, it is a major pest of a wide range of exotic fruits and some vegetables. In the north-west of the state *D. jarvisi* attacks mangoes.

Of 18 tephritid species recorded from the Northern Territory (Area F, Fig. 2.5.1) 3 are considered as current or potential pests. In recent years there has been a considerable expansion of horticultural crops and the economic importance of fruit flies may be expected to increase as a consequence. *D. jarvisi* is recognised as a pest of mangoes (which is not true for Queensland) and this results in quarantine restrictions to southern markets. Recently this species has been recorded from citrus and again this could complicate the export of citrus, particularly limes. *Dacus aquilonis* (May), which is closely related to *D. tryoni*, also attacks citrus as well as other fruit such as peach and guava. *D. cucumis* was detected in 1980 but its effect on cucurbit crops has yet to be defined. Delimiting its range is difficult because it does not respond to any of the male lures (Table 2.5.1).

On 2 occasions (1976/77 and 1981/82) infestations of *C. capitata* were detected in Alice Springs, and were eradicated by collection and destruction of host fruits, and protein bait and cover sprays.

Tasmania does not have endemic tephritid pest species and has no fruit fly problem. Although *C. capitata* was detected there at the turn of this century it did not establish.

2.3 New Caledonia (Area G, Fig. 2.5.1)

Dacus psidii (Froggatt) has been reared from citrus, mango, granadilla and guava, and potentially could become a serious pest if these hosts were developed commercially. However, it is likely that the introduced *D. tryoni* would be the more important species. In any event, both species could be managed by the same control measures. Another species, *Dacus curvipennis* Froggatt, is restricted to New Caledonia and infests citrus.

2.4 Fiji, Samoa, Tonga (Area H, Fig. 2.5.1)

Three local species have been recorded from Fiji and Samoa attacking a variety of fruit crops including peach, mango, citrus and guava and vegetable crops such as tomato and capsicum, viz. *Dacus xanthodes* Broun, *Dacus kirki* Froggatt and *Dacus passiflorae* (Froggatt). On Tonga *Dacus facialis* Coquillett is a serious pest of tomatoes and has been recorded from a variety of other fruits and vegetables.

2.5 Cook Islands (Area I, Fig. 2.5.1)

Dacus melanotus Coquillett has been recorded only from this region and is considered a very serious pest of citrus. It also attacks mango and guava. *D. xanthodes* has recently been recorded on these islands (M.A. Bateman, pers. comm.).

2.6 Society Islands (Area J, Fig. 2.5.1)

One endemic species, *D. kirki*, has been recorded from several hosts, viz. peach, mango, guava, and capsicum. However, the exotic species *D. tryoni* is likely to be a greater problem in fruit and vegetable crops.

3 CONCLUSION

This chapter would not be complete without addressing the problem of the introduction of exotic species. That this threat is a serious one is amply illustrated by (i) spread of *D. tryoni* to islands of the South Pacific, (ii) the invasion of northern Australia by *D. frauenfeldi*, and (iii) the establishment of *D. cucurbitae* in Papua New Guinea.

The establishment by *D. dorsalis* in New Guinea and/or Australia would be a very serious matter. However, there are more than 20 species within the *"dorsalis* complex", which all respond to methyl eugenol, and which are difficult to separate on classical taxonomic criteria. The magnitude of the problem became clear after the detection in the Northern Territory in 1976 of flies that at the time were believed to be *D. dorsalis*. Although morphologically they could not be distinguished from *D. dorsalis* existing in south-east Asia and the Hawaiian Islands, they were subsequently defined as a new species probably endemic to the area, *Dacus opiliae* Drew and Hardy, on ecological and cytological grounds. This species has been recorded only from one non-economic native host.

The existence of extensive populations of *D. cucurbitae* in Papua New Guinea, a scant 150 km from the Australian mainland is a matter of continuing concern, given the level of local maritime commerce across Torres Strait. Similarly the spread of *D. xanthodes, D. facialis, D. kirki* and *D. passiflorae* from their current ranges to other islands in the Samoan, Fijian and Tongan island groups should be monitored.

The history of past movements of tephritid fruit flies coupled with an increase in tourism and landfall by itinerant yachts within the area and between this area and south-east Asia, demonstrates the necessity for efficient quarantine services. In 1984 Australia-wide quarantine confiscated 7.1 tonnes of fruit from incoming plane passengers and crew, and this arose from a 5% inspection

Chapter 2.5 references, p. 72

rate. In addition 3.8 tonnes of fruit were discarded in "honesty bins" prior to quarantine inspection (A. Allwood, pers. comm.). While there is no information on the rate of fruit fly infestation of such fruit from Australia, New Zealand does record such data and they have reared seven *Dacus* spp. from produce originating from within the Pacific region, *D. cucurbitae* and *Dacus ciliatus* Loew from India, *C. capitata* from Spain, and *Anastrepha fraterculus* (Wiedemann) from Peru (Keall, 1981). Clearly quarantine efforts in all areas should be backed by extensive and intensive monitoring by appropriate traps to detect incipient invasions of exotic species.

4 ACKNOWLEDGEMENTS

We are grateful for information provided by A. Allwood, T.G. Amos, J.G. Gallatley, P.E. Madge, and A.N. Sproule.

5 REFERENCES

Drew, R.A.I., 1974. The response of fruit fly species (Diptera: Tephritidae) in the South Pacific area to male attractants. Journal of the Australian Entomological Society, 13: 267–270.

Drew, R.A.I., 1982. Taxonomy. In: R.A.I. Drew, G.H.S. Hooper and M.A. Bateman (Editors), Economic Fruit Flies of the South Pacific. 2nd ed. Department of Primary Industries, Brisbane, pp. 1–97.

Drew, R.A.I. and Hooper, G.H.S., 1981. The responses of fruit fly species (Diptera: Tephritidae) in Australia to male attractants. Journal of the Australian Entomological Society, 20: 201–205.

Keall, J.B., 1981. Interceptions of insects, mites and other animals entering New Zealand 1973–1978. New Zealand Ministry of Agriculture and Fisheries, Levin, pp. 126–129.

Chapter 2.6 Hawaiian islands and North America

ERNEST J. HARRIS

1 INTRODUCTION

Insects are considered to be pests when they damage crops, commodities and stored products, transmit diseases, or annoy beneficial animals (Anonymous[a], 1969). In this chapter the primary interest is in the pest status, past and current, of fruit flies in Hawaii and North America. For introduced species, such status begins when the insects are introduced into new areas where, unhindered by climate and natural enemies, they increase and cause economic damage and loss of a crop or commodity. Pest status may also be given to an insect of minor importance when it changes its habits and attacks a crop or commodity of economic importance. Regardless of how pest status is achieved, the impact of the presence of an insect on control, management, quarantine, and eradication methods is a primary consideration. The objectives of this discourse are to describe and elucidate the following events: (1) pest introduction; (2) initial attempt to eliminate the problem; (3) initial spread, distribution, and host range; (4) subsequent occurrence, spread, distribution and host range; (5) influence of biological, environmental, and social factors on future host status; and (6) population stability.

2 HAWAIIAN ISLANDS

The pest status of fruit flies in the Hawaiian islands can be seen in Fig. 2.6.1.

2.1 The melon fly, *Dacus cucurbitae* Coquillett and the Mediterranean fruit fly, *Ceratitis capitata* (Wiedemann)

Back and Pemberton (1917) reported that the melon fly, *D. cucurbitae* was introduced from Japan as early as 1895. By 1917 the melon fly was then infesting 15 species of vegetables and 6 species of fruits. Unencumbered by any control activities, the fly spread quickly to the major islands in Hawaii. Its spread necessitated quarantine because upon arrival in California the insect was found alive as larvae or pupae in produce brought aboard ship in Hawaii by passengers or stored on board cargo ships. It greatly damaged cucurbits by stinging and depositing eggs in young fruit and blossoms and in vines. Thus in Hawaii farmers stopped growing cucumbers, squash, and tomatoes. Later when production of these vegetables was resumed, the size and number of plantings was greatly reduced and chemical control efforts were intensified.

On June 21, 1910 Back and Pemberton (1918) discovered the medfly, *C. capitata* in Honolulu, Hawaii following introduction via ship from Australia.

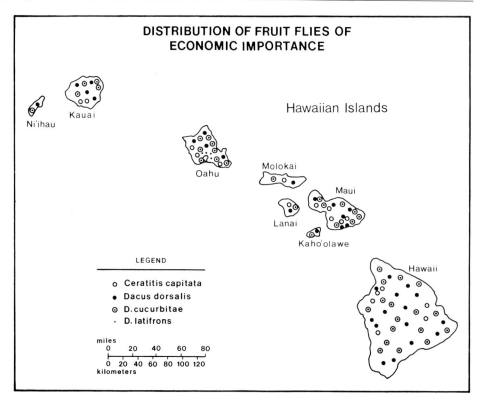

Fig. 2.6.1. Distribution of fruit flies of economic importance in the Hawaiian islands.

The presence of both melon fly and the medfly in Hawaii dealt a serious blow to agricultural development just at the time when such development was increasing. The medfly attacked some 72 host species including fruits, nuts, and vegetables. The quarantines against all fruits, particularly avocado and mango, seriously affected small farmers. The favorable climate and abundant host plants provided an ideal setting for both fruit flies.

2.2 The oriental fruit fly, *Dacus dorsalis* Hendel

In May 1946, during routine observations of insect collections, a member of the entomology staff of the Territorial Board of Agriculture and Forestry (TBAF) discovered the presence of *D. dorsalis*, the oriental fruit fly (Carter, 1952). Within two years this insect became a major pest of almost every variety of fruit grown commercially in Hawaii (Carter, 1952). It also attacked wild fruits, domestic fruits, and ornamental berries and even invaded fruit stands inside stores and laid its eggs in imported fruit. The losses this fly caused were of three kinds: (1) reduction of fruit grade caused by stings that spoiled the appearance of the fruit, even in many fruits unfavorable for larval survival; (2) damage of the entire fruit by larvae and resulting decay; and (3) indirect damage from quarantining the Islands to prevent dissemination of the fly (Carter, 1952).

From 1947 to 1948 action against the oriental fruit fly was promptly taken by the TBAF along with private agencies. Since control of insect pests in Hawaii by means of parasites and predators was well established, the biological control method received early attention. Temperature and host availability were found to be the major natural factors controlling populations. In 1946, prior to the arrival of the oriental fruit fly the medfly had been known to infest guava and other fruits throughout the Hawaiian islands. Simultaneously with the spread of the oriental fruit fly throughout Hawaii, infestations of guavas by the medfly, especially in lowland areas, disappeared because of displacement by the oriental fruit fly (Bess, 1953). By December 1949 less than 5% of guava samples from Oahu were infested by the medfly. These samples came from a few locations with a cool climate. From 1966 to 1968 populations of the oriental fruit fly and the medfly were much lower than prior to 1950 because the size of guava tree stands were reduced by land development and commercialization (Haramoto and Bess, 1970).

2.3 *Dacus latifrons* (Hendel)

In March 1983, a fourth economically important fruit fly, *D. latifrons* was discovered on peppers, *Capsicum annum* L., and eggplant, *Solanum melongena* L. in Honolulu Country. Immediately preceding this discovery, a package from Hawaii with infested peppers was intercepted in a southern Californian post office. Subsequently *D. latifrons* was found by Hawaii State and Federal entomologists infesting solanaceous (nightshade) vegetables. It was widely distributed on the island of Oahu attacking tomato, eggplant, and peppers. A survey has shown that to date *D. latifrons* has such a wide distribution only on the island of Oahu (Vargas and Nishida, 1985) (Fig. 2.6.1). The ability of *D. latifrons* to live in small urban rural vegetable gardens has permitted this species to fill a niche not occupied by the melon fly or the oriental fruit fly.

3 NORTH AMERICA

3.1 *Rhagoletis* spp.

The fruit flies established in North America (Fig. 2.6.2) consist primarily of *Rhagoletis* spp. (Bush, 1966; Biggs, 1972). Published studies of *Rhagoletis* spp. have focused primarily on those of economic importance. The *pomonella* group consists of four species: apple maggot fly, *Rhagoletis pomonella* (Walsh); blueberry maggot fly; *Rhagoletis mendax* Curran, and the dogwood berry fly *Rhagoletis cornivora* Bush, all of which are sympatric, and the snowberry fruit fly *Rhagoletis zephyria* Snow. The last named species was once believed to be parapatric with apple maggot only in the extreme western part of the latter's range, but is now known to reach New York State. The apple maggot attacks apples and in some localities cherries, plums, apricots, and pears. The eastern and western cherry fruit flies, *Rhagoletis cingulata* (Loew), *Rhagoletis indifferens* Curran, and the black cherry fruit fly, *Rhagoletis fausta* (Osten Sacken), attack sweet and sour cherry; the blueberry maggot fly attacks blueberries; the walnut husk fly, *Rhagoletis completa* Cresson, and *Rhagoletis sanvis* Loew, attack walnuts and in some localities peaches; and *Rhagoletis basiola* (Osten Sacken) attacks rose hip. These species are widely distributed in both southern Canada and the United States (Fig. 2.6.2).

Most *Rhagoletis* spp. are univoltine and oligophagus. They feed on the pulp of developing fruit, pupate in the soil, diapause through the winter, and emerge as adults over a period of several weeks during the spring and summer. These insects feed for several weeks, mate, and oviposit in host fruits. The apple maggot fly is an indigenous North American insect whose principal native host is hawthorn (*Crategus* spp.) (Bush, 1966). For about 120 years, the apple maggot has expanded its host range to include fruits of the following introduced plants; *Malus* spp. Walsh, domestic plum (Herrick, 1920), sour cherry (Shervis et al., 1970), apricot (Lienk, 1970), pear (Prokopy and Bush, 1972), and most recently rose hips (Prokopy and Berlocher, 1980). The apple maggot has been more successful than *R. zephyria*, and *R. cornivara*, in shifting to introduced plants.

Chapter 2.6 references, p. 80

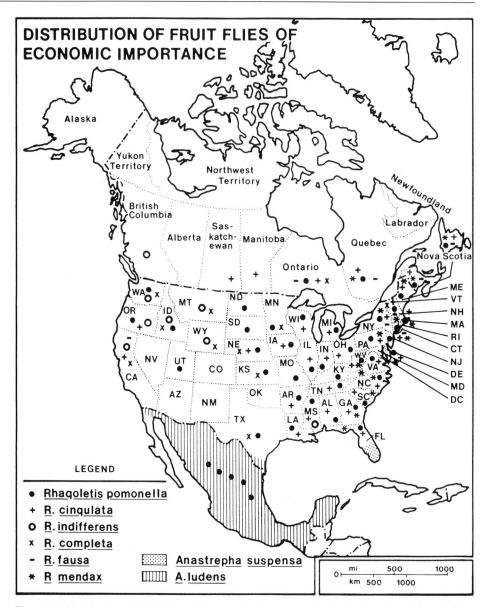

Fig. 2.6.2. Distribution of fruit flies of economic importance in North America (adapted from Arnett, 1985).

Visual and chemical traps are used to detect *Rhagoletis* spp. Control of the apple maggot is directed against the adults through trapping and applications of insecticides to prevent oviposition.

3.2 The Mexican fruit fly, *Anastrepha ludens* Loew

This is a subtropical tephritid species whose native home is thought to be in north-eastern Mexico (Ebeling, 1959). Its present distribution (Fig. 2.6.2) extends throughout much of Mexico and as far south as northern South America. In the United States, the Mexican fruit fly occurred for the first time in April 1927 in Mission, Texas, primarily in the Lower Rio Grande Valley. Occasionally it appears in southern California, most recently in the Los Angeles area where an infestation of this pest was established and subsequently eradicated in 1983. The Mexican fruit fly is multivoltine and polyphagus attacking more than 14 different hosts. Its hosts include oranges, grapefruit, apples, peaches, pear, pomegranates, quinces, and yellow sapote, which is the principal wild host in Mexico (Stone, 1942; Hoidale, 1952). This plant is generally distributed for a distance of > 320 km along the Sierre Madre Orientalis mountains where moisture conditions are favorable. It provides an important reservoir from which fruit flies originate to establish infestations in Texas citrus groves. It is also of great concern to citrus producers in southern California and the Lower Rio Grande Valley of Texas. As it is not a tropical species, it can adapt to areas that are dry, rainy, mountainous, or nearby coastal plains. Since the mexican fruit fly is native to Mexico, the problem of controlling it is more acute in central to southern Mexico than in areas bordering the United States. When ethylene dibromide (EDB) plus vapor heat was accepted as a fumigant by APHIS, Texas citrus growers began to regard this fly primarily as a nuisance. Policy dictated that fumigation was not required for Texas fruit until the fly appeared in traps on the U.S. side of the border. Since populations in Texas are generally very low throughout the year, much speculation takes place concerning the source of the flies that are trapped; were they newly emerged migrants from Mexico or did they survive in low numbers and then rebuild populations when the host situation became more favorable (William G. Hart, personal communication)?

Current records from APHIS and ARS show that host availability strongly influences the abundance of the Mexican fruit fly and the weather strongly influences the abundance of the host. However, market demand determines the harvest practices of growers. During 1983 when the market was poor, many grapefruit were left on trees causing a very large increase in the Mexican fruit fly, a condition that persisted into the summer and fall. In the winter of 1983 a freeze occurred in Weslaco, Texas. Immediately afterward, large numbers of active larvae were found in grapefruit, and flies also emerged from pupae in the soil, an indication that this fly can overwinter in southern Texas. This finding is strong evidence of the adaptive ability of the fly and explains its sporadic appearance annually in southern Texas (William G. Hart, personal communication).

3.3 The Caribbean fruit fly, *Anastrepha suspensa* Loew

This species has been introduced into Florida on three occasions since the first infestation in 1931. In November 1936 hog plums, (*Spandias mombin* L.) were found infested in Key West, Florida. Adults of three fruit flies were reared from collected fruits and identified as *Anastrepha fraterculus* (Wiedemann), *A. suspensa*, and *Anastrepha acidusa* Walker. Difficulty was encountered in correctly identifying the fruit fly species causing the damage. Although eradication was attempted these fruit flies disappeared spontaneously and reappeared in 1959 when two adults were trapped at Key West (Swanson and Baranowski, 1972). An infestation detected in Miami Spring in April 1965 (Weems, 1966) was determined to be *A. suspensa*. By December 1965, *A. suspensa* was found infesting 15 hosts. A decision was made not to conduct an eradication campaign because it was believed that the fly would disappear as it had in the 1930's. Instead of disappearing *A. suspensa* became well established in Florida.

Swanson and Baranowski (1972) produced a host list for *A. suspensa* in Florida which includes 84 host fruits belonging to 23 families. The major hosts included rose apple, cattley guava, Surinam cherry, tropical almond, guava, and loquat. In 1968 *A. suspensa* was found for the first time attacking dooryard grapefruit in Florida, and in 1974 was found in commercial grapefruit exported to Japan (A.K. Burditt, personal communication). Export of grapefruit was made possible by fumigating the fruit with EDB. The banning of the use of EDB by EPA has made it necessary to treat grapefruit with methyl bromide fumigation and cold treatment and to create zones free of fruit flies for export of fruits from Florida and Texas. The shift of *A. suspensa* to infesting commercial citrus has identified this pest as a target for eradication.

4 CURRENT EXPRESSION OF PEST STATUS: HAWAII AND THE MAINLAND UNITED STATES

After the melon fly, medfly, and oriental fruit fly became established in Hawaii, the severity of the damage they caused to host fruits declined following the establishment of an equilibrium between these insects and the habitats in which they live. Haramota and Bess (1970) called attention to the influence of the environment on the reduction of population levels of Hawaiian fruit flies. Strong evidence of the effect of changes in the environment between 1910 and 1985 is shown in Table 3.6.1 (Harris, unpublished data). In June 1910, shortly after the medfly was established in the Punchbowl district of Honolulu, 4,610 host fruit trees were present in this area. By June 1985, fewer than 100 host trees were found in the same area because of commercial and residential development of downtown Honolulu. In areas adjacent to Honolulu, habitat changes have been less severe and populations of the three fruit flies have shifted their distribution to these locations (Harris and Lee, 1986).

Apple and hawthorne infested with races of *Rhagoletis* spp. are generally distributed over the eastern half of the continental United States, extending south from New York to Florida and east to North Dakota and eastern Texas (Fig. 3.6.2). A separate population is known to occur in the highlands of central Mexico. Apple production concentrated in the Pacific coastal regions of the U.S. in the states of California, Oregon, and Washington has now become threatened by the establishment of the apple maggot (Fig. 2.6.2) (Ali Niazee and Penrose, 1981). The immense potential for damage by this insect, if not controll- ed, has been documented by Glass and Lienk (1971). In addition, the potential for host race formation by *Rhagoletis* spp. exploiting new habitats is revealed by the occurrence of these insects in Oregon, California, and Utah. Establish- ment of *Rhagoletis* spp. has already resulted in the application of quarantine measures to prevent spread into uninfested areas. In the tropical and subtropi- cal areas of mainland United States, there exist many favorable locations in which tephritid fruit flies potentially can become established. Most adult finds and new establishments of the oriental fruit fly have occurred in California. The greatest diversity in species introduced into the U.S. have occurred in that State. Four recent medfly infestations were discovered in California, with the most severe occurring from 1980 to 1982 (see Chapter 10.8). Infestations of the Mexican fruit fly were also found in California. Undoubtedly, the high annual tourist traffic entering that State both by air and sea has increased the pressure on the quarantine barrier by concomitantly increasing the risk of introduc- tions. Moreover, commercial and residential development has increased the number of favorable habitats, because of homeowners planting a diversity of fruit fly host plants around their homes.

5 IMPLICATIONS OF PEST STATUS: PRESENT AND FUTURE

The current and future pest status of fruit flies is closely related to their intrinsic rate of increase, *r* (Bateman, 1972; Price, 1975). *Rhagoletis* spp. (the K strategists) breeding primarily in one host or in only a few hosts, generally do not greatly extend their range of distribution as readily as *Dacus* spp. If the K strategists increase their range of distribution, they do so very slowly over a

TABLE 2.6.1

Name and numbers of host species of the Mediterranean fruit fly found in the Punchbowl District of Honolulu in June 1910 when the fly was first established compared with host species found in the same location in June 1985

Host	Number	
	1910	1985
Apricot, *Prunus armeniaca* L.	1	0
Avocado, *Persea americana* Mill.	653	4
Breadfruit, *Artocarpus incis communis* Forst.	58	1
Carambola, *Averrhoa carambola* L.	48	0
Chinese inkberry, *Cestrum diurnum* L.	6	0
Chinese orange, *Citrus japonica* Swingle	148	0
Coffee, *Coffea arabica* L.	298	0
Coffee, Liberiam, *Coffea liberica* Bull.	8	14
Cotton, *Gossypium tomentosum* Nutt.	11	0
Custard apple, *Annova reticulata* L.	1	0
Damson plum, *Chrysophyllum oliviforme* L.	4	0
Fig, *Ficus carica* L.	201	1
Guava, common *Psidium guajava* L.	94	0
Guava, strawberry, *Psidium cattleianum* Sabine	73	18
Java plum, *Eugenia cuminii* (L.) Druce	80	0
Kamani, ball, *Calophyllum inophyllum* L.	4	1
Kamani, winged, *Terminalia catappa* L.	13	15
Kumquat, *Fortunella japonica* Swingle	4	0
Lemon, *Citrus medica* L.	22	0
Lichee, *Litchi chinensis* Sonn	40	4
Lime, *Citrus limon* (L) Burm.f.	10	0
Loquat, *Eriobotrya japonica* (Thunb.) Lindl.	33	1
Mandarin, *Citrus nobilis* Swingle	28	0
Mango, *Mangifera indica* L.	1154	23
Mangosteen, *Garcinia mangostema* L.	7	0
Mountain apple, *Jambosa malaccensis* L.	41	0
Mock orange, *Murraya exotica* L.	33	6
Orange, sweet, *Citrus aurantium* L.	372	3
Papaya, *Carica papaya* L.	687	6
Peach, *Prunus persica* L. Sieb, & Zucc.	69	0
Pear, Bartlett, *Pryus communis* L.	2	0
Pomegranite, *Punica granatum* L.	128	0
Pomelo, *Citrus paradisis* Macf.	15	0
Rose apple, *Eugenia jambos* L.	25	0
Sapota, *Casimiroa edulis* Llave & Lex	30	0
Sour sop, *Annona muricata* L.	57	0
Spanish cherry, *Eugenia dombeyi* (Spreng.) Skeels	1	0
Star apple, *Chrysophyllum cainito* L.	4	2
Surinam cherry, *Eugenia uniflora* L.	63	0
Wi, *Spondias ducis* Forst.f.	19	0
Total	4610	99

period of several years; changes in the host species they infest also occur slowly. The r strategists (*Dacus* spp. and the medfly) breed in many hosts, including fruits and vegetables. Perhaps the most efficient species in host utilization are the melon fly and the medfly. In Hawaii the melon fly breeds in papaya and other fruits when cucurbits, its preferred hosts, are unavailable. A high *r* value is associated with great adaptive ability. The global distribution of the medfly which is indicative of a high *r* value enables the medfly to adapt to diverse habitats in subtropical and temperate areas. Among *Dacus* spp., the oriental fruit fly is widely distributed and well adapted to infest hosts in tropical rain forests as well as in arid, and subtropical climates. In contrast, *D. latifrons* is quite restricted in its distribution in Hawaii and is more like a K

Chapter 2.6 references, p. 80

strategist. *Anastrepha* spp. are intermediate in their biology and ecology, with some characteristics of both K and r strategists. For example, *Anastrepha* spp. can be found across the border in Mexico near the Rio Grande Valley of Texas and near San Diego, California; but few outbreaks of these species have occurred on the mainland U.S. as compared with those involving *Dacus* spp. and the medfly.

r strategists are more effective at becoming established in new areas (Chambers et al., 1974) as evidenced by recent outbreaks in Florida and California. The medfly and the oriental fruit fly are introduced with increasing frequency into the continental United States. Trade between nations and mass movement of their peoples by automobile, boat, and airplane will continue to increase the risk of introduction of fruit flies into new areas.

6 REFERENCES

AliNiazee, M.T. and Penrose, R.L., 1981. Apple maggot in Oregon: a possible new threat to the northwest apple industry. Entomological Society of America Bulletin, 27 (4): 245–246.

Anonymous[a], 1969. Principles of plant and animal pest control Vol. 3. Insect Pest Management and Control Pub. 1695 National Academy of Sciences, Washington D.C.

Arnett, Jr. R.H., 1985. American insects: a handbook of the insects of America North of Mexico, Van Nostrand Reinhold co. New York.

Back, E.A. and Pemberton, C.E., 1917. The melon fly in Hawaii USDA. Bull. 491, 63 pp.

Back, E.A. and Pemberton, C.E., 1918. The Mediterranean fruit fly. USDA. Bull. 640, 44 pp.

Bateman, M.A., 1972. The ecology of fruit flies. Annual Review of Entomology, 17: 493–518.

Bess, H.A., 1953. Status of *Ceratitis capitata* in Hawaii following the introduction of *Dacus dorsalis* and its parasites. Proceedings of the Hawaiian Entomological Society, 15: 221–234.

Biggs, J.D., 1972. Aggressive behavior in the apple maggot (Diptera: Tephritidae) Canadian Entomologist, 104: 349–353. Scotia Bull. Nova Scotia Dept. Agric. No. 9, pp. 1–70.

Bush, G.L., 1966. The taxonomy, cytology and evolution of the genus *Rhagoletis* in North America (Diptera: Tephritidae). Bulletin of the Museum of Comparative Zoology, Harvard University, 134: 431–562.

Carter, W., 1952. The oriental fruit fly. In: the Yearbook of Agriculture. Insects. USDA, Washington D.C., pp. 551–559.

Chambers, D.L., Cunningham, R.T. and Thrailkill, R.B., 1974. Pest control by attractants: A case study demonstrating economy specificity, and environmental acceptability. Bio Science, 24 (3): 150–152.

Ebeling, W., 1959. Subtropical fruit pests. Div. Agric. Sci. Univ. Calif. Berkeley, Calif., 436 pp.

Glass, E.H. and Lienk, S.E., 1971. Apple insect and mite populations developing after discontinuance of insecticides: 10-year record. Journal of Economic Entomology, 64: 23–26.

Haramoto, F.R. and Bess, H.H., 1970. Recent studies on the abundance of the oriental and Mediterranean fruit flies and the status of their parasites. Proceedings of the Hawaiian Entomological Society, 20 (3): 551–566.

Harris, E.J. and Lee, C.Y.L., 1986. Seasonal and annual occurrence of Mediterranean fruit flies (Diptera: Tephritidae) in Makaha and Waianae Valleys, Oahu, Hawaii. Environmental Entomology, 15: 507–512.

Herrick, G.W., 1920. The apple maggot in New York. Cornell Univ. Agr. Exp. Sta. Bull. 402: 89–101.

Hoidale, P.A., 1952. The Mexican fruit fly. In: the Yearbook of Agriculture. Insects. United States Department of Agriculture, Washington D.C., pp. 559–562.

Lienk, S.E., 1970. Apple maggot infesting apricot. Journal of Economic Entomology, 63: 1684.

Newell, W., 1936. Progress report on the Key West (Florida) fruit fly eradication project. Journal of Economic Entomology, 29 (1): 116–120.

Price, P.W., 1975. Insect ecology. John Wiley and Sons, New York, pp. 124–165.

Prokopy, R.J. and Bush, G.L., 1972. Apple maggot infestation of pear. Journal of Economic Entomology, 65: 597.

Prokopy, R.J. and Berlocher, S.H., 1980. Establishment of *Rhagoletis pomonella* (Diptera: Tephritidae) on rose hips in southern New England. Canadian Entomologist, 112 (12): 1319–1320.

Shervis, L.J., Boush, G.M. and Koval, C.F., 1970. Infestation of sour cherries by apple maggot: Confirmation of a previously uncertains host status. Journal of Economic Entomology, 63: 294–295.

Stone, A., 1942. The fruit flies of the genus, *Anastrepha*. USDA. Misc. Publ. No. 439, 112 pp.

Swanson, R.W. and Baranowski, R.M., 1972. Host range and infestation by the Caribbean fruit fly,

Anastrepha suspensa (Diptera: Tephritidae), in South Florida. Proceedings of the Florida State Horticultural Society, 85: 271–274.

Vargas, R.I. and Nishida, T., 1985. Survey for *Dacus latifrons* (Hendel) (Diptera: Tephritidae): A recently discovered fruit fly in Hawaii. Journal of Economic Entomology, in press.

Weems, H.V., Jr., 1966. The caribbean fruit fly in Florida. Proceedings of the Florida State Horticultural Society, 79: 401–403.

Chapter 2.7 Mexico, Central and South America

DIETER ENKERLIN, LAURA GARCIA R. and FIDEL LOPEZ M.

1 INTRODUCTION

It is an extremely difficult task to adequately address the question of the pest status of so many different species of fruit flies belonging to five genera: *Dacus*, *Toxotrypana*, *Anastrepha*, *Rhagoletis* and *Ceratitis* (Diptera: Tephritidae) with a vast variety of hosts in many different countries in Mexico, Central and South America, including the Caribbean area.

The pest status differs in each country and has changed in the last decade. For many years cotton pests were considered to be much more important than fruit flies. The present importance of fruit flies is linked principally to established or potential export markets for fruits in the tropics and sub-tropics (mainly citrus and mango), but in temperate regions, northern Mexico or Chile, also for apples, peaches and other rosaceae fruit. More recently, due to increasing immigration from developing countries in the tropics to developed countries, the market in the latter for exotic fruit is increasing. Furthermore, many countries in need of foreign currency are expanding the area dedicated to the production of fruit for export.

With the phasing out of ethylene dibromide (EDB) in the late 1970s, as a practical commodity treatment, fruit fly importance has increased enormously and the export market has been severely affected. Therefore indirect losses due to fruit flies have risen sharply, in countries with an established fruit export.

The average losses in many Latin American countries (under this term we include for practical purposes all countries south of the USA) and the Caribbean area, are considered to be around 25%; while they may be low at the beginning of the fruiting season for the different hosts, losses can rise to 80% or more late in the season.

Many of the tropical fruits in most countries are very important from the socio-economic point of view: peasant farmers, village and hillside dwellers are practically dependent on different fruit which they harvest all year from trees along the roadsides and paths. The fruits may also be sold in local markets but much is discarded because of fruit fly damage. Hence while these crops are important for these people, it is extremely difficult to develop reliable loss figures in most countries as no one knows how much mango, guava, chapote and other tropical fruit grows wild in the countryside. Further, chemical control of fruit flies is much less expensive than for other pests as the different recommended insecticides are mixed with protein to give insecticide-baits. These can be applied in bands, on one side of the tree only, or as spot treatments giving savings of at least 50% of total material and thereby lowering the cost and undesirable side effects.

In this chapter we will discuss and/or present data related to the five fruit

Chapter 2.7 references, p. 90

fly genera mentioned above, including the most important species and their hosts in different areas as well as fruit production data for some countries, when available. We hope, that in the future, as more knowledge on fruit flies is accumulated in the Americas, more accurate information on their pest status can be presented. In this discussion we will follow the taxonomic classification of Foote (1980).

2 GENUS *DACUS* FABRICIUS

This genus belongs to the subfamily Dacinae and probably contains most of the important fruit fly species on a world wide basis. *Dacus dorsalis* Hendel, the oriental fruit fly and *Dacus cucurbitae* Coquillett, the melon fly, have been accidentally introduced into North America from larvae presumably originating from Acapulco, Mexico in 1976, but each time they were eradicated. No other specimens have been reported from the remainder of Latin America except for a recent report that *D. dorsalis* was found in Surinam and probably had been there since 1980 (G.G. Rohwer, pers. comm., 1986). *D. cucurbitae* would be specially devastating if it became established as its life cycle is much shorter than that of *Anastrepha* spp. and it has a high fecundity (Aluja, 1984). Therefore continuous trapping should be maintained in all possible ports of entry in Latin America and the Caribbean area.

3 GENUS *TOXOTRYPANA* GERSTAECKER

This genus of the subfamily Dacinae contains a very large fruit fly, viz. the papaya fruit fly *Toxotrypana curviacauda* Gerstaecker with a wide distribution in the Americas and the Caribbean area. Other species have been identified in South America, but apparently are of no economic importance. Even though *T. curviacauda* is reported as a serious pest of papaya we have only been able to find adults in traps, including liver baited screw worm traps in the Rayones Cañon, N.L., but have never found larvae in the locally produced and/or shipped in fruit from all of Mexico. In Florida, USA it apparently causes some damage but we have no figures at hand. In South America, Brazil has the largest production of papaya, 470 thousand tons (Table 2.7.3) but Gallo et al. (1970) do not mention *T. curviacauda* as a pest. Wille (1952) in his book of agricultural entomology refers to fruit flies as very important pests but he does not mention *Toxotrypana* spp. Overall, we feel that the papaya fruit fly, although mentioned as an important pest is not very abundant in most countries. Occasionally it has been reported as attacking mangoes.

4 GENUS *ANASTREPHA* SHINER

Anastrepha is a New World genus and its origin most likely is tropical and subtropical, although some species have adapted themselves to colder climates, especially *Anastrepha fraterculus* (Wiedemann). Some 155 economically important species are known but there are some problems with synonymy and detailed accurate host lists are required. One good example of host range problem is given by *Anastrepha suspensa*, the Caribbean fruit fly; in the Greater Antilles several species of Citrus are reported as hosts; however, according to Van Whervin (1974), to local agronomists and entomologists and our own experience, *Citrus* spp. have not been found to be infested. Fruits of the Myrtaceae family are the preferred hosts, also tropical almonds, this may be the explanation that citrus "escapes". Finally for many of the species their hosts or host preference is not known; nor, therefore, is their pest status.

While *Anastrepha* spp. are by far the most damaging fruit flies in Latin America, few concrete figures on economic losses are available; losses are calculated as being 25% on the average but this figure fluctuates depending on species, crops, time of the year, etc. For instance, late in the season in areas like Chiapas, Mexico, Jamaica and most likely all tropical areas, over 80% of hosts are infested if no control measures are taken. As stated before, the recent increase of the economic importance of the *Anastrepha* complex is due to the phasing out of EDB. In the State of Nuevo Leon, Mexico there are around 40 fruit packing plants; most of the commodities processed are for the export market, mainly USA. Citrus and mangoes are the most important crops but much of this fruit comes from different parts of Mexico, some being areas heavily infested with fruit flies. The total yearly amount of fruit packed on the average was 1.75 million tons with an installed capacity of 3 million. If the border is closed the indirect losses caused by the fruit flies would be very high, as packing for the local market is uneconomic. About 2500 workers would be without a job, the investment in the packing plants would be partially lost, the trucking business would be affected and a chain of other losses would be caused. This is the reason why Mexico and the USA are working out a protocol to establish "fly free" orchards or larger blocks of orchards. However, the establishment of such areas implies insecticide-bait applications which will increase the overall production cost.

In the following paragraphs we intend to discuss the most important species of the *Anastrepha* complex in relation to their hosts in different countries of Latin America and the Caribbean area.

For Mexico Aluja (1984) mentions 17 species of which we consider only five of economic importance. Of these *Anastrepha ludens* (Loew) is the principal pest being reported from 24 different hosts. *Citrus* species, especially *C. grandis* and *C. paradisi* are probably the most preferred. In north eastern Mexico, "Yellow Chapote" *Sargentia greggii* I. Watts of the Family Rutaceae is the most important alternative host, and according to some authors *A. ludens* may have evolved on these shrubs or trees which generally can be found along small rivers and creeks, fruiting late in the spring and during the summer when no susceptible citrus fruit is present. Pears are also alternative hosts in north eastern Mexico but our research has shown that infestation of these fruits is very low; furthermore orchard owners never use chemical control in pear orchards. Further south and in the tropics mangoes and guavas are very important hosts, together with different *Citrus* species. We believe that all other fruit provides occasional hosts in the tropics as *Citrus* and mangoes are available mostly throughout the year area wise, and for the export market Citrus, followed by mangoes and guavas, are the most important hosts and as stated before an overall loss of 25% is generally acknowledged. This figure can be related to Table 2.7.1 for calculating possible losses in U.S. dollars.

Probably the second most important species is the West Indian fruit fly *Anastrepha obliqua* (Macquart), as mango is the main host and a very important crop for the local and export market. We believe, according to observations in Chiapas and Jamaica that this pest is the most abundant in mango and tropical plum of the genus *Spondias* spp. Although reported in some *Citrus* species, in guava and rose apple, its damage seems to be of little importance other than from the quarantine point of view. *A. fraterculus*, the South American fruit fly, is the most widely distributed species and has been reported from Texas, USA down to Chile and Argentina in South America, where it is the most important fruit fly pest. However, in Mexico it is less important than the two preceding species except for the export market mainly to the USA,

TABLE 2.7.1

Production of some fruits in metric tons, for Mexico, Central America and Panama[1]

Country	Oranges	Lemons	Grapefruit	Avocado	Mangoes	Papaya	Coffee beans
Mexico	1600	600	105	440	670	300	415
Guatemala	–	–	–	36	–	–	140
El Salvador	100	21	–	17	15	3	166
Costa Rica	78	–	–	10	–	4	124
Panama	66	–	–	7	28	–	9
Nicaragua	55	–	–	37	–	–	46
Honduras	47	–	–	37	14	–	73
Belize	–	–	28	–	1	–	–

[1] FAO Annuary of Production, 1983.

where it is considered the second most important "exotic" species after the Mediterranean fruit fly, *Ceratitis capitata* (Wiedemann) from mango producing countries. It can be confused with *A. obliqua*.

Anastrepha serpentina (Wiedemann) can also be found in North, Central and South America including Panama and the Caribbean area; at least 16 hosts are reported. In Mexico it is a more tropical species. Its main hosts are of the family Sapotaceae, especially mamey sapote *Calocarpum sapota* Merr. which is a well liked fruit by local people, but the acreage is of very little economic importance. Citrus and mango can also be attacked but there is little hard data.

The other two species of some importance for tropical fruit are *Anastrepha striata* Schiner, mainly attacking guava and *Anastrepha distincta* Green, which can attack oranges and mango.

In summary only *A. ludens* and *A. obliqua* are considered to be species which require chemical control, but extensive studies are needed to actually determine the pest status of the species mentioned here and other species which may be very important potential pests as the others are controlled and/or eradicated.

After studies in six countries of Central America, the *Anastrepha* complex was considered as a pest of minor importance. This has been changing in relation to the increasing export market. For instance Belize has large mango plantations and an effort was made to implement the "fly free" concept, but apparently there was no success in reaching the probit 9 level of infestation. Further, as the CAP-MED project progresses, trying to eliminate the medfly from Central America and Panama, movement of fruit even within some of the countries will have to follow very strict quarantine and this, of course, will much increase the importance of *Anastrepha* spp. As stated above *A. obliqua* and *A. ludens* are the most important species but it is suggested that all Central American countries and Panama initiate and/or strengthen research on their own fruit flies. Costa Rica has the most developed fruitculture in the area (Anonymous, 1985). In Panama more than 40 species are reported but their importance seems to be negligible, at least according to the literature. Again, it must be pointed out that the socio-economic importance of fruit flies to all of these tropical countries is very hard to put into figures.

In the Caribbean area, consisting of some 22 countries and/or islands divided into two groups, the Greater Antilles and the Lesser Antilles, fruitculture is becoming more and more important. One reason is the low value of sugarcane due to world over-production. For instance, in Jamaica many cane fields are ploughed in and planted with citrus and mangoes, both crops grown mainly for the export market. Therefore fruit flies are becoming important pests; 18 *Anastrepha* spp. have been reported, very similar to those of Mexico and Central America as well as to their importance. However one important species *A.*

TABLE 2.7.2

Production of some fruits in metric tons; in countries of the Caribbean area[1]

Country	Oranges	Lemons and Limes	Grapefruit and Pomelo	Avocado	Mangoes	Papayas	Coffee beans
Cuba	374	48	151	17	15	42	21
Dominican R.	75	14	3	25	185	9	49
Grenada	1	–	2	–	2	–	–
Guadeloupe	1	1	–	2	1	–	–
Haiti	32	26	12	2	340	–	38
Jamaica	32	23	21	7	4	2	2
Martinique	1	2	–	23	1	–	–
Puerto Rico	29	3	4	36	5	5	16
Trinidad-T.	7	1	7	3	–	–	2

[1] FAO Annuary of Production, 1983.

suspensa, the Caribbean fruit fly is reported only from Jamaica, Cuba, Haiti, Dominican Republic and Puerto Rico; this species has established itself also in Florida, USA. *A. ludens* has not been collected. Therefore the two most important species are *A. obliqua*, causing heavy losses in mango, up to 80% or more when no chemical control is applied, and *A. suspensa*, attacking different Myrtaceae species. We did some calculations for Jamaica (Enkerlin, 1986) for 2000 acres of commercial varieties of mangoes. Considering a 25% loss due to fruit flies, mainly *A. obliqua*, the yearly loss would be equivalent to U.S. $ 800,000 when not using chemical control which according to one mango farmer, reduces losses to 5% but at a very high cost as he applies weekly malathion cover sprays from early fruiting to harvest. Estimations on one million mango trees of all varieties, at a density of 70 trees per acre and a 25% loss due to fruit flies in Jamaica, would give losses equivalent to 5.5 million U.S. $ in damage yearly. If we take this example for other crops attacked by fruit flies and estimate only 5% losses on the average, to tropical and subtropical fruit, their importance for Latin America and the Caribbean area is still very high. *A. fraterculus*, *A. serpentina* and *Anastrepha striata* are restricted mainly to Trinidad and Tobago, according to the literature and are in general less important although *A. fraterculus* is a very important quarantine problem for the USA. A summary of fruit production can be seen in Table 2.7.2.

For South America we do not have much data available other than different species reported in some of the countries, but as stated by Stone (1942), Wille (1952) and Gallo et al. (1970) the South American fruit fly *A. fraterculus* is probably the most important with a very wide number of hosts ranging from tropical to temperate fruit species. It is also the most widely distributed species from Chile to North America. In Chile *A. fraterculus* is the only species reported as an important pest of apples, peach, plum and other stone fruits. However *A. fraterculus* was eradicated from this country in 1964 (Olalquiaga, 1986; pers. comm.). Other species generally found throughout South America are *A. serpentina*, *A. distincta* and *A. striata*. In Brazil close to 60 species are reported but for many of them their hosts are either not known or are of no economic importance. *A. grandis*, which exists in several of the countries in South America, attacks fruits of the family Cucurbitaceae and in the last few years has become a rather important pest; together with *A. consobrina* and *A. pseudoparallela* it also attacks the passion fruit. However *A. fraterculus* is by far the most important species (Malavasi, 1986; pers, comm.). Nuñez (pers. comm.) in Colombia has been working for many years on the fruit flies but her work is not available to us at present. In general the studies in the last 20 years or so have been concentrated on the medfly rather than on *Anastrepha*. For fruit production see Table 2.7.3.

Chapter 2.7 references, p. 90

TABLE 2.7.3

Production of some fruits in metric tons, for South America[1]

Country	Apples	Pears	Prunes	Peaches	Grapefruit and Pomelo	Apricots	Avocado	Mangoes	Papaya	Coffee beans
Argentina	934	155	53	241	140	29	3	2	1	–
Bolivia	10	5	5	25	47	–	4	4	5	24
Brazil	95	34	–	20	–	–	115	525	470	2452
Chile	410	54	25	122	–	14	27	–	–	–
Colombia	–	–	–	–	–	–	18	23	67	945
Ecuador	35	8	1	3	41	–	30	20	23	330
Guyana	–	–	–	–	–	–	–	3	–	2
Paraguay	1	–	3	4	61	–	4	15	15	14
Peru	71	9	1	29	5	–	67	81	49	143
Surinam	–	–	–	–	1	–	–	–	–	–
Uruguay	43	10	2	20	7	–	–	–	–	–
Venezuela	–	–	–	8	–	–	43	102	32	256

[1] FAO Annuary of Production, 1983.

5 GENUS *CERATITIS* MACLEAY

According to Foote (1980) the Mediterranean fruit fly, *Ceratitis capitata* (Wiedemann) is the only species of this genus found in the Americas, being of course introduced. Gallo et al. (1970) reports the first record of its presence in Brazil as 1905 but other authors mention 1901. The medfly, as it is commonly known, has its origin probably in East Africa, spreading from there to many countries in most of the continents.

We believe that, next to household insects and those attacking stored products it is the species of the widest distribution and host range of all phytophagous insects, feeding on many fruit and horticultural crops, belonging to about 200 species. For instance in the southern valleys of Peru, medfly can be found attacking olives, hot peppers, *Capsicum* spp., or squash varieties; also, when coffee berries are available in Central America, the fly stays on this host, moving to *Citrus* spp. or other hosts in the off season. In coffee, losses have been calculated between 5% to 15% damage, as the coffee berries "mature" faster when attacked by medfly larvae, falling to the ground, where they still can be picked; however increasing labour costs and reducing the quality of the beans. The establishment of the medfly in Mexico would have been an economic disaster for the agriculture, not only because of the actual damage to the fruit and horticultural crops but also due to very strict export limitations.

In Central America the medfly was first reported from Costa Rica in 1955, spreading to Nicaragua in 1960 and to Panama in 1963, but insufficient funds were made available by these countries and by the rest of Central America, Mexico, the USA and International Organizations to eradicate *C. capitata* by the sterile insect technique (Enkerlin, 1984). As a consequence the medfly spread further, being detected in 1975 and 1976 in El Salvador and Guatemala, respectively, moving towards Mexico along the coffee belt where the first fly was detected on 31 January 1977 in spite of all quarantine efforts. After six years of very costly control efforts including the building of the largest laboratory in the world for producing sterile insects and a total expenditure close to 100 million U.S. dollars the medfly was suppressed in Mexico (Gutierrez Samperio, 1976; Hendrichs et al., 1982; Enkerlin, 1984).

As the Central American countries and Panama are becoming more interested in exporting different kinds of fruit, the medfly, like all other fruit flies, is becoming more important and therefore an international program has been established, the CAP-MED project aiming to eradicate *C. capitata* from all of Central America and Panama. This project will also protect Mexico and the USA from possible re-invasions of this pest.

In the Caribbean area, the medfly has not been reported but constant trapping efforts should be undertaken. In South America, fruit producing countries like Argentina, Chile (where presently no *C. capitata* has established itself), Peru and to some extent Brazil and Venezuela have control programs, quarantines and in the case of Peru, an effort to eradicate medfly at least from the southern valleys. We must state that *C. capitata* probably is the most important fruit fly species in the Americas, but *Anastrepha* with its numerous species of which some six are major pests, as a whole is more important.

6 GENUS *RHAGOLETIS* LOEW

Even though Foote (1980) states that 21 species occur in Mexico, the West Indies and throughout most of America, we have found only little information on this genus. For Mexico *Rhagoletis cingulata* has been reported as attacking wild cherries, *Prunus capuli* Cav. and *Rhagoletis pomonella* in *Crataegus mexicana* (García Martell, 1981), both fruits of very little economic importance and no control measures are applied. For Central America, Panama and the Caribbean area we have no reports.

For South America, Wille (1952) reports *Rhagoletis ochraspis* Wiedemann as infesting tomatoes and *Solanum pimpinellifolium*; also in fruits of potato, where *Rhagoletis psalida* is reported as a recent pest. According to Olalquiaga (1986, personal communication) *R. ochraspis* also exists in Chile as a pest of tomatoes but recently it was divided into *Rhagoletis nova* and *Rhagoletis tomatis*.

7 REFERENCES

Aluja Schunemann, M., 1984. Manejo Integrado de las Moscas de la Fruta. Programa Mosca del Mediterraneo. Dir. Gen. San. Vegetal, SARH. Mexico, pp. 241.

Anonymous, 1985. Compendio Estadistico del Area de OIRSA. Organismo Internacional Regional de Sanidad Agropecuaria, pp. 48–55.

Enkerlin, S. D., 1984. Success and Problems in the Use of the Sterile Insect Technique for the Eradication of the Medfly and the Screwworm in Mexico. In: Advances in Invertebrate Reproduction 3. Proc. Third Internal. Symp., Internal. Soc. Invertebrate Reproduction. Elsevier Science Publishers, New York, pp. 505–513.

Enkerlin, S. D., 1986. Final Report. Project TCP/Jam/4507-Fruit Fly Eradication. Food and Agricultural Organization, Kingston, Jamaica.

Foote, R.H., 1980. Fruit Fly Genera South of the United States (Diptera: Tephritidae), U.S. Dept. Agric. Tech. Bull. 1600.

Gallo, D.N.O., Wiendel, F.M., Silveira Neto, S. and Ricardo, P.L.C., 1970. Manual de Entomología, Edit. Agronomica Ceres, Sao Paulo.

García Martell, C., 1981. Lista de Insectos y Acaros Perjudiciales a los cultivos en México. Dir. Gen. San. Veg. SARH, Fitofilo 86, México, pp. 186.

Gutierrez Samperio, J., 1976. La Mosca del Mediterraneo, *Ceratitis capitata* (Wiedemann) y los Factores que favorecerían su establecimiento y propagación en México. Dir. Gen. San. Vegetal, SARH. México, pp. 121.

Hendrichs, J., Ortíz, J., Schwarz, A. and Liedo, P., 1982. Six Years of Successful Fight Against the Mediterranean Fruit Fly in Mexico. Proc. Internal. Symp. on Fruit Flies of Economic Importance, Athens, Greece, pp. 13.

Stone, A., 1942. The Fruit Flies of the Genus Anastrepha. U.S. Dept. Agric. Misc. Public. 439.

Van Whervin, L., 1974. Some Fruit Flies (Tephritidae) in Jamaica. PANS 20:1, pp. 11–19.

Wille, T.J.E., 1952. Entomología Agrícola del Perú. Junta San. Veg., Dir. Gral. San. Veg., Ministerio de Agricultura, Lima, Perú.

Chapter 2.8 Temperate Europe and West Asia

P. FISCHER-COLBRIE and E. BUSCH-PETERSEN

1 INTRODUCTION

The Tephritid family is rather large, with about 4000 species distributed throughout the temperate, subtropical and tropical areas of the world (Christenson and Foote, 1960). The frugivorous species are abundant in most of the countries of temperate Europe and West Asia and are often considered of high economic importance. However, non-frugivorous fruit fly species, which feed on other structures of plants, such as leaves, shoots, roots or inflorescences, may also be of economic importance. The pest status of some important representatives of both types of the family Tephritidae will be summarized in this chapter.

2 FRUGIVOROUS TEPHRITID FLIES

2.1 *Rhagoletis cerasi* (Linnaeus)

R. cerasi, also known as the "European cherry fruit fly", is a member of the subfamily Trypetinae in the large genus of *Rhagoletis*. It belongs to the group of specialized frugivorous fruit flies which infest all varieties of cherries. Alternative host fruits include the *Caprifoliaceae* and some species of *Lonicera*. It is distributed throughout most of the cherry growing area of Europe (Bezzi, 1927). *R. cerasi* is univoltine and the pupae diapause in the soil in the immediate vicinity of the host. The species is well equipped to survive extreme environmental conditions. Adult emergence is synchronized with the ripening of the host fruits and is brought about by a diapause system closely adjusted to climate and photoperiod (Boller and Prokopy, 1976).

The fly attacks mainly medium- and late-ripening cherries at a time when the fruit colour changes from green to yellow. Early ripening varieties are also infested with eggs, but rarely allow larval development prior to harvest. Thus protective measures are usually focussed on medium- and late-varieties. Infestation rates, particularly in late-varieties, may reach more than 90%. The degree of infestation is influenced mainly by climatic conditions during the egg-laying period (Boller, 1966) as well as by the ratio of flies to host fruits. Dispersive movements of *R. cerasi* are very strong immediately after emergence, especially in cases of insufficient supply of host fruits (Boller et al., 1970). Therefore, steady sources for reinfestation of cultivated cherry are provided from wild cherry trees and alternative hosts, which are abundant in many types of landscapes. Observations to determine adult population density are possible by the use of yellow sticky traps (Prokopy and Boller, 1971; Russ et al., 1973).

It is impossible to detect and eliminate infested fruits during harvest. This often makes it very difficult to meet the tolerance limits of commercial markets, which commonly allow only 0 to 4% infestation. As harvest costs for cherries are high, price reductions due to infestation rates beyond tolerance limits may result in severe financial losses. Plant protection measures are therefore obligatory in most areas where cherries are grown on a commercial basis. *R. cerasi* is distributed throughout most of the areas covered by this report. In the **Soviet Union** it is an important pest of medium- and late-varieties of cherries, particularly in the south of Ukraine (Bogdan, 1968), including Crimea, Soviet Moldavia, North Caucasus and Transcaucasia, extending north up to Leningrad (Smol'yannikov, 1977). Routine control measures are regularly applied.

R. cerasi is found in most parts of **Bulgaria** and **Romania** (Beratlief et al., 1981; Chorbadzhiev, 1939; Popov, 1956), leading to damage levels of 80 to 90% of late ripening fruits. It also causes severe damage in **Hungary, Czechoslovakia** and **Poland**. Reports from Hungary (Toth, 1970; Saringer, 1972; Jenser and Toth, 1978) quote infestation rates of up to 90%. In **Czechoslovakia**, fruit damage is reported from Moravia (Rozsypal, 1942) as well as from western and central Slovakia (Seda, 1938).

The first outbreaks of *R. cerasi* in **Poland** occurred in 1923 and 1924 (Leski, 1969). Since then, infestation rates vary from very low (Kagan and Lewartowsky, 1978) up to 90%; especially in central regions of Poland where light soils and climatic conditions are favourable (Leski, 1963).

In the **German Democratic Republic** Masurat and Stephan (1963) found 9.3% of cherry fruit trees to be heavily infested (average of 51% of the fruit) in the district of Potsdam. No major activity of the pest was noted in 1985 and control measures were only necessary in three counties in the southern part of the country (Ramson et al., 1986). In **Austria** *R. cerasi* is present in all cherry growing areas of the country and often causes more than 80% infestation especially in untreated host trees in the eastern part of the country, (Böhm, 1949; Böhm and Gläser, 1975). It is considered to be the major pest in Austrian cherry growing.

High infestation pressures and annually applied control measures are reported particularly from southern parts of the **Federal Republic of Germany** (F.R.G.) (Engel, 1976), where *R. cerasi* is firmly established. In the Upper Rhine region it may cause infestations of up to 100% in late varieties of cherries, such as "Hedelfinger" and "Langstiel". Routine chemical control in commercial orchards is applied every year, thus effectively reducing infestations to below the economic threshold of 4% (Engel, 1976). The presence of *R. cerasi* further north was mentioned by Haisch (1978), who reported the presence of unidirectional race incompatibility between populations collected north and south of a line running approximately between Frankfurt and Franconia (see Chapter 5.2). No report on the pest status in northern F.R.G. has been found, suggesting that the economic impact in this region is minimal or non-existent. Although Thiem (1933) reports that no infestation occurs in northern F.R.G. *R. cerasi* is known to be present in **Denmark** (P. Esbjerg, personal communication), especially on the islands of Zeeland and Funen, and in the south-eastern parts of the mainland. However, no control measures are taken, and the fly is not considered to be a pest.

R. cerasi is the only fruit fly present in **Norway**. Its distribution falls east of the 7° longitude and south of the 61° latitude, whereas 80% of all commercial fruit growing areas are located west of the 7° longitude. No control measure is undertaken (T. Edland, personal communication).

The distribution of the cherry fruit fly in **Sweden** is also limited to areas south of the 61° latitude. However, cherry is very uncommon as a commercial crop in Sweden (C. Solbreck, personal communication). Infestation is usually of only local importance, but was fairly severe in parts of the south in 1961 (Von Rosen, 1965), with an average of 20% and a maximum of 70% of cherries being attacked in certain localities.

The cherry fruit fly is an important pest in **Switzerland**, infesting up to 53% of middle and late ripening varieties in uncontrolled orchards in 1976 (Katsoyannos and Boller, 1980). Much research is being carried out in attempts to reduce the number of chemical applications by improving the efficiency and timing of these applications (Kundert et al., 1982) and by incorporating alternative control measures (Boller and Remund, 1982).

The pest status of *R. cerasi* in southern **France** is outlined in Chapter 2.1. The fly is located throughout France (Verguin, 1927), but its major economic importance is focussed in the Mediterranean area and in Elsass-Lothringen (Bezzi, 1927). Its distribution extends also into **Belgium** where it has been verydamaging in the past, especially on sweet cherry (G. Sterk, personal communication). The damage is no longer important because of efficient control applications, but close supervision is assured by the authorities every year in order to prevent difficulties arising during export (G. Sterk, personal communication). De Meijere (1916) reported the presence of *R. cerasi* in **The Netherlands**; however, Thiem (1933) found no evidence of infested fruit, and no subsequent report of its presence has been found, thus suggesting that the cherry fruit fly is of no economic importance in this country. Also, no important fruit flies appear to occur in the **United Kingdom** (T. Lewis, personal communication).

Although the cherry fruit fly is endemic throughout **Portugal** (CIE, 1956), commercial cherry orchards are largely restricted to three main regions in the northern and central parts of the country. Pinto de Matos (1976) reported high infestations in the Lamego and Fundao regions whereas the infestation in Alenquer was rather low.

In **Turkey**, *R. cerasi* is considered a major pest of sweet cherries, causing considerable damage every year, especially in the Aegean (Zümreoglu et al., 1981) and Marmara regions (Birkardesler, 1971).

Very severe infestations of cherries and barberries are reported from **Iran** (Farahhakhsh, 1961), mainly in Teheran and vicinity, in Esfahan, Khorassan, Azarbaijan and Hamadan.

No report of infestation by *R. cerasi* in other **West Asian** countries has been found.

2.2 *Ceratitis capitata* (Wiedemann)

C. capitata, the Mediterranean fruit fly, also commonly called the medfly, is a cosmopolitan pest, probably originating in East Africa. It was reported from Mediterranean countries, such as Spain, Algeria, Italy and Tunesia, already in the second half of the last century (De Breme, 1842).

During the present century the medfly has spread to all five continents of the world. Because of its exceptionally broad host spectrum of more than 200 plant species (US/AID, 1977; Tejeda, 1980) the medfly is rated as one of the most destructive and widespread fruit pests of the world.

Potential main hosts include the *Prunus* spp. (peach, nectarine, apricot, plum, cherry, almond) *Pyrus communis* (pear), *Malus sylvestris* (apple) and some *Citrus* spp. (orange, mandarin orange) and secondary host, like *Asparagus* spp., *Ficus* spp., *Lycopersicon exulentum, Morus* spp., *Olea europea, Solanum* spp., *Sorbus* spp., *Vitis* spp. and others.

Climatic conditions are generally not favourable for an endemic establishment of this pest in temperate Europe. However, sporadic outbreaks occur regularly. Strict quarantine measures are therefore often implemented to regulate the movement of infested host materials in an attempt to prevent the entry and dispersal of this pest.

Within the geographical region covered by this chapter *C. capitata* is considered an established pest only in **Jordan, Turkey**, parts of **Saudi Arabia** and **Portugal** (Karpati, 1983). In **Jordan**, medfly infestation rates in 1961–62 of 20–25% in citrus, 91% in peach, 55% in apricot and 15% in plum were recorded (Abu Yaman, 1966), while Tuncyurek (1972) recorded damage by the medfly in **Turkey** of 6.1 to 7.8% in the western parts of the country in 1961, and 3.5 to 6.2% in the south in 1961–65. The medfly was found in asiatic Turkey in 1904 and has since then been rated as a serious pest of various fruit trees (Giray, 1979).

In **Saudi Arabia** the medfly is established only in the region of Taif (FAO, 1970). According to Karpati (1983), the medfly is not established in **Iran**; however, Jafari (1982) reported the presence of an infestation in the Mazandera Province in 1977, which, by 1982, had increased in distribution, population density and infestation rate. Orange, peach, pear, mandarin and persimmons were infested. The medfly was first reported in **Portugal** in 1898 (Coutinho, 1898). It is now distributed throughout the country (Karpati, 1983) and is rated as a pest of major economic importance (EPPO, 1979).

C. capitata is not established in the countries mentioned below. The following part of this section is therefore concerned only with occasional infestations due to imported flies, as they have been reported in the literature. In spite of quarantine measures, *C. capitata* has several times been found in the **Soviet Union**, probably imported in infested fruits (Kryachko, 1979). Foci of infestation were detected in Odessa in 1937, 1964 and 1966 and in Sebastopol from 1964 to 1966, where it was found heavily infesting some 75 acres of apricot, peaches and pears, and to a lesser extent apples (Mel'nikova and Pyshkalo, 1965). Tereshkova (1981) reported, that this species has evidently persisted despite many quarantine measures, including fumigation, being taken.

No report was found of medfly infestations in **Romania** and **Bulgaria**. However, one may assume that the occurrence of the pest in these countries would be similar to that of the Crimea and Hungaria. The medfly was first reported from **Hungaria** in 1928 (Bako, 1928), but did not establish itself. Protection measures prevented the establishment of this pest in later incidents where infested fruit was imported (Martinovich, 1967). The climatic conditions in the eastern fruit growing areas of **Austria** are suitable for the development of *C. capitata* during the growing season, but no evidence has been found for the overwintering of this pest (Böhm, 1958). Infestations usually occur in the neighbourhood of fruit importing centres, such as railway stations and wholesale fruit markets, thus suggesting that infestations are due to imported fruits (especially citrus), despite strict quarantine measures. When introduced in early spring the medfly may reproduce in the field on native crops. Infestation rates, especially on peaches and apricots, may reach 40% in endangered gardens (own observations). These are controlled by chemical treatments or by the release of sterile flies, reared at the IAEA-Laboratories in Seibersdorf.

In **Germany** *C. capitata* has been found repeatedly since 1934 (Thiem, 1937; Maassen, 1979) and has occasionally caused very high damage on peaches (Baas, 1960). Repeated infestations were observed on peaches in Frankfurt and in Baden-Württemberg between 1950 and 1959 (Rump, 1959; Baas, 1960) as well as in 1951 in Trier (Drees, 1955). In 1955, infestation rates of up to 100% were found in peaches in Frankfurt (Baas, 1960). The establishment of the fly in these areas is unlikely because of the lack of possibilities for overwintering in the field (Mayer, 1970). Although it cannot be excluded that the fly may overwinter in frost-free warehouses (Rump, 1959), the import of infested fruits was considered to act as a permanent infestation source (Mayer, 1970). Strict quarantine measures, implemented in 1955 (Drees, 1955), resulted in the rejection of about 50% of fruit imports between 1960 and 1968 due to infestation by *C. capitata* (Mayer, 1970). Despite these measures an infestation in Baden-Baden in 1964 caused 100% damage in peaches (Maassen, 1979).

In **Switzerland**, the medfly was first observed in 1935 (Baas, 1955). It exists only locally and is of little economic importance (EPPO). The medfly was first observed in southern **France** in 1772 and, according to Maassen (1979), in 1979 covered large parts of France as far north as the English Channel. Other reports (Milaire, 1964; Trouillon, 1976, 1977) restrict the medfly distribution to the southern parts of the country. However, Chancogne et al. (1961) reported an outbreak near Paris in 1960. For further details see Chapter 2.1.

The medfly is occasionally observed in **Belgium** (EPPO) but is of no economic importance (G. Sterk, personal communication). In **The Netherlands** a severe out-break was reported in 1955 (Van de Pol, 1957; Anonymous, 1956) and in 1959. Since 1960, continuous monitoring of insect pests in fruit orchards throughout the country has never shown any new infestations by *C. capitata*. It is concluded that the organism is not present. Necessary quarantine measures are taken to prevent introduction (EPPO Plant Protection Service, RSE-479).

No observation has been made on the presence of medflies in the **United Kingdom** and **Scandinavia**, although infested fruit is commonly imported (Mehl, 1980).

2.3 *Dacus* spp.

The olive fruit fly, *Dacus oleae* (Gmelin), is found in regions of western Asia and in **Portugal** (Christenson and Foote, 1960; CIE, 1957).

In **Portugal**, much damage is caused to olives due to infestations by this fly, and control measures are necessary (Rosa de Azevedo, 1943).

Until very recently *Dacus cucurbitae* Coquillett, the melon fly, had not been reported in regions covered by this report. Farahhakhsh (1961), in listing the economically important insects in **Iran**, did not include this species. However, in a report on a new pest of cucurbitaceous plants in Iran, Arghand (1983) reported the destruction of up to 40% of the total harvest by this fly.

Dacus ciliatus Loew, the lesser pumpkin fly, is the most serious pest of cucurbitaceous plants in **Saudia Arabia**, preferentially attacking various varieties of melon. It is controlled by routine applications of insecticides. A related pest, *D. frontalis* Becker is restricted to the eastern provinces of Saudia Arabia (Talhouk, 1983).

2.4 *Myiopardalis pardalina* (Bigot)

M. pardalina, the Baluchistan melon fly, is distributed throughout western **Asia** (CIE, 1961). It is a host of melon, muskmelon and other varieties, where the larvae feed on the flesh of the fruit. It is reported to cause very severe damage in **Iran** (Farahhakhsh, 1961).

3 NON-FRUGIVOROUS TEPHRITID FLIES

3.1 *Platyparea poeciloperta* (Schrank)

P. poeciloperta, the asparagus fly, is a serious pest of asparagus in Central Europe (Hendel, 1927; Christenson and Foote, 1960). The larvae feed on the

Chapter 2.8 references, p. 96

stem marrow of the plant. Masurat and Stephan (1963) reported infestation in the **German Democratic Republic** in 1962 of 32.4% on the total crop area and up to 49% in the Potsdam area. Similar levels of infestation were observed in 1961. The insect is regarded as a pest also in **Hungary** (Bodor, 1966), the **Federal Republic of Germany** (Eckle, 1966) and **Austria** (unpublished). Its northernmost distribution is reached in Southern **Sweden** and Gotland where, however, its pest status is poorly known (C. Solbreck, personal communication). The pest is less common in Southern Europe (Hendel, 1927).

3.2 *Acanthiophilus* spp. and *Chaetorellia* spp.

The larvae of both *Acanthiophilus* spp. and *Chaetorellia* spp. feed on seeds within the flower heads of safflower (*Carthamus tinctarius* L.). Their distributions cover large parts of **Western Asia** and **South-eastern** and **Central Europe** (Martinovich, 1967; Dirlbek and Dirlbekova, 1974; Selim, 1977).

In **Iraq** infestation rates of up to 10% have been observed by *A. helianthi* Rossi, reducing the oil content of the infested safflower seeds from 37.8% to 21.8% (Al-Ali et al., 1977). Usually, however, these species are not reported to be of economic importance (Al-Ali et al., 1978/79).

4 SUMMARY

The pest status of the most important Tephritid flies in temperate Europe and western Asia has been reviewed. *Rhagoletis cerasi* was the most widely distributed species and, with some exceptions, was regarded to be of economic importance throughout its range. *Ceratitis capitata* is endemic south of the 41° northern latitude, where it is commonly of high economic importance. *Dacus* spp. are mainly found in western Asia where individual species may locally be of high importance. Non-frugivorous Tephritids are found throughout the area reviewed but, with the exception of *Platyparia poeciloperta*, are rarely of economic importance.

5 REFERENCES

Abu Yaman, I., 1966. Jahreszeitliche Populationsschwankungen von *Ceratitis capitata* Wied. in Jordanian, Anzeiger für Schädlingskunde, 39: 136–140.

Al-Ali, A., Al-Neamy, I., Abbas, S. and Abdul-Masih, A., 1977. On the life-history of the safflower fly *Acanthiophilus helianthi* Rossi (Dipt., Tephritidae) in Iraq. Zeitschrift für Angewandte Entomologie, 83: 216–223.

Ali-Ali, A., Abbas S., Al-Neamy, I. and Abdul-Masih, A., 1979. On the biology of the yellow safflower fly *Chaetorellia carthami* Stack. (Dipt., Tephritidae) in Iraq. Zeitschrift für Angewandte Entomologie, 87: 439–445.

Anonymus, 1956. De Middellandse Zeevlieg. Plantenziektenkundige Dienst Wageningen, Vlugschrift Nr. 75.

Arghand, B., 1983. A new pest on cucurbitaceous plants in Iraq (engl. Summary). Entomologie et Phytopathologie Appliquees, 51: 11.

Baas, J., 1955. Die Mittelmeerfruchtfliege *Ceratitis capitata* Wied. in Mitteleuropa. Gartenbauwissenschaft, 1: 340–365.

Baas, J., 1960. Die Mittelmeerfruchtfliege in Hessen-Nassau im Jahre 1959, Gesunde Pflanzen, 12: 106–110.

Bakó, G., 1928. Narancslégy; Gyümölcsöseink új ellensége. Növenyvedelem, 4: 210–214.

Beratlief, C., Ionescu, C. and Mustatea, D., 1981. Observations sur l'efficacite des pieges visuels dans l'avertissement contre la mouche des cerises (*Rhagoletis cerasi* L.). Bulletin de l'Academie des Sciences Agricoles et Forestieres, 10: 93–102.

Bezzi, M., 1927. Sulla distributione geografica della mosca delle ciliege (*Rhagoletis cerasi* L., Dipt.). Bollettino Lab. Zool. Gen. Agr. Sc. sup. Agric., 20: 7–16.

Birkardesler, H., 1971. Marmara bölgesinde kraz sinegi (*Rhagoletis cerasi* L.) ne karsi bazi insektisitlerin etkileri üzerinde instirmalar. Bitki Koruma Bülteni, 11 (1): 15–32.

Bodor, J., 1966. A sparga legykartevöi. Kerteszet es szöleszet, 15/6.

Bogdan, L., 1968. *Rhagoletis cerasi* L. in der Waldsteppe und im Waldgebiet der Ukraine. XIII Int. Congr. Ent. Moskau. p. 34.

Böhm, H., 1949. Untersuchungen über die Lebensweise und Bekämpfung der Kirschfliege (*Rhagoletis cerasi* L.). Pflanzenschutzberichte, 3: 177–185.

Böhm, H., 1958. Zum Vorkommen der Mittelmeerfruchtfliege *Ceratitis capitata* Wied. im Wiener Obstbaugebiet. Pflanzenschutzberichte, 21: 129–158.

Böhm, H. und Gläser, G., 1975. Bericht über das Auftreten wichtiger Krankheiten und Schädlinge an Kulturpflanzen in Österreich im Jahre 1974. Pflanzenschutzberichte, 45: 7–12.

Boller, E., 1966. Der Einfluss natürlicher Reduktionsfaktoren auf die Kirschenfliege *Rhagoletis cerasi* L. in der Nordwestschweiz, unter besonderer Berücksichtigung des Puppenstadiums. Schweizerische Landwirtschaftliche Forschung, 5: 153–210.

Boller, E. and Prokopy, R., 1976. Bionomics and management of Rhagoletis. Annual Review of Entomology, 21: 223–246.

Boller, E. and Remund, U., 1982. Field feasibility study for the application of SIT in *Rhagoletis cerasi* L. in Northwest Switzerland (1976–1979). CEC/IOBC Symposium, Athens, pp. 366–370.

Boller, E., Haisch, A. and Prokopy, R., 1970. Ecological and behavioural studies preparing the application of the sterile-insect-release-method (SIRM) against *Rhagoletis cerasi*; Presented at IAEA Symp. Athens: Sterility Principle for Insect Control or Eradication.

Chancogne, M., Court, D., Cantuel, J. and Destruel, C., 1961. Essais en vergers de poires et de pechers de produits contre *Ceratitis capitata* W. Phytiatrie–Phytopharmacia, 9: 227–232.

Chorbadzhiev, P., 1939. Materialien über die schädlichen Insekten und anderen Feinde der Kulturpflanzen in Bulgarien, (German summary). Mitteilungen bulg. Entomologischen Gesellschaft, 10: 55–72.

Christenson, L. and Foote, R., 1960. Biology of fruit flies. Annual Review of Entomology, 5: 171–192.

Commonwealth Institute of Entomology, 1956. Distribution maps of insect pests. Series A, Map Nr. 65.

Commonwealth Institute of Entomology, 1957. Distribution maps of insect pests. Series A, Map Nr. 74.

Commonwealth Institute of Entomology, 1961. Distribution maps of insect pests. Series A, Map Nr. 124.

Coutinho, A., 1898. A Mosca da laranja e do pesego. Agricultura contemporanea, 9: pp. XXX.

De Breme, F., 1842. Note sur le genre Ceratitis de M. Mac Leay (Diptera). Annales de la Societe Entomologique de France, 11: 183–190.

De Meijere, J., 1916. Tijdschr. Ent. Nederl. 59: 309 in: Commonwealth Institute of Entomology, 1956. Distribution maps of insect pests. Series A, Map Nr. 65.

Dirlbek, J. and Dirlbekova, O., 1974. Vorkommen der Bohrfliegen aus der Gattung Chaetorellia (Diptera, Trypetidae) in der CSSR. Folia Facultatis Scientiarum Naturalium Universitatis Purkynianae Brunensis, 15: 83–84.

Drees, H., 1955. Verordnung zur Verhütung der Einschleppung der Mittelmeerfruchtfliege. Gesunde Pflanzen, 7: 69–70.

Eckle, 1966. Schädiger im Spargelanbau. Süddeutsche Erwerbsgärten, 20: 296–298.

Engel, H., 1976. Untersuchungen über die Besatzdichte der Kirschfruchtfliege (*Rhagoletis cerasi* L.). Zeitschrift für Pflanzenkrankheiten und Pflanzenschutz, 83: 53–58.

EPPO, 1979. Data sheets on quarantine organisms, List A2: *Ceratitis capitata* Wied. EPPO, Paris.

FAO, 1970. Report to the governments of the countries of the Near East and North Africa on Citrus production problems in the Near East and North Africa based on the work of H. Chapot. FAO Report Nr. TA 2870 AGP:TA/184.

Farahhakhhsh, G., 1961. A checklist of economically important insects and other enemies of plants and agricultural products in Iran. Plan. Organization, Pren, Teheran.

Giray, H., 1979. Türkiye Trypetidae (Diptera) faunasina ait ilk liste. Türkiye Bitki Koruma Dergisi, 3: 35–46.

Haisch, A., 1978. Crossing experiments with different populations of *Rhagoletis cerasi* L. in Germany. (Abstract). OILB-Proceedings Sassari, p. 150.

Hendel, F., 1927. Trypetidae. Aus: Lindner, Palaearktische Fliegen, Bd. 49. E. Schweizerbart'sche Verlagsbuchhandlung, Stuttgart.

Jafari, M. and Sabzewari, A., 1982. Bioecology of Mediterranean fruit fly in Mazandaran province. Proceedings of VII Plant Prot. congr. of Iran. Teheran, pp. 61–65.

Jenser, G. and Tóth, E., 1978. Analysing applicability of colour traps for *Rhagoletis cerasi* forecast in an orchard consisting of several varieties of cherries. Research Institute for Horticulture, Budapest, Hungary, pp. 235–239.

Kagan, F. and Lewartowski, R., 1978. Charakterystyka rozwoju, nasilenia wysrepowania i szkodliwosci wazniejszych szkodnikow drzew i krzewow owocowych w Polsce w roku 1976. Biuletyn Instytutu Ochrony Rosilin, 62: 331–421.

Karpati, J., 1983. The mediterranean fruit fly. (Its importance, detection and control). FAO, Rome.

Katsoyannos, B. and Boller, E., 1980. Second field application of oviposition-deterring pheromone of the European cherry fruit fly, *Rhagoletis cerasi* L. (Diptera: Tephritidae). Zeitschrift für Angewandte Entomologie, 89: 278–281.

Kryachko, Z., 1979. The mediterranean fruit-fly. (Engl. abstract). Zashchita Rastenii, 12: 59.

Kundert, J., Remund, U. and Potter, C.A., 1982. Pflanzenschutz im Kirschenanbau. Schweizerische Zeitschrift für Obst und Weinbau, 118: 354–356.

Leski, R., 1963. Studia nad biologia i ekologia nasionnicy trzesniowki *Rhagoletis cerasi* L. (Dipt., Trypetidae). Polskie Pismo Entomologiczne, Seria B: 153–240.

Leski, R., 1969. Population studies of the cherry fruit fly, *Rhagoletis cerasi* L. Proceedings of a panel on "insect ecology and the sterile-male technique" 1967, IAEA, Vienna.

Maassen H., 1979. Zur Befallssituation von importierten Zitrusfrüchten mit der Mittelmeerfrucht-fliege (*Ceratitis capitata* Wied.) und ihre Bedeutung für den südwestdeutschen Obstbau. Gesunde Pflanzen, 3: 57–64.

Martinovich, V., 1967. Adatok a Kárpátmedence furólegyeinek ismeretéhez (Diptera, Trypetidae). Folia Entom. Hung., 22: 459–473.

Masurat, G. and Stephan, S., 1963. Das Auftreten der wichtigsten Krankheiten und Schädlinge der landwirtschaftlichen und gärtnerischen Kulturpflanzen im Jahre 1962 im Bereich der Deutschen Demokratischen Republik. Nachrichtenblatt für den Deutschen Pflanzenschutzdienst, 12: 185–212.

Mayer, K., 1970. Die Mittelmeerfruchtfliege *Ceratitis capitata* Wied., ein gefährlicher Quarantäne-schädling. Zeitschrift für Angewandte Entomologie, 3: 357–363.

Mehl, R., 1980. Appelsinflua *Ceratitis capitata* i importert Frukt. Fauna, 33: 155–156.

Mel'nikova, R. and Pyshkalo, G., 1965. A dangerous quarantine pest (engl. abstract). Zashchita Rastenii, 5: 49–51.

Milaire, H., 1964. Observations sur la mouche mediterraneenne des fruits en 1963. Phytoma, 16: 26–27.

Pinto de Matos, A., 1976. Estudo de prospeccao da mosca da cereja (*Rhagoletis cerasi* L.) em Portugal entre 1974 e 1976. Reparticao de servicos fitopatologicos, RSF (D)-21/76.

Popov, P., 1956. Studies on the flight, ecology and prognosis of the cherry fly (*Rhagoletis cerasi* L.) in Bulgaria (engl. summary). Plant Protection Institute, Sofia, 3: 297–326.

Prokopy, R. and Boller, E., 1971. Response of European cherry fruit flies to colored rectangles. Journal of Economic Entomology, 64: 1441–1447.

Ramson, A., Arlt, K., Hänsel, M., Herold, H., Plescher, A., Reuter, E. and Sachs, E., 1986. Das Auftreten der wichtigsten Schaderreger in der Pflanzenproduktion der Deutschen Demokratis-chen Republik im Jahre 1985 mit Schlussfolgerungen für die weitere Arbeit im Pflanzenschutz. Nachrichtenblatt für den Pflanzenschutz in der DDR, 40: 89–112.

Rosa de Azevedo, A., 1943. O valor da colheita precoce da azeitona como metodo preventivo de combate ao *Dacus oleae* Gmel. Agronomia Lusitana, 5: 83–89.

Rosen, H. von, 1965. Beobachtungen über die Kirschfruchtfliege *Rhagoletis cerasi* (L.) in Schweden (Dipt., Trypetidae). Statens Vaxtskydsanstalt Meddelanden (National Swedish Institute for Plant Protection), 13: 149–167.

Rozsypal, J., 1942. Zprava o skodlivych cinitelich kulturnich plodin (vyjma oves, brambory, len) ve vegetacnim obdobi 1940–41 na Morava. Ochrona Rost. 18: 17–24.

Rump, L., 1959. Die Mittelmeerfruchtfliege in Rheinland-Pfalz im Sommer 1959. Gesunde Pflanzen, 11: 239–241.

Russ, K., Boller, E., Vallo, V., Haisch, A. Sezer, S., 1973. Development and application of visual traps for monitoring and control of populations of *Rhagoletis cerasi* L. Entomophaga, 18: 103–116.

Sáringer, G., 1972. Adatok a cseresznyelégy (*Rhagoletis cerasi* L., Diptera: Trypetidae) diapauzájá-nak ismeretéhez. Növényvédelem Idoszeru Kerdesei, 8: 152–157.

Seda, A., 1938. Zprava o skodlivych cinitel'och kulturnych plodin na vychod. Slovensku a Podkar-pat. Rusi za hospodarsky rok 1936–37. Ochrona Rost. 14: 16–23.

Selim, A., 1977. Insect pests of safflower (Carthamus tinctorius) in Mosul Northern Iraq. Mesopota-mia Journal of Agriculture, 12: 75–78.

Smol'yannikov, V.V., 1977. Pests of cherry (engl. abstract). Zaschita Rastenii, 8: 60–61.

Talhouk, A., 1983. Wichtige Schadinsekten der Cucurbitaceae in Saudi-Arabien. Anzeiger für Schädlingskunde, Pflanzen und Umweltschutz, 56: 6–9.

Tejada, L., 1980. Estudio sobre las hospederas potenciales de la mosca del Mediterraneo, *Ceratitis capitata* Wied. SARH, Sanidad Vegetal, Mexico, 95 pp.

Tereshkova, E., 1981. Mediterranean fruit-fly; methods of demonstration and control (engl. abs-tract). Zashchita Rastenii, 1: 42–43.

Thiem, H., 1933. Beitrag zur Epidemiologie und Bekämpfung der Kirsch-fruchtfliege (*Rhagoletis cerasi* L.). Nachrichtenblatt für den Deutschen Pflanzenschutzdienst, 13. Jahrgg., 5: 33–35.

Thiem, H., 1937. Auftreten der Mittelmeerfruchtfliege (*Ceratitis capitata* Wied.) in Deutschland. Nachrichtenblatt Deutscher Pflanzenschutzdienst, Berlin, 17: 45.

Tóth, G., 1970. Klórozott szénhidrogének helyettesitése a esereszny-elégy (Rhagoletis cerasi L.) elleni küzdelemben. Növényvédelem, 6. szám.: 145–148.

Trouillon, L., 1976. Un succes du controle phytosanitaire: la disparition en France de la mouche des fruits. Phytoma, 28: 27–29.

Trouillon, L., 1977. A nouveau la mouche des fruits. Phytoma, 29: 17–18.

Tuncyurek, M., 1972. Recent advances in the Bio-ecology of Ceratitis capitata Wied. in Turkey. OEPP/EPPO Bull. 6: 77–81.

US/AID, 1977. The Mediterranean fruit fly and its economic impact on Central American countries and Panama. US/AID pest management and related environmental protection project, US Agency for International Development.

Van de Pol, 1957. De Middellandse-Zeevlieg. Mededelingen Directie van de Tuinbouw, 20: 36–38.

Verguin, M., 1927. La mouche des cerises (*Rhagoletis cerasi* L.); etat actuel de la question. Annales des Epiphyties, 13: 31–41.

Zümreoglu, A., Akman, K., San, S. and Ulu, O., 1981. Investigations on the possibilities to apply the bait-spraying technique against the european cherry fruit fly (*Rhagoletis cerasi* L.) in Aegean region. (engl. Summary). Aralik, XX: 218–230.

PART 3

BIOLOGY AND PHYSIOLOGY

Chapter 3.1 Nutrition

3.1.1 Requirements

JOHN A. TSITSIPIS

1 INTRODUCTION

Food is one of the main environmental factors determining the abundance
and distribution of insects. Its dual function is to induce feeding by insects by
possessing appropriate characteristics (phagostimulatory, arrestant) and to
offer the necessary nutrients that provide fuel for meeting their energy requirements and the necessary building blocks for substance acquisition and progeny
production. Additionally, in insects where food is their habitat, it also offers
those physical conditions allowing life activities to go on. Insects, however,
have further specific nutritional requirements that should be fulfilled by the
food available.

A concise reference to the nutrition terminology, used in the present review,
is given in this section. House (1974) defined nutritional requirements as the
chemical factors that are essential for the adequacy of the absorbed nutritive
material. This was a modification of the definition of Beck (1956) in which the
idea of absorbed, instead of ingested nutritive material, was introduced thus
covering the contribution of symbiotic microorganisms in nutrients to the
insects. Nutritional requirements can be *qualitative* and *quantitative*. The
former pertain to the different classes of nutrients serving as nutrients, i.e.
proteins, sugars, vitamins, the latter refer to the relative amounts thereof in
the diet. The nutrients required for optimal growth and development are
designated as *essential*.

Dadd (1985) discusses in length the terms *essential* and *required*. He considers a nutrient as essential when its deletion from a diet results in the
termination of growth, development or continued reproduction. A nutrient is
termed as required when its deletion slows down the rates of growth and
development though optimal final size and form are attained. Artificial diets
used for nutritional studies or for rearing insects, depending on the degree of
the chemical determination of their constituents, have been called by Dougherty (1959) *holidic*, *meridic* or *holigidic*. In holidic, or chemically-defined diets,
all constituents are chemically known. Meridic diets contain mostly chemically defined chemicals with the presence of at least one material not chemically
defined. Holigidic are the diets that are mainly composed of ingredients of
unknown chemical structure. For a more comprehensive treatment of nutrition
terminology the review of Dadd (1985) should be consulted.

Insect nutrition and in particular determination of nutritional requirements
in insects has shown a remarkable progress in the last twenty-five years.
Several methodological achievements have contributed to this goal such as:
the development of chemically defined diets, the axenic culture of insects, the
utilization of conventional methodologies i.e. the deletion technique, the

---

Chapter 3.1.1 references, p. 116

chemical analysis of natural foods, the comparison of chemical analysis of food tissues and insect excreta, or more sophisticated ones,i.e. the use of the isotope technique or the use of antimetabolites to study nutrient essentiality (Dadd, 1985). Some of the most recent reviews on insect nutrition are by House (1972, 1974), Rodriguez (1972), Dadd (1973, 1977, 1985). Knowledge of insect nutrition has been formulated from extensive work on species belonging mainly to the taxonomic orders of Homoptera, Orthoptera, Lepidoptera, Diptera, Coleoptera. A considerable amount of research has been done on the nutritional requirements of Diptera and has been reviewed by Friend (1968). The nutritional requirements of the fruit flies, insects belonging to the family Tephritidae, have been briefly treated by Bateman (1972) and Steiner and Mitchell (1966) in reviews on the ecology of fruit flies and their mass rearing respectively.

In the family Tephritidae about 4000 species have been described, some of them being of great economic importance, living in temperate, subtropical and tropical world-areas (Christenson and Foote, 1960). The larvae of most of the species live and feed in plant stalks, leaves, fruit, flower-heads or seeds. Certain species form galls in stems, crowns, rhizomes (Novak and Foote, 1980). The adults feed on glandular secretions of plants, nectar, plant sap exudates, decaying fruit, pollen, honeydew secretions of homopteran insects, bird droppings, leaf and fruit surface bacteria (Drew et al., 1983 and Chapter 3.1.3 of this volume).

Some species are oligophagous, breeding on a limited host range, i.e. *Dacus oleae* (Gmelin), *Rhagoletis cerasi* (Linnaeus), while others are polyphagous, i.e. *Dacus dorsalis* Hendel, *Dacus tryoni* (Froggatt), *Ceratitis capitata* (Wiedemann) (Drew, 1978).

The determination of nutritional requirements, sensu stricto, in insects implies that investigations are carried out with holidic diets under axenic conditions. In fruit flies very little work has been done at that level, being limited only to adult nutrition. Most of the existing information derives from research with insect species of economic importance on holigidic diets and partly on natural foods. This preference has resulted from a desire to understand the biology and ecology of those species being pests of many agricultural crops. As a consequence demand for availability of large numbers of insects necessitated active research on rearing methods on a large scale. The need was more intense in species that were considered to be controlled by the Sterile Insect Technique (Steiner and Mitchell, 1966). Efficient methodologies for mass production of the species *Anastrepha ludens* (Loew), *Anastrepha suspensa* (Loew), *C. capitata*, *Dacus cucurbitae* Coquillett, *D. dorsalis*, *D. oleae*, *D. tryoni* have been developed and their rearing methods have been reviewed by Singh (1977). The necessity for immediate results and the development of economically cheap diets emphasized research on holigidic diets. All this work was based on the findings of Hagen (1953), Maeda et al. (1953), Finney (1956), Hagen et al. (1963).

Very little effort was given to holidic diet development. Nutritional studies of fruit fly larvae are generally impeded by difficulties such as the feeding behavior of larvae tunnelling in the diet, and not being therefore easily accessible for measurements (Friend, 1968), the extreme difficulty in collecting excretory material (Friend, 1968) and the dependence of larvae on extracellular symbiotes for the provision of essential nutrients to a great extent (see Chapter 3.1.2 of this volume). Hagen (1966) suggested that utilization of olives by *D. oleae* larvae is made possible by the action of the symbiote, *Pseudomonas savastanoi* Smith.

The nutritional requirements of tephritids are fulfilled by nutrients from the diet they ingest, and/or by transfer of nutrients from earlier life stages, and/or by the activity of symbiotic microflora. In the section that follows, research data on tephritid nutritional requirements will be reviewed derived from work with natural foods, holigidic and holidic diets, since there is insufficient information on holidic diets alone. Emphasis will be given on very basic work and on new published information from which principles and areas for further work will be pointed out.

2 NUTRITIONAL REQUIREMENTS OF LARVAE

2.1 Natural food nutrition

The nutrition of tephritid fruit flies in the larval stage is considered very important, since nutrients are required, qualitatively and quantitatively,not only to provide energy and building material for survival, growth, development, but also for storage material to be utilized in the pupal stage. In pupae, nutrients are required for metamorphosis and a part is transferred to the adult stage. Many species utilize a very wide host-range, while others a very limited one, these are also monophagous species. The biological performance of larvae, measured as growth, development, survival, effect on adult fecundity, of one species can vary not only from host to host but also within the same host. Thus, although species can breed for many generations on the same host, still the above diversity indicates existence of differences in the quality and/or quantity of the nutrients provided. These differences usually are not at the level of essential nutrients. Christenson and Foote (1960) list a number of fruit fly species in which the numerical host-range is indicated. The medfly, *C. capitata* can grow on 200 hosts, *D. cucurbitae* on over 80 hosts, *D. oleae*, practically monophagous, on species of the genus *Olea*. *D. dorsalis* has been reported to attack 173 species (Drew, 1978) and a number of species of the genera *Dioxyna* and *Paroxyna* have a limited host range between one and 17 species (Novak, 1974). Further work will probably expand the host range either because of existing unrecorded hosts, or because of adaptation of species to new ones. The larvae of the olive fruit fly, *D. oleae*, could develop successfully in mature tomatoes in the laboratory (Tzanakakis, 1974). It would be interesting to test the nutritional adequacy of tomato xenically, by trying to rear the insect for consecutive generations, excluding in this way possible essential nutrient transfer via the egg. Christenson and Foote (1960) and Bateman (1972) refer to data showing differences in the development of *A. ludens* and *C. capitata* in different hosts. In *D. cucurbitae* host suitability, in descending order, was: cucurbits, tomatoes, egg-plants (Rahamannar, 1962). In *C. capitata* larval developmental time was found in peaches, mespils and tomatoes to be 6, 8 and 11.8 days respectively (Mourikis, 1965). In *D. dorsalis*, of eight fruits tested, papaya was the most suitable with developmental period of 11.3 days, pupal weight of 11.6 mg and adult emergence 44.5%, while pineapple was the least suitable with respective values of 8.9 days, 4.9 mg and 8.9% (Ibrahim and Rahman, 1982). In *C. capitata*, larval development in immature peaches lasted 8 days compared with 6 days in mature ones. Respective values for apricots were 10 and 8 days (Mourikis, 1965). Krainacker et al. (1987) have examined the performance of *C. capitata* larvae on 24 different hosts. They found that there was considerable variation in the developmental time, growth, survival, eventual adult fecundity, and survival of the larvae on the various hosts. They concluded that the Mediterranean fruit fly is a successful generalist frugivore because it maintains a relatively high intrinsic rate for population increase (*r*), this being the net result of two counteracting tendencies for host specificity: one tending to lower *r* and one tending to increase it.

Chapter 3.1.1 references, p. 116

TABLE 3.1.1A

Composition of holigidic larval diets for the rearing of certain species of tephritid fruit flies[1]

	Dacus dorsalis *Dacus cucurbitae* *Ceratitis capitata*		*Anastrepha ludens* *Anastrepha suspensa*		*Ceratitis capitata*	*Dacus tryoni*
	(1)		(2)		(3)	(4)
Water	100[2]	100	70	80.92	50	90
Agar	1.1	1.3				
Sucrose			6.22		13	
Glucose	4.0	4.9				
Cellulose powder						
Shredded tissue paper						
Corn cob grits			13.0			
Granulated carrot (dehydrated)				7.0		
Carrot powder			5.0	7.0		9.17
Wheat germ		0.175				
Wheat germ oil						
Brewer's yeast	3.25	3.0		4.0	8.5	9.17
Torula yeast			5.0			
Yeast hydrolyzate				0.3		
Soy hydrolyzate enzymatic						
Olive oil						
Cholesterol	0.35	0.175				
Choline chloride	0.07	0.07				
Tween-80						
VanderZant Vitamin mix.						
Wheat flour						
Wheat shorts						
Wheat bran					26	
Salt mixture (U.S.P. XIII)	0.35	0.35				
HCl			0.6	0.6	2.3	0.5
Propionic acid						
Nipagin		0.12	0.1	0.1	0.1	0.08
Butoben (Tegosept B)			0.08	0.08	0.1	
Sodium benzoate						
Potassium sorbate						

[1] (1) Maeda et al. (1953); (2) Lopez (1970); (3) Nadel (1970); (4) Monro and Osborn (1967). [2] Figures in g for solids and ml for liquids.

TABLE 3.1.1B

Composition of holigidic larval diets for the rearing of certain species of tephritid fruit flies[1]

	Dacus oleae		Dacus zonatus	Rhagoletis cerasi	Rhagoletis pomonella	Anastrepha obliqua
	(1)	(2)	(3)	(4)	(5)	(6)
Water	62.5[3]	55.0[2]	80[3]	80.4	100	100[3]
Agar	1.0		0.7	4.2	1.0	1.0
Sucrose		2.0	6.6	4.0	4.0	
Glucose		30.0				
Cellulose powder					2.0	
Shredded tissue paper						
Corn cob grits						
Granulated carrot (dehydrated)						
Carrot powder	12.5					
Wheat germ				4.0		
Wheat germ oil						
Brewer's yeast	7.5	7.5	3.4		2.5	2.67
Torula yeast				5.6		
Yeast hydrolyzate						
Soy hydrolyzate enzymatic	3.0	3.0				
Olive oil	7.5	2.0				
Cholesterol	0.025				0.35	
Choline chloride		0.75			0.7	
Tween-80	2.5			1.0		
VanderZant Vitamin mix.						
Wheat flour						2.67
Wheat shorts			20			
Wheat bran						
Salt mixture (U.S.P. XIII)					0.35	
HCl	3.5 (2N)	3.0 (2N)	4.0 (1N)	0.4 (4N)		
Propionic acid				0.4		
Nipagin		0.2	0.1			1.0 (20% sol.)
Butoben (Tegosept B)	0.025					
Sodium benzoate	0.15					
Potassium sorbate		0.05				
Formalin					(0.05)[4]	

[1] Hagen et al. (1963); (2) Tsitsipis (1977a); (3) Quereshi et al. (1974); (4) Haisch (1975); (5) Neilson (1973); (6) Message and Zucoloto (1980). [2] Figures converted to approximate % composition. [3] Figures in g for solids and ml for liquids. [4] A similar diet without formalin.

2.2 Holigidic diet nutrition

Remarkable progress has been made in larval nutrition of tephritids with holigidic diets. Nutritionally adequate diets have been developed for polyphagous and oligophagous species,the most important of which are: *Anastrepha fraterculus* (Wiedemann), *A. ludens, A. suspensa, D. dorsalis, D. cucurbitae, Dacus cucumis* French, *D. oleae, D. tryoni, Dacus zonatus* (Saunders), *C. capitata, Ceratitis rosa* (Karsch), *R. cerasi, Rhagoletis pomonella* (Walsh) (diets compiled by Singh, 1977), *Anastrepha obliqua* (Macquart) (Message and Zucoloto, 1980). The developed diets, coupled with efficient rearing techniques, allowed production of flies, in certain species, by the million. Steiner and Mitchell (1966) discuss in considerable detail the stepwise progress made in the development of larval diets for tephritids. Finney (1953, 1956) and Maeda et al. (1953) laid the foundations for the development and further improvement of diets for many polyphagous species such as *D. dorsalis, D. cucurbitae, D. tryoni, C. capitata, A. ludens*. The breakthrough that allowed achievement of high yields of pupae from larval diets was the introduction of dehydrated plant materials (i.e. carrot powder) and dry yeasts. Less efficient diets have been developed for oligophagous species, such as *D. oleae, R. cerasi, R. pomonella*. The first attempt to rear *D. oleae* under axenic conditions was made by Moore (1959, 1962), but the results were not deemed satisfactory. The introduction of soya hydrolyzate enzymatic, by Hagen et al. (1963), made rearing of the fruit fly possible. Hagen (1966) reasoned that the presence of a hydrolyzed protein was indispensable to the olive fruit fly larvae, since under the laboratory rearing conditions, where antibiotics were used in the adult diets and preservatives in the larval diet at a low pH, the symbiotic bacterium, presumably *Pseudomonas savastanoi*, was eliminated. The function of the bacterium in nature is to hydrolyze the fruit protein, making it available in such form to the larvae, and also to provide them with the essential amino acids methionine and threonine, in which olive fruit is deficient (Narasaki and Katakura, 1954). A review of the methods for the rearing of the olive fruit fly has been made by Tzanakakis (1971).

The development of holigidic diets is an attempt to find substitutes for natural foods. These diet components should provide the larvae with all classes of essential nutrients, qualitatively and quantitatively, i.e. amino acids, vitamins, sugars, minerals, growth factors. In order to be successful, certain prerequisites are necessary. The nutritional and physical requirements of the diets of larvae of particular species ought to be fulfilled. Additionally, diets should contain preservatives that keep them free, for appropriate periods of time, from spoilage by microorganisms. Categories of diet ingredients for the above purpose are shown in Tables 3.1.1.1A, 3.1.1.1B, where the composition of larval diets for certain species is shown. The list of species and the specific diets cited are intended to give an indication of the variety of ingredients used. To meet the nutritional requirements a large array of materials have been used: dry carrot, wheat products (germ, bran, flour, shorts), yeasts (brewer's, torula, extract), hydrolyzed proteins (soy), oils (olive, wheat germ), sucrose, cholesterol, choline chloride, vitamin mixtures and salt mixtures. Ingredients used to give a proper texture to the diet are: agar, cellulose powder, tissue paper, corn cob grits (could also provide nutrients). Some of the materials providing nutrients contribute also to diet texture. Tween-80 is used as an emulsifier for oils. Finally, as antimicrobial agents there are substances such as: nipagin (methyl *p*-hydroxybenzoate), butoben (butyl *p*-hydroxybenzoate), potassium sorbate, sodium benzoate, propionic acid, formalin. Hydrochloric acid is used to adjust the pH of the media to levels (~ 4) unfavorable for microbial growth. The ingredients given make up only a part in a large variety of materials used for the development and improvement of diets. VanderZant (1974) in a review paper discusses at length the utilization and the role of plant materials, yeasts, stabilizing agents and antimicrobials in the development of practical diets. Tables 3.1.1.1A and 3.1.1.1B show that all diets contain yeast, an important constituent providing both macro- and micro-nutrients. Most of the diets contain a carbohydrate, sucrose, as an energy source, and HCl for pH adjustment. The other ingredients, nutrients, inerts or antimicrobials, vary in kind and concentration. This indicates possible differences among the species for nutritional, feeding, physical requirements and sensitivity of the larvae to antimicrobial agents.

Larval nutrition research in many tephritids, in which production of large numbers is required, aims at improving already known diets through an optimization procedure. This is pursued by either improving their efficiency, producing thus more pupae per gram-diet, or by substituting ingredients by cheaper ones locally available. A change of an ingredient or the concentration thereof, upsets a multidimensional and delicate equilibrium. The ratios, solids to liquids, carbon to nitrogen and relationships between similar classes of nutrients change, affecting the nutritional and physical characteristics of the diet. Tzanakakis (1971) discusses this problem and maintains that general conclusions of components being indispensable cannot be drawn. What can be said is that in a diet an ingredient is needed at a certain concentration. Another uncertainty entering the work with holigidic diets is the instability of the qualitative and quantitative characteristics of certain ingredients. Different brands of brewer's yeast and different batches of the same brand may differ. The above are illustrated in work on diet improvement in *D. oleae*. Elimination of agar (0.5 g/55 ml H_2O) and increase of cellulose powder content by 50% (from 20 to 30 g/55 ml H_2O) (Tsitsipis, 1977b) in the diet of Tzanakakis et al. (1970) and replacement of brewer's yeast by a different brand (Tsitsipis, 1977a) increased diet efficiency, measured as pupal recovery, more than fivefold (Tsitsipis, 1982). The first alteration drastically changed diet texture from pastelike to a more favorable granular form and the second improved pupal weight and yield. In different batches of the yeast brand previously used, there had been found great variation in the non-protein nitrogen content (Manoukas, 1974). Manoukas (1977) studied the effect of brewer's yeast, soy hydrolyzate, olive oil and sugar on larval performance of *D. oleae*. He concluded that all ingredients, except sugar, are essential for normal growth and development. Respective optimal levels of these components for larval performance were 7.5 g, 2 g and 4 ml per 55 ml water of diet. Tsitsipis (1983), in a series of experiments, designed to optimize larval diet constituent levels in *D. oleae*, found that brewer's yeast, a different brand from that of Manoukas (1977), was indispensable for growth and development. Soy hydrolyzate, sucrose, olive oil and Tween-80 were considered "required" for optimal performance at 4 g, 2 g, 3 ml, 0.5–1.0 ml per 55 ml water of diet respectively. Optimum level of brewer's yeast was 10 g. The differences in the results of the authors, despite differences in methodology, indicate the difficulty in establishing nutritional requirements even at the ingredient level.

An approach that has been used to obtain limited information on nutritional requirements in species reared in holigidic diets is to substitute or to supplement crude materials by known chemicals. Chawla (1966) devised a semisynthetic diet for *D. cucurbitae* by supplementing it with amino acids, nucleotide bases, vitamins etc. The diet could not support development to completion and was inferior to the control. Manoukas (1986) substituted soy hydrolyzate in the larval diet of *D. oleae* with three different 18 amino acid mixtures, or methionine or lysine, the last two in the presence or absence of non-essential amino acids. All diets with amino acid mixtures were as good as the control, and one

Chapter 3.1.1 references, p. 116

of them gave higher survival rates than the control. Methionine and lysine were detrimental to the larvae regardless of the presence of non-essential amino acids or not. The results showed that soy hydrolyzate could be substituted by defined chemicals. The role of the individual amino acids is not likely to be determined, since brewer's yeast, contained in the diet provides amino acids although not at adequate levels for optimal performance.

An antipode of minimal nutritional requirements determination is the knowledge on the tolerance levels of nutrients. Such data provide invaluable information on diet development. Tolerance levels of 18 amino acids, supplementing a basal diet containing brewer's yeast, were established in *D. oleae* larvae (Manoukas, 1981). Methionine, lysine and glycine affected adversely survival, growth and adult emergence at levels above 0.1–0.4 g, aspartic and glutamic acid at levels above 4.0–32.0 g, and histidine, tyrosine, threonine, tryptophane, cysteine, serine, valine, arginine, isoleucine, leucine, phenylalanine, proline and alanine at levels over 0.4–4.0 g/55 ml dietary water (amino acid ranking in ascending order). Tolerance levels were also determined for inorganic salts in *D. oleae* larvae (Manoukas, 1982) with a holigidic diet of known mineral content. Potassium chloride affected larvae at 4.0 g and copper sulfate at 0.005 g/100 g diet. Sodium chloride, calcium chloride, magnesium sulfate, manganese sulfate, zinc sulfate and iron sulfate influenced adversely larval performance at concentrations higher than the ones in the range 0.4–0.025 g/100 g diet for the respective salts.

The nutritional requirements of tephritids can be viewed from the standpoint of their ability to exploit their particular hosts. In oligophagous, and in particular in polyphagous species, nutritional requirements of larvae show great variability, since they are obliged to live on hosts with varying nutritional characteristics. The property of wide host-range utilization should be controlled genetically, so that its maintenance is ensured. Cavicchi and Zaccarelli (1974) investigated polyphagy in the larvae of *C. capitata* in laboratory studies. They came to the conclusion that the genetic variability underlying the control of larval nutritional requirements is a characteristic of the entire population (evolutionary plasticity) as well as one of the individual (individual plasticity). ity).

2.3 Meridic diet nutrition

Based on a presumably meridic diet (Srivastava, 1975, original not seen), the requirements for vitamins, fatty acid and sterols, and carbohydrates in *D. cucurbitae* larvae have been examined. From the vitamin B complex (Srivastava et al., 1977) thiamin, riboflavin, nicotinic acid and pantothenic acid were considered essential. Pyridoxine and choline chloride severely affected growth, while *p*-aminobenzoic acid (a constituent of folic acid), inositol, biotin and folic acid did not affect growth and development. The optimal levels of the first six vitamins, in the order cited above, were: 30, 30, 20, 10, 20 and 200 µg/ml diet respectively. Srivastava and Pant (1979) reported that cholesterol was an essential requirement for *D. cucurbitae* larvae, that could be substituted by sitosterol, ergosterol and to a lesser extent by stigmasterol. Linoleic acid, melonic acid and arachidic acid alone did not support growth and development. A series of carbohydrates tested by Srivastava et al. (1978) on *D. cucurbitae* larvae could be utilized at various degrees. Upon examination of the data it is seen that glucose, fructose, mannose, sorbose, ribose, sucrose, maltose, lactose, raffinose, inulin, mannitol, sorbitol and dulcitol improved survival, although with some of them developmental period was prolonged. Growth was equal or better than the control, except with lactose where it was lower. Trehalose, arabinose, xylose and starch were about as good as the control, but melezitose affected performance adversely. Absence of variability and statistical treatment of data do not allow extraction of accurate conclusions.

3 NUTRITIONAL REQUIREMENTS OF ADULTS

Nutrients in the adult stage of tephritids are required for survival and reproduction. The nutritional requirements of adults are met by food ingestion, by nutrient transfer from the larval stage via the pupa and by nutrient supply by symbiotic microorganisms. It is principally the first source that will be treated in the following section. Considerable amount of work has been done covering this subject with natural foods, holigidic and holidic diets.

3.1 Natural food nutrition

Tephritid adults in nature acquire necessary nutrients from a large variety of food sources (Christenson and Foote (1960), Steiner and Mitchell (1966), Bateman (1972) and Drew et al. (1983)). Honeydew, containing carbohydrates, amino acids, vitamins and minerals, was first pointed out by Hagen (1958) as being an important food source for many tephritids. It cannot be considered, however, as providing all necessary nutrients, since honeydews from certain homopterans have been reported (Craig, 1960) to lack certain amino acids such as tryptophan, cystine, cysteine, or to contain them at quantities lower than required. Tsiropoulos (1977a) showed that *D. oleae* adults were able to survive and reproduce on honeydews of *Saissetia oleae* Bern, *Filippia oleae* Costa and *Euphyllura olivina* Costa, and on pollens, supplementing a supply of a sucrose solution, of *Olea europaea* Linnaeus, *Vitis vinifera* Linnaeus, *Pistacia lentiscus* Linnaeus, *Pinus* sp. and *Quercus coccifera* Linnaeus. The pollen of the first two plants gave longer survival and better fecundity than that of the last two ones.

3.2 Holigidic diet nutrition

Work on holigidic diets has shown that adults require a protein source for improved survival and egg production. Fluke and Allen (1931) first found that inclusion of an unhydrolyzed yeast in the diet of *R. pomonella* prolonged survival and increased fecundity. Later, Hagen and Finney (1950) were the first ones to use a protein hydrolyzate in the diet of *D. dorsalis*, *D. cucurbitae* and *C. capitata*. As a result, fecundity increased greatly. Protein hyrolyzates are a good source for vitamins and minerals, besides being rich in amino acids. Since the fundamental discovery of the importance of protein hydrolyzates in adult tephritid nutrition, development of diets for a large number of species has occurred. The composition of diets for few tephritids is given in Table 3.1.1.2. There is a variation in the nutritional requirements among the different species as it is evidenced by the optimum ratios of the diet components used. The main components that are found in diets are: water, a carbohydrate source, usually sucrose, a protein source, either unhydrolyzed, i.e. brewer's yeast, or hydrolyzed, i.e. yeast hydrolyzate, soy hydrolyzate, or both. The protein source contains also mineral salts and vitamins. Certain species require additional constituents usually found in the other components of the diet in insufficient amounts, i.e. mineral salts, chicken egg yolk. The ratio of the main components, excluding water, (carbohydrate:protein) ranges from less than one, as in *R. pomonella* (Boush et al., 1969), to one and a half, as in *D. dorsalis* (Taguchi, 1966), to two as in *C. capitata* (Tanaka et al., 1970), to five as in *C. rosa* (Barnes, 1976). The other components are found at much lower levels. Water is a constituent indispensable for the utilization of other nutrients. In *D. oleae*

Chapter 3.1.1 references, p. 116

TABLE 3.1.1.2

Composition of holigidic diets for certain adult tephritids

Species	Diet form	Enzymatic hydrolyzate				Other constituents	References[1]
		Water	Sucrose	Yeast	Soy hydrolyzate		
Anastrepha ludens	Solid		3			1-Orange juice crystals	1
Anastrepha suspensa	Solid	3	+[2]	+			2
Ceratitis capitata	Solid	3	+[2]	+			3
Ceratitis capitata	Solid	3	3–5	1			4
Ceratitis capitata	Solid		2	1			5
Ceratitis rosa	Solid		5	1			6
Dacus cucumis	Solid		+[2]	+			7
Dacus cucurbitae	Solid		+[2]	+			3
Dacus dorsalis	Solid		+[2]	+			3
Dacus dorsalis	Solid		6	2	2	0.4-Salt mixture	8
Dacus oleae	Liquid	100	80	20		20-Brewer's yeast	9
Dacus oleae	Liquid	100	80	20		0.1-Choline chloride	10
Dacus oleae	Solid		80	20		40-Brewer's yeast	11
Dacus oleae	Liquid	100	80	20		14-Fresh egg yolk	12
Dacus oleae	Solid		80	20		6-Dry egg yolk	13
Dacus oleae	Solid		80	30		6.6-Dry egg yolk	14
Dacus tryoni	Solid		+[2]	+			7
Rhagoletis cerasi	Liquid	100	80	20		40-Brewer's yeast	15
Rhagoletis cerasi	Solid		80	20			16
Rhagoletis pomonella	Solid				20	20-Fructose + 10-Brewer's yeast + 5-Salt mixture	17

[1] Data obtained from: 1, Rhode and Spishakoff (1965); 2, Burditt et al. (1974); 3, Mitchell et al. (1965); 4, Nadel (1970); 5, Tanaka et al. (1970); 6, Barnes (1976); 7, Monro and Osborn (1967); 8, Taguchi (1966); 9, Moore (1962); 10, Hagen et al. (1963); 11, Cavalloro (1967); 12, Economopoulos and Tzanakakis (1967); 13, Tsitsipis (1975); 14, Tsitsipis and Kontos (1983); 15, Haisch (1968); 16, Prokopy and Boller (1970); 17, Boush et al. (1969).

[2] Diet components offered separately.

[3] Water offered on 0.75–1% agar blocks.

(Tsitsipis and Voyatzoglou, unpublished data) adults lived for 2.4 days without water, 2.6 days with water only, 10 days with sucrose alone, presumably utilizing the metabolic water produced during sucrose catabolism, and 7.6 days with solid diet (Tsitsipis, 1975) only. They lived, however, for 47 (♀) or 23 (♂) days with sucrose plus water and for 58 (♀) or 56 (♂) days with solid diet plus water.

The form of diet, solid or liquid, and the way it is presented to the flies, ingredients mixed or separate, vary with the species. The efficiency of the diet does not depend only on the presence of the nutrients at appropriate levels but also on their phagostimulatory characteristics. Possession of such properties plays an important role in feeding behavior by ensuring ingestion. Some species, or even ecotypes of a single species perform better on liquid than on solid diets. Higher numbers of female *D. oleae* (Tzanakakis et al., 1967), derived from a humid or a dry area, oviposited when fed on a liquid than on a solid diet. Flies from the humid area, feeding on a liquid diet, laid more eggs and had a shorter preoviposition period than those feeding on solid diet. Females from the dry area responded differently by laying more eggs on solid than on liquid diet. In regard to diet presentation, in *D. dorsalis*, *D. cucurbitae* and *C. capitata* (Mitchell et al., 1965) sucrose, protein hydrolyzate and water are given separately. Insects perform better on that than on a diet with mixed components presumably thereby ingesting appropriate needed quantities of each ingredient. In *D. oleae* (Tzanakakis et al., 1967), from a dry area, a solid diet with mixed components was better than one with its constituents given separately when the number of ovipositing flies, fecundity and preoviposition period were considered. Besides sucrose, a known phagostimulant and arrestant, hydrolyzed protein was found to be phagostimulatory to fruit flies. In *C. capitata* both sexes ingested more casein hydrolyzate (CH) plus sucrose (S) liquid diet than sucrose solution alone (Galun et al., 1980). In further studies Galun et al. (1981) showed that preference for CH + S ceased when the flies were previously fed with a protein-containing diet. Irradiation of flies in the pupal stage, at a dose causing gonad degeneration, affected quantity of diet imbibed. Although preference of both sexes of normal and irradiated flies was higher for CH + S than for S, absolute amounts of ingested diets were lower in irradiated than in normal ones. In *A. suspensa* Sharp and Chambers (1984) found an increased CH or yeast hydrolyzate (YH) plus sucrose diet consumption than sucrose solution alone, but similar quantities of the two diets were consumed when the CH or YH concentration was one tenth as high (0.25%) as the previous one. The consumption of protein hydrolyzates was higher in laboratory-reared flies at higher concentrations of the protein than at lower concentrations when offered at alternate concentrations (low-high or vice versa). There were no differences at the lowest concentrations tested, (2.5, 5.0%). Normal females of *C. capitata* and *A. suspensa* (Galun et al., 1985) consumed more CH + S and YH + S respectively than irradiated insects. The same trend with males of *A. suspensa* but not with *C. capitata*, where there were no differences observed. Irradiation seems to have in some way affected nutritional needs which are presumably lower than in normal ones because of undeveloped ovaries.

The phagostimulatory properties of protein hydrolyzates as well as the olfactory responses of flies to them (Galun et al., 1985) were conceived by Steiner (1952), who first used them successfully as poisoned baits for fruit fly control. Even today this technique is widely used.

3.3 Holidic diet nutrition

Holidic diets have been developed for few species and major nutrient categories have been identified, with respect to their effect on survival, fecundity and fertility. In the nutrient categories are included amino acids, carbohydrates, vitamin mixtures, salt mixtures and additional factors such as sterols and RNA. Some work has been also done on utilization of carbohydrates. In insect nutrition, utilization of or requirements for a nutrient is closely associated with feeding behaviour and in particular with the ability of the insect to ingest and maintain feeding. This procedure is accomplished by the presence of phagostimulants, i.e. substances either non-nutritive, "token" stimulants, or common nutrients. In certain tephritids some amino acids and carbohydrates were found to function as such. The basic nutritional work on adult nutritional requirements has concentrated on the development of diets that support survival and egg production. Some further work has also been done on the determination of qualitative nutritional requirements. No "minimal" diet has been developed yet and practically no investigation of quantitative requirements has been done. Work in these areas is needed to understand not only the nutritional requirements of fruit flies, but also to better appreciate their nutritional ecology. Such knowledge could be utilized for the management of the pest species. Holidic diets have been developed by Hagen (1952, 1958) and Hagen and Tassan (1972) for *D. dorsalis, D. cucurbitae, C. capitata*, by Boush et al. (1969) for *R. pomonella*, by Tsiropoulos (1977b, 1978) for *D. oleae* and *Rhagoletis completa* (Cresson). They all contained an amino acid mixture based on *Aphis pomi* De Geer honeydew analysis for *R. pomonella*, or on yeast hydrolyzate analysis, a protein hydrolyzate used successfully in meridic diets, for *R. completa* and *D. oleae*. They contained fructose or sucrose (*D. oleae* only) and a vitamin and salt mixture. Additionally, all, except *R. pomonella*, contained RNA and cholesterol. Hagen (1952, 1958) and Hagen and Tassan (1972) showed that protein was indispensable for egg production in *D. dorsalis* and *D. cucurbitae*. In *C. capitata* some eggs could be laid without protein, although those eggs would not hatch. Protein greatly improved fecundity and it was required by the males for mating, but it could not be utilized unless mineral salts were present. Vitamins were indispensable for egg fertility. By adding L-tryptophan, L-cysteine and L-cystine to the amino acid complement based on the apple aphid honeydew analysis, Boush et al. (1969) transformed an inadequate diet for *R. pomonella* to a successful one comparable to a meridic diet. The authors reasoned that in nature the missing amino acids could be provided for the flies by either microorganisms, contaminants of the honeydew, being able to synthesize amino acids, or from the activity of a symbiotic microorganism of the fly, and concurrently a pathogen to the plant, capable of amino acid synthesis. In *D. oleae* and *R. completa* (Tsiropoulos, 1977b, 1980a, 1978) a complete diet with 10 essential amino acids was as good as one with 19 amino acids as far as survival, fecundity and fertility were concerned. Survival could be supported by a carbohydrate source alone, but fecundity was not possible (*R. completa*) or it was very little (*D. oleae*) even in the presence of protein without mineral salts. Fecundity was greatly improved with the addition of vitamins in *D. oleae* but not in *R. completa*. Fertility was greatly enhanced with B-vitamins in both species, but it was not affected by Vitamin E, cholesterol and RNA in *R. completa*. Individual amino acids when given as sole nitrogen source were found to affect adversely *D. oleae* adult survival and fecundity (Tsiropoulos, 1983). The degree of influence varied with the amino acid. Considering the effect of individual amino acids in single amino acid deletions it was found that the 10 essential ones plus valine were indispensable for fecundity, while survival in males was also affected by alanine, hydroxyproline and tryptophan omission. Differences were noted in the survival between the two sexes in alanine, aspartic acid, cystine, glycine and tryptophan deficient diets. Of 15 vitamins tested plus RNA on the adults of *D. oleae* (Tsiropoulos, 1980b), i.e. vitamins A, C, D, E, K, B_{12}, biotin, calcium pantothenate, choline chloride, folic acid, inositol, nicotinic acid, pyridoxine, riboflavin and thiamin, survival of both sexes was affected by pyridoxine, riboflavin, or all vitamins omission, while female survival was additionally affected by omission of both folic acid and RNA. Egg production and hatchability were significantly reduced when vitamin E, biotin, choline chloride, inositol, nicotinic acid, riboflavin were individually omitted, or all vitamins, or all B-complex vitamins, or folic acid plus RNA were absent. Hatchability was additionally reduced with single omissions of calcium pantothenate, folic acid, pyridoxine and thiamine. Utilization of various sugars encountered in natural foods that adult fruit flies are known to feed on has been examined by Tsiropoulos (1980c) in *D. oleae*. The sugars tested were: rhamnose, ribose, xylose, sorbose, mannose, glucose, fructose, sucrose, cellobiose, melibiose, lactose, trehalose, maltose, raffinose, melezitose, dextrin, inulin, glucogen, erythrytol, dulcitol, sorbitol and inositol. Of those sugars, mannose, glucose, fructose, sucrose, melibiose, trehalose, maltose, melezitose and sorbitol could support survival and a low fecundity. Utilization of other sugars cannot be precluded, though at low rates, in case some of them elicit phagodeterrent stimuli to the flies, or they do not induce feeding. D-arabinose and L-arabinose could not support survival in *D. oleae* (Tsitsipis and Voyatzoglou, unpublished data) (2.5 and 3.5 days survival respectively), but when they were mixed with sucrose in the ratio arabinose/sucrose 70/30, survival increased to 6.5 and 13 days respectively. Amino acids, when complementing a sucrose solution, elicit a phagostimulatory or inhibitory response. In *C. capitata* (Galun et al., 1980) phenylalanine, glutamic acid, arginine, glycine, serine and methionine have been reported as highly stimulatory, valine, aspartic acid, lysine, alanine, proline, hydroxyproline and threonine were found less stimulatory, and cysteine induced phagorepellent response to both sexes. Differences were found between the sexes and in some amino acids with age. Sharp and Chambers (1984) found phagostimulatory action of amino acids in *A. suspensa*. Inhibitory response gave the amino acids cysteine and hydroxyproline to females, and proline to males. Certain sugars were also tested in *A. suspensa* (Sharp and Chambers, 1984). Sucrose was the most phagostimulatory, second were fructose and maltose, third dextrose and fourth glycogen with water. Females drunk equal amounts of glycogen and water, but males imbibed less glycogen.

4 CONCLUDING REMARKS

Work on the nutritional requirements in tephritids has not followed the progress accomplished in certain other insect taxa. The available information on the nutritional requirements of larvae stems from the search for diet ingredients, usually crude, that would improve survival, growth and development. The major contribution in this endeavour has been the utilization in the diets of dry plant material, dry yeasts, and in one case, of hydrolyzed protein (*D. oleae*), beyond the ameliorations brought about by improvement of diet texture. In adult nutrition more progress has been achieved through the development of holidic diets for a few species. Despite limitations of the research done, the qualitative nutritional requirements have been studied of certain nutrient classes such as amino acids, vitamins, growth factors, and the utilization of certain sugars and have been found generally similar to those of other insect groups. Profound nutrition work, however, is lagging behind that of other insects, a major inherent difficulty being the inability to rear larvae on holidic diets. Rearing for two consecutive generations on holidic diets is, therefore, not feasible and the problem of nutrient transfer cannot be controlled. It is of priority then that research proceeds to this direction so that more

Chapter 3.1.1 references, p. 116

concrete data on qualitative and quantitative nutrient requirements are obtained.

Research on tephritid species is very active. An important reason is their great economic importance. Some species are reared by the million to be used for Sterile Insect Technique. A better understanding of the nutritional requirements of tephritids could mean cheaper production cost. It would lead to a better understanding of their ecology and possibly more rational and efficient pest management.

5 ACKNOWLEDGMENT

Typing of the manuscript by Mrs. A. Kanoussi is gratefully acknowledged.

6 REFERENCES

Barnes, B.N., 1976. Mass rearing the Natal fruit fly *Pterandrus rosa* (Ksh.) (Diptera: Trypetidae). Journal of the Entomological Society of South Africa, 39: 121–124.

Bateman, M.A., 1972. The ecology of fruit flies. Annual Review of Entomology, 17: 493–518.

Beck, S.D., 1956. The european corn borer, *Pyrausta nubilalis* (Hubn.), and its principal host plant. II. The influence of nutritional factors on larval establishment and development on the corn plant. Annals of the Entomological Society of America, 49: 582–588.

Boush, G.M., Baerwald, R.J. and Miyazaki, S., 1969. Development of a chemically defined diet for adults of the apple maggot based on amino acid analysis of honeydew. Annals of the Entomological Society of America, 62: 19–21.

Burditt, A.K., Jr., Lopez, D.F., Steiner, L.F., Windeguth, D.L., von, Baranowski, R. and Anwar, M., 1974. Application of sterilization techniques to *Anastrepha suspensa* Loew in Florida, United States of America. In: Sterility Principle for Insect Control. Proc. Panel Organized by the Joint FAO/IAEA Division of Atomic Energy in Food and Agriculture, pp. 93–101.

Cavalloro, R., 1967. Orientamenti sull'allevamento permanente di *Dacus oleae* Gmelin (Diptera, Trypetidae) in laboratorio. Redia, 50: 337–344.

Cavicchi, S. and Zaccarelli, D., 1974. Lo studio della *Ceratitis capitata* Wied. in laboratorio: Fattori che controllano le esigenze nutrizionali allo stadio di larva. Rivista di Biologia N.S., 27: 139–151.

Chawla, S.S., 1966. Studies on the nutritional requirements of the fruit fly, *Dacus cucurbitae* Coquillet (Diptera: Trypetidae). III. Development of a synthetic medium for *D. cucurbitae* larvae. Research Bulletin Punjab University, N. Series, 17: 243–250.

Christenson, L.D. and Foote, R.H., 1960. Biology of fruit flies. Annual Review of Entomology, 5: 171–192.

Craig, R., 1960. The physiology of excretion in the insect. Annual Review of Entomology, 5: 53–68.

Dadd, R.H., 1973. Insect nutrition: current developments and metabolic implications. Annual Review of Entomology, 18: 381–420.

Dadd, R.H., 1977. Qualitative requirements and utilization of nutrients: insects. In: M. Rechcigl (Editor), Handbook Series in Nutrition and Food, Section D, Vol. 1. Nutritional Requirements. CRC Press, Cleveland, pp. 305–346.

Dadd, R.H., 1985. Nutrition: Organisms. In: G. Kerkut and L. Gilbert (Editors), Comprehensive Insect Physiology, Biochemistry and Pharmacology, Vol. 4. Pergamon Press, Oxford, New York, Toronto, Frankfurt, Tokyo, Sydney, pp. 313–390.

Dougherty, E.C., 1959. Introduction to axenic culture of invertebrate metazoa: A goal. Annals of New York Academy of Sciences, 77: 27–54.

Drew, R.A.I., 1978. Taxonomy. In: Economic Fruit Flies of the South Pacific Region. Plant Quarantine, Dept. of Health, Canberra, A.C.T., Australia, pp. 2–94.

Drew, R.A.I., Courtice, A.C. and Teakle, D.S., 1983. Bacteria as a natural source of food for adult fruit flies (Diptera: Tephritidae). Oecologia, 60: 279–284.

Economopoulos, A.P. and Tzanakakis, M.E., 1967. Egg yolk and olive juice as supplements to the yeast hydrolyzate-sucrose diet for adults of *Dacus oleae*. Life Sciences, 6: 2409–2416.

Finney, G.L., 1953. A summary report on the mass-culture of fruit flies and their parasites in Hawaii. Special Report on the Control of the Oriental Fruit Fly (*Dacus dorsalis*) in the Hawaiian Islands. 3rd Senate of State of California, pp. 77–83.

Finney, G.L., 1956. A fortified carrot medium for mass culture of the oriental fruit fly and certain other tephritids. Journal of Economic Entomology, 49: 134.

Fluke, C.L. and Allen, T.C., 1931. The role of yeast in life history studies of the apple maggot *Rhagoletis pomonella* Walsh. Journal of Economic Entomology, 24: 77–80.

Friend, W.G., 1968. The nutritional requirements of Diptera. In: Radiation, Radioisotopes and Rearing Methods in the Control of Insect Pests. Proc. IAEA, Vienna, STI/PUB 185, pp. 41–57.

Galun, R., Nitzan, Y., Blondheim, S. and Gothilf, S., 1980. Responses of the Mediterranean fruit fly *Ceratitis capitata* to amino acids. Proceed. Symposium on Fruit Fly Problems, Kyoto and Naha, National Institute of Agricultural Sciences, Japan, pp. 55–63.

Galun, R., Gothilf, S., Blondheim, S. and Lachman, A., 1981. Protein and sugar hunger in the Mediterranean fruit fly *Ceratitis capitata* (Wied.) (Diptera: Tephritidae). In: Determination of Behaviour by Chemical Stimuli, Proceed. 5th European Chemoreception Research Organization Symposium, pp. 245–251.

Galun, R., Gothilf, S., Blondheim, S., Sharp, J.L., Mazor, M. and Lachman, A., 1985. Olfactory and gustatory responses of normal and irradiated fruit flies *Ceratitis capitata* (Wied.) and *Anastrepha ludens* (Loew) to nutrients. Environmental Entomology, 14: 726–732.

Hagen, K.S., 1952. Influence of adult nutrition upon fecundity, fertility, and longevity of three tephritid species. Ph.D. Thesis, University of California, Berkeley.

Hagen, K.S., 1953. Influence of adult nutrition upon the reproduction of three fruit fly species. In: Third Special Report on the Control of the Oriental Fruit Fly *Dacus dorsalis* in the Hawaiian Islands. pp. 72–76. Senate, State of California, Sacramento, California.

Hagen, K.S., 1958. Honeydew as an adult fruit fly diet affecting reproduction. Proceedings 10th Intern. Congress of Entomology, Montreal, 1956, Vol. 3, pp. 25–30.

Hagen, K.S., 1966. Dependence of the olive fly, *Dacus oleae*, larvae on symbiosis with *Pseudomonas savastanoi* for the utilization of olive. Nature (London), 209: 423–424.

Hagen, K.S. and Finney, G.L., 1950. A food supplement for effectively increasing the fecundity of certain tephritid species. Journal of Economic Entomology, 43: 735.

Hagen, K.S. and Tassan, R.L., 1972. Exploring nutritional roles of extracellular symbiotes on the reproduction of honeydew feeding adult Chrysopids and Tephritids. In: J.G. Rodriguez (Editor), Insect and Mite Nutrition. North-Holland-American Elsevier, Amsterdam and London, pp. 323–351.

Hagen, K.S., Santas, L. and Tsekouras, A., 1963. A technique for culturing the olive fly *Dacus oleae* Gmelin on synthetic media under xenic conditions. Proceedings Symposium on Radiation and Radioisotopes Applied to Insects of Agricultural Importance, Athens, Greece, 1963, International Atomic Energy Agency, Vienna, pp. 333–356.

Haisch, A., 1968. Preliminary results in rearing the cherry fruit-fly (*Rhagoletis cerasi* L.) on a semi-synthetic medium. In: Radiation, Radioisotopes and Rearing Methods in the Control of Insect Pests. Proc. Panel Organized by the Joint FAO/IAEA Division of Atomic Energy in Food and Agriculture, pp. 69–78.

Haisch, A., 1975. Raising and multiplying of the European cherry fruit fly, *Rhagoletis cerasi* L. In: Controlling Fruit Flies by the Sterile-Insect Technique. Proceed. Panel Organized by Joint FAO/IAEA Division of Atomic Energy in Food and Agriculture, pp. 120–121.

House, H.L., 1972. Insect nutrition. In: N.T.W. Fiennes (Editor), Biology of Nutrition, International Encyclopaedia of Food and Nutrition. Vol. 18. Pergamon Press, Oxford and New York, pp. 513–573.

House, H.L., 1974. Nutrition. In: M. Rockstein (Editor), The Physiology of Insecta. Vol. V. Academic Press, New York and London, pp. 1–62.

Ibrahim, A.G. and Rahman, M.D.A., 1982. Laboratory studies of the effects of selected tropical fruits on the larvae of *Dacus dorsalis* Hendel. Pertanika, 5: 90–94.

Krainacker, D.A., Carey, J.R. and Vargas, R.I., 1987. Effect of larval host on life history traits of the Mediterranean fruit fly, *Ceratitis capitata*. Oecologia (In press).

Lopez, D.F., 1970. Sterile insect technique for eradication of the Mexican and Caribbean fruit flies. Review of current status. In: Sterile Male Technique for Control of Fruit Flies. Proc. Panel Organized by Joint FAO/IAEA Division of Atomic Energy in Food and Agriculture, pp. 111–117.

Maeda, S., Hagen, K.S. and Finney, G.L., 1953. The role of microorganisms in the culture of fruit fly larvae. In: Third Special Report of the Control of the Oriental Fruit Fly, *Dacus dorsalis*, in the Hawaiian Islands, Senate, State of California, Sacramento, California, pp. 84–86.

Manoukas, A.G., 1974. Protein hydrolysate-free larval diets for rearing of the olive fruit fly, *Dacus oleae*. Proc. Symp. Sterility Principle for Insect Control, International Atomic Energy Agency, SM-186, pp. 219–228.

Manoukas, A.G., 1977. Biological characteristics of *Dacus oleae* larvae (Diptera, Tephritidae) reared on a basal diet with variable levels of ingredients. Annales de Zoologie et Ecologie Animale, 9: 141–148.

Manoukas, A.G., 1981. Effect of excess levels of individual amino acids upon survival, growth and pupal yield of *Dacus oleae* (Gmel.) larvae. Zeitschrift für Angewandte Entomologie, 91: 309–315.

Manoukas, A.G., 1982. Effect of excess levels of inorganic salts upon survival, growth and pupal yield of *Dacus oleae* (Gmel.) larvae. Zeitschrift für Angewandte Entomologie, 93: 208–213.

Manoukas, A.G., 1986. Biological aspects of the olive fruit fly grown in different larval diets. In:

R. Cavalloro (Editor), Fruit Flies of Economic Importance. Balkema, Rotterdam, The Nether-lands, pp. 81–87.

Message, C.M. and Zucoloto, F.S., 1980. Valor nutritivo do levedo de cerveja para *Anastrepha obliqua* (Diptera, Tephritidae). Ciência e Cultura, 32: 1091–1094.

Mitchell, S., Tanaka, N. and Steiner, L.F., 1965. Methods for mass culturing oriental, melon, and Mediterranean fruit flies. U.S. Dept. of Agriculture, ARS 33-104, pp. 1–22.

Monro, J. and Osborn, A.W., 1967. The use of sterile males to control populations of Queensland fruit fly, *Dacus tryoni* (Frogg.) (Diptera: Tephritidae). I. Methods of mass-rearing, transporting, irradiating, and releasing sterile flies. Australian Journal of Zoology, 15: 461–473.

Moore, I., 1959. A method for artificially culturing the olive fly (*Dacus oleae* Gmel.) under aseptic conditions. Ktavim, 9: 295–296.

Moore, I., 1962. Further investigations on the artificial breeding of the olive fly – *Dacus oleae* Gmel. – under aseptic conditions. Entomophaga, 7: 53–57.

Mourikis, P.A., 1965. Data concerning the development of the immature stages of the Mediter-ranean fruit fly (*Ceratitis capitata* (Wiedemann) (Diptera: Trypetidae)) on different host fruits and on artificial media under laboratory conditions. Annals of the Benaki Phytopathological Institute N.S., 7: 59–105.

Nadel, D.J., 1970. Current mass-rearing techniques for the Mediterranean fruit fly. In: Sterile Male Technique for Control of Fruit flies. Proc. Panel Organized by Joint FAO/IAEA Division of Atomic Energy in Food and Agriculture, pp. 13–19.

Narasaki, T. and Katakura, K., 1954. Fundamental studies on the utilization of olive fruits. II. Identification of the amino acids in the protein hydrolyzate of ripe olive flesh by paper chro-matography. Technical Bulletin of the Kagawa Agricultural College, 6: 194–198.

Neilson, W.T.A., 1973. Improved method for rearing apple maggot larvae on artificial media. Journal of Economic Entomology, 66: 555–556.

Novak, J.A., 1974. A taxonomic revision of *Dioxyna* and *Paroxyna* (Diptera: Tephritidae) for America North of Mexico. Melanderia, 16: 1–53.

Novak, J.A. and Foote, B.A., 1980. Biology and immature stages of fruit flies: The genus *Eurosta* (Diptera: Tephritidae). Journal of the Kansas Entomological Society, 53: 205–219.

Prokopy, R.J. and Boller, E.F., 1970. Artificial egging system for the European cherry fruit fly. Journal of Economic Entomology, 63: 1413–1417.

Quereshi, Z.A., Ashraf, M., Bughio, A.R. and Hussain, S., 1974. Rearing, reproductive behaviour and gamma sterilization of fruit fly, *Dacus zonatus* (Diptera: Tephritidae). Entomologia Ex-perimentalis et Applicata, 17: 504–510.

Rahamannar, N., 1962. Growth, orientation and feeding behaviour of the larva of melon fly, *Dacus cucurbitae* Coq. on various plants. Proceedings of National Institute of Science, India, Pt. B. Biological Sciences, 28: 133–142.

Rhode, R.H. and Spishakoff, L.M., 1965. Tecnicas usadas en la cultivo de *Anastrepha ludens* (Loew). II. Memorias del dia Parasitologia, Departo de Parasitologica. Escuola Nacionale Agronomico Chapingo, (1964): 23–28.

Rodriguez, J.G. (Editor), 1972. Insect and Mite Nutrition. North Holland-American Elsevier, Amsterdam and London.

Sharp, J.L. and Chambers, D.L., 1984. Consumption of carbohydrates, proteins, and amino acids by *Anastrepha suspensa* (Loew) (Diptera: Tephritidae) in the laboratory. Environmental Entomol-ogy, 13: 768–773.

Singh, P., 1977. Artificial Diets for Insects, Mites, and Spiders. IFI/Plenum, New York, Washin-gton, London, 594 pp.

Srivastava, B.G., 1975. Nutritional behaviour of *Dacus cucurbitae* (Coquillett) larvae. Doctoral Thesis, University of Gorakhpur, India, 87 pp.

Srivastava, B.G. and Pant, G.C., 1979. Fatty acids and sterol requirements of *Dacus cucurbitae* (Coquillett) maggots under aseptic conditions. Bulletin of Entomology, 20: 124–127.

Srivastava, B.G., Pant, N.C. and Choudhary, H.S., 1977. Vitamin B requirement of *Dacus cucurbitae* (Coquillett) maggots (Diptera: Trypetidae). Indian Journal of Entomology, 39: 308–318.

Srivastava, B.G., Pant, N.C. and Chaudry, H.S., 1978. Carbohydrate requirement of *Dacus cucur-bitae* (Coquillett) maggots under aseptic conditions. Indian Journal of Entomology, 40: 150–155.

Steiner, L.F., 1952. Fruit fly control in Hawaii with poison-bait sprays containing protein hydroly-zates. Journal of Economic Entomology, 45: 838–843.

Steiner, L.F. and Mitchell, S., 1966. Tephritid fruit flies, In: C.N. Smith (Editor), Insect Coloniza-tion and Mass Production. Academic Press, New York and London, pp. 555–583.

Taguchi, T., 1966. Influence of different diets upon the longevity of the adult of oriental fruit fly, *Dacus dorsalis* Hendel. Research Bulletin Plant Protection Service of Japan, 20: 16–19 (in Japanese with English summary).

Tanaka, N., Okamoto, R. and Chambers, D.L., 1970. Methods of mass rearing the Mediterranean fruit fly currently used by the U.S. Department of Agriculture. In: Sterile Male Technique for Control of Fruit Flies. Proc. Panel Organized by the Joint FAO/IAEA Division of Atomic Energy in Food and Agriculture, pp. 19–23.

Tsiropoulos, G.J., 1977a. Reproduction and survival of the adult *Dacus oleae* feeding on pollens and honeydews. Environmental Entomology, 6: 390–392.

Tsiropoulos, G.J., 1977b. Survival and reproduction of *Dacus oleae* (Gmel.) fed on chemically defined diets. Zeitschrift für Angewandte Entomologie, 84: 192–197.

Tsiropoulos, G.J., 1978. Holidic diets and nutritional requirements for survival and reproduction of the adult walnut husk fly. Journal of Insect Physiology, 24: 239–242.

Tsiropoulos, C.J., 1980a. Major nutritional requirements of adult *Dacus oleae*. Annals of the Entomological Society of America, 73: 251–253.

Tsiropoulos, G.J., 1980b. The importance of vitamins in adult *Dacus oleae* (Diptera: Tephritidae) nutrition. Annals of the Entomological Society of America, 73: 705–707.

Tsiropoulos, G.J., 1980c. Carbohyrate utilization by normal and γ-sterilized *Dacus oleae*. Journal of Insect Physiology, 26: 633–637.

Tsiropoulos, G.J., 1983. The importance of dietary amino acids on the reproduction and longevity of adult *Dacus oleae* (Gmelin) (Diptera: Tephritidae). Archives Internationales de Physiologie et de Biochimie, 91: 159–164.

Tsitsipis, J.A., 1975. Mass rearing of the olive fruit fly, *Dacus oleae* (Gmel.) at "Democritos". In: Controlling Fruit Flies by the Sterile-Insects Technique. Proc. Panel Organized by the Joint FAO/IAEA Division of Atomic Energy in Food and Agriculture, STI/PUB/392, pp. 93–100.

Tsitsipis, J.A., 1977a. An improved method for the mass rearing of the olive fruit fly, *Dacus oleae* (Gmel.) (Diptera, Tephritidae). Zeitschrift für Angewandte Entomologie, 83: 419–426.

Tsitsipis, J.A., 1977b. Larval diets for *Dacus oleae*: The effect of inert materials cellulose and agar. Entomologia Experimentalis et Applicata, 22: 227–235.

Tsitsipis, J.A., 1982. Mass rearing of the olive fruit fly: Recent improvements. In: Sterile Insect Technique and Radiation in Insect Control, International Atomic Energy Agency, ST/PUB/595, pp. 425–427.

Tsitsipis, J.A., 1983. Optimization of a holigidic diet for the larvae of the olive fruit fly. In: R. Cavalloro (Editor), Fruit Flies of Economic Importance. Balkema, Rotterdam, The Netherlands, pp. 423–428.

Tsitsipis, J.A. and Kontos, A., 1983. Improved solid adult diet for the olive fruit fly, *Dacus oleae*. Entomologia Hellenica, 1: 24–29.

Tzanakakis, M.E., 1971. Rearing methods for the olive fruit fly *Dacus oleae* (Gmelin). Annals School of Agriculture and Forestry of the University of Thessaloniki, 14: 293–326.

Tzanakakis, M.E., 1974. Tomato as food for larvae of *Dacus oleae* (Diptera: Tephritidae). Annals of School of Agriculture and Forestry of University of Thessaloniki, 17: 549–552.

Tzanakakis, M.E., Tsitsipis, J.A. and Steiner, L.F., 1967. Egg production of olive fruit fly fed solids or liquids containing protein hydrolyzate. Journal of Economic Entomology, 60: 352–354.

Tzanakakis, M.E., Economopoulos, A.P. and Tsitsipis, J.A., 1970. Rearing and nutrition of the olive fruit fly. I. Improved larval diet and simple containers. Journal of Economic Entomology, 63: 317–318.

VanderZant, E.S., 1974. Development, significance, and application of artificial diets for insects. Annual Review of Entomology, 19: 139–160.

3.1.2 The Symbionts of *Rhagoletis*

DANIEL J. HOWARD

1 INTRODUCTION

Before embarking on a discussion of the symbionts of *Rhagoletis* it is necessary to define a few terms; in particular, symbiosis and symbiont. The word symbiosis is used as it was originally defined by De Bary (1879), to refer to the *living together of dissimilar organisms*. According to this definition, the association does not have to be beneficial to any of the organisms involved to be considered symbiotic. Thus, parasitic, commensalistic, and mutualistic relationships that involve the permanent or semi-permanent association of different forms of life come under the heading of symbiosis.

The term symbiont can properly be applied to all members of a symbiotic association (Steinhaus, 1967). However, it has become customary to use symbiont to describe the smaller member of a symbiosis and to refer to the larger member as the host. The usage of symbiont will be limited to the microorganisms intimately associated with *Rhagoletis*.

2 THE MICROORGANISMS ASSOCIATED WITH *RHAGOLETIS* — SYMBIOTIC OR NON-SYMBIOTIC?

The initial description of the microorganisms associated with *Rhagoletis pomonella* (Walsh) was furnished by Allen and his colleagues (Allen, 1931; Allen and Riker, 1932; Allen et al., 1934). They were not interested in an exhaustive description; rather, they wanted to determine whether a microbial associate of the apple maggot, *R. pomonella*, was responsible for the rot which commonly occurs in maggot-infested apples. They discovered that a bacterium identified as *Pseudomonas (Phytomonas) melophthora* could cause rot in apples and that this bacterium was frequently associated with various life stages of *R. pomonella*. However, the relationship was by no means permanent. Forty-five percent of all flies studied did not harbor the bacterium (Allen et al., 1934) and Allen and his co-workers refrained from calling the relationship a symbiosis. It is not clear from their reports whether *P. melophthora* was the predominant microorganism they isolated from *R. pomonella*.

The next paper relevant to *Rhagoletis*-microorganism symbiosis appeared more than twenty years later when Hellmuth (1956) published the results of a survey of bacteria associated with 30 species of tephritids, including *Rhagoletis cerasi* Linnaeus, *Rhagoletis alternata* (Fallen), and *Rhagoletis meigeni* (Loew), but not *R. pomonella*. She found that the majority of the insects studied (including these species of *Rhagoletis*) harbored a fluorescent pseudomonad, which she named *Pseudomonas mutabilis*. The high frequency of the association between *P. mutabilis* and tephritids, and Hellmuth's inability to isolate this bacterium from larval habitats are suggestive of a symbiotic relationship. However, because Hellmuth did not provide an adequate biochemical description of the bacterium, *P. mutabilis* has not been recognized as a species by Bergey's Manual of Determinative Bacteriology (1974), and the true identity of the bacterium is open to question. Hellmuth's results have not been verified by other investigators, and at this point the relationship between *P. mutabilis* and *Rhagoletis* is incompletely characterized.

In the last twenty years the pace of the work on *Rhagoletis*-microorganism symbiosis has increased dramatically, with *R. pomonella* attracting most of the attention. Based on the investigations of the 1930s (Allen, 1931; Allen and Riker, 1932; Allen et al., 1934), Boush and his co-workers interpreted the relationship between *R. pomonella* and *P. melophthora* as an obligate symbiosis and began a series of studies designed to elucidate the functional significance of the bacterium to the fly by describing the bacterium's metabolic capabilities (Boush and Matsumura, 1967; Miyazaki et al., 1968). They reportedly obtained pure cultures of *P. melophthora* by streaking directly from macerated digestive tracts of larvae of the apple maggot, and they demonstrated that the bacterium could degrade pesticides and synthesize methionine and cystine, two essential amino acids. However, the picture of a close relationship between *R. pomonella* and *P. melophthora* generated by this work has not been supported by subsequent investigations.

Dean and Chapman (1973) performed an extensive series of isolations from the digestive tracts of field-collected and laboratory-reared adult *R. pomonella* and failed to obtain a single colony of *P. melophthora*. Most of their isolates represented bacterial species that are widely distributed in soil, water, and the digestive tracts of animals. In addition, they isolated one yeast-like organism. The microorganism found most frequently in both laboratory-reared and field-collected flies was *Klebsiella oxytoca*, occurring in 33 of the 40 individuals studied.

The work of Huston (1972) generated further doubt about the status of *P. melophthora* as a symbiont of *R. pomonella*. He characterized 193 bacterial cultures isolated from 159 individuals of *R. pomonella*. Most of the insects were field-collected, and all life stages were represented. None of the cultures matched the description of *P. melophthora* given by Allen and Riker (1932). Although the numerical taxonomic approach taken by Huston precluded him from attaching species names to his isolates, according to his biochemical descriptions the predominant group of isolates belonged in the genus *Klebsiella*. Referred to by Huston as taxon XII, representatives of this group were isolated from 41% of the flies.

Huston (1972) also studied American Type Culture Collection (A.T.C.C.) culture number 23460, a representative of *Pseudomonas melophthora* isolated by Boush in 1967. This culture displayed only a very low level of similarity with two other *Pseudomonas* species, it was however very closely related to *Serratia marcescens*, and Huston identified A.T.C.C. 23460 as an achromogenic strain of *S. marcescens*. Only one of the cultures that he obtained from *R. pomonella* exhibited a high level of similarity with A.T.C.C. 23460.

Huston (1972) concluded that *R. pomonella* is not involved in a symbiotic relationship with *P. melophthora* or any other microorganism, and that the microflora associated with *R. pomonella* is determined by the habitat of the insect.

The importance of environmental factors was also stressed by Tsiropoulos (1976), who studied the bacteria associated with the walnut husk fly, *Rhagoletis completa* (Cresson). Tsiropoulos isolated microorganisms from all life stages of the fly and from rotted walnut husk, and he obtained 15 morphologically different isolates. Only two morphological isolates were associated with all life stages, but even these were not found in all individuals. Tsiropoulos identified these two common morphological types as a *pseudomonas* species and a *Xanthomonas* species. However, because he characterized only one isolate of each morphological type, his identifications must be considered tentative.

The most recent surveys of the bacteria associated with *Rhagoletis* have concentrated on identifying the inhabitants of the oesophageal bulb (EB) (Rossiter et al., 1983; Howard et al., 1985). The EB is a muscular evagination of the dorsal wall of the esophagus found only in adult tephritids (Girolami, 1973; Ratner and Stoffolano, 1982). The major functions of this organ appear to be the regulation of water and/or salt balance (Ratner and Stoffolano, 1984, 1986) and the maintenance of extracellular gut flora (Girolami, 1973; Ratner and Stoffolano, 1982). The gut microflora of adult *R. pomonella* make their first appearance in the EB and reach their highest densities there (Ratner, 1981). The high concentration of bacteria in the EB may be attributable to entrapment by the fibrous material in the EB lumen (Ratner and Stoffolano, 1982). Recent bacterial surveys have focused on this organ because of its apparent mycetomal function.

The surveys of the EB of *R. pomonella* have yielded results similar to those of Huston (1972) and Dean and Chapman (1973). Typically, more than one bacterial species was isolated from an EB, and the predominant species in an EB varied even among *R. pomonella* individuals originating from a single population (Rossiter et al., 1983; Howard et al., 1985). Yet, despite the variability, the most common EB inhabitant was *Klebsiella oxytoca*. This was the most common bacterium found in the digestive tract of adult *R. pomonella* by Dean and Chapman (1973), and it is probably the bacterial species represented by taxon XII of Huston (1972).

Howard et al. (1985) also characterized the EB inhabitants of six other *Rhagoletis* species (*Rhagoletis mendax* Curran, *Rhagoletis cornivora* Bush, *Rhagoletis tabellaria* Fitch, *Rhagoletis electromorha* Berlocher, *Rhagoletis suavis* Loew, and *Rhagoletis completa*). They found substantial microbial species diversity, although *K. oxytoca* turned out to be the predominant inhabitant of the EB in every fly species except *R. tabellaria*. The bacterium most frequently isolated from *R. tabellaria* was *Enterobacter agglomerans*, a close relative of *K. oxytoca*.

In summary, substantial evidence now supports the view that species of *Rhagoletis* do not enter into symbiotic relationships with microorganisms. The vast majority of the microorganisms that have been isolated from *Rhagoletis* are enteric bacteria that are widely distributed in nature. Even the microbial species most frequently associated with *Rhagoletis (K. ocytoca)* does not occur in all individuals. Confidence in the conclusion of no symbiosis is enhanced by the similarity in the findings of Huston (1972), Dean and Chapman (1973), Rossiter et al. (1983), and Howard et al. (1985) with regard to the microbial associates of *R. pomonella*. The similarity in findings is especially compelling when one considers that these investigators differed in their use of media, in their isolation techniques, in the life stages and body parts they studied, and in the geographic origin of their flies.

Yet one is still left with the uncomfortable task of explaining the disparity between the results of early surveys of the microorganisms associated with *R. pomonella* (Allen and Riker, 1932; Allen et al., 1934; Boush and Matsumura, 1967) and the results of later surveys (Huston, 1972; Dean and Chapman, 1973; Rossiter et al., 1983; Howard et al., 1985). The most likely explanation, given the clear misidentification of A.T.C.C. 23460 by Boush, is that Allen and his co-workers misidentified their isolates and that some subsequent investigators were misled by using only a small number of diagnostic biochemical tests to characterize cultures. Unfortunately, Allen did not establish a type strain, making it impossible to verify his work.

The cosmopolitan nature of the bacteria found in the digestive tract of *R. pomonella* suggests that these bacteria are picked up from the environment. This suggestion is bolstered by Huston's (1972) observation that 8 of the 20 eggs and 48 of the 70 larvae he examined harbored no bacteria, and by the discovery of Howard and Bush (unpublished data) that *K. oxytoca* occurs on the surface of unpicked apples in the field. All adult *R. pomonella* ever studied have contained bacterial cells (Huston, 1972; Dean and Chapman, 1973; Rossiter et al., 1983; Howard et al., 1985), indicating that an eventual association between *R. pomonella* and bacteria is virtually inevitable.

3 THE ECOLOGICAL INTERACTION BETWEEN *RHAGOLETIS* AND MICROORGANISMS

Even though species of *Rhagoletis* do not appear to be involved in symbiotic relationships with bacteria, the certainty that the flies will encounter bacteria makes it important to understand the nature of the interaction between the two groups. Ecologists recognize three fundamentally different types of interactions among organisms: competition, predation, and mutualism. Attention here is devoted to assessing the evidence for and against mutualism, not because the possibility of predation or competition can be dismissed out of hand, but because the interactions between *Rhagoletis* and bacteria have generally been regarded as mutualistic. Consequently, most of the relevant research has focused on uncovering the benefits provided to a fly by its microflora. The emphasis on the fitness of the fly rather than the fitness of the microflora reflects the entomological inclinations of most of the researchers interested in *Rhagoletis*-microorganism symbiosis.

Three major factors have contributed to the prevailing view that the *Rhagoletis*-microorganism interaction is mutualistic. The first factor was an experiment carried out by Allen and Riker (1932) in which the larvae from surface-sterilized eggs of *R. pomonella* died a couple of days after being placed in apple tissue, whereas larvae from untreated eggs developed normally. Many investigators took these results as evidence that bacterial associates are crucial to the survival and development of larvae of *R. pomonella*. However, a close examination of Allen and Riker's experiment reveals that their surface disinfectant killed 23 out of 25 eggs, so that only 2 larvae from treated eggs were followed in the experiment. Not only can one reasonably attribute the early deaths of the two larvae to the effects of the surface disinfectant, but the small sample size removes any meaning from the experiment. The second factor was results from research on another tephritid, *Dacus oleae* (Gmelin) (the olive fruit fly), which suggested that this fly enjoys an intimate relationship with the olive knot bacterium, *Pseudomonas savastanoi*, and that the presence of the bacterium is essential for normal development, survival, and reproduction (Petri, 1910; Stammer, 1929; Fytizas and Tzanakakis, 1966; Hagen, 1966). The third and perhaps most important factor was a pervasive conviction among students of insect-microorganism symbioses that if a microorganism is universally present in an insect species it can be inferred that the relationship is a mutually dependent symbiosis (Brooks, 1963). This was a position forstered by Paul Buchner (e.g., Buchner, 1965), the leader in insect-microorganism symbiosis research for the first half of the 20th century, and his students. Although this attitude is no longer as widespread as it once was, it certainly influenced the interpretation of the interaction between *Rhagoletis* and microbial symbionts.

3.1 The provisioning of essential nutrients

The larvae of *Rhagoletis* are highly host-specific frugivores (Boller and Prokopy, 1976), whereas the adults are believed to feed primarily on homopterous honeydews (Hagen and Tassan, 1972). Because both types of food are low in nitrogen (Burroughs, 1970; Hagen and Tassan, 1972), and reportedly lack one or more essential amino acids (Miyazaki et al., 1968; Boush et al., 1969; Salama and Rizk, 1969; Burroughs, 1970), much attention has focused on the nutritional significance of bacteria associated with *Rhagoletis*. Three types of experiments have been used to assess the nutritional dependency of a fly on its microflora.

The first type of study attempts to establish the metabolic capabilities of the microorganism(s) and the nutrient composition of the host's food materials. The functional significance of the microbe(s) is then explained in terms of its ability to provide missing nutrients or to break down refractile compounds. In the first study employing this approach, Miyazaki et al. (1968) demonstrated that a bacterium isolated from *R. pomonella*, and identified as *P. melophthora*, could synthesize methionine and cystine, two amino acids that the investigators could not find in the flesh of ripe McIntosh apples. Miyazaki et al. (1968) tentatively suggested, based on these results, that *P. melophthora* might provide amino acids to *R. pomonella*. The relevance of these findings is undermined by the uncertain identity of their bacterium and by reports from other investigators that methionine and cystine do occur in apple flesh (Davis et al., 1949).

Howard et al. (1985) assessed whether bacteria associated with *Rhagoletis* break down pectic substances in fruit to metabolites that larvae are capable of digesting, by investigating the abilities of a large number of bacterial isolates from *Rhagoletis* to digest polygalacturonic (pectic) acid. Their results appear to rule out such a function. Although some isolates exhibited pectolytic activity, more than a quarter of the flies studied, including all of the *R. pomonella* larvae, harbored only bacteria lacking pectolytic activity.

Howard et al. (1985) also evaluated the capacity of bacteria associated with *R. pomonella* to fix atmospheric nitrogen, by using the acetylene reduction assay. The assay was sensitive enough to detect nitrogen fixing activity supplying as little as 0.5% of a fly's nitrogen requirements over the course of its life, and the results were entirely negative. If the bacteria associated with *R. pomonella* do fix atmospheric nitrogen, they are doing so at a rate meaningless to the fly.

The assessment of the metabolic capabilities of a microorganism associated with an insect can provide valuable information on the metabolic limitations of the microorganism and thus define what functions it cannot perform for its host. However, the demonstration that a microorganism possesses a certain capability does not establish that the capability has any significance to its host, and must be followed up with additional studies. Unfortunately, the follow-up studies are rarely carried out.

The second approach that has been used to examine the nutritional dependency of *Rhagoletis* on microorganisms involves removing the microorganisms and evaluating the effect of the elimination on various fitness components of flies reared on a series of chemically-defined diets. Taking this approach, Tsiropoulos (1981) studied the interaction between *R. completa*, the walnut husk fly, and its microflora. He reported that the incorporation of antibiotics into a number of chemically-defined diets adversely affected adult survival, the pre-oviposition period, or fecundity or fertility. From these results, he inferred that microbial associates provide vitamins, amino acids, and possibly minerals to *R. completa*. However, Tsiropoulos' inferences must be viewed with some

Chapter 3.1.2 references, p. 128

caution. In his paper, he attributes the effect of the antibiotics to the elimination of the fly's microflora; however, he had no control in his experiment for toxic reactions to the antibiotics. It is possible that the antibiotics themselves adversely affected flies fed certain diets. Moreover, Tsiropoulos neglected to demonstrate that flies reared on diets containing antibiotics had lost their microflora. He assumed that bacteria were eliminated because bacteria isolated from the fly were sensitive to the antibiotics utilized in the experiment. Several cases are known in which antibiotics have failed to produce aposymbiotic insects (Pant and Dang, 1972).

A more general problem with the use of chemically-defined media in removal experiments is the difficulty of extrapolating the results to what is going on in the field. Chemically-defined diets rarely, if ever, mimic the texture, aroma, or nutrient composition of an insect's natural food substrate. Thus, they cannot be used as substitutes for natural foods when one is attempting to determine whether microorganisms are necessary for an insect's development, survival, or reproduction in nature. Natural substrates must be used to address these issues.

The third approach is to analyze the nutritional significance of bacteria associated with *Rhagoletis* by employing bacterial removal techniques and a natural food (Huston, 1972). Huston surface-disinfected eggs of *R. pomonella* to eliminate bacteria and then monitored the development and survival of larvae reared in apple, either axenically or with a variety of bacteria isolated from the fly. He found that apple maggot larvae reared axenically survived as well or better than larvae reared with bacteria. The presence or absence of bacteria also did not influence development times to pupation or the weights of pupae. For unknown reasons pupal survival was very low in all of Huston's experiments, making it difficult to determine the impact of bacteria on pupal survival and development, and on adult longevity, fecundity, and fertility. Huston's efforts were exemplary and clearly demonstrate that the presence of bacteria is not essential for the survival and development of apple maggot larvae. Unfortunately, in all of his re-infection treatments he used bacteria that are only infrequently associated with *R. pomonella*, rather than the commonly associated bacteria, such as his taxon XII (probably Klebsiella oxytoca). If the commonly associated bacteria have a quantitative effect on larval fitness, that is, if they are not essential for development and survival but they do enhance the fitness of their hosts, Huston would not have detected their effect.

The possibility that bacteria have a quantitative effect on *R. pomonella* fitness is eliminated by recent unpublished work of Howard and Bush. These investigators performed a detailed experiment comparing the development and survival of untreated larvae, axenic larvae, and larvae reinfected with *K. oxytoca*, reared in apples. They found no significant differences in development rates or in the numbers of larvae that survived to pupation among the three groups. Moreover, there were no significant differences among the three groups in development time to adulthood or in the proportion of pupae reaching the adult stage.

It seems safe to conclude that the bacteria associated with *R. pomonella* do not enhance the fitness of the fly by providing nutrients important for larval and pupal survival and development. Whether bacteria provide nutrients vital for adult longevity and reproductive success cannot be answered at this time.

3.2 Bacteria as food

Drew et al. (1983) and Chapter 3.1.3 have recently advocated the idea that the digestion of ingested bacteria represents an important source of nutrition for adult tephritids. These investigators have shown in laboratory experiments that on a diet of bacteria, sugar, and water, *Dacus tryoni* (Froggatt) adults have equal longevity and increased fecundity compared with flies fed the conventional diet of autolyzed brewer's yeast, sugar, and water. Additional evidence marshalled by Drew et al. (1983) to support their contention is (1) that bacteria isolated from the crops and stomachs of *D. tryoni* are morphologically similar to bacteria isolated from the surfaces of mulberry fruit and leaves (the natural food substrate of *D. tryoni*); (2) that *D. tryoni* and *Dacus cacuminatus* (Hering) showed a strong feeding response to bacteria isolated from the digestive tract of flies collected in the mulberry tree; and (3) that bacteria occurred in large quantities in crops and stomachs but in much lower quantitites in feces, indicating that they were being digested. Dean and Chapman (1973) found a greater number of bacteria at the beginning of the alimentary canal than at the end in *R. pomonella* and they too suggested that bacteria may be used as food by the fly.

Although tephritids may indeed digest ingested bacteria, the evidence that the bacteria are an important source of nutrition remains unconvincing. Laboratory studies have established the ability of *Rhagoletis* to live on many substances that they are unlikely to encounter in nature (Dean and Chapman, 1973). A laboratory demonstration that a substance is ingested and supplies enough nutrients for survival and reproduction does not provide much information on natural adult foods. A more meaningful experiment would be to compare the longevity and reproductive success of adult flies reared axenically and non-axenically on natural substrates. Such experiments with the larvae of *R. pomonella* have clearly shown their lack of dependence on bacteria as food (Huston, 1972; Howard and Bush, unpublished data).

3.3 Bacteria as detoxification agents

Boush and Matsumura (1967) demonstrated that a bacterial culture, identified as *P. melophthora*, could degrade a variety of pesticides, apparently through the hydrolytic action of strong esterases. They went on to suggest that these esterases might break down certain unwanted components of the host fly's food, thereby protecting the fly. The relevance of the findings of Boush and Matsumura (1967) is weakened by the uncertainty over the identity of the bacterium they were studying. Nevertheless, the possibility that the bacteria associated with *Rhagoletis* help to detoxify harmful host plant compounds is intriguing. It should be noted that because Huston (1972) and Howard and Bush (unpublished data) used picked, mature apples, which are low in phenolics in their removal experiments, their results are not particularly pertinent to this issue.

Howard and Bush (unpublished data) recently addressed the question of food detoxification by examining the impact of bacteria removal on the survival and development of *R. suavis* larvae reared on their natural food. The larvae of *R. suavis* eat the husks of *Juglans* (walnut) species, a food that is extremely rich in phenolics and quinones (Leistner, 1981), such as juglone. Howard and Bush reasoned that if any species of *Rhagoletis* needs help with food detoxification, it is *R. suavis*. Their experiment clearly demonstrated that the larvae of *R. suavis* do not depend on bacteria for food detoxification. Larvae reared axenically developed and survived as well as larvae reared with *K. oxytoca*, a common bacterial associate. Thus, it is unlikely that the bacteria associated with *Rhagoletis* help with food detoxification.

3.4 Bacteria as protection against potential pathogens

Because *Rhagoletis* larvae infest fruit, they must share their food with a diverse array of microbes, some of them perhaps pathogenic. Therefore, it is

Chapter 3.1.2 references, p. 128

possible that there has been selection on *Rhagoletis* to form associations with benign bacteria that protect a fly from invasion by microbial pathogens. As yet there has been no work on this topic.

4 CONCLUSIONS

Although many aspects of the relationship between *Rhagoletis* and microorganisms remain to be studied, the available evidence discloses that *Rhagoletis* do not enter into symbiotic relationships with microorganisms. It also appears that a fly does not benefit nor suffer injury from its loose association with a variety of microorganisms, at least at the larval stage of development. If the adult fly relies on microorganisms as a food substance or to perform some necessary function, the needs of the fly can be satisfied by a number of different bacterial species.

These conclusions have important implications for biologists studying *Rhagoletis*. One of the major motivations behind *Rhagoletis*-microorganism symbiosis research has been a desire to understand host specificity and host shifts among species of *Rhagoletis*. The reasoning has been that if microbial symbionts play a role in larval nutrition or in host plant detoxification, then it may be an alteration or a replacement of the symbionts which permits a fly to expand or alter its source of nutrition. However, with the premise of larval dependence on bacterial symbionts increasingly less tenable, we can concentrate on other areas in our efforts to comprehend the host shifts and food limitations of *Rhagoletis*.

Another driving force behind *Rhagoletis*-microorganism symbiosis research has been the need to devise more efficient and ecologically acceptable means of pest management. The hope has been to control the fly by interrupting its association with a single microbial species. The absence of a tight relationship between a specific microorganism and species of *Rhagoletis* and the seeming lack of larval dependence on microorganisms render this a remote hope, indeed.

I do not advocate abandoning research into *Rhagoletis*-microorganism interactions. Much interesting biology has yet to be worked out, especially the effect of microorganisms on adult fitness. But other concerns, such as the refinement of adult rearing media, will have to motivate this research.

5 ACKNOWLEDGEMENT

The writing of this chapter was supported by NSF grant BSR-8211153.

6 REFERENCES

Allen, T.C., 1931. Bacteria producing rot of apple in association with the apple maggot, *Rhagoletis pomonella*. Phytopathology, 21: 338.
Allen, T.C. and Riker, A.J., 1932. A rot of apple fruit caused by *Phytomonas melophthora*, n. sp., following invasion by the apple maggot. Phytopathology, 22: 557–571.
Allen, T.C., Pinckard, J.A. and Riker, A.J., 1934. Frequent association of *Phytomonas melophthora* with various stages in the life cycle of the apple maggot, *Rhagoletis pomonella*. Phytopathology, 24: 228–238.
Bergey's Manual of Determinative Bacteriology, Eighth Edition. 1974. The Williams & Wilkins Company, Baltimore, 1268 pp.
Boller, E.F. and Prokopy, R.J., 1976. Bionomics and management of *Rhagoletis*. Annual Review of Entomology, 21: 223–246.
Boush, G.M. and Matsumura, F., 1967. Insecticidal degradation by *Pseudomonas melophthora*, the bacterial symbiote of the apple maggot. Journal of Economic Entomology, 60: 918–920.

Boush, G.M., Baerwald, R.J. and Miyazaki, S., 1969. Development of a chemically defined diet for adults of the apple maggot based on amino acid analysis of honeydew. Annals of the Entomological Society of America, 62: 19–21.

Brooks, M.A., 1963. The microorganisms of healthy insects. In: E.A. Steinhaus (Editor), Insect Pathology, Vol. 1. Academic Press, New York, pp. 215–250.

Buchner, P., 1965. Endosymbiosis of Animals with Plant Microorganisms. Interscience Publishers, New York.

Burroughs, L.F., 1970. Amino Acids. In: A.C. Hulme (Editor), The Biochemistry of Fruits and Their Products. Academic Press, London, pp. 119–146.

Davis, S.C., Fellers, C.R. and Esselen, W.B., 1949. Composition nature of apple protein. Food Research, 14: 417–428.

Dean, R.W. and Chapman, P.J., 1973. Bionomics of the apple maggot in eastern New York. Search Agric. Entomol. Geneva 3(10), 62 pp.

De Bary, A., 1879. Die Erscheinung der Symbiose. Karl J. Trübner, Strassburg, 300 p.

Drew, R.A.I., Courtice, A.C. and Teakle, D.S., 1983. Bacteria as a natural source of food for adult fruit flies (Diptera: Tephritidae). Oecologia, 60: 279–284.

Fytizas, E. and Tzanakakis, M.E., 1966. Some effects of streptomycin, when added to the adult food, on the adults of *Dacus oleae* (Diptera: Tephritidae) and their progeny. Annals of the Entomological Society of America, 59: 269–273.

Girolami, G., 1973. Reperti morfo-istologici sulle batteriosimbiosi del *Dacus oleae* Gmelin e di altri Ditteri Tripetidi, in natura e negli allevamenti su substrati artificiali. Estratto da REDIA, 54: 269–294.

Hagen, K.S., 1966. Dependence of the olive fly, *Dacus oleae*, larvae on symbiosis with *Pseudomonas savastanoi* for the utilization of olive. Nature, 209: 423–424.

Hagen, K.S. and Tassan, R.L., 1972. Exploring nutritional roles of extracellular symbiotes on the reproduction of honeydew feeding adult chrysopids and tephritids. In: J.G. Rodriguez (Editor), Insect and Mite Nutrition. North-Holland Publishing Company, Amsterdam, pp. 323–351.

Hellmuth, H., 1956. Untersuchungen zur Bakteriensymbiose der Trypetiden (Diptera). Zeitschrift für Morphologie und Oekologie der Tiere, 44: 483–517.

Howard, D.J., Bush, G.L. and Breznak, J.A., 1985. The evolutionary significance of bacteria associated with *Rhagoletis*. Evolution, 39: 405–417.

Huston, F., 1972. Symbiotic association of bacteria with the apple maggot, *Rhagoletis pomonella* (Walsh) (Diptera: Tephritidae). M.S. thesis. Acadia Univ., Wolfville, NS, Can.

Leistner, E., 1981. Biosynthesis of plant quinones. In: E.E. Conn (Editor), The Biochemistry of Plants, Vol. VIII. Secondary Plant Products. Academic Press, New York, pp. 403–423.

Miyazaki, S., Boush, G.M. and Baerwald, R.J., 1968. Amino acid synthesis by *Pseudomonas melophthora*, bacterial symbiote of *Rhagoletis pomonella* (Diptera). Journal of Insect Physiology, 14: 513–518.

Pant, N.C. and Dang, K., 1972. Physiology and elimination of intracellular symbiotes in some stored product beetles, In: J.G. Rodriguez (Editor), Insect and Mite Nutrition. North-Holland Publishing Company, Amsterdam, pp. 311–322.

Petri, L., 1910. Untersuchung über die Darmbakterien der Olivenfliege. Zentralblatt für Bakteriologie, Parasitenkunde, Infektionskrankheiten und Hygiene, Abteilung 2, 26: 357–367.

Ratner, S.S., 1981. Structure and function of the esophageal bulb of the apple maggot fly, *Rhagoletis pomonella* Walsh. Ph.D. Diss. Univ. Massachusetts, Amherst, MA.

Ratner, S.S. and Stoffolano, J.G., 1982. Development of the esophageal bulb of the apple maggot, *Rhagoletis pomonella* (Diptera: Tephritidae): morphological, histological, and histochemical study. Annals of the Entomological Society of America, 75: 555–562.

Ratner, S.S. and Stoffolano, J.G., 1984. Ultrastructural changes of the esophageal bulb of the adult female apple maggot, *Rhagoletis pomonella* (Walsh) (Diptera: Tephritidiae). International Journal of Insect Morphology and Embryology, 13: 191–208.

Ratner, S.S. and Stoffolano, J.G., 1986. *In vitro* measurement of movement of water and solutes out of the esophageal bulb of *Rhagoletis pomonella* Walsh (Diptera, Tephritidae). Estratto da REDIA, in press.

Rossiter, M.A., Howard, D.J. and Bush, G.L., 1983. Symbiotic bacteria of *Rhagoletis pomonella*. In: R. Cavalloro (Editor), Fruit Flies of Economic Importance. A.A. Balkema, Rotterdam, pp. 77–84.

Salama, H.S. and Rizk, A.M., 1969. Composition of the honeydew in the mealy bug, *Saccchariococcus sacchari*. Journal of Physiology, 15: 1873–1875.

Stammer, H.J., 1929. Die Bakteriensymbiose der Trypetiden (Diptera). Zeitschrift für Morphologie und Oekologie der Tiere, 15: 481–523.

Steinhaus, E.A., 1967. Insect Microbiology. Hafner Publishing Company, New York, 763 pp.

Tsiropoulos, G.J., 1976. Bacteria associated with the walnut husk fly, *Rhagoletis completa*. Environmental Entomology, 5: 83–86.

Tsiropoulos, G.J., 1981. Effect of antibiotics incorporated into defined adult diets on survival and reproduction of the walnut hustk fly, *Rhagoletis completa* Cress. (Dipt., Trypetidae). Zeitschrift für Angewandte Entomologie, 91: 100–106.

3.1.3 Bacteria associated with Fruit Flies and their Host Plants

R.A.I. DREW and A.C. LLOYD

1 INTRODUCTION

Although it has been recognized for many years that microorganisms may have various non-pathological associations with insects (Steinhaus, 1954), the true biological significance of many of these relationships remains poorly understood. As stressed by Steinhaus (1960), the role of microorganisms in the ecology of insects has been, and still is, largely neglected by entomologists and microbiologists alike.

Recent studies by Drew et al. (1983), Courtice and Drew (1984), Drew (1987), Drew and Lloyd (in press), with tropical and subtropical Dacinae suggest that the bacteria associated with fruit flies and with their host plants play an essential role in the ecological behaviour of these insects.

2 HISTORICAL DEVELOPMENT

Most specific studies on the relationship of bacteria to fruit flies relate back to the early work of Petri (1910) in which the association of the organism *"Bacillus (Pseudomonas) savastanoi"* Smith (the known cause of olive knot disease) with *Dacus oleae* (Gmelin) (the olive fruit fly) was defined as 'symbiosis'. Most studies since then have continued to refer to bacteria associated with fruit flies as obligate or probable symbiotic organisms.

In studying the alimentary tract bacteria of *D. oleae* Petri also described another bacterium, *"Ascobacterium luteum,"* as occurring in association with *"B. savastanoi"*. In this work it was stated that the bacteria are smeared onto the egg from rectal glands which open near the oviduct, the bacteria then enter the egg through the micropyle and are thus incorporated into the newly hatched larva. In the larvae, the bacteria develop colonies in four blind sacs at the anterior end of the mid-gut and from here are transferred into the adult when some bacteria are incorporated into the oesophageal bulb which develops in 5-day old pupae. The biological significance of the bacteria was not clearly explained but it was thought that they may be important in larval digestion of olive fruit tissue. Petri presented inconclusive experimental micro-biological data to support his interpretation of symbiosis yet it is upon this study that subsequent workers have built their theories. Stammer (1929) isolated bacteria from 37 species of Tephritidae and further described the bacterial transfer system through each stage of the life cycle of *D. oleae*.

Yamvrias et al. (1970) could not isolate either *"Pseudomonas savastanoi"* or *"Ascobacterium luteum"* from eggs and oesophageal bulbs of field collected adults of *D. oleae* and emphasised that Petri also failed to isolate these bacteria from the flies even though he proceeded to call them symbionts. This finding of Yamvrias et al. (1970) also brings into question the conclusions of Hagen (1966) who claimed that the addition of streptomycin in the adult fly diet destroyed the symbiotic bacterium "*Pseudomonas savastanoi.*" Hagen based his conclusions on those of Petri without providing experimental evidence.

Luthy et al. (1983) presented results of scanning electron microscope studies of the oesophageal bulb of *D. oleae* and concluded that one bacterial symbiont was present although the identity of the bacteria was not established conclusively.

Girolami (1973, 1983) defined two different types of symbiosis in Tephritidae, one with bacteria in the adult oesophageal bulb and the other with bacteria in the blind sacs at the anterior end of the larval mid-gut. Girolami did not give any bacterial identification of these symbiotic organisms although the release of compact masses of bacteria from the oesophageal bulb into the mid-gut is described.

Allen and Riker (1932) described the bacterium "*Pseudomonas (Phytomonas) melophthora*" as causing apple rot in association with the larval feeding of *Rhagoletis pomonella* (Walsh) (apple maggot). Allen et al. (1934) identified the same bacterium associated with all stages of *R. pomonella*, oviposition punctures, larval burrows and exit holes in apple fruit. Rossiter et al. (1983) also stated that essential symbiotic bacteria are present and necessary for normal development in most tephritid species. Their work on *R. pomonella* was significant in providing identification of two bacteria, viz., *Klebsiella oxytoca* and *Enterobacter cloacae* and showing that they are capable of causing rot in apples. They were unable to isolate "*P. melophthora*" as previously identified by Allen and Riker (1932).

The bacteria associated with *R. pomonella* were further studies by Baerwald and Boush (1968) and Miyazaki et al. (1968) who stated that "*P. melophthora*" was an obligate symbiont of the fly but they based their conclusion on the work of Allen et al. (1934) who did not mention symbiosis. No experimental evidence was provided.

The work of Dean and Chapman (1973) on *R. pomonella* was the first real attempt to explain the biological significance of bacteria associated with adult fruit flies. *Klebsiella oxytoca* was the most common bacterium isolated and larger numbers of isolates were obtained from the crop of the fly than from the mid-gut and rectal sac. They showed that the only proteinaceous material in the crop was in the form of bacterial cells and that the numbers of these cells decreased through the alimentary canal from the crop to the rectum. If the crop contents are representative of the substances which the fly ingests then such bacterial cells on the plant surface would appear to be an important food for adult fruit flies in nature.

Ratner and Stoffolano (1982) studied the development of the oesophageal bulb in *R. pomonella* and concluded that this organ may contain a feeder culture of bacteria for slow release to the gut. They did not, however, provide any direct evidence for this.

Tsiropoulos (1976) found 15 morphologically different bacteria as a result of isolations from all stages of *Rhagoletis completa* (Cresson) (the walnut husk fly) and rotted walnut pulp. Only a *Pseudomonas* sp. and a *Xanthomonas* sp. were associated with all stages of the fly. He stated that it was hypothetical as to whether or not any of these bacteria were symbiotic in any stage of the fly. In a latter study of the bacteria associated with oesophageal bulbs and ovipositors of wild and laboratory reared adult olive flies, Tsiropoulos (1983) found a wide variety of bacterial species in both types and concluded that a "more complicated system of relationships" existed than was originally assumed. Fitt and

O'Brien (1985) have suggested a lack of strict symbiosis between *Dacus* species and bacteria.

Although there is a growing literature on bacteria and fruit flies, there is very little that provides identification of bacterial species and experimental evidence giving an understanding of the biological significance of their relationships with the flies. Bateman (1972) accepts symbiosis in fruit flies as an established phenomenon, but Prokopy (1977) notes that the role of symbionts and other microorganisms in larval development "remains poorly understood."

3 THE MICROBIOLOGY OF THE PHYLLOPLANE

The phyllosphere or phylloplane is the term used by many workers for the layer of microorganisms growing on the leaf and fruit surfaces. Ruinen (1961) emphasised that this was an "ecologically neglected milieu". Since that time numerous publications on phylloplane microflora have been produced, many of which have concentrated on phytopathogenic organisms. Some properties of non-pathogenic (saprophytic) microorganisms colonizing green leaves have been reviewed (Last and Warren, 1972) and the interaction of such organisms with pathogens has been studied (Blakeman and Brodie, 1976; Blakeman and Fokkema, 1982).

The resident bacterial populations on aerial plant parts consist primarily of Gram-negative rod shaped organisms in the genera *Erwinia, Pseudomonas, Xanthomonas* and *Flavobacterium* and less frequently Gram-positive organisms such as *Lactobacillus, Bacillus* and *Corynebacterium* (Billing, 1976).

The most extensively studied epiphytic microorganisms belong to the ubiquitous *Erwinia herbicola* group. These organisms have attracted attention because of their common association with plant pathogenic *Erwinia* species. Organisms in the *herbicola* group known as *Enterobacter agglomerans* have been recognised as frequently associated with pathological conditions in man and animals (Starr and Chatterjee, 1972; Gibbons, 1978; Perombelon, 1981; Slade and Tiffin, 1984).

Most published works on epiphytic microflora of aerial plant parts are of little direct relevance to our understanding of fruit fly ecology as fruit fly host trees in particular have rarely been the subject of microbiological investigations. The studies of phylloplane bacteria of *Lolium perenne* (rye grass) by Dickinson et al. (1975) described and compared useful quantitative and qualitative methods for the isolation of plant surface microorganisms. The predominant organisms detected were *Chromobacterium, Corynebacterium, Pseudomonas* and *Xanthomonas* but few such studies have been reported on fruit fly host fruit and leaves.

The microflora of olive leaves (the host tree for *D. oleae*) were studied in detail and the predominant organism found to be "*Pseudomonas savastanoi*" (Ercolani, 1978). This bacterium is the cause of olive knot and as mentioned has been reported by some workers as the "symbiont organism" in *D. oleae*, although no mention of this insect association was made by Ercolani.

Leaves and fruit are bounded by an environment rich in microbial propagules and the surrounding atmosphere exerts diurnal, seasonal and random fluctuations influencing the growth of these microorganisms (Dickinson, 1971). Many, if not all, chemicals within the plant are leached from the plant onto the leaf and fruit surfaces. Such leachates include inorganic nutrients e.g. all essential minerals, organic substances, e.g. free sugars, pectic substances, sugar alcohols, all amino acids found in plants, many organic acids, vitamins, alkaloids and phenolic substances (Tukey, 1971; Godfrey, 1976). Soft fruits such

Chapter 3.1.3 references, p. 138

as mulberry are susceptible to leaching especially just prior to harvest. These leachates provide significant nutrients for the growth of phylloplane micro-organisms.

Fruit flies, particularly Dacinae, require a diet rich in amino acids, B complex vitamins, minerals, sugars and water for survival and reproduction (Hagen, 1953). A high concentration of free amino acids is essential for egg production. Honeydew, believed for many years to be the natural source of food for adult fruit flies, is both nutritionally unsuitable and unavailable for tropical Dacinae in Queensland (Drew et al., 1983). Recent studies on adult fruit fly feeding have shown that certain bacteria isolated from the crops and stomachs of actively feeding field flies provided virtually a complete breeding diet (Drew et al., 1983). The question now raised is what is the availability of these bacteria in the field?

4 RECENT INVESTIGATIONS OF THE ECOLOGICAL RELATIONSHIPS OF BACTERIA TO FRUIT FLIES

The distribution and abundance of bacteria, isolated from both fruit flies and fruit fly host trees and the significance of these organisms in the life cycle of the fly is now the subject of a major research programme by the authors (Lloyd et al., 1986).

Bacterial isolations have been made from a total of 80 adult flies, representing four species [*Dacus tryoni* (Froggatt), *Dacus neohumeralis* Hardy, *Dacus cacuminatus* (Hering) and *Dacus musae* (Tryon)] collected in the field from host trees (wild tobacco, guava, mulberry, peach and banana) during an eighteen month period. All flies examined have been wax-plugged at anal and oral openings and surface sterilized prior to aseptic dissection in which the crop and mid-gut or oesophageal bulb were removed for culturing. Bacteria have also been isolated from faeces of field collected flies, as well as from host fruit surfaces, oviposition sites and larvae-infested tissue in host fruit.

Results of our isolations and subsequent identifications (Table 3.1.3.1) indicate that the predominant microflora within the tropical/subtropical Dacinae comprise bacteria belonging to the family *Enterobacteriaceae*. Mid-guts of field collected flies almost always contained large numbers of bacteria but crops, faeces and oesophageal bulbs were found to contain variable, and sometimes very low, numbers of bacteria.

Of 135 bacterial isolates obtained from adult fly specimens, 120 were identified as *Enterobacteriaceae*. The species *Erwinia herbicola*, *Enterobacter cloacae* and *Klebsiella oxytoca* were the most frequently isolated, with *Citrobacter freundii*, *Proteus* and *Providencia* species, *Escherichia coli* and *Serratia* species being less frequently found. The three most common species found in

TABLE 3.1.3.1

Bacteriological examination of the alimentary tract and faeces of field collected *Dacus* species in Queensland

	Crop (No.)	Mid-gut (No.)	Oesophageal bulb (No.)	Faeces (No.)
Specimens examined	52	53	16	30
Specimens with				
> 5 colonies per plate	34	52	12	13
Bacterial isolates obtained	38	65	12	20
Isolates identified as Enterobacteriaceae	30	59	11	20
(as %)	(79%)	(91%)	(92%)	(100%)

TABLE 3.1.3.2

Bacteria isolated from *Dacus tryoni* (Froggatt) and from peach fruit on the tree in which the flies were collected (+ = bacteria present)

Bacterium	D. tryoni		Oviposition site and fruit rot	Fruit surface
	Crop	Mid-gut		
Erwinia herbicola	+	+	+	+
Enterobacter cloacae	+	+	+	+
Klebsiella oxytoca	+	+	+	+
Proteus spp.	+		+	+

flies were also the most frequently isolated from oviposition sites and larval induced rot in field stung fruit. Fly species, sex, host tree and seasonal time did not appear to affect the types of bacteria found in field collected flies. No one species of bacteria was found to be consistently associated with any particular section of the alimentary canal or with any one species of fly. Hence the term "symbiosis" does not appear to be appropriate for the association of bacteria with these species of Dacinae and we believe that a more extensive ecological relationship is involved.

One important result of our investigations has been the recovery of the same bacteria from fruit surfaces, fruit rots and within crops and stomach of adult flies when all samples were collected from the same host tree (Table 3.1.3.2).

When surfaces of caged (fruit flies excluded) and uncaged (exposed to fruit flies) guava fruit were examined for presence of *Enterobacteriaceae*, few were present on immature fruit in both situations. However, in more mature fruit, once oviposition has commenced in uncaged (exposed) fruit, many more "fruit fly type" *Enterobacteriaceae* were detected on those fruit surfaces than on caged fruit surfaces.

A similar result was obtained when the microflora of a caged nectarine tree was compared with that of an uncaged nectarine tree exposed to wild *D. tryoni* and *D. neohumeralis*. Uncaged fruit showed a rapid increase in surface bacterial counts from 2.7×10^2 organisms/cm^2 of fruit surface (when ovipositions were first observed) to a maximum of 6×10^6 organisms/cm^2 of fruit surface 15 days later (one day prior to total fruit drop). The large surface populations on stung fruit were almost entirely "fruit fly type" *Enterobacteriaceae*, representative isolates being identified as *E. cloacae, E. herbicola, K. oxytoca* and *C. freundii*. These bacteria were not detected on uncaged leaves or on caged fruit or leaves and these results suggest that flies are introducing their own specific bacteria onto host fruit surfaces.

Further evidence to support this theory was obtained using adult *D. tryoni* fed on a culture of "marked" *E. cloacae* (double antibiotic-resistant mutant) as the only protein source and then released into a cage containing nectarine trees bearing mature green fruit. One week later, large numbers of *E. cloacae* were recovered from fruit surfaces and oviposition sites in stung fruit. Other observations using *D. tryoni* fed on labelled *K. oxytoca* have shown that flies regurgitate alimentary tract bacteria in droplets of spittle and even in the absence of obvious regurgitated fluid, the proboscis was constantly spreading these same bacteria when individual flies were allowed to forage on agar medium.

Another significant finding is that fruit flies are strongly attracted to species of *Enterobacteriaceae*. The attraction of *D. tryoni* to two different protein suspensions with and without *Providencia rettgeri* inoculum was tested in a field cage experiment and replicated 3 times. The protein suspensions attracted a total of 41 and 83 flies while the bacteria innoculated suspensions attracted

TABLE 3.1.3.3

Number of *Dacus tryoni* (Froggatt) attracted to bait solutions sprayed on avocado trees at Mt. Tamborine, Queensland

Bait	No. females	No. males	Total
Bacteria bait	489	386	875
Mauri protein autolysate	290	229	519
Millers Nu-lure	177	118	295

244 and 248 flies respectively. Also the attraction was tested to protein suspension innoculated with *P. rettgeri* used as spot baits in an avocado orchard in south-east Queensland (Table 3.1.3.3). While this is an area of continuing research in Queensland, these initial results indicate that actively growing bacteria isolated from the flies are powerful attractants. Once attracted the flies feed vigorously on the bacteria. These attractant and feeding responses may play a key role in the behaviour of the flies.

Flies collected in mulberry trees bearing ripening fruit have purple fruit colour visible in the alimentary canal if the flies have been feeding in the tree (Drew et al., 1983). Four species of flies were collected in one such tree, only two of these species being known to oviposit in mulberries. Gravid females of these two species, *D. neohumeralis* and *D. tryoni*, had virtually empty crops and very pale purple colour in the mid-gut, indicating that they had fed at least 24 h before collection. Immature females had distended crops and mid-guts with deep purple contents. Males of both species had brown coloured crop and mid-gut contents indicating that they were feeding on different substances from the females. Females of the other species, viz., *D. cacuminatus* and *Dacus brunneus* (Perkins and May) had deep purple crop and stomach contents, all specimens being immature and actively feeding.

From this it can be concluded that the mulberry fruit surfaces were a significant feeding site for 4 fly species. Mature gravid flies which oviposit in the fruit did not feed on these surfaces after egg maturation (consistent with the fact that flies do not require protein after egg maturation), immature females of all species were actively feeding on the fruit and males of all species were feeding on other substrates. The fruit were at early stages of ripening at the beginning of the fruiting season and there was no fruit breakdown and fluid loss. Our bacterial isolation studies have revealed large quantities of 'fruit fly type' bacteria on the fruit surfaces, indicating that such host fruit surfaces are significant sources of protein and other nutrients for developing flies. It is also possible that proliferating bacteria on the host tree form a powerful attractant drawing more flies into the tree.

The protein requirements of *D. tryoni* have been assessed by feeding 50 males and 50 females separately for 15 days from emergence, then combining the sexes for three days prior to a 5-h egg production test (Table 3.1.3.4). In another experiment the egg production of *D. tryoni* was assessed when cages of 50 males and 50 females were given a single feed of excess protein hydrolysate for 6 h on only one day and egged 15 days after emergence (Table 3.1.3.5). These studies have shown that (a) males require little or no protein to fulfil their role of fertilising eggs, (b) a single protein feed on one day is adequate for a female to produce up to 40 eggs in one oviposition period 3 days after feeding, (c) females accrue most benefit from protein after 4 days from emergence.

From these studies it appears that immature females in the post-teneral dispersal phase are attracted by bacterial odours and these could be an important source of attraction to a host tree. The flies could then utilise bacterial food available in the tree, apparently on the surfaces of maturing fruit. Feeding

TABLE 3.1.3.4

Influence of protein hydrolysate fed to *Dacus tryoni* (Froggatt) males and females on egg production and fertility (all cages fed sugar and water; nil diet = sugar and water only)

Protein fed to ♂♂	Protein fed to ♀♀	No. eggs/♀/h	Egg hatch (%)
Nil	Excess	3.1	72
Excess	Nil	0	0
Excess	Excess	3.6	87
0.01 g/week	Excess	6.8	81
Excess	0.01 g/week	0	0

flies remain in the tree and oviposit when their eggs become mature (within 3 days of feeding). For this behaviour pattern to develop the flies would have to establish a reasonably permanent population within the host plant. This indeed is what happens. Recent observations on *D. tryoni* in peach trees and *D. cacuminatus* in *Solanum mauritianum* have revealed that the flies feed, mate and oviposit in the same host plant.

The biological and ecological significance of these *Enterobacteriaceae* species in the life cycle of the fruit fly remains to be explained. We have shown that at times they are prolific on host fruit surfaces and it appears that the flies carry the bacteria and introduce them to the host plant. In this way, colonies of bacteria are established on specific plant parts, particularly host fruit, when required. This would be particularly significant when rain and resulting plant surface leachates are available and it is the wet summer period when the tropical Dacinae are most active. This system would ensure that, weather conditions permitting, adult food is not limiting. It would also be more economical with respect to the energy budget of the fly to search for oviposition sites prior to maturation and fertilization of eggs. In this respect the bacteria are important in ensuring reproductive success within fruit fly populations.

5 DISCUSSION

It appears that the association of bacteria and fruit flies is complex. It is well known that microorganisms are associated with insects in ways other than as

TABLE 3.1.3.5

Influence of a single feed of protein hydrolysate for 6 h on egg production and fertility in *Dacus tryoni* (Froggatt) of different ages (all flies fed sugar and water)[1]

Age of flies when fed protein (days)	No. eggs/♀/h	Egg hatch (%)
2	0.07	100
4	0.94	78
6	2.86	86
8	5.05	88
10	7.54	92
12	7.95	87
14	0	–

[1] Note: (a) Flies were egged on day 15, and the 14 day old flies were egged again on day 16. (b) 0 eggs produced 1 day after feeding. (c) Small egg production (0.37 eggs/♀/h) 2 days after feeding. (d) Large egg production (7.95 eggs/♀/h) 3 days after feeding.

Chapter 3.1.3 references, p. 138

pathogens but entomologists have tended to neglect the role of microorganisms in the ecology of insects (Steinhaus, 1960).

The term "symbiosis" was first used by De Bary (1879) who defined it as "a phenomenon in which dissimilar organisms live together". De Bary originally coined the term "symbiont" to refer to either partner of an association and numerous workers including Dindal (1975) have followed this. Brooks (1963) preferred the term "symbiontes" defined as "two dissimilar organisms which live together" and avoids using the term symbiosis for many microorganism/ insect relationships because one of the key points of the original meaning "being together as in a partnership" is often missing. Brooks also emphasises that much literature on symbiosis implies mutually dependent symbiosis without rigorous proof. De la Torre-Bueno (1962) defined symbiosis as "a living together in more or less intimate association, or even close union, or organisms of different species". He includes all possible relationships (even parasitism) as symbiosis. Definitions describing interactions between microorganisms and insects have been misused, misinterpreted and confused and before a particular relationship can be classified as symbiosis, ecological criteria should be considered (Dindal, 1975). Under the classification of "Interspecific Relationships" by Dindal (1975), the relationship of bacteria and fruit flies could possibly be called Trophobiosis, where the bacteria supply food and the fly provides protection for the bacteria and moves them to and from the feeding site.

Of the 10 possible relationships between insects and microorganisms outlined by Steinhaus (1954), the known bacterial associations in fruit flies could fall into several groups — free living bacteria serving directly as food; fruit flies and bacteria existing separately but in association, the fly acting as a carrier and specially culturing bacteria to ingest as food; fruit flies acting as hosts to commensal bacteria in the alimentary canal. While some aspects of the fruit fly/bacteria association could come under the original broad definition of symbiosis (De Bary, 1879), viz., "a phenomenon in which dissimilar organisms live together", it appears from present knowledge that the system consists of several diverse aspects and it is therefore not possible to place it under one definitve category. Most workers have inferred an intimate mutually dependent symbiosis essential for the survival of the fly yet to date there is no satisfactory published evidence for this and few have considered that the bacteria could be involved in the ecosystem of the fly.

Considering that the bacteria comprise proteinaceous food for adult fruit flies (and probably larvae), entirely new areas of research become exposed. Studies on the identity, distribution and abundance of the bacteria in the field, the definition and biological significance of the fruit fly/bacteria association, and the attractancy of adult fruit flies to bacteria, will require much attention and are sure to prove most rewarding. Such work would make a valuable contribution to our understanding of fruit fly biogeography, speciation, ecology and control.

6 REFERENCES

Allen, T.C. and Riker, A.J., 1932. A rot of apple fruit caused by *Phytomonas melophthora*, n. sp., following invasion by the apple maggot. Phytopathology, 22: 557–571.

Allen, T.C., Pinckard, J.A. and Riker, A.J., 1934. Frequent association of *Phytomonas melophthora*, with various stages in the life cycle of the apple maggot, *Rhagoletis pomonella*. Phytopathology, 24: 228–238.

Baerwald, R.J. and Boush, G.M., 1968. Demonstration of the bacterial symbiote *Pseudomonas melophthora* in the apple maggot, *Rhagoletis pomonella*, by fluorescent-antibody techniques. Journal of Invertebrate Pathology, 11: 251–259.

Bateman, M.A., 1972. The ecology of fruit flies. Annual Review of Entomology, 17: 493–518.

Billing, E., 1976. The taxonomy of bacteria on the aerial parts of plants. In: C.H. Dickinson and T.F. Preece (Editors), Microbiology of aerial plant surfaces. Academic Press, London, pp. 223–273.

Blakeman, J.P. and Brodie, I.D.S., 1976. Inhibition of pathogens by epiphytic bacteria on aerial plant surfaces. In: C.H. Dickinson and T.F. Preece (Editors), Microbiology of aerial plant surfaces. Academic Press, London, pp. 529–557.

Blakeman, J.P. and Fokkema, N.J., 1982. Potential for biological control of plant diseases on the phylloplane. Annual Review of Phytopathology, 20: 167–192.

Brooks, M.A., 1963. The micro-organisms of healthy insects In: E.A. Steinhaus (Editor), Insect Pathology, An Advanced Treatise, Vol. 1. Academic Press, New York, pp. 215–250.

Courtice, A.C. and Drew, R.A.I., 1984. Bacterial regulation of abundance in tropical fruit flies (Diptera: Tephritidae). Australian Zoologist, 21: 251–268.

Dean, R.W. and Chapman, P.J., 1973. Bionomics of the apple maggot in eastern New York. Search Agriculture Entomology, Geneva, 3, 62 pp.

De Bary, A., 1879. Die Erscheinung der Symbiose. K.J. Trubner, Strassburg.

De la Torre-Bueno, J.R., 1962. A Glossary of Entomology. Brooklyn Entomology Society, Brooklyn, 336 pp.

Dickinson, C.H., 1971. Cultural studies of leaf saprophytes. In: T. S. Preece and C.H. Dickinson (Editors), Ecology of leaf surface micro-organisms. Academic Press, London, pp. 129–137.

Dickinson, C.H., Austin, B. and Goodfellow, M., 1975. Quantitative and qualitative studies of phylloplane bacteria from *Loloium perenne*. Journal of General Microbiology, 91: 157–166.

Dindal, D.L., 1975. Symbiosis: Nomenclature and proposed classification. The Biologist, 57: 129–142.

Drew, R.A.I., 1987. Behavioural strategies of fruit flies of the genus *Dacus* (Diptera: Tephritidae) significant in mating and host-plant relationships. Bulletin of Entomological Research, 77: 73–81.

Drew, R.A.I. and Lloyd, A.C., 1987. The relationship of fruit flies (Diptera: Tephritidae) and their bacteria to host plants. Annals of the Entomological Society of America, in press.

Drew, R.A.I., Courtice, A.C. and Teakle, D.S., 1983. Bacteria as a natural source of food for adult fruit flies (Diptera: Tephritidae). Oecologia (Berlin), 60: 279–284.

Ercolani, G.L., 1978. *Pseudomonas savastanoi* and other bacteria colonizing the surface of olive leaves in the field. Journal of General Microbiology, 109: 245–257.

Fitt, G.P. and O'Brien, R.W., 1985. Bacteria associated with four species of *Dacus* (Diptera: Tephritidae) and their role in the nutrition of the larvae. Oecologia (Berlin), 67: 447–454.

Gibbons, L.N., 1978. *Erwinia herbicola*: a review and perspective. Proceedings 4th International Conference on Plant Pathogenic Bacteria, Angers, 1978, pp. 403–431.

Girolami, V., 1973. Reperti morfo-istologici sulle batteriosimbiosi del *Dacus oleae* Gmelin e di altri ditteri tripetidi, in natura e negli allevamenti su substrati artificiali. Redia, 54: 269–294.

Girolami, V., 1983. Fruit fly symbiosis and adult survival: General aspects. In: R. Cavalloro (Editor), Fruit flies of economic importance. A.A. Balkema, Rotterdam, pp. 74–76.

Godfrey, B.E.S., 1976. Leachates from aerial parts of plants and their relation to plant surface microbial populations. In: C.H. Dickinson and T.F. Preece (Editors), Microbiology of aerial plant surfaces. Academic Press, London, pp. 433–439.

Hagen, K.S., 1953. Influence of adult nutrition upon the reproduction of three fruit fly species. Special report on control of the oriental fruit fly (*Dacus dorsalis*) in the Hawaiian Islands, 3rd Senate of the State of California, 72–76.

Hagen, K.S., 1966. Dependence of the olive fly, *Dacus oleae*, larvae on symbiosis with *Pseudomonas savastanoi* for the utilization of olive. Nature (London), 209: 423–424.

Last, F.T. and Warren, R.C., 1972. Non-parasitic microbes colonizing green leaves: their form and functions. Endeavour, 31: 143–150.

Lloyd, A.C., Drew, R.A.I., Teakle, D.S. and Hayward, A.C., 1986. Bacteria associated with some *Dacus* species (Diptera: Tephritidae) and their host fruit in Queensland. Australian Journal of Biological Sciences, 39: 361–368.

Luthy, P., Struder, D., Jacquet, F. and Yamvrias, C., 1983. Morphology and in vitro cultivation of the bacterial symbiote of *Dacus oleae*. Mitteilungen der Schweizerischen Entomologischen Gesellschaft, Bulletin de la Societe Entomologique Suisse, 56: 67–72.

Miyazaki, S., Boush, G.M. and Baerwald, R.J., 1968. Amino acid synthesis by *Pseudomonas melophthora* bacterial symbiote of *Rhagoletis pomonella* (Diptera). Journal of Insect Physiology, 14: 513–518.

Perombelon, M.C.M., 1981. The ecology of Erwinias on aerial plant surfaces. In: J.P. Blakeman (Editor), Microbial Ecology of the Phylloplane. Academic Press, London, pp. 411–431.

Petri, L., 1910. Untersuchung über die Darmbakterien der Olivenfliege. Zentralblatt fur Bakteriologie, Parasitenkunde, Infektionskrankheiten und Hygiene, 26: 357–367.

Prokopy, R.J., 1977. Stimuli influencing trophic relations in Tephritidae. Colloques Internationaux de C.N.R.S., 265: 305–366.

Ratner, S.S. and Stoffolano, J.G., 1982. Development of the esophageal bulb of the apple maggot, *Rhagoletis pomonella* Diptera: Tephritidae); morphological, histological, and histochemical study. Annals of the Entomological Society of America, 75: 555–562.

Rossiter, M.A., Howard, D.J. and Bush, G.L., 1983. Symbiotic bacteria of *Rhagoletis pomonella*. In: R. Cavalloro (Editor), Fruit flies of economic importance. A.A. Balkema, Rotterdam, pp. 77–82.

Ruinen, J., 1961. The phyllosphere. I. An ecologically neglected milieu. Plant and Soil, 15: 81–109.

Slade, M.B. and Tiffin, A.I., 1984. Biochemical and serological characterization of *Erwinia*. Methods in Microbiology, 15: 227–293.

Stammer, H.J., 1929. Die Bakteriensymbiose der Trypetiden (Diptera). Zeitschrift für Morphologie und Ökologie der Tiere, 15: 481–523.

Starr, M.P. and Chatterjee, A.K., 1972. The genus *Erwinia*: Enterobacteria pathogenic to plants and animals. Annual Review of Microbiology, 26: 389–426.

Steinhaus, E.A., 1954. The effects of disease on insect populations. Hilgardia, 23: 197–261.

Steinhaus, E.A., 1960. Symposium: selected topics in microbial ecology. II. The importance of environmental factors in the insect-microbe ecosystem. Bacteriological Reviews, 24: 365–373.

Tsiropoulos, G.J., 1976. Bacteria associated with the walnut husk fly, *Rhagoletis completa*. Environmental Entomology, 5: 83–86.

Tsiropoulos, G.J., 1983. Microflora associated with wild and laboratory reared adult olive fruit flies, *Dacus oleae* (Gmel.). Zeitschrift für Angewandte Entomologie, 96: 337–340.

Tukey, H.B., 1971. Leaching of substances from plants. In: T.F. Preece and C.H. Dickinson (Editors), Ecology of leaf surface micro-organisms. Academic Press, London, pp. 67–80.

Yamvrias, C., Panagopoulos, C.G. and Psallidas, P.G., 1970. Preliminary study of the internal bacterial flora of the olive fly (*Dacus oleae* Gmelin). Annals of the Institute of Phytopathology Benaki, 9: 201–206.

Chapter 3.2 Oogenesis and Spermatogenesis

D. LEROY WILLIAMSON

1 INTRODUCTION

The study of gonadal development in the cyclorraphous Tephritidae is fundamental to understanding the nature of reproduction among these insects. Development of the oocyte generated by the female will proceed only after fusion with a spermatozoan from the male. Ovaries of the female, testes of the male, and associated gonoducts in each sex are paired. Contents are discharged into a median duct of cuticular origin forming the vagina in the female and the ejaculatory duct in the male. Gonadal development apparently does not govern sexual activity (Wigglesworth, 1965), but the nutritional state which directly affects activity of the corpus allatum is essential to maturation of gametes and promotion of reproductive stimuli (Bonhag, 1958; Englemann, 1968).

Anatomical organization of genitalia is generally similar among the tephritid fruit flies although some differences are of taxonomic value. Within species, details of sexual maturation are useful for ecological studies, particularly where the distinction between teneral and sexually mature insects is important (Drew, 1969). Renewed interest in cytological and histochemical descriptions of reproduction as it relates to control of economically important species of fruit flies is timely and necessary for successful application of modern techniques such as genetic control.

2 FEMALE REPRODUCTIVE SYSTEM

Structure and function of the tephrited female reproductive system received attention in the early part of this century as pest species increased in economic importance and global distribution (Back and Pemberton, 1917; Miyake, 1919; Illingworth, 1912). Since eggs are oviposited beneath the fruit surface and larvae develop inside the protective covering of the host, conventional insecticide applications were aimed at destroying adults before sexual maturity and oviposition began. Thus the pre-oviposition period, the period from emergence of adult females from puparia to maturation of eggs preparatory to oviposition, was, and remains, a matter of considerable importance.

2.1 Genital glands and ducts

Internal structures of ectodermal origin are the common oviduct, spermathecae, accessory glands, bursa copulatrix, and ventral receptacle (Fig. 3.2.1a, b).

The common oviduct, or vagina, unites the lateral oviducts from the ovaries

---

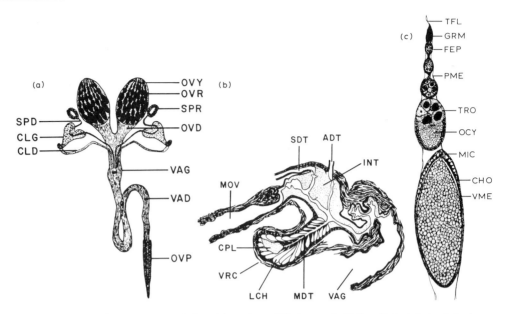

Fig. 3.2.1 (a) Reproductive system of *Ceratitis capitata* (Wiedemann). CLD, colleterial gland duct; GLG, colleterial gland; OVD, oviduct; OVP, ovipositor; OVR, ovariole, OVY, ovary; SPD, spermathecal duct; SPR, spermatheca; VAD, vaginal duct; VAG, vagina (after Hanna, 1938). (b) Saggital section of bursa copulatrix and ventral receptacle, *Rhagoletis pomonella* (Walsh). ADT, accessory gland duct; CPL, chitinous plate; INT, chitinous intima; LCH, lateral chamber; MDT, median duct; MOV, median oviduct; SDT, spermathecal duct; VAG, vagina; VRC, ventral receptacle (after Dean, 1935). (c) Longitudinal section of ovariole from mature *Rhagoletis pomonella* (Walsh). CHO, chorion; FEP, follicular epithelium; GRM, germarium; MIC, micropyle; OCY, oocyte; PME, petitoneal membrane; TFL, terminal filament; TRO, trophosome; VME, vitelline membrane (after Dean, 1935).

anteriorly and extends distally into the ovipositor. The ventral opening on the tephritid ovipositor is functionally both vulva and anus. Hence, within the basal portion of the ovipositor two tubular ducts unite. The smaller, dorsal duct is the rectum and the larger ventral one is the proximal part of the vagina.

Accessory glands are paired, pyriform sacs, and larger than the spermathecae. These glands are found near the spermathecae and each communicates to the vagina through a slender, chitinised duct. The gland appears as a translucent sac containing an opaque, whitish mass of coarsely granular secretion. The inner wall of the sac consists of a layer of large epithelial cells with a vacuole that opens into the lumen through an intercellular duct. The activity of these glands in fruit flies is not clearly known. Based on functions in other insects, they may be "colleterial" in supplying adhesive substance to the egg surface or serve some glandular function such as activating stored spermatozoa (Dean, 1935).

Spermatheca are darkly pigmented, glandular bodies where the spermatozoa are stored. Each small, irregularly shaped spermathecal gland, that lies near the base and dorsal of the ovaries, has a duct leading to the dorsal wall of the bursa copulatrix. Depending on species, 2 or 3 spermatheca may be present. Three spermathecae are found in *Rhagoletis* spp. and *Anastrepha* spp. and two spermathecae in *Dacus* spp. and *Ceratitis* spp. According to Wigglesworth (1965) three spermatheca are characteristic of higher Diptera, i.e. fewer would indicate a more primitive form.

The bursa copulatrix is a distinct organ of the vagina indicated by its greater thickness (Fig. 3.2.1b). At this site enter ducts from the spermathecae and accessory glands, as well as bands of muscle. The ventral receptacle is a protrusion from the bursa copulatrix. The flattened structure is lined with lateral chambers that unite in a median duct. The distal end of the duct terminates in a flat, pigmented plate. Dean (1935) found coiled spermatozoa in the lateral chambers of the ventral receptacle. Whether the ventral receptacle receives sperm used first in fertilizing eggs has not been determined in Tephritidae, but this function in other insects suggests the possibility.

2.2 Ovaries

In a sexually mature female, each heavily tracheated ovary consists of a series of ovarioles. Each ovariole contains a linear series of progressively developing oocytes, with a mature egg distally. The ovarioles are encased in a sac-like peritoneal membrane composed of a double sheath of endothelium and connective tissue which continues into a supportive terminal filament attached to a dorsal sclerite. The number of ovarioles in tephritid fruit flies may vary between individual females as well as between ovaries of the same individual. *Rhagoletis pomonella* (Walsh) has an average of 22 ovarioles per ovary with a range of 18 to 26 (Dean, 1935). *Ceratitis capitata* (Wiedemann) has about 28 ovarioles in each ovary (Hanna, 1938) and *Dacus tryoni* (Froggatt) approximately 44 ovarioles per ovary (Anderson and Lyford, 1965). Ovarioles are polytrophic (Grosz, 1903 cited by De Wilde, 1964) in which endofollicular trophocytes (nutritive cells) are contained with the oocyte inside a follicle formed by a single layer of cuboid cells (Fig. 3.2.1c). Hanna (1938) described ovarioles of *C. capitata* as panoistic, however, Williamson et al. (1985) shows photographically the polytrophic ovarioles in this species. Drew (1969) incorrectly recorded panoistic ovarioles in *D. tryoni* that had been shown by Anderson and Lyford (1965) to be polytrophic. Trophocyte cells are located at the anterior of the follicular cell with the oocyte at the posterior pole, characteristic of polytrophic ovarioles. The numerical relationship of the division products forming trophocytes and the oocyte is common to cyclorraphous Diptera at 15:1. Dean (1935) found 15 trophocytes associated with each developing oocyte in *R. pomonella* as did Anderson and Lyford (1965) in *D. tryoni*.

2.3 Oogenesis

Within the ovary between the terminal filament and the lateral oviduct, activity of each ovariole is marked by three distinct zones: the germarium, vitellarium and pedicel. The germarium contains oogonia enveloped in a layer of mesodermal cells. At the anterior end of the germarium differential oogonial mitoses occurs. Sixteen sister cells produced by differential divisions produce an oocyte and companion trophocytes. These cell clusters or cysts become enveloped by prefollicular tissue (Bonhag, 1958) which gives rise to the epithelial follicle. The vitellarium is the most extensive zone of the ovariole and in sexually mature female Tephritids contains a linear series of progressively developing follicles. A mass of epithilium plugs the terminal end of the vitellarium. The pedicel or ovariole stalk unites each ovariole with one of the oviducts. When ovulation occurs, the epithelial plug breaks down and a mature chorionated oocyte passes into the pedicel.

Tephritid female adults under optimum conditions usually require several days to produce mature eggs. Though all developmental divisions do not occur simultaneously in ovarioles within the ovary, a mature ovariole can be categorized into approximately 6 divisions of developing oocytes. A generalized pattern of oogenesis related to age of the fly is described for illustrative purpose as follows: beginning from the germarium of an ovariole, 2 or 3 pre-vitellogenic follicles are formed in the first day of adult life. The distinguishable oocyte cytoplasm is less granular and less basophilic than that of the trophocyte cells.

Chapter 3.2 references, p. 151

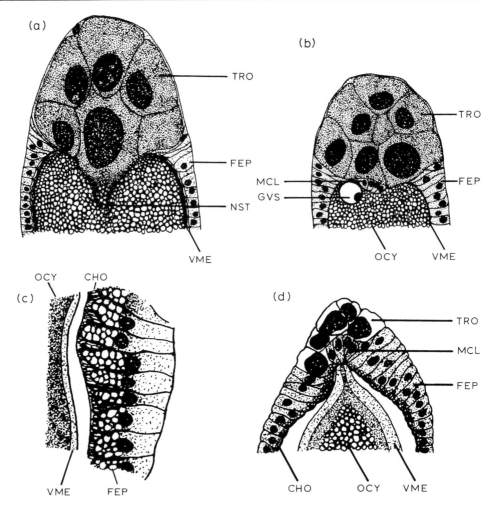

Fig. 3.2.2 (a) Anterior portion of follicle in longitudinal section showing invagination of trophocyte cells into the oocyte and discharge of nutrient. (b) Longitudinal section of anterior follicle showing position of micropyle cells between trophosome and oocyte. (c) Development of micropyle and degenerate trophocytes in anterior, longitudinal section. (d) Longitudinal section showing secretion of chorion by a group of follicular epithelial cells. CHO, chorion; FEP follicular epithelium; GYS, germinal vesicle; MCL, micropyle cells, NST, nutrient stream; OCY, oocyte; TRO, trophosome; YME, vitelline membrane (after Dean, 1935).

The oocyte nucleolus would show elongation and convolution of chromosomes. The chromosomes are Fuelgen-negative but stain intensely with mercuric bromophenol blue indicating differentiation of the germinal vesicle (Anderson and Lyford, 1965, for *D.tryoni*). By day 3, the oldest follicle, i.e. the most distal, is enlarged having completed the pre-vitellogenic growth stage denoted by large-sized trophocytes, and increase in diameter of the germinal vesicle. By day 4, a fourth follicle appears in the anterior ovariole. A rapid increase in growth of the oldest, most distal follicle occurs as vitellogenesis proceeds. Follicle cells undergo differentiation; those surrounding the trophocytes remain cuboidal. The nutritive function of the trophocyte cells can be observed in the most advanced follicle. Trophocyte cells nearest the oocyte can be observed protruding into the oocyte cytoplasm (Fig. 3.2.2a). Discharge of yolk granules occurs resulting in rapid expansion of the oocyte. The rapid process of trophocyte emptying and a flowing of contents into the oocyte is repeated, for each cell, with corresponding shrinkage. By day 5, a fourth new follice appears in the ovariole. Columnar follicle cells around the most mature oocyte deposit the vitelline membrane. Also, a distant group of follicular cells, termed micropyle cells, at the anterior of the oocyte will have invaginated between the degenerate trophocytes and the oocyte (Fig. 3.2.2b). These columnar cells contribute to the unique structure of the micropyle and the impress of these cells is retained on the secreted chorion (Fig. 3.2.2c). Relic trophocyte nuclei can be found at the anterior pole of the oocyte (Fig. 3.2.2d). By about day 6, there are 5 follicles in the ovariole, the oldest being a mature oocyte with a fully formed chorion and vitelline membrane. The egg nucleus would be in the metaphase of the first maturation division.

This generalized pattern is representative of the sequence of oogenesis observed in *R. pomenella* (Dean, 1938), *D. tryoni* (Anderson and Lyford, 1965), *C. capitata* (Hanna, 1938) and *Anastrepha ludens* (Loew). The effects of nutrition and temperature in producing wide variations, including the phenomenon of oocyte resorption are discussed under specific sections. The rates of gonadal maturation between self perpetuating laboratory colonies and feral populations is also a matter of practical importance. Mazomenos et al. (1977) noted maturation of *Anastrepha suspensa* (Loew) laboratory stock flies preceeded the wild strain by about 5 days. Although only the germarium is expected in newly emerged tephritid females (Dean, 1935; Fletcher, 1975), the presence of previtellogenic follicles in female puparia near eclosion may not be uncommon in laboratory stock flies. In *C. capitata* colonized for more than 330 generations, there was rapid development of pre-vitellogenic follicles in females prior to emergence from the puparium.

Further investigation is needed to determine the biochemical implications of selection for earlier gonadal development in laboratory colonies, particularly the female. During early growth phases of the oocyte and companion trophosome, prior to yolk deposition, mitotic division of follicular cells, for example, is most important whereas later, during vitellogenesis, cellular activity is more of change in shape and growth (Bonhag, 1958). Hence, with the initiation of vitellogenesis in an individual ovariole the germinal divisions abate and the events of oocyte enlargement become more analogous to somatic cell activity. Selection for accelerated gametic maturation can occur in artificial mass rearing systems. Vitellogenesis might commence prior to emergence from the puparium coincident with the time when gamma sterilization techniques are usually applied. Under such circumstances, it is conceivable that radiation susceptibility of vitellogenic follicles would be reduced. A result would be the appearance of potentially viable eggs in released female fruit flies to the detriment of a sterile insect program. Williamson et al. (1985) describes the recovery of viable eggs from female *C. capitata* following male sterilizing dose of ^{60}Co in a nitrogen atmosphere. The effects of gamma irradiation in both air and hypoxic atmosphere aimed at inhibiting further development of germarial products needs to be elucidated among those pest species colonized for such purposes to assure that target levels of sterility are achieved.

The linear sequence of follicular development in tephritid polytrophic ovarioles affords a sequential picture of vitellogenesis. Many aspects of the chemical nature of protein yolk bodies and whether follicular cells are involved along with the trophosome in transfer to the oocyte remain unresolved. These periodic acid-shiff reactive materials are positive in trophocytes and follicular cells during pre-vitellogenesis. Histochemical test for desoxyribonucleic (DNA) and ribonucleic acid (RNA), such as Fuelgen and Brachet's methods respectively, allow study of the yolk synthesis by trophocyte cells and passage to the oocyte. It has been suggested (Bonhag, 1958) that DNA transforms into a Fuelgen-negative substance that passes to the growing oocytes. Anderson and Lyford (1965) found no evidence of Fuelgen-positive DNA form transferred from the trophocytes to the oocyte in *D. tryoni*. The manner and temporal sequence in which proteins, lipids, DNA and RNA are synthesized and passed from the nutritive cells to the oocyte warrants further investigation. Histochemical and physiological determination would greatly benefit the understanding of sexual maturation and adaptive mechanisms.

Additional study of the physiological processes of vitellogenesis is also needed to understand the pre-ovipositional maturation time. *C. capitata* fed on protein diet in the laboratory began oviposition 4 days after emergence (Hanna, 1947). Single pairs of *A. ludens* confined in a small cage with fruit had a mean pre-oviposition of 11 days in the summer months and 24.6 days during January to March at Cuernavaca, Mexico (McPhail and Bliss, 1933). *Dacus ciliatus* Loew in Egypt required about 6 days for pre-ovipositional development in summer (27.2°C and 52.7% RH) and 10 to 14 days in winter (22.1°C and 65% RH) (El-Nahal et al., 1970). *R. pomonella* in eastern New York was found to have a pre-oviposition period of 9 to 10 days average with a minimum of 7 to 8 days (Dean, 1935). Temperature-regulated change in ovarian maturation is also a mechanism being elucidated in the Tephritidae (see Chapter 3.10 p. 273).

2.4 Mature egg

The final product of oogenesis is the mature fertilized egg which is the beginning of independent life. De Wilde (1964) noted that activity of the follicle

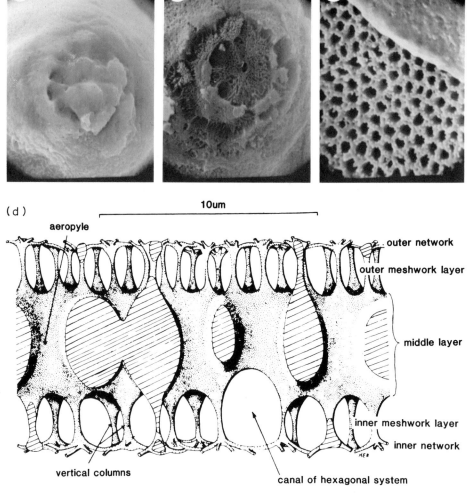

Fig. 3.2.3 (a–b) Apical end of *Dacus cucurbitae* Coquillett egg with centrally located micropyle orifice bedore and after removal of outer chorionic surface. (c) Meshwork layer on lateral region of *D. cucurbitae* egg beneath chorionic surface layer. (d) Schematic diagram of a muscid fly chorion to illustrate structure diversity (after Hinton, 1981).

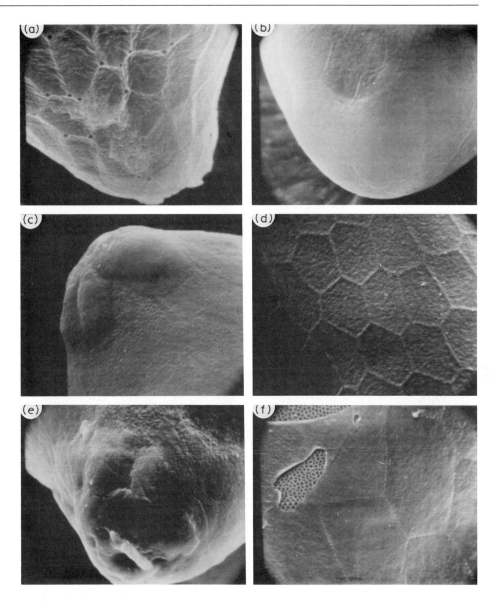

Fig. 3.2.4 (a–b) Micropyle end and distal end of *Ceratitis capitata* (Wiedemann) egg. (c–d) Micropyle end and lateral surface sculpturing of *Dacus dorsalis* Hendel egg. (e–f) Micropyle end and lateral surface area of *Dacus cucurbitae* Coquillett egg. Scanning electron microscopy X1900.

cells in secreting the variety of sculpturing and substances found in insect egg chorion is undoubtedly one of the most complicated features of oogenesis. In a voluminous work on the biology of insect eggs, Hinton (1981) commented on the remarkable lack of investigative reports on the egg stage of economically important tephritid fruit flies. Part of the difficulty appears to be in improving techniques to study this delicate stage. Although fruit fly eggs are well adapted to their natural, almost aquatic, environment, they are a fragile life form for experimentation. Intricate chorionic structural adaptations exist which combine the need for rigidity of the egg with porous, gas-filled networks (Fig. 3.2.3a–d). These structures allow diffusion of respiratory products through the egg surface while regulating moisture and other factors. Also, maternal secretions from colleterial glands could be involved in the physical communication of the developing embryo with its environment, all of which merit investigation. The economics of mass colonization of tephritid fruit flies, required for sterile insect technique programs, and improvement of insect quality should derive immediate benefit from increased knowledge of egg stage development. Further, the taxonomic value of tephritid egg morphology is virtually unexplored although differences exist amoung species (Fig. 3.2.4a–f).

3 MALE REPRODUCTIVE SYSTEM

The male tephritid initiates delivery of the spermatozoan to the mature oocyte. The female, however, controls the process by several means of non-response to mating signals from the male, tactile rejection or escape. Earlier sexual maturation of males than females of the same age, often by several days, is evident from laboratory observations of attempted matings by males with unreceptive females. The male abdominal cavity is generally smaller in each species and lacks an elongated eighth abdominal segment that sheaths the ovipositor and readily distinguishes the female.

3.1 Testes

The paired testes of the male are separated organs that lie longitudinally within the abdominal cavity. They are readily identified by the pyriform shape and yellow pigmentation of the outer connective tissue cells. The apical end of each testis is usually bent slightly, and the main body surrounded by a thickened layer of fat body (Fig. 3.2.5a). Unlike the female ovary which undergoes enormous expansion during maturation of oocytes, the male testis is a more rigid structure with contents in various developmental zones is contingent on developmental stage and sexual activity. Tubular follicles of the testes, analogous to ovarioles in the female and correspondingly pronounced in some insects (see Wigglesworth, 1965) are unrecognized in the male tephritid. A succession of developmental zones is discernable, however, in whole mounts of testes. These distinct zones apparent in the intact testis from the apical end to

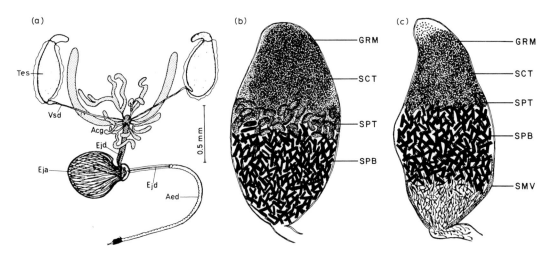

Fig. 3.2.5 (a) Male reproductive system of *Dacus tryoni* (Froggatt) Acg, accessory glands; Aed, aedeagus; Eja, ejaculatory apodeme; Ejd, ejaculatory duct; Tes, testes; Vsd, vas deferens (after Drew, 1969). (b) Testis from 1-day old male *Ceratitis capitata* (Wiedemann) showing sperm in sperm bundle stage of development. (c) Testis from sexually mature 8-day old *Ceratitis capitata* (Wiedemann) showing free sperm in lower end of testis. GRM, germarium; SCT, spermatocytes; SMV, seminal vesicle; SPB, sperm bundles; SPT, spermatids (after Anwar et al., 1971).

the vas deferens are: the germarium, the sperm bundles and free swimming spermatozoa. Spermatophores are not formed.

The germarium or zone of spermatogonia consists of densely packed primordial germ cells surrounded by somatic mesodermal cells. Within this zone are found spermatocytes undergoing differential mitoses and reduction divisions into spermatocytes and spermatids. The zone of sperm bundles is discernible in the mid-area of a mature testis and lastly a zone of flagellated spermatozoa are visible at the base of the testis where they are contained in the seminal vesicle.

3.2 Accessory glands and ejaculatory organ

The accessory glands are 4 pairs of tubular structures, varying in length. Usually one pair is elongated and may be thicker than the others, either slightly curved or convoluted. The remaining pairs are shorter and may be variously branched in some species. All are joined by ducts to the apical end of the seminal vesicle. Ducts from the testes, the vasa deferentia, are rather long ducts that join the apical part of the seminal vesicle.

A common ejaculatory duct or vas efferens leads from the seminal vesicle to the sclerotized and muscled erecting and pumping organ. As the nomenclature indicates, the function of this distinct sac-like organ is to erect the long, coiled adeagus and to force, by changes in sac volume, the spermatozoa and seminal fluid through the duct of the adeagus.

3.3 Spermatogenesis

At the apical end of the testis, primary spermatogonial cells differentiate and after successive mitotic divisions produce a cyst of spermatocyte cells enclosed in an epithelial sheath (Anwar et al., 1971). Formation of nutritive cells, analagous to the trophocyte cells in the female, have not been shown in male fruit fly testes, but it is possible that such a mechanism exists within the spermatocyte cluster (see De Wilde, 1964). Primary spermatocytes divide meiotically and resulting secondary spermatocytes divide into 2 pre-spermatids. The total complement of each sperm bundle originates from a single spermatogonial cell.

Spermatogenesis is frequently in the final stages of spermatozoan transformation by the time the adult male emerges from the puparium. This cyclical process continues through the reproductively active life of the fly. *C. capitata* and *Dacus dorsalis* Hendel have a similar pattern of sexual maturation in relation to the adult ages as follows: At day one, the mature spermatogonial cells in the germarium have differentiated into spermatocyte cells and spermatid cysts. This growth zone comprises about 1/3 of the testicular sheath content. Sperm bundles fill the remaining testicular cavity (Fig. 3.2.5b). At day two, the proportion of the testes occupied by the germarium and sperm bundles remains fairly constant although more sperm bundles have been generated but become more densely packed. Loosening of sperm bundles can be observed although no free sperm may be present (Anwar et al., 1971). By day four, free sperm appear at the distal end of the testes, decreasing the space occupied by sperm bundles. The number of spermatocytes and spermatids increase in the germinal zone as a result of continuing spermatogenesis. From day 5 or 6 onwards in the sexually mature male testes (Fig. 3.2.5c) the germarial, sperm bundle and spermatozoan zones are fairly proportionate depending on the quantity of spermatozoa ejaculated from the seminal vesicle. Sperm and seminal plasma are opaque, cream colored fluid. The exact chemical nature in terms of pH, relative proportions of DNA, proteinaceous material and carbohydrates has not been determined.

Chapter 3.2 references, p. 151

4 HORMONAL REGULATION OF GONADAL DEVELOPMENT

Hormonal stimuli fuel the reproductive process. It is a cyclical process throughout the adult life for which mechanisms of restraint exist as may be imposed by periods of nutritional or climatic stress. Since reserves for sexual development come from the larva, then larval nutrition can affect egg production in the female and perhaps insemination capacity of the male. Insufficient food or low quality of diet will reduce the number of eggs oviposited. However the Dacinae do not produce eggs unless fed protein at the adult stage. Also, in the adult, when the corpus allatum again becomes active, nutrition plays an important role in the activity of this gland in governing oocyte maturation as well as male accessory gland development. Wigglesworth (1964) summarizes the complexity of the study of hormonal regulation. A system of neurosecretory cells in the protocerebrum and sub-oesphageal ganglion, the corpus allatum, corpus cardiacium and the reproductive organs are all involved. These components interact by nervous stimuli, by neurosecretions, and by humoral factors in the circulating blood. The relationships of these elements which vary among insects have not been defined in tephritid fruit flies. An understanding of the endocrine system and its response to chemical stimuli would be a productive and useful area of study.

5 SYMBIOTIC RELATIONSHIPS

The mutual adaptation of microorganisms with some fruit flies e.g. *Dacus oleae* (Gmelin) and mechanisms for transmission to offspring has long been known (Stammer, 1929 cited in Wigglesworth, 1964) *D. oleae* has developed specific integumental invaginations for harboring such organisms. Steinhaus (1967) illustrates and describes structures in *D. oleae* female reproductive system whereby bacteria are transmitted to the egg surface and to the embryo via the micropyle. Organisms can be passed from the hindgut into the ovipositor where they either adhere to the chorion or enter the micropyle during oviposition. Earlier chapter 3.1.2 describes these relationships.

6 CONCLUSION

It is clear that a detailed knowledge of the processes of oogenesis and spermatogenesis among tephritid fruit flies is of immediate value scientifically and practically. Methods that interrupt reproductive functions within the insect, such as the sterile insect technique, offer viable containment and control opportunities. Major areas of research would include: any environmental influence on oogenesis and spermatogenesis and adaptive mechanisms, such as oocyte resorption; definition of requirements for the delicate, immobile egg stage both in its natural host environment as well as under large-scale rearing conditions; the role of microorganisms in reproduction; nutritional influence on the biochemical and physiological processes of oogenesis and spermatogenesis; functional definition and physiological analysis of ovipositor sensillae; implications of selection pressure toward more rapid sexual maturation of colonized species used for sterilization and release. Progress can be expected to accelerate through application of modern technology. Scanning electron microscopy, electrophoretic procedures, and advanced biochemical analyses offer excellent opportunities to expand the knowledge of hormonal regulation, cellular function, and biochemistry of fruit fly reproduction. Such fundamental knowledge may also have an important role in fruit fly systematics and genetics as well as in elucidating mechanisms of spread and adaptation to food cultivars.

7 REFERENCES

Anderson, D.T. and Lyford, G.C., 1965. Oogenesis in *Dacus tryoni* (Frogg.) (Diptera: Trypetidae). Australian Journal of Zoology, 13: 423–435.

Anwar, M., Chambers, D.L., Ohinata, K. and Kobayashi, R.M., 1971. Radiation sterilization of the Mediterranean fruit fly (Diptera: Tephritidae): Comparison of spermatogenesis in flies treated as pupae or adults. Annals of the Entomological Society of America, 64: 627–633.

Back, E.A. and Pemberton, C.E., 1917. The melon fly in Hawaii. U.S. Dept. Agric. Bull. No. 491.

Bonhag, P.F., 1958. Ovarian structure and vitellogenesis in insects. Annual Review of Entomology, 3: 137–160.

De Wilde, J., 1964. Reproduction In: M. Rockstein (Editor), The Physiology of Insects. Academic Press, New York, pp. 9–90.

Dean, R.W., 1935. Anatomy and postpupal development of the female reproductive system in the apple maggot fly, *Rhagoletis pomonella* Walsh. New York State Agric. Exper. Sta. Tech. Bull. No. 229. 31 p.

Drew, R.A.I., 1969. Morphology of the reproductive system of *Strumeta tryoni* (Froggatt) (Diptera: Trypetidae) with a method of distinguishing sexually mature adult males. Journal of the Australian Entomological Society, 8: 21–32.

El-Nahal, A.K.M., Azab, A.K. and Swailem, S.M., 1970. Studies on the biology of the melon fruit fly, *Dacus ciliatus* Loew. Bulletin Société Entomologie Egypte, 54: 231–241.

Englemann, F., 1968. Endocrine control of reproduction in insects. Annual Review of Entomology, 13: 1–26.

Fletcher, B.S., 1975. Temperature-regulated changes in the ovaries of overwintering females of the Queensland fruit fly, *Dacus tryoni*. Australian Journal of Zoology, 23: 91–102.

Hanna, A.D., 1938. Studies on the Mediterranean fruit fly: *Ceratitis capitata* Wied. I. The structure and operation of the reproductive organs. Bulletin de la Societé Foread Ier D'Entomologie, 22: 39–52.

Hanna, A.D., 1947. Studies on the Mediterranean fruit fly, *Ceratitis capitata* Wied. II. Biology and Control. Bulletin de la Société Foread Ier D'Entomologie, 31: 251–285.

Hinton, H.E., 1981. The Biology of Insect Eggs. Pergamon Press, Oxford, Vols. I–III, 1125 pp.

Illingworth, J.F., 1912. A study of the biology of the apple maggot *(Rhagoletis pomonella)*, together with an investigation of methods of control. Cornell Univ. Agric. Expt. Sta. Bull. No. 324.

Mazomenos, B., Nation, J.L., Coleman, W.J., Dennis, K.C. and Esponda, R., 1977. Reproduction in Carribbean fruit flies: Comparisons between a laboratory strain and a wild strain. The Florida Entomologist, 60: 139–144.

McPhail, M. and Bliss, C.I., 1933. Observations on the Mexican fruit fly and some related species in Cuernavaca, Mexico, in 1928 and 1929. U.S. Dept. Agric. Cir. No. 255, 24 pp.

Miyake, T., 1919. Studies on the fruit flies of Japan: I. Japanese orange fly. Imp. Centr. Agric. Expt. Sta. Bull. No. 2 pp. 85–165.

Steinhaus, E.A., 1967. Insect Microbiology. Hafner Publishing Co. 763 p.

Wigglesworth, V.B., 1964. The hormonal regulation of growth and reproduction in insects. In: J.W.L. Beament, J.E. Treherne and V.B. Wigglesworth (Editors), Academic Press, New York, pp. 247–336.

Wigglesworth, V.B., 1965. The Principles of Insect Physiology. Butler and Tanner Ltd, Frome, 741 pp.

Williamson, D.L., Mitchell, S. and Seo, S.T., 1985. Gamma irradiation of the Mediterranean fruit fly (Diptera: Tephritidae): Effects of puparial age under induced hypoxia on female sterility. Annals of the Entomological Society of America, 78: 101–106.

Chapter 3.3 The Effect of Ionizing Radiation on Reproduction

G.H.S. HOOPER

1 INTRODUCTION

Over the past two decades there has been considerable interest in the application of the sterile insect technique (SIT) for the eradication and/or control of pest tephritid species. The ability to induce high levels of sterility, with a minimum deterioration in fitness, in the target species is a prerequisite for any SIT program. When reproductive cells are exposed to ionizing radiation (X and γ radiation, neutrons, alpha particles) dominant lethal mutations (DLM) result (Muller, 1927). These can be defined as chromosomal changes which need be present in only one gamete to prevent maturation of the zygote. The occurrence of DLM need not prevent the maturation of the gamete, nor affect zygote formation. In Diptera DLM can result from loss of chromosomal fragments and/or the formation of acentric and dicentric chromosomes. The net result is chromosomal imbalance in the zygote which, in Diptera, usually dies during the early cleavage divisions. While induction of DLM may be the primary and preferred basis for sterility in irradiated insects, sterility can arise from other causes, e.g. aspermia or sperm inactivation in male insects and infecundity in females. In tephritids spermatogenesis is well advanced by the late pupal stage when irradiation is usually carried out and hence aspermia is unlikely to occur. However, aspermia may occur secondarily as a result of depletion of sperm (carrying DLM) after a variable number of matings, if the radiation dose was sufficient to eliminate early spermatogonial cells from which sperm could subsequently develop. Females of most tephritid species are more radio-sensitive than males and hence, at doses required for the induction of DLM in sperm, females are usually infecund if mature pupae are irradiated.

The incidence of DLM in sperm is usually determined by measuring the hatch of eggs from untreated (U) females mated with irradiated (I) males. The dose–egg hatch response curves are similar in all tephritid species so far studied and suggest that "one-hit" ionising events only are required for the induction of DLM (La Chance and Graham, 1984). Figure 3.3.1 shows this situation for two species. While initially the dose-response curve is linear, it becomes non-linear when the percentage of DLM reaches ca. 80%. This arises because as the dose increases, progressively more sperm are exposed to more than one ionizing event and any DLM beyond the first is superfluous. With females the situation may be different. Limited data (Fig. 3.3.1) suggest that more than one ionizing event may be required for induction of DLM in oocytes.

While early work on irradiation utilised X-radiation, more recent studies have utilised γ-radiation, usually from Co 60 (energy 1.17 and 1.33 MeV) and less frequently from Cs 137 (energy 0.66 MeV) sources. Dose data in the literature are reported in "roentgen" and "rad'". The latter is the more useful for

Chapter 3.3 references, p. 162

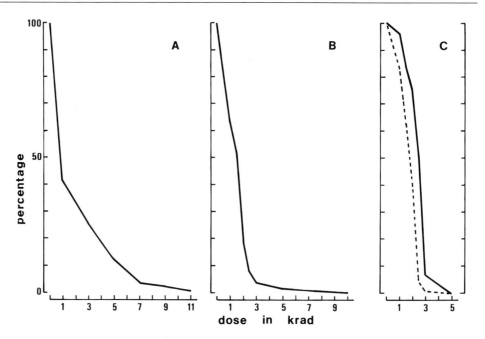

Fig. 3.3.1. Sterility-dose response relationships for: (A) irradiated male *C. capitata* (Hooper, 1971c);
(B) irradiated male *D. tryoni* (Bhatti, 1970); (C) irradiated female *D. tryoni*, fecundity - solid line,
fertility - dotted line (Bhatti, 1970).

radiobiological work since it refers to the energy absorbed by tissue, and with
soft tissue the roentgen and rad are roughly equivalent. In this paper all doses
have been expressed in rads since that is the unit most commonly reported even
though now the current international unit of absorbed dose is the gray – 100
rad being equivalent to 1 Gy.

2 BIOLOGICAL CONSIDERATIONS

While irradiation of larvae and young pupae will result in sterility in the
subsequent adult stage, such treatment usually also causes somatic damage
which may manifest itself as prolonged developmental times, decreased eclosion and survival of adults, and possibly aspermia in males. Hence, it is prefer-
able to irradiate mature pupae or young adults to optimalise the yield of sterile
adults. With this approach there is a margin between the dose required to
induce sterility and that which causes severe deleterious effects, e.g. with
Mediterranean fruit fly, *Ceratitis capitata* (Wiedemann), while doses of 8–
10 krad will induce a high level of sterility, eclosion from irradiated mature
pupae and longevity of the emerging adults is not significantly affected until
doses of 20 krad are used (Katiyar and Valerio, 1964). However, even with doses
which induce sterility in the 95–100% range there are effects on the competitiveness of males and these effects are dose-dependent (Hooper and
Katiyar, 1971).

Spermatogenesis is well advanced in pupae 1–2 d before adult eclosion with
spermatids and spermatozoa already differentiated. With doses that induce a
high level of DLM in spermatids and spermatozoa, death of pre-meiotic stages
and atrophy of the germinal tissue is usual, and thus regeneration of unda-
maged spermatozoa from spermatogonia, which could be manifested as a recovery of fertility, is rare. Young adults will contain more spermatozoa than
mature pupae and irradiation at the former stage will produce males capable
of transferring more sperm (Ohinata et al., 1971), and to an increased number
of females, before aspermia intervenes. The female reproductive system of

tephritids is of the polytrophic type. When mature pupae are irradiated the number of eggs laid by females is reduced with increasing dose, and dominant lethal mutations may be present in those eggs which are produced. At doses causing high levels of sterility in males, females are usually infecund. Ovipositional drive is significantly reduced, but not eliminated so that "blind stings" can occur on fruit. Both developing oocytes and trophocytes could be damaged, but once endomitosis has occurred the trophocytes are more radio-resistant. As the time of adult eclosion approaches, oogenesis proceeds rapidly (Williamson et al., 1985) so that following irradiation of young adult females a few eggs derived from mature oocytes are laid before infecundity occurs.

While most tephritid females are not monogamous the incidence of a second mating is usually low if the initial mating has been satisfactory in terms of amount of sperm and/or accessory material transferred. With *C. capitata* the urge to remate appears to be correlated with the amount of sperm within the spermathecae (Nakagawa et al., 1971), and females mated with irradiated males are more likely to remate than those females mated by untreated males (Katiyar and Ramirez, 1970). Further work established that the amount of sperm transferred by males is related to the duration of mating and that a copulation period of ca. 90 min was necessary for untreated males to transfer sufficient sperm to fill the spermathecae of females (Farias et al., 1972). Hence, irradiated males which are to function appropriately in SIT programs must have sufficient sperm (carrying DLM) to fill the spermathecae of their female partners, and the act of mating must be long enough to transfer the required number of sperm. Ideally an irradiated male should have sufficient sperm to successfully mate a number of females.

There was no evidence for any sterility in F_1 offspring of irradiated male and untreated female matings when the males were given sub-sterilising doses (Ohinata et al., 1971; Teruya, 1983). Neither is there any evidence that mating with a sterile male affects the number of eggs produced by an untreated female.

With tephritids male pheromones play an essential role in mating in the field. Fletcher and Giannakakis (1973) found that doses up to 20 krad delivered to mature pupae did not affect pheromone production by male *Dacus tryoni* (Froggatt). However, in *C. capitata* Zumreoglu et al. (1979) found that 10 krad to mature pupae significantly reduced pheromone production, while 14 krad in a nitrogen atmosphere did not. In both these cases pheromone production and/or release was assayed by monitoring the response of females to the calling males.

The competitiveness of irradiated males, i.e. their ability to effectively interact with, and reduce the reproductive potential of, the target population, is clearly of considerable importance. The preferred laboratory approach to evaluate this parameter involves "ratio tests" in which irradiated males are combined with untreated males and females in various ratios and the resultant egg hatch is measured. Haisch (1970) and Fried (1971) developed identical formulae to derive a competitiveness value (which can vary from 0–1) from the experimental data, and Hooper and Horton (1981) reported a procedure to determine the variance associated with the competitiveness value. The egg hatch derived from these ratio tests reflects the interaction of a number of factors: mating ability, sperm complement, sperm competitiveness, multiple mating, sequence of mating. Other factors which would be of importance in the field, e.g. courtship behaviour, lekking behaviour, and flight ability clearly cannot operate in such a laboratory procedure. Nevertheless laboratory ratio tests are, and will continue to be, used to give an indication of the competitiveness of irradiated male insects. Evaluation of the irradiated males against wild insects in field cages is an essential subsequent step but it must be recognised that in this situation the competitiveness of the males will be determined by two factors: laboratory colonisation as well as the irradiation procedure.

Chapter 3.3 references, p. 162

It has been clearly established that competitiveness of irradiated males is inversely related to the irradiation dose, while the level of sterility is of course directly related to dose (Hooper and Katiyar, 1971; Hooper, 1972). Hence it is essential that if the competitiveness of males sterilised by different treatments, or of males irradiated at different developmental stages, is to be compared then those males must exhibit equivalent levels of sterility. There is evidence, with *C. capitata* pupae, that competitiveness is affected less as the time between irradiation and eclosion decreases (Hooper, 1971c). Further, there is evidence, again with *C. capitata*, that irradiation of young adults gives a more competitive male than does irradiation of mature pupae (Hooper, 1971a; Anwar et al., 1975). However, this was not so for *Dacus cucurbitae* Coquillett (Ohinata et al., 1971).

While sperm inactivation usually only occurs with doses much greater than those causing high levels of sterility, the possibility of reduced competitiveness of sperm must be considered since this could affect the mating response of females. This aspect can be explored by mating females sequentially with irradiated and untreated males and recording the hatch of eggs after each mating. Sperm precedence of first and subsequent matings would also be revealed. With *Rhagoletis cerasi* Linnaeus, *C. capitata*, *Dacus cucumis* French, and *D. cucurbitae* at least, sperm precedence does not occur; the results obtained after 2 matings indicate rather that sperm mixing occurs in the spermatheca. However, the impact on egg hatch of an irradiated male mating following an initial mating with an untreated male is usually less than the impact of a mating with an untreated male following a mating with an irradiated male. This suggests that while sperm mixing occurs, and that while sperm from both matings are subsequently used, the sperm from irradiated males are somewhat less competitive than those from untreated males.

3 SPECIES INVESTIGATED

3.1 *Dacus dorsalis* Hendel

In early field tests US workers irradiated mature pupae with 10 krad. Since Wong et al. (1982) report that irradiation of pupae 2 days before eclosion with 10 krad (Co 60, 4–5 krad/min) in an atmosphere of nitrogen caused 99% sterility in males, a 10 krad dose in air would appear excessive. Data from the Philippines indicated that when pupae 2 days pre-eclosion were irradiated (Co 60) a dose of 5 krad caused infecundity in females and 99% sterility in males (E.C. Manoto, personal communication, 1976). However, recent taxonomic research has indicated that the major species attacking mango in the Philippines is *Dacus occipitalis* (Bezzi) and queries whether *D. dorsalis* occurs there (Drew, 1983). Hence the relevance of the above data to *D. dorsalis* must be suspect. In connection with an eradication in the Ogasawara Islands, Japanese workers used a 9 krad dose which was fractionated into doses of 5 and 4 krad delivered to pupae 3 and 2 days before eclosion (Habu et al., 1984). This treatment completely sterilised both sexes. Taiwanese workers found that males irradiated (Co 60, 1.78 krad/min) 1–2 days pre-eclosion in nitrogen with 15 krad were essentially sterile (0.2% egg hatch) and were much more competitive than males irradiated with 13 krad in air (zero egg hatch) – competitiveness value of 0.26. Even at 21 krad which gave 0.05% egg hatch, competitiveness was possibly greater (0.35) than in the air treatment (Chang and Lee, 1984).

3.2 *Dacus tryoni* (Froggatt)

Early work with *D. tryoni* reported the sterilising dose to be from 4–12 krad when mature pupae were treated. However, detailed work by Bhatti (1970) showed that when 8-day old pupae, i.e. 2 days pre-eclosion, were irradiated, females were rendered sterile, through infecundity, at 5 krad, while corrected egg hatch from irradiated males was > 1% at 7.5 krad and zero at 10 krad.

3.3 *Dacus cucurbitae* Coquillett

Japanese workers found that, following irradiation (Co 60, 333 rad/min) of pupae 1–2 days before adult eclosion, 6 krad produced ca. 99% sterility in males and females were infecund (Teruya et al., 1975). Using a large number of doses Anwar et al. (1975) found that when pupae were irradiated 2 days before eclosion (Co 60, 2.5 krad/min) 6 krad caused complete sterility in both males and females. However, when 2 day old adults were irradiated the sterilising dose for males had to be increased to 8 krad. Both groups of workers have found that at the 98–99% level of sterility males were fairly competitive (C values of 0.67–1.00).

3.4 *Dacus oleae* (Gmelin)

When irradiated (Co 60, 700 rad/min) 2–3 days before adult eclosion sterility of ca. 99% in males was achieved with 8 krad, while females were infecund after receiving 6 krad (Tzanakakis et al., 1966). Economopoulos (1972) concluded that, with a dose of 8 krad, adults irradiated 2 days after eclosion were more competitive than adults arising from pupae irradiated 1–2 days before emergence. However, since (a) data in this paper suggest that 8 krad gave somewhat less sterility in adult irradiation than quoted for pupal irradiation by Tzanakakis et al. (1966); and (b) it has been established that competitiveness is related to the level of sterility (Hooper, 1972) this conclusion should be re-tested. Adults eclosing from "advanced" pupae irradiated (Co 60, 3.98 krad/min) with 8 krad were able to transfer sperm for 5–11 matings before aspermia occurred (Tsiropoulos and Tzanakakis, 1970).

3.5 *Dacus ciliatus* Loew

A dose of 8.5 krad (Co 60, 2.82 krad/min) is reported as causing complete sterility in males, and in females (through infecundity), when late stage (7 day old) pupae were irradiated (Huque and Ahmad, 1969).

3.6 *Dacus zonatus* Saunders

Huque and Malik (1967) found when pupae were irradiated (Co 60, 3.55 krad/min) just before eclosion, 9 krad caused complete sterility in both sexes with no drastic effect on longevity. These results were confirmed in a further study by Qureshi and Bughio (1969) who found that irradiation (Co 60, dose rate stated to be 560 krad/min, which seems very high) of 6 day old pupae with 8 krad gave complete sterility in males. In females 2 krad gave 94% reduction in hatch of eggs and no effect on fecundity, 4 krad permitted production of a few infertile eggs and 6 krad caused infecundity. However, at ratios of 3–5/1/1 (I ♂/U ♂/U ♀) male competitiveness was low (ca. 0.2) at 9 krad and this may relate to the maturity of the pupae at irradiation. On the other hand when 1 day old adults were irradiated (Co 60, 80 rad/min) 12 krad was required to produce less than 1% egg hatch from treated males, and longevity was affected at all doses (9–12 krad) (Qureshi et al., 1974). The reason for the much larger sterilising dose in this work on adult vs. the earlier pupal irradiation, may lie in the dose rate; at 80 rad/min a dose of 9 krad would require 112 min and 12 krad would require 150 min. Unless precautions were taken, and this is not indicated, the

Chapter 3.3 references, p. 162

development of anoxia during such long exposures could result in a significant increase in the dose required for sterility.

3.7 *Dacus cucumis* French

When 1-day old adult *D. cucumis* were irradiated (Co 60, 9.3 krad/min) ca. 90% sterility in males was obtained with 9 krad. With females, hatch of eggs was eliminated at 5 krad and infecundity occurred at 6 krad. The competitiveness of males receiving 9 krad was only 0.3, mating capability was significantly decreased, but longevity was not adversely affected (Hooper, 1975a).

3.8 *Ceratitis capitata* (Wiedemann)

Irradiation of pupae 1–2 days prior to eclosion with 3–4 krad causes infecundity in females (Hooper, 1971c; Katiyar, 1962). As the interval between irradiation and eclosion decreases there is more likelihood of some eggs being produced (Williamson et al., 1985). When 2–6 h old adults were irradiated (Co 60, 8 krad/min) a reduced number of eggs with low hatch was produced even with a dose of 9 krad (Hooper, 1971b). With males, irradiation (Co 60, 8 krad/min) of pupae 2 days pre-eclosion with a dose of 9–11 krad was required to produce 99% sterility, and a similar situation exists if young adults are irradiated (Hooper, 1971a). Due to the wide distribution and pest status of *C. capitata*, there is a considerable body of data on radiation induced sterility. While the doses required to produce high levels of sterility in males do not differ dramatically, the slope of the sterility-dose response curve does differ in various reports (Fisher, 1984). Whether this difference is due to dosimetry, dose rate, rearing procedures, or strain differences is unknown.

3.9 *Ceratitis rosa* Karsch

A dose of 8 krad gave 98% sterility in males when pupae were irradiated (Cs 137, 1.5 krad/min) 1 day prior to eclosion, and a dose of 4 krad caused females to be infecund (Anon., 1982).

3.10 *Rhagoletis cerasi* Linnaeus

Early work with *R. cerasi* involved pupal irradiation. However, there is a pupal diapause in this species and emergence of adults is prolonged once the diapause is broken. Hence it is difficult to irradiate pupae at a defined developmental stage. Nevertheless Haisch and Boller (1971) have reported work on pupal irradiation (50% of the flies emerged within 4 days of irradiation). Irradiation (Cs 137, 70 rad/min) of males gave a corrected egg hatch of 1.7% at 8 krad. With females a 98% reduction in fecundity was achieved by doses of 2–8 krad and zero hatch of eggs was achieved at 8 krad. The competitiveness of males (pupae irradiated with Co 60 source, dose rate ca. 1.4 krad/min) at a sterility level of 93% (5 krad) was 0.63 and this level did not change when pupae were irradiated from 2–8 days prior to eclosion.

Subsequent work has involved irradiation of young adults to overcome the protracted eclosion from diapause pupae and because a level of fecundity is desirable to ensure appropriate mating behaviour in the released females. With doses of 2–8 krad (Co 60, 5 krad/min) to 2 day old females fecundity was reduced by 60% and increasing the dose to 20 krad did not cause any further reduction. *R. cerasi* is unusual in that in terms of egg hatch, males are more radiosensitive than females; sterility of > 99% was achieved by 6 krad in males but by 8 krad in females. With a dose of 10 krad male competitiveness was 0.95; at 12.5 krad and above competitiveness was drastically reduced. At 10 krad, aspermia in irradiated males did not occur in eight successive matings, neither longevity nor flight ability were impaired, but mating capacity and speed of mating decreased somewhat (Boller et al., 1975).

3.11 *Rhagoletis pomonella* (Walsh)

When *R. pomonella* pupae 1 day prior to eclosion were irradiated (Cs 137) a dose of 3 krad gave 99.9% sterility in males and infecundity in females (Myers et al., 1976).

3.12 *Anastrepha ludens* (Loew)

Rhode et al. (1961) found that when pupae were irradiated 4 days before emergence (Co 60, < 100 rad/min) 4 krad produced complete sterility in males and zero hatch of eggs produced by irradiated females. At 5 krad irradiated females were infecund. However, at 5 krad the data indicate that the males were very poorly competitive. On the other hand Velasco and Enkerlin (1982) found that 4 krad (Co 60) induced 90% sterility while 10 krad was required to produce ca. 99% sterility. When both irradiated males and females were combined with untreated adults in a 9/1 ratio, 4 krad gave the lowest egg hatch of a series of doses ranging from 1–12 krad. An approximate calculation again suggests that competitiveness was poor at the 90% sterility level (ca. 0.18).

3.13 *Anastrepha suspensa* (Loew)

When 0–6-day old adults were irradiated > 99% sterility was achieved in males and females with 5 krad. Fecundity of irradiated females was reduced by 96 and 99% when irradiated with 3–5 and 6 krad respectively, and hatch of any eggs produced was prevented by 6 krad. However, longevity of both sexes was markedly reduced by doses of 3–6 krad. In subsequent work infecundity was achieved with doses of 4–8 krad when females were irradiated as mature pupae or young adults. When irradiated as 10-day old pupae (the pupal period has been reported as 12 days at 27°C and 14 days at 26°C) 4 krad gave complete sterility in males, while 8 krad was required for 12 day old pupae and 1 day old adults (Burditt et al., 1975).

4 IRRADIATION PARAMETERS

4.1 Dose rate

Earlier it was stated that the sterility-dose response curves for irradiated males indicated that DLM are due to one-hit ionizing events. If this conclusion is correct the rate at which the dose is delivered might not be expected to affect the level of sterility produced by a given dose. However, the evidence on this point is equivocal. A review of reported work with *C. capitata* did not indicate any major effect of dose rates which varied by factors of 4 for Co 60 and 13 for Cs 137 (Hooper, 1970). In *D. cucumis* there was no significant effect on sterility of dose rates which varied by a factor of 11 (0.87–9.51 krad/min) (Hooper, 1975b). However, Wakid (1982) reported a significant effect of dose rate which varied by a factor of ca. 14-fold at doses of 7 and 11 krad with *C. capitata* (the higher dose rate being more effective), and work with *A. ludens* (Rhode et al., 1961) suggests a similar phenomenon (dose rate varied 10-fold). However, when low strength irradiation sources are used the problem of anoxia developing during

Chapter 3.3 references, p. 162

the long exposures involved with low dose rates must be borne in mind; in the
A. ludens study the irradiation times for the highest and lowest dose rates were
22 and 200 min, respectively. Unless special precautions were taken anoxia
could have been involved in the reduced sterilising effect of the low dose rate
treatments. In any event any dose rate effect would be of value only if there was
a worthwhile effect on the competitiveness of irradiated males; there is no
evidence from studies on two species that this is so (Hooper, 1975, Wakid et al.,
1982).

4.2 Dose fractionation

While with mammalian cells fractionation of a given dose leads to less
damage than if the same dose is given at one time, presumably due to repair
mechanisms operating during the intra-dose interval, no such effect of fractionation on sterility in *C. capitata* was found. While male competitiveness was
lower when a 9 krad dose was fractionated over 3 days (delivered 4, 3 and 2 days
before eclosion) than when the 9 krad dose was given 2 days before eclosion,
this effect could occur because pupae were irradiated at an early age (Wakid
et al., 1982). Mayas (1975) reported exactly the reverse effect of fractionation
with the same species.

4.3 Oxygen effect

The influence of oxygen tension on the biological effect of exposure to
ionizing radiation has been known for many years. When pupae, 2 days before
eclosion, of *C. capitata* were irradiated in a nitrogen atmosphere, it was necessary to increase the dose to obtain a level of sterility equivalent to that obtained
if pupae were irradiated in air; ca. 98% sterility being achieved by 9 krad in air
and 11 krad in nitrogen (Hooper, 1971a). However, males irradiated in nitrogen
were more (3-fold) competitive than their counterparts irradiated in air. This
effect has been confirmed in a number of species (Fig. 3.3.2). At equivalent
levels of sterility male *C. capitata* irradiated in nitrogen transferred more

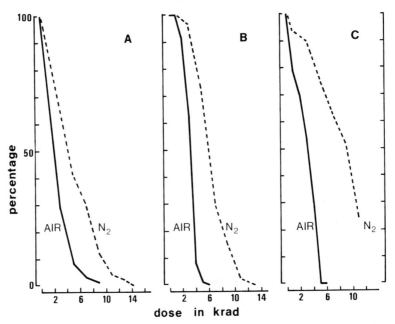

Fig. 3.3.2. Effect of irradiation in air and nitrogen on dose-response relationships: (A) sterility of
irradiated male *D. cucumis*; (B) fecundity of irradiated female *D. cucumis*; (C) fertility of irradiated
female *D. cucumis*; data from Hooper (1975c).

sperm than males irradiated in air (Zumreoglu et al., 1979). Ohinata et al. (1971) found that irradiation of mature pupae of *C. capitata* in a partial vacuum, and in atmospheres of carbon dioxide and helium gave results similar to that obtained with a nitrogen atmosphere. Irradiation in an environment which reduces the oxygen level in tissues probably lessens the effect of free radicals and peroxide, which arise from the irradiation of aqueous material, on the chromosomal DNA. Why, at equivalent levels of sterility, the competitiveness of males irradiated in nitrogen is less affected than that of males irradiated in air remains an intriguing question!

Since the solubility of oxygen in water increases as temperature decreases one might expect that increased damage i.e. greater induction of DLM, would occur if pupae were irradiated at low temperatures. With *C. capitata*, Langley and Maly (1971) found some evidence for such an effect, but Wakid et al. (1982) did not. Teruya (1984) found no effect of temperature with *D. cucurbitae*.

4.4 Neutron irradiation

Although it is unlikely that high energy radiation such as neutrons will be commonly used to sterilise insects, it is appropriate to review the small amount of work that has been conducted for its radiobiological significance. Fast neutrons are more effective in producing mutations in insects because of their increased biological effectiveness vis-a-vis X or γ radiation. This arises because of their high rate of transfer of linear energy (LET). The mean LET of protons arising from irradiation with fast neutrons is of the order of 8–43 keV/μm compared with a mean LET for γ radiation of ca. 0.4 keV/μm of track in tissue. Thus high LET radiation would be expected to have a high relative biological effectiveness (RBE) compared with low LET radiation such as γ radiation. With *C. capitata* an RBE of ca. 4–5 for induction of a high level of sterility in males, and 3–4 for infecundity in females has been found (Hooper, 1971b). This is of the order previously found for *C. capitata* and for other insects e.g. *Anthonomus grandis* Boheman and *Bombyx mori* (Linnaeus). There were indications that, for equivalent levels of sterility, the high LET fast neutron treatment reduced *C. capitata* male competitiveness less than did low LET γ radiation.

5 CONCLUSIONS

After reviewing the published radiobiological data a number of points became obvious. Firstly, the dose rate of the source used to irradiate the insects should always be stated. There is conflicting evidence as to whether sterility and/or competitiveness is dose rate dependent. Data assembled by Fisher (1984) for *C. capitata* indicate that the slopes of the sterility-dose response lines reported by various authors differ. This suggests that dose rate could be important. On the other hand, as indicated in the text, very low dose rates would involve lengthy irradiation exposures and unless specific precautions are taken anoxia could occur which would affect the level of sterility.

Secondly, since it is known that the levels of sterility and competitiveness obtained with a given dose are dependent on the age of the pupae when irradiated, it is essential that the pupal age be stated. Preferably authors should indicate when the irradiation is performed in relation to the time of adult eclosion. It is of little value to state that "*x*" day old pupae were irradiated if the length of the pupal period, which is temperature dependent, is not also stated.

Thirdly, the possibility that populations of a given species from different

---

Chapter 3.3 references, p. 162

parts of its geographical range could differ in radiosensitivity should be kept in mind. This has been alluded to by some authors e.g. Thomou (1963).

Reproduction in its broadest sense involves behavioural and ecological parameters as well as reproductive physiology. The released sterilised males must be able to disperse, find food and survive, find appropriate mating arenas, participate in lek formation in those species where that is involved, and perform the appropriate courtship ritual. The extent to which the released males can perform these functions will determine the extent to which they impact on the reproductive potential of the target population. The challenge for the future will be to devise appropriate field tests to quantify the impact of irradiation procedures on the overall reproductive impact of sterilised insects. This will be difficult as other factors e.g. transport and release stresses and the intrinsic "quality" of the mass-produced insects also will be involved.

6 REFERENCES

Anon., 1982. Annual Report of the Ministry of Agriculture and Natural Resources and the Environment. Government Printer, Mauritius, pp. 106–128.

Anwar, M., Chatha, N., Ohinata, K. and Harris, E.J., 1975. Gamma irradiation of the melon fly: laboratory studies of the competitiveness of flies treated as pupae 2 days before eclosion or as 2-day-old adults. Journal of Economic Entomology, 68: 733–735.

Bhatti, M.A., 1970. Sterility and sexual competitiveness of Queensland fruit fly (Diptera: Tephritidae). Ph.D. thesis, University of New South Wales, Sydney.

Boller, E.G., Remund, V. and Zehnder, J., 1975. Sterilization influence of the quality of the European cherry fruit fly, *Rhagoletis cerasi*. In: IAEA Sterility Principle for Insect Control 1974. IAEA, Vienna, pp. 179–189.

Burditt, A.K., F. Lopez D., Steiner, L.F. and Von Windeguth, D.L., 1975. Application of sterilization to *Anastrepha suspensa* Loew in Florida, United States of America. In: Sterility Principle for Insect Control 1974. IAEA, Vienna, pp. 93–101.

Chang, T. and Lee, W., 1984. Effect of gamma irradiation on the laboratory strain of Oriental fruit fly (*Dacus dorsalis* Hendel) in nitrogen. Bulletin of the Institute of Zoology, Academia Sinica, 3: 193–197.

Drew, R.A.I., 1983. Report on the pest status of fruit flies in mango fruit in the Philippines. Queensland Department of Primary Industries, pp. 1–12.

Economopoulos, A.P., 1972. Sexual competitiveness of X-ray sterilised males of *Dacus oleae*. Mating frequency of artificially reared and wild females. Environmental Entomology, 1: 490–497.

Farias, G.J., Cunningham, R.T. and Nakagawa, S., 1972. Reproduction in the Mediterranean fruit fly: abundance of stored sperm affected by duration of copulation, and affecting egg hatch. Journal of Economic Entomology, 65: 914–915.

Fisher, K.T., 1984. The methodology of mass rearing, sterilisation and release of Mediterranean fruit fly (*Ceratitis capitata* (Wiedemann) in Western Australia. M. Sc. thesis, University of Western Australia, Perth.

Fletcher, B.S. and Giannakakis, A., 1973. Sex pheromone production in irradiated males of *Dacus (Strumeta) tryoni*. Journal of Economic Entomology, 66: 62–64.

Fried, M., 1971. Determination of sterile-insect competitiveness. Journal of Economic Entomology, 64: 869–872.

Habu, N., Iga, M. and Numazawa, K., 1984. An eradication program of the Oriental fruit fly, *Dacus dorsalis* Hendel (Diptera: Tephritidae) in the Ogasawara (Bonin) Islands. I. Eradication field test using a sterile fly release method on small islets. Applied Entomology and Zoology, 19: 1–7.

Haisch, A., 1970. Some observations on decreased vitality of irradiated Mediterranean fruit fly. In: Sterile-Male Technique for Control of Fruit Flies. IAEA, Vienna, pp. 71–75.

Haisch, A. and Boller, E.F., 1971. Genetic control of the European cherry fruit fly, *Rhagoletis cerasi* L. In: Sterility Principle for Insect Control or Eradication. IAEA, Vienna, pp. 67–76.

Hooper, G.H.S., 1970. Sterilization of the Mediterranean fruit fly. A review of laboratory data. In: Sterile-Male Technique for Control of Fruit Flies. IAEA, Vienna, pp. 3–12.

Hooper, G.H.S., 1971a. Competitiveness of gamma-sterilized males of the Mediterranean fruit fly: effect of irradiating pupal or adult stage and of irradiating pupae in nitrogen. Journal of Economic Entomology, 64: 1364–1368.

Hooper, G.H.S., 1971b. Sterilization and competitiveness of the Mediterranean fruit fly after irradiation of pupae with fast neutrons. Journal of Economic Entomology, 64: 1369–1372.

Hooper, G.H.S., 1971c. Gamma sterilization of the Mediterranean fruit fly. In: Sterility Principle for Insect Control or Eradication. IAEA, Vienna, pp. 87–95.

Hooper, G.H.S., 1972. Sterilization of the Mediterranean fruit fly with gamma radiation: effect on male competitiveness and change in fertility of females alternately mated with irradiated and untreated males. Journal of Economic Entomology, 65: 1–6.

Hooper, G.H.S., 1975a. Sterilization of *Dacus cucumis* French (Diptera: Tephritidae) by gamma radiation. I. Effect of dose on fertility, survival and competitiveness. Journal of the Australian Entomological Society, 14: 81–87.

Hooper, G.H.S., 1975b. Sterilization of *Dacus cucumis* French (Diptera: Tephritidae) by gamma radiation. II. Effect of dose rate on sterility and competitiveness of adult males. Journal of the Australian Entomological Society, 14: 175–177.

Hooper, G.H.S. and Horton, I.F., 1981. Competitiveness of sterilised male insects: a method of calculating the variance of the value derived from competitive mating tests. Journal of Economic Entomology, 74: 119–121.

Hooper, G.H.S. and Katiyar, K.P., 1971. Competitiveness of gamma-sterilised males of the Mediterranean fruit fly. Journal of Economic Entomology, 64: 1068–1071.

Huque, H. and Ahmad, C.R., 1969. Studies on the control of *Dacus ciliatus* Loew (Tephritidae: Diptera) by sterile male release technique. International Journal of Applied Radiation and Isotopes, 20: 791–795.

Huque, H. and Malik, Q.R., 1967. Control of fruit flies *Dacus zonatus* Saunders by gamma rays. International Journal of Applied Radiation and Isotopes, 18: 658–661.

Katiyar, K.P., 1962. Possibilities of eradication of the Mediterranean fruit fly, *Ceratitis capitata* Wied., from Central America by gamma-irradiated males. Fourth Inter-American Symposium on the Peaceful Applications of Nuclear Energy, Mexico City, pp. 211–217.

Katiyar, K.P. and Ramirez, E., 1970. Mating frequency and fertility of Mediterranean fruit fly females mated with normal and irradiated males. Journal of Economic Entomology, 63: 1247–1250.

Katiyar, K.P. and Valerio, J., 1964. Further studies on the possible use of sterile male release technique in controlling or eradicating the Mediterranean fruit fly, *Ceratitis capitata* Wied. from Central America. Fifth Inter-American Symposium on the Peaceful Applications of Nuclear Energy, Valpariaso (9–12 March, 1964) pp. 197–202.

Langley, P.A. and Maly, H., 1971. Control of the Mediterranean fruit fly (*Ceratitis capitata*) using sterile males: effects of nitrogen and chilling during gamma-irradiation of puparia. Entomologia Experimentalis et Applicata, 14: 137–146.

La Chance, L.E. and Graham, C.K., 1984. Insect radiosensitivity: dose curves and dose-fractionation studies of dominant lethal mutations in the mature sperm of 4 insect species. Mutation Research, 127: 49–59.

Mayas, I.A., 1975. Effets du fractionnement de la dose sterilisante de rayons gamma sur l'emergence, la fertilitie et la competitivite de la mouche Mediterraneenne des fruits, *Ceratitis capitata* Wied. In: Sterility Principle for Insect Control 1974. IAEA, Vienna, pp. 229–235.

Muller, H.J., 1927. Artificial transmutation of the gene. Science, 66: 84–87.

Myers, H.S., Barry, B.D., Burnside, J.A. and Rhode, R.H., 1976. Sperm precedence in female apple maggots alternately mated to normal and irradiated males. Annals of the Entomological Society of America, 69: 39–41.

Nakagawa, S., Farias, G.J., Suda, D., Cunningham, R.T. and Chambers, D.L., 1971. Reproduction of the Mediterranean fruit fly: frequency of mating in the laboratory. Annals of the Entomological Society of America, 64: 949–950.

Ohinata, K., Ashraf, M. and Harris, E.J., 1977. Mediterranean fruit flies: sterility and sexual competitiveness in the laboratory after treatment with gamma irradiation in air, carbon dioxide, helium, nitrogen or partial vacuum. Journal of Economic Entomology, 70: 165–168.

Ohinata, K., Chambers, D.L., Fujimoto, M., Kashiwai, S. and Miyabara, R., 1971. Sterilization of the Mediterranean fruit fly by irradiation: comparative mating effectiveness of treated pupae and adults. Journal of Economic Entomology, 64: 781–784.

Qureshi, Z.A. and Bughio, A.R., 1969. Sterilization and competitive ability of gamma sterilised males to normal males of *Dacus zonatus* (Saunders). International Journal of Applied Radiation and Isotopes, 20: 473–476.

Qureshi, Z.A., Ashraf, M., Bughio, A.R. and Hussain, S., 1974. Rearing, reproductive behaviour and gamma sterilization of fruit fly, *Dacus zonatus* (Diptera: Tephritidae). Entomologia Experimentalis et Applicata, 17: 504–510.

Rhode, R.H., F. Lopez D. and Eguisa, F., 1961. Effect of gamma radiation on the reproductive potential of the Mexican fruit fly. Journal of Economic Entomology, 53: 202–203.

Teruya, T., 1983. Sterilization of the melon fly *Dacus cucurbitae* Coquillett (Diptéra: Tephritidae), with gamma-radiation: fertility of F_1 progeny of flies treated with sub-sterilizing doses. Applied Entomology and Zoology, 18: 335–341.

Teruya, T., 1984. Sterilization of the melon fly, *Dacus cucurbitae* Coquillet (Diptéra: Tephritidae), with gamma-radiation: sterility of flies irradiated under a low temperature condition. Applied Entomology and Zoology, 19: 109–111.

Teruya, T., Zukeyama, H. and Ito, Y., 1975. Sterilization of the melon fly, *Dacus cucurbitae* Coquillet with gamma radiation: effect on rate of emergence, longevity and fertility. Applied Entomology and Zoology, 10: 298–301.

Thomou, H., 1963. Sterilization of *Dacus oleae* by gamma radiation. In: Radiation and Radioisotopes Applied to Insects of Agricultural Importance. IAEA, Vienna, pp. 413–424.

Tsiropoulos, G.J. and Tzanakakis, M.E., 1970. Mating frequency and inseminating capacity of radiation-sterilized and normal males of the olive fruit fly. Annals of the Entomological Society of America, 63: 1007–1010.

Tzanakakis, M.E., Tsitsipis, J.A., Papageorgehiou, M. and Fytizas, E., 1966. Gamma radiation-induced dominant lethality to the sperm of the olive fruit fly. Journal of Economic Entomology, 59: 214–216.

Velasco, H. and Enkerlin, D., 1982. Determinacion de la doses optima de irradiacion relativa a la competitividad del macho esteril de *Anastrepha ludens* (Loew); su atraccion a trampas de color y al atrayente sexual. In: Sterile Insect Technique and Radiation in Insect Control. IAEA, Vienna, pp. 323–339.

Wakid, A.M., Amin, A.H., Shoukry, A. and Fadel, A., 1982. Factors influencing sterility and vitality of the Mediterranean fruit fly, *Ceratitis capitata* Wiedemann. In: Sterile Insect Technique and Radiation in Insect Control. IAEA, Vienna, pp. 379–386.

Williamson, D.L., Mitchell, S. and Seo, S.T., 1985. Gamma irradiation of the Mediterranean fruit fly (Diptera: Tephritidae): effects of puparial age under induced hypoxia on female sterility. Annals of the Entomological Society of America, 78: 101–106.

Wong, T.T.Y., Couey, H.M. and Nishmoto, J.I., 1982. Oriental fruit fly: sexual development and mating response of laboratory-reared and wild flies. Annals of the Entomological Society of America, 75: 191–194.

Zumreoglu, A., Ohinata, K., Fujimoto, M., Higa, H. and Harris, E.J., 1979. Gamma irradiation of the Mediterranean fruit fly: effect of treatment of immature pupae in nitrogen on emergence, longevity, sterility, sexual competitiveness, mating ability, and pheromone production of males. Journal of Economic Entomology, 72: 173–176.

Chapter 3.4 Mating Pheromones

3.4.1 Tropical Dacines

J. KOYAMA

1 INTRODUCTION

The study of the mating pheromones of fruit flies is important in order to understand their role in the mating behavior of the flies, and to determine whether they could be utilized as practical control measures. In this section the pheromones of tropical Dacines are discussed from the above points of view.

2 MATING BEHAVIOR

Most species of tropical Dacines of economic importance including the Queensland fruit fly, *Dacus tryoni* (Froggatt), oriental fruit fly, *Dacus dorsalis* Hendel, and melon fly, *Dacus cucurbitae* Coquillett, mate at dusk (Fletcher, 1968, 1969; Tychsen and Fletcher, 1971; Fletcher and Giannakakis, 1973a; Tychsen, 1977; Kobayashi et al., 1978; Suzuki and Koyama, 1981; Kuba and Koyama, 1982; Kuba et al., 1984; Arakaki et al., 1984).

Tychsen (1977) studied the mating behavior of *D. tryoni* in simulated natural conditions in a field cage containing a peach tree. Under natural lighting conditions mating was entirely limited to a period of about half an hour as the light intensity fell at dusk. All males in the cage aggregated into a loose flying swarm which soon settled on the windward side of the tree. Within the settled swarm, each male defended a small area on a leaf where he stridulated and released pheromone. Females flew into the aggregation individually. Each female landed near a male and then walked up to him. The male's behavior did not change until the female entered his visual field. Then mounting occurred immediately and stable copulation was achieved within a few seconds.

Similar observations were made for *D. cucurbitae* in a field cage containing a coral tree by Kuba et al. (1984) and Kuba and Koyama (1985). When the light intensity decreased toward dusk, males aggregated on the tree and stationed themselves on the bottom surface of leaves. In this species, a flying swarm as in *D. tryoni* was not observed. Each male individually occupied a leaf as a territory and defended it against other males which approached. Males remained stationary and engaged in stridulation and released a sex pheromone to which females were attracted. When a female approached a male to within a distance of 2–3 cm, the male approached the female and attempted copulation. Although field cage observations have not been made, the courtship behavior of *D. dorsalis* appears to be very similar to that of *D. cucurbitae* (Arakaki et al., 1984).

The aggregation behavior of the males of these species during courtship is a strong indication that they engage in lek behavior (Emlen and Oring, 1977).

In the lek, pheromone released by the males plays an important role by attracting females for mating. Males frequently wipe (*D. tryoni*) or beat (*D. cucurbitae*) the posterior part of their abdomen with their hind legs when engaged in rapid wing vibration (Fletcher, 1969; Kuba and Koyama, 1982). This behavior may help to spread the pheromone on the body surface, which with the local air currents made by wing vibration, may enhance evaporation of the pheromone and increase female response. When sex pheromone release occurs in the group situation of the lek, it may be more effective in attracting females than release from single males.

When both antennae were removed from females of *D. cucurbitae* they rarely copulated with males (Suzuki and Koyama, 1981), which suggests that females perceive the pheromone by receptors on the antennae.

3 PHEROMONE GLAND

Males of *D. tryoni*, *D. dorsalis* and *D. cucurbitae* have sex pheromone glands associated with the rectum (Fletcher, 1968, 1969; Schultz and Boush, 1971; Nation, 1981) (Fig. 3.4.1.1).

Fletcher (1968, 1969) and Fletcher and Giannakakis (1973a, 1973b) studied the structure and function of the pheromone glands of *D. tryoni*. The pheromone was secreted by a gland complex which developed from the posterior ventral wall of the rectum. In mature males this consisted of a secretory sac and a ventral reservoir in which the pheromone was stored before release. Pheromone was first detected in flies 2 days after emergence, but in the majority the reservoir did not become full until 12 or 14 days after emergence. There was a close correlation between the amount of secretion in the reservoir and the onset of sexual activity. Sexually receptive females responded to extracts of reservoir by characteristic types of behavior including preening and probing with the ovipositor just like their response to males. Females responded maximally when their ovaries were mature, and around the dusk period, which is the normal time of mating in this species. After mating, females become unresponsive to the pheromone. Four weeks after the first mating some females had regained their responsiveness. Exposure of the pupae to 8,000 or 20,000 rad on the 9th day after pupation had no detectable effects on pheromone production.

Kobayashi et al. (1978) found that the rectal gland complexes of male *D. dorsalis* were highly attractive to females of the same species. Similar glands of *D. cucurbitae* males were only weakly attractive to conspecific females.

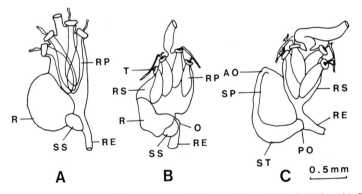

Fig. 3.4.1.1. Rectal gland complex of three adult male tephritids: (**A**) *Dacus tryoni*, (**B**) *Dacus dorsalis*, (**C**) *Dacus cucurbitae*. AO, anterior opening; O, orifice opening into reservoir; PO, posterior opening; R, reservoir; RE, "anal tube" (posterior rectum); RP, rectal papilla; RS, rectal sac; SP, secretory pouch; SS, secretory sac; ST, secretory tube; T, trachea. (Redrawn from Fletcher (1968) (**A**) and Schultz and Boush (1971) (**B**, **C**)).

However, Koyama (unpublished) found that fluid released from anus of male *D. cucurbitae* was highly attractive to conspecific females. The response behavior of females of these two species to the pheromones of conspecific males were similar to those in *D. tryoni*.

Kobayashi et al. (1978) observed that the glands of *D. cucurbitae* males elicited a strong response from females of *D. dorsalis*, which also responded positively to live males of *D. cucurbitae*. They suggested that the components of the sex pheromone of these two species may be structurally related. When Arakaki et al. (1984) caged males of *D. cucurbitae* with females of *D. dorsalis*, several males of *D. cucurbitae* attempted copulation with females of *D. dorsalis*. However, when a male jumped on a female's back he immediately departed. These results suggest that males are unable to distinguish the species until the time of tactile contact. Tactile stimulus may be one of the mechanisms involved in species discrimination.

4 CHEMICAL STRUCTURE OF PHEROMONES

Bellas and Fletcher (1979) identified 6 aliphatic amides from the rectal gland secretions of *D. tryoni* and a closely related species, *Dacus neohumeralis* Hardy (Table 3.4.1.1). The mixture of these amides showed only short-range attraction for females. The proportions of the various amides in the two species were similar, but *D. neohumeralis* mated at high light intensities which inhibited the sexual activity of *D. tryoni*. This difference in mating time was thought to be a barrier to hybridization between the two species in the field, although they produced fertile hybrids in the laboratory (Tychsen and Fletcher, 1971).

Baker et al. (1982) identified 3 amides, 3 pyrazine derivatives and 2-ethoxy benzoic acid from the rectal gland secrection of *D. cucurbitae* (Table 3.4.1.1). The *N*-2-methylbutylacetamide and *N*-3-methylbutylacetamide are common among *D. tryoni*, *D. neohumeralis* and *D. cucurbitae*. The latter compound had been shown to elicit activation and increased flight activity of *D. cucurbitae* females, but the exact role of the other components remains to be defined.

As described above, the chemical composition of sex pheromones in Dacines

TABLE 3.4.1.1

Chemical components of the rectal gland secretions in *Dacus tryoni*, *Dacus neohumeralis* and *Dacus cucurbitae*

Species	Components	Authors
Dacus tryoni and *Dacus neo-humeralis*	(1) *N*-3-methylbutylpropanamide (2) *N*-3-methylbutylacetamide (3) *N*-(3-methylbutyl)-2-methyl-propanamide (4) *N*-2-methylbutylpropanamide (5) *N*-2-methylbutylacetamide (6) *N*-(2-methylbutyl)-2-methyl-propanamide	Bellas and Fletcher (1979)
Dacus cucurbitae	(1) tetramethylpyrazine (2) methylpyrazine (3) 2,3,6-trimethylpyrazine (4) *N*-3-methylbutylacetamide (5) *N*-2-methylbutylacetamide (6) 2-methoxy-*N*-3-methylbutyl-acetamide (7) 2-ethoxy benzoic acid	Baker et al. (1981)

has not yet been fully analysed and their roles in sexual functions are not well understood. More investigations are needed before it can be determined if sex pheromones can be used for the practical control of these fruit flies.

5 REFERENCES

Arakaki, N., Kuba, H. and Soemori, H., 1984. Mating behavior of the oriental fruit fly, *Dacus dorsalis* Hendel (Diptera: Tephritidae). Applied Entomology and Zoology, 19: 42–51.

Baker, R., Herbert, R.H. and Lomer, R.A., 1982. Chemical components of the rectal gland secretions of male *Dacus cucurbitae*, the melon fly. Experientia, 38: 232–233.

Bellas, T.E. and Fletcher, B.S., 1979. Identification of the major components in the secretion from the rectal pheromone glands of the Queensland fruit flies *Dacus tryoni* and *Dacus neohumeralis* (Diptera: Tephritidae). Journal of Chemical Ecology, 5: 795–803.

Emlen, S.T. and Oring, L.W., 1977. Ecology, sexual selection and the evolution of mating systems. Science, 197: 215–223.

Fletcher, B.S., 1968. Storage and release of a sex pheromone by the Queensland fruit fly, *Dacus tryoni* (Diptera; Tephritidae). Nature, 219: 631–632.

Fletcher, B.S., 1969. The structure and function of the sex pheromone glands of the male Queensland fruit fly, *Dacus tryoni*. Journal of Insect Physiology, 15: 1309–1322.

Fletcher, B.S. and Giannakakis, A., 1973a. Factors limiting the response of females of the Queensland fruit fly, *Dacus tryoni*, to the sex pheromone of the male. Journal of Insect Physiology, 19: 1147–1155.

Fletcher, B.S. and Giannakakis, A., 1973b. Sex pheromone production in irradiated males of *Dacus (Strumeta) tryoni*. Journal of Economic Entomology, 66: 62–64.

Kobayashi, R.M., Ohinata, K., Chambers, D.L. and Fujimoto, M.S., 1978. Sex pheromone of the oriental fruit fly and the melon fly: Mating behavior, bioassay method, and attraction of females by live males and by suspected pheromone glands of males. Environmental Entomology, 7: 107–112.

Kuba, H. and Koyama, J., 1982. Mating behavior of the melon fly, *Dacus cucurbitae* Coquillett (Diptera: Tephritidae): Comparative studies of one wild and two laboratory strains. Applied Entomology and Zoology, 17: 559–568.

Kuba, H. and Koyama, J., 1985. Mating behavior of wild melon flies, *Dacus cucurbitae* Coquillett (Diptera: Tephritidae) in a field cage: Courtship behavior. Applied Entomology and Zoology, 20: 365–372.

Kuba, H., Koyama, J. and Prokopy, R.J., 1984. Mating behavior of wild melon flies, *Dacus cucurbitae* Coquillett (Diptera: Tephritidae) in a field cage: Distribution and behavior of flies. Applied Entomology and Zoology, 19: 367–373.

Nation, J.L., 1981. Sex-specific glands in Tephritid fruit flies of the genera *Anastrepha, Ceratitis, Dacus* and *Rhagoletis* (Diptera: Tephritidae). Journal of Insect Morphology and Embryology, 10: 121–129.

Schultz, G.A. and Boush, G.M., 1971. Suspected sex pheromone glands in three economically important species of *Dacus*. Journal of Economic Entomology, 64: 347–349.

Suzuki, Y. and Koyama, J., 1981. Courtship behavior of the melon fly, *Dacus cucurbitae* Coquillett (Diptera: Tephritidae). Applied Entomology and Zoology, 16: 164–166.

Tychsen, P.H., 1977. Mating behavior of the Queensland fruit fly, *Dacus tryoni* (Diptera: Tephritidae), in field cages. Journal of Australian Entomological Society, 16: 459–465.

Tychsen, P.H. and Fletcher, B.S., 1971. Studies on the rhythm of mating in the Queensland fruit fly, *Dacus tryoni*. Journal of Insect Physiology, 17: 2139–2156.

3.4.2 *Dacus oleae*

B.E. MAZOMENOS

1 INTRODUCTION

The olive fruit fly, *Dacus oleae* (Gmelin) is one of the most serious pests of olive fruits. The larvae feed exclusively on olive fruits, and adults have several food sources such as honey dew, nectar or pollen. Olive fruit fly occurs in most of the countries surrounding the Mediterranean sea and it cause an average of 30% crop losses annually.

Chemicals modifying insect behaviour and especially chemicals involved with the mating process such as sex pheromones are very promising tools to be used for the control of insect pests.

Behavioural studies conducted mainly under laboratory conditions have shown that many species of fruit flies in addition to other stimuli (visual, auditory) use a sex pheromone for their mating communication. Although sex pheromone seem to be present in many fruit fly species, the chemical structure of the pheromone has only been characterized in few cases.

The studies conducted during the last decade on the mating pheromone of the olive fruit fly, *D. oleae*, are the subject of this chapter.

2 MATING BEHAVIOUR

Most of our knowledge on the mating behaviour of the olive fruit fly is based on studies of artificially reared flies under laboratory conditions. Female olive fruit flies are oligogamous and mate 1–3 times during their life, (Tzanakakis et al., 1968; Zouros and Krimbas, 1970; Cavalloro and Delrio, 1971). Male olive fruit flies on the other hand are polygamous and they can mate daily if receptive females are available (Zervas, 1982).

During the mating period sexually active males stridulate by fanning their wings over a pair of combs which forms bristles on the third abdominal tergite, the high frequency sound that is produced has been interpreted by Feron and Andriew (1962) as a mating stimulus.

Economopoulos et al. (1971), reported that odours emitted by males during the active period were different from the odours emitted by females, and that the odour concentration was higher during the last hour of the sexually active period. DeMarzo et al. (1978) showed that female olive fruit flies were attracted to males in olfactometer tests.

Anatomical and hystological studies showed that the male have an evagination (sac) at the lower part of the posterior rectum. Secretory cells are present in the sac and in the posterior rectum. No such evagination and area with secretory cells were observed in the female rectum (Economopoulos et al., 1971;

Schultz and Boush, 1971; DeMarzo et al., 1978). These authors suspected that the male olive fruit fly produces a pheromone which facilitates mating and that the rectum is the site of production.

Haniotakis (1974), demonstrated in laboratory studies that virgin, sexually mature females attracted males during the mating period and he suggested that virgin females release an airborne sex pheromone which attracts males. These results were verified in field tests, with traps baited with virgin females, (Haniotakis, 1977). However, no female attraction to traps baited with live males was observed either in laboratory or field tests. Laboratory behavioural tests indicated that virgin females attracted sexually mature males from the second day post emergence, while males responded to the released female sex pheromone after the third day post emergence. Sexual attraction was observed in the last 3 h of 12:12 h light-dark regime, indicating that there is a daily rhythm in female release of pheromone and in male response. Field observations revealed that matings and male response to virgin females and pheromone extracts occur in the late afternoon or at dusk. Courtship behaviour of the olive fruit fly has not been studied in depth. Females during calling did not assume a characteristic posture as occurs in many other fruit fly species. Female probing is observed during the day, but it is not certain if the pheromone is released during probing. Pheromone cause short and long range orientation of males to females. The different behavioural steps of males are not easily recognized. Mature male locomotor activity, wing vibration and attempts to copulate with other males are common behaviour patterns during the day.

3 PHEROMONE COLLECTION ISOLATION AND IDENTIFICATION

3.1 Pheromone collection

There are two efficient methods used for pheromone collection from the virgin female olive fruit flies as follows.

Air volatiles emitted by sexually mature females during the sexually active period are trapped by passing air over the caged females and then into a glass trap immersed in liquid nitrogen. The liquid volatiles collected after 3 h are allowed to evaporate under low temperature and the trapped volatiles are taken up in diethyl ether. The cold trap method has the advantage that it allows continuous use of the same females to collect pheromone for many days and the material is free from non volatile lipids, (Haniotakis et al., 1977; Gariboldi et al., 1982).

The rectum of mature virgin females can be excised and extracted in diethyl ether for 24 h (Vita et al., 1979; Mazomenos and Haniotakis, 1981), or sealed in glass capillaries and analyzed on gas chromatograph (GC) using a solid sample injector system (Baker et al., 1980; Gariboldi et al., 1983).

3.2 Pheromone isolation

The ether extracts of both these methods are of sufficient purity to be introduced directly into the GC after concentration of the solution. Sequential GC fractionation on non-polar and polar columns revealed that the crude extract obtained with the cold trap contained four peaks, which have various degrees of male attraction (Mazomenos and Haniotakis, 1981). The Kovat's indices (Kovats, 1961) for the four active peaks in two different columns OV-101 and Carbowax 20M are shown in Table 3.4.2.1. The ratio of the active components in the crude sample was determined by measuring the peak area of

TABLE 3.4.2.1

Kovat's indices for active peaks of olive fruit fly sex pheromone isolated from cold trap (CT) and rectal gland extracts (GE)

Compound	Column			
	OV-101 (2 m × 1.8 mm (ID))		Carbowax 20M (2 m × 1.8 mm (ID))	
	CT	GE	CT	GE
A	956	–	1065	–
B	1063	–	1380	–
C	1139	1145	1380	1380
D	1527	1534	1864	1858

each component and was found to be approximately (1/0.1/3/1) for the components A, B, C and D respectively. Biological studies of the four components in laboratory and field cage bioassays indicated that all the components tested individually elicited mature male sexual excitation and attraction at the concentration of 1 female equivalent (FE) in the laboratory and 10 FE in a field cage. Component C showed the major attraction. Combinations of the three secondary components with component C at the concentration of 0.1 FE, the lowest concentration we found in laboratory bioassays to cause male upwind orientation to the pheromone source for component C and the appropriate ratio of the secondary components, showed that the secondary components have an additive effect on male attraction and that a combination of the four components constantly gave higher male attraction. The three secondary components when combined attracted less males than the major component.

GC fractionation of the pheromone gland extract indicated the presence of two active components with Kovat's indices which coincided with two of the components isolated from the cold trap extract (Table 3.4.2.1). The two components which were not present in the pheromone gland extract are probably synthesized and stored elsewhere in the insect body.

3.3 Pheromone identification

The chemical structure of the major pheromone component was elucidated by Baker et al. (1980). GC-MS analysis of the gland volatiles gave the mass spectrum of this component with the atomic composition $C_9H_{16}O_2$. Comparing the mass spectrum of this component with those published for alkyl-1,6-dioxaspiro[5.5]undecanes (Francke et al., 1979), the chemical structure of the component was assigned as 1,7-dioxaspiro[5.5]undecane.

This structure for the major female pheromone component was confirmed by Mazomenos et al. (1981) and Gariboldi et al. (1983). 1,7-Dioxaspiro[5.5]undecane corresponds to component C isolated from female volatiles trapped with liquid nitrogen and the gland extracts. In addition to 1,7-dioxaspiro[5.5]undecane, Mazomenos et al. (1981), identified the other three secondary components by GC-MS spectroscopy. The chemical structure of these components were assigned as C_6H_{20} a-pinene, $C_9H_{20}O$ n-nonanal and $C_{14}H_{30}O_2$ ethyl dodecanoate, which correspond to components A, B, and D respectively in Table 3.4.2.1 and Fig. 3.4.2.1. The mass spectra of the natural components were compared with those of synthetic samples and were found to be identical.

The biological activity of the synthetic components was tested both in the laboratory and the field. In laboratory tests all the components possess the same activity as the corresponding natural components. 1,7-Dioxaspiro[5.5]undecane was the most active (Mazomenos and Haniotakis, 1985).

Chapter 3.4.2 references, p. 176

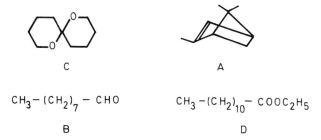

C   A

$CH_3-(CH_2)_7 - CHO$ $CH_3-(CH_2)_{10}- COOC_2H_5$

B   D

Fig. 3.4.2.1. Chemical structures of the sex pheromone components of the female olive fruit fly *Dacus oleae* (Gmelin).

Electroantennogram (EAG) studies showed that all four components elicited the same male antennal response as the natural ones (Van de Pers et al., 1985). 1,7-Dioxaspiro[5.5]undecane and n-nonanal elicited higher EAG responses than a-pinene and ethyl dodecanoate. Also of interest is the finding that 1,7-dioxaspiro[5.5]undecane and n-nonanal are detected by independent sensory systems.

In a series of field experiments at different periods of the year male olive fruit fly attraction to the synthetic pheromone components, various combinations of the four components and crude female extract released from polyethylene vials or rubber stoppers, showed that 1,7-dioxaspiro[5.5]undecane attracted high number of males, the other three secondary components showed no significant attraction when tested individually. Combinations of 1,7-dioxaspiro[5.5]undecane with the other components except ethyl dodecanoate increased male attraction. Various combinations of the secondary components without the spiroketal were not attractive. When the complete mixture of the four components was dispensed from polyethylene vials it attracted consistently higher number of males than the spiroketal alone. In comparative attraction studies between female crude extract and the complete mixture of the synthetic components approximately the same number of males was attracted (Mazomenos and Haniotakis, 1985).

Rossi et al. (1978) and Gariboldi et al. (1982), reported that the components p-cymene and (E)-6-nonen-1-ol, which were isolated from female volatiles trapped with liquid nitrogen, showed biological activity to males in laboratory bioassays, while in field tests both sexes were attracted. We were not able to isolate these components from the female volatiles and our laboratory test with the synthetic components revealed no such activity. Jones et al. (1983) tested the same components in the field and observed no fly activity. The same authors tested (E)- and (Z)-6-nonen-1-ol in combination with 1,7-dioxaspiro[5.5]undecane and found that (E)-6-nonen-1-ol depressed male catches, while (Z)-6-nonen-1-ol had no significant effects.

In further studies Baker et al. (1982) using the same procedure identified two additional spiroketals 3- and 4-hydroxy-1,7-dioxaspiro[5.5]undecanes. The two components were present at levels of ca. 10 ng per female. Both have been synthesized by Baker et al. (1982) and Kocienski and Yeates (1983), but the biological activity of these components has not been tested.

3.4 Isolation of 1,7-dioxaspiro[5.5]undecane from males

1,7-Dioxaspiro[5.5]undecane was characterized by Baker et al. (1980) as a female specific component. The isolation of the same component from the rectal gland of males (Mazomenos and Pomonis, 1983), indicates that it is not only produced by the females. Synthesis of the same component by both sexes modifying different aspects of their behaviour has been reported for many bark beetle species (Wood, 1982). We postulated that 1,7-dioxaspiro[5.5]undecane produced by females and males may act to control many aspects of the insects behaviour in nature. In laboratory bioassays the component isolated from males was as attractive to mature males as the synthetic spiroketal. The same component isolated either from males or females showed no biological activity to females. However, EAG studies of the natural and synthetic component elicited a similar response by both male and female antennae (Van de Pers et al., 1985). Field observations (Haniotakis, 1984) indicated that in olive orchards or individual trees in which 1,7-dioxaspiro[5.5]undecane was released the female population was higher than the controls. It seems possible that mature females aggregate near the pheromone source, where the concentration is high.

4 SYNTHESIS OF 1,7-DIOXASPIRO[5.5]UNDECANE

1,7-Dioxaspiro[5.5]undecane is a white oil with a pleasant heavy odour and a boiling point of 77–80°C at 13 mm Hg. The compound has been described and synthesized by Stetter and Rauchut (1958), in studies investigating spiro centers among organic compounds. It was synthesized by cyclisation of 1,9-dihydroxynonan-5-one. Baker et al. (1980) and Gariboldi et al. (1983), synthesized the racemic mixture of 1,7-dioxaspiro[5.5]undecane (see Fig. 3.4.2.2). The above method proved to be a convenient method for large scale preparation.

The 1,7-dioxaspiro[5.5]undecane exists as enantiomers. Recently Mori et al. (1984) and Redlich and Francke (1984), have synthesized S-($+$)- and R-($-$)-1,7-dioxaspiro[5.5]undecane starting from D-glucose.

Biological tests with pure R-($-$)- and S-($+$)-enantiomers would answer which of the two isomers is more attractive to males. Also the finding of the absolute configuration of the naturally produced 1,7-dioxaspiro[5.5]undecane by females and males olive fruit flies would prove whether both sexes produce the same or different enantiomers.

5 FACTORS AFFECTING PHEROMONE PRODUCTION

Age of insects, mating history, internal rhythm and environmental conditions have been found to play important roles in pheromone production in many insect species. We studied the effects of the above factors on the production of the major pheromone component in olive fruit fly.

5.1 Effect of age

Pheromone production in female olive fruit flies seems to follow a different pattern from that known for other insect species. Under laboratory conditions females begin to produce pheromone on the third day post emergence to the amount of 46 ng/female. Pheromone production is rather stable for the following two days and then increases rapidly. 7-day-old females produced an average of 321 ng/female. The quantity of pheromone produced remains high in 8-day-

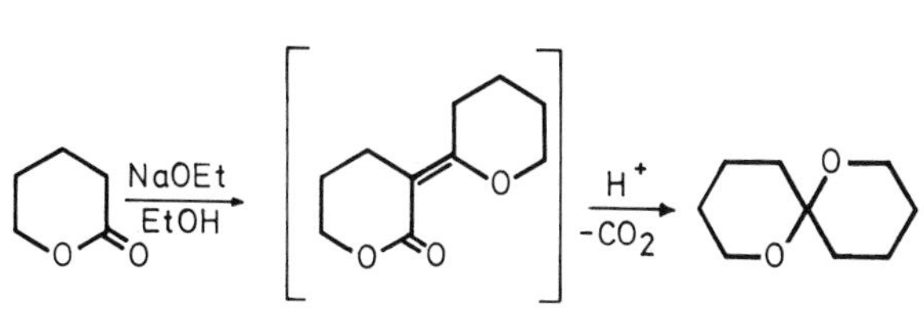

Fig. 3.4.2.2. Synthetic route of 1,7-dioxaspiro[5.5]undecane (Baker et al., 1980).

Chapter 3.4.2 references, p. 176

old females and then decreases to ca. 80 ng/female in 12 to 15 day-old females. Subsequently pheromone production increases to ca. 240 ng/female in 17 to 19 day-old females and increases once again in 27 to 29 day-old females (Mazomenos, 1984). Initiation of pheromone production coincides with sexual maturation of the female. Females under laboratory conditions become receptive on the third day post emergence. Although females produce relatively small amounts of pheromone during the third to fifth day post emergence, 90% of the matings occur in that period. Possible differences in pheromone concentration are of less importance in revealing variation in propensity to mate of different age classes because of the close synchrony of the flies imposed by laboratory conditions. In contrast, wild females obtained from infested olive fruits and maintained under laboratory conditions became receptive 6–15 days after emergence (Zervas, 1982). Females reared as larvae on olive fruits reach the maximal level of pheromone production at 12 day post emergence (Mazomenos, unpublished data). In both types of females pheromone production appears to be cyclical with peaks of production recurring at about 10 day intervals which coincide with female receptivity, each peak lasts 2–3 days. As the female ages the quantity of pheromone produced decreases.

Synthesis of the same component by male flies is completely different. Male pheromone glands contain small quantities ca. 8 ng/male, of 1,7-dioxaspiro[5.5]undecane, during the first day after emergence. The quantity of the pheromone increases and 9 to 15-day-old males produce ca. 50 ng/male. The production of pheromone decreases as the male ages. The quantity produced by males is lower than the quantity produced by females of the same age (Mazomenos, 1983).

5.2 Effect of mating

Mated females produce less pheromone than virgin females of the same age (Mazomenos, 1984). Females 2 and 5 days after mating produce an average of 32.2 and 30.2 ng/female, respectively. Virgin females of the same age produce ca. 136.5 and 120.3 ng/female. Twelve days after mating the pheromone production of mated females increases and they produce ca. 168.7 ng/female. The way in which mating influences pheromone production is not known. Tompkins and Hall (1981) reported that substances transferred from males during copulation diminished pheromone production in *Drosophila melanogaster* Meigen. Tzanakakis et al. (1968) also suggested that olive fruit fly female receptivity is inhibited for a certain period by substance(s) transferred by males during copulation. Possibly, mating controls female receptivity by diminishing pheromone production.

5.3 Effect of internal rhythm

Male response to the pheromone source is restricted to the last 3 h of the photophase in the laboratory or the dusk period in nature. This is the result of an internal rhythm, which coincides with that of the mating period. Light intensity which has been found to play a definitive role in male responsiveness in other insect species, is not critical in the case of olive fruit fly; males responded to the pheromone source even at high light intensity. Females start releasing pheromone at a low rate with the onset of the light period. The low rate of pheromone release is maintained until the initiation of the sexual active period when pheromone release rate is increased. With the beginning of the dark period, release rate drops significantly and ceases in a few hours (Haniotakis, 1979).

6 BIOSYNTHESIS OF 1,7-DIOXASPIRO[5.5]UNDECANE

The 1,7-dioxaspiro[5.5]undecane isolated from female and male olive fruit flies as the major pheromone component has a structure which is unique among those of many pheromone previously reported; this suggests a unique pathway for biosynthesis.

We studied the biosynthesis of 1,7-dioxaspiro[5.5]undecane using ^{14}C-labeled substrates, such as sodium 2-^{14}C-acetate (57 mCi/mol), sodium 2-^{14}C-malonate (55.5 mCi/mol), sodium 2-^{14}C-propionate (58 mCi/mol), sodium 2,3-^{14}C-succinate (50 mCi/mol) and L-U-^{14}C-glutamate (250 mCi/mol) (Mazomenos, 1983).

6.1 Incorporation of ^{14}C-label by feeding

The radiolabeled substrates were incorporated at the rate of either 50 or 100 μCi into a sugar-water syrup or the normal adult diet (Tsitsipis, 1975), upon which the adult flies (100–200) were allowed to feed ad libidum for 4–5 days post emergence. The pheromone and the other volatile substances were extracted with 1 ml of ether for 24 h.

The pheromone glands of mature virgin females were removed and placed in 1 ml of phosphate buffer (0.05 M, pH 7.2) and then treated with the described ^{14}C-labeled substrates at the rate of 2 μCi per treatment. Samples were incubated for 24 h at 25°C in stoppered pyrex tubes with agitation in a water bath fitted with a shaking platform. The solution of the incubated glands were extracted with equal volume of ether.

The extracts were fractionated on GC and the fractions were transferred to scintillation vials containing standard scintillation fluid and counted.

The radiolabeled substrates with the exception of acetate, were highly incorporated into the major pheromone component. Thus 14, 18, 22 and 9.3% respectively, of the total radio-activity of the chromatographic fractions for propionate, succinate, malonate and glutamate were found in the fraction containing the pheromone component 1,7-dioxaspiro[5.5]undecane.

Acetate was incorporated into the pheromone at the rate of only 2.6%. The low rate of incorporation of acetate relative to other substrates indicated that acetate is probably diverted to several anabolic processes which dilute the availability for incorporation into the pheromone molecule. Alternatively, acetate may not be used in the biosynthetic process directly such as use in primer or "starter piece" formation, but may be converted to malonate, succinate etc. Glutamate cannot be used directly for the biosynthesis of fatty acids but it can enter the tricarboxylic acid cycle (TCA) by oxidative deamination to α-ketoglutarate. When the glands were incubated with malonate and glutamate only malonate was incorporated into the pheromone molecule. Apparently the isolated gland cannot convert glutamate to usable intermediates for synthesis of the pheromone molecule.

Malonate on the other hand is a common building block for fatty acids providing two carbon fragments for chain elongation utilising the fatty acid synthetases. The isolation of 5-oxo-nonadioate from the rectal gland of wild males (Mazomenos and Pomonis, 1983) suggested that biosynthesis proceeds by the fatty acid synthetase system utilizing activated propionate as the starter unit. Recent studies indicated that when wild males were fed with labeled propionate and malonate the incorporation into the 1,7-dioxaspiro[5.5]undecane molecule was 8.3 and 14.1%, respectively. A high percentage was incorporated into the fraction containing the 5-oxo-nonanedioate. When explanted female glands were incubated with radiolabeled 5-oxo-nonanedioate isolated

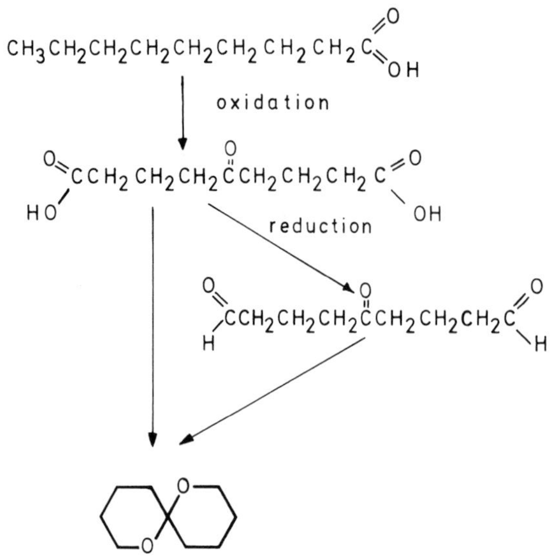

$$CH_3CH_2CH_2CH_2CH_2CH_2\ CH_2\ CH_2\ C\underset{OH}{\overset{O}{\diagdown}}\qquad (2)$$

oxidation

$$\underset{HO}{\overset{O}{\diagdown}}CCH_2\ CH_2CH_2\overset{O}{\overset{\|}{C}}CH_2CH_2CH_2\ C\underset{OH}{\overset{O}{\diagup}}\qquad (3)$$

reduction

$$\underset{H}{\overset{O}{\diagdown}}CCH_2CH_2CH_2\overset{O}{\overset{\|}{C}}CH_2CH_2CH_2C\underset{H}{\overset{O}{\diagup}}\quad (4)$$

(1)

Fig. 3.4.2.3. Proposed pathway for the biosynthesis of 1,7-dioxaspiro[5.5]undecane by olive fruit fly *Dacus oleae* Gmelin.

from males the female glands converted 5-oxo-nonanedioate to 1,7-dioxa-spiro[5.5]undecane (Mazomenos, unpublished data).

We proposed the scheme shown in Fig. 3.4.2.3 for the biosynthesis of 1,7-diox-aspiro[5.5]undecane. Starting with propionate and utilising malonate for the two carbon units, nonanoate (2) is synthesized by fatty acid synthetase system, which is terminally oxidized to nonanedioate and mid-chain oxidation leads to 5-oxo-nonanedioate (3). The keto-dicarboxylic acid could then undergo cycliza-tion or reduction to the unstable 5-oxo-1,9-nonadiol (4) which easy cyclizes to the 1,7-dioxaspiro[5.5]undecane (1).

7 CONCLUSIONS

The olive fruit fly *D. oleae*, has a very complex pheromone system and our knowledge on its effect on the behaviour of the fly is still limited. Since 1980, research has concentrated primarily on the identification of the pheromone components. Detailed quantitative behavioural studies should be undertaken in the laboratory with the synthetic pheromone components to establish the role of each component on the mating behaviour of the fly.

The major female sex pheromone component 1,7-dioxaspiro[5.5]undecane was found to be produced also by males. Research is needed to elucidate the absolute configuration of the component produced by each sex and to clarify, using behavioural studies the effect of pure enantiomers on the behaviour of both sexes.

Field experiments conducted during the last four years with the synthetic pheromone showed that the pheromone will be a major component in future integrated pest management programs for control of the olive fruit fly.

8 REFERENCES

Baker, R., Herbert, R.H., Howse, P.E., Jones, O.T., Francke, W. and Reith, W., 1980. Identification and synthesis of the major sex pheromone of the olive fly, *Dacus oleae*, Journal of the Chemical Society, Chemical Communications, 1: 52–54.

Baker, R., Herbert, R.H. and Parton, A.H., 1982. Isolation and synthesis of 3 and 4-hydroxy-1, 7-dioxaspiro[5.5]undecane from the olive fly, *Dacus oleae*, Journal of the Chemical Society, Chemical Communications, 11: 601–603.

Cavalloro, R. and Delrio, G., 1971. Rivelli sul comportamenta sessuale di *Dacus oleae* Gmelin (Diptera: Trypetidae) in laboratorio. Redia, 52: 201–230.

DeMarzo, L., Nuzzari, L. and Solinas, M., 1978. Studio anatomico, istologico, ultrastrutturale, e fisiologico del retro ed osservazione etologiche alla possible produzione di feromone sessuali nel maschio di *Dacus oleae* Gmel. Entomologica XIV Barl. pp. 203–266.

Economopoulos, A.P., Gianakakis, A., Tzanakakis, M.E. and Voyatzoglou, A., 1971. Reproductive behavior and physiology of the olive fruit fly. 1. Anatomy of the adult rectum and odors emitted by adults. Annals of the Entomological Society of America, 64: 1112–1116.

Feron, M. and Andriew, A.J., 1962. Etude des signaux acoustiques du male dans le comportement sexual des *Dacus oleae* Gmel. (Diptera: Trypetidae). Annales des Epiphyties, 13: 269–276.

Francke, W., Hindorf, G. and Reith, W., 1979. Alkyl-1,6-dioxaspiro[4.5]decanes. A new class of pheromones. Naturwissenschaften, 66: 618–619.

Gariboldi, P.G., Jommi, G., Rossi, R. and Vita, G., 1982. Studies on the chemical constitution and sex pheromone activity of volatile substances emitted by *Dacus oleae*. Experientia, 38: 441–444.

Gariboldi, P., Verotta, L. and Fanelli, R., 1983. Studies on the sex pheromone of *Dacus oleae*. Analysis of the substances contained in the rectal glands. Experientia, 39: 502–505.

Haniotakis, G.E., 1974. Sexual attraction in the olive fruit fly, *Dacus oleae* (Gmelin). Environmental Entomology, 3: 82–86.

Haniotakis, G.E., 1977. Male olive fly attraction to virgin females in the field. Annales de Zoology-Ecologie Animale, 9: 273–276.

Haniotakis, G.E., 1984. Control of the olive fruit fly, *Dacus oleae* Gmelin. Mass trapping: Present status-Prospects. Proc. VIII Circum-Mediterranean plant protection meeting of European and Mediterranean Plant Protection Organization, Chanea, Grete, Greece Sept. 24–28, 1984. (in press).

Haniotakis, G.E., Mazomenos, B.E. and Tumlinson, J.H., 1977. A sex attractant of the olive fruit fly, *Dacus oleae*, and its biological activity under laboratory and field conditions. Entomologia Experimentalis et Applicata, 21: 81–87.

Jones, O.T., Lisk, S.C., Longurst, G., Howse, P.E., Ramos, P. and Campos, M., 1983. Development of a monitoring trap for the olive fly *Dacus oleae* (Gmelin) (Diptera: Tephritidae) using a component of its pheromone as lure. Bulletin of Entomological Research, 73: 97–106.

Kocienski, Ph. and Yeates, C., 1983. A new synthesis of 1,7-dioxaspiro[5.5]undecanes. Application to a rectal gland secretion of the olive fruit fly, *Dacus oleae*. Tetrahedron Letters, 24: 3905–3906.

Kovats, E., 1961. Zusammenhänge zwischen Struktur und Gaschromatographishen Daten Organischer Verbindungen. Zeitschrift für Analytische Chemie, 181: 351–360.

Mazomenos, B.E., 1983. Biosynthesis of a sex pheromone of the olive fruit fly, *Dacus oleae* (Gmel.). Ph. D. Thesis University of Gent, Belgium. 137 pp.

Mazomenos, B.E., 1984. Effect of age and mating on pheromone production in the female olive fruit fly, *Dacus oleae*. Journal of Insect Physiology, 30: 765–769.

Mazomenos, B.E. and Haniotakis, G.E., 1981. A multicomponent female sex pheromone of *Dacus oleae* (Gmelin), isolation and bioassay. Journal of Chemical Ecology, 7: 437–443.

Mazomenos, B.E. and Pomonis, J.G., 1983. Male olive fruit fly pheromone: Isolation, identification and lab-bioassays. In: R. Cavalloro (Editor), Fruit flies of Economic Importance. Proc. CEC/IOBC Intern. Symp. Athens/Nov., 1982. A.A. Balkema, Rotterdam, pp. 96–103.

Mazomenos, B.E. and Haniotakis, G.E., 1985. Male olive fruit fly attraction to synthetic sex pheromone components in laboratory and field tests. Journal of Chemical Ecology, 11: 397–405.

Mazomenos, B.E., Haniotakis, G.E., Tumlinson, J.H. and Ragousis, N., 1981. Isolation, identification, synthesis and bioassays of the olive fruit fly sex pheromone. Proc. Panhellenic, Congress of Agricultural Research, Kalithea, Halkidikis, Greece May 5–8, 1981, pp. 96–97.

Mori, K., Vematsu, T., Watanade, H., Yanaqi, K. and Manode, M., 1984. Synthesis of the enantiomers of 1,7-dioxaspiro[5.5]undecane, the components of the olive fly pheromone. Tetrahedron Letters, 25: 3875–3878.

Redlich, H. and Francke, W., 1984. Synthesis of enantiomerically pure 1,7-dioxaspiro[5.5]undecanes, pheromone components of the olive fly, *Dacus oleae*. Angewandte Chemie, International Edition in English, 23: 519–520.

Rossi, R., Carpita, A. and Vita, G., 1978. (*Z*)-6-nonen-1-ol and related compounds as attractant of the olive fruit fly, *Dacus oleae* (Gmelin) (Diptera: Trypetidae). Gazzetta Chimica Italiana, 108: 709–712.

Schultz, C.A. and Boush, G.M., 1971. Suspected sex pheromone glands in three economically important species of *Dacus*. Journal of Economic Entomology, 64: 347–349.

Stette, H. and Rauhut, H., 1958. Über spirocyclische Ketale verschiedener Ringgrösse. Chemische Berichte, 91: 2543–2548.

Tompkins, L. and Hall, J., 1981. The different effects on courtship of volatile compounds from mated and virgin *Drosophila* females. Journal of Insect Physiology, 27: 17–21.

Tsitsipis, J.A., 1975. Mass-rearing of the olive fruit fly *Dacus oleae* (Gmel.) at "Democritos". In:
    Controlling Fruit Flies by Sterile Insect Technique. IAEA Vienna STI/PUB/392, pp. 93–100.

Tzanakakis, M.E., Tsitsipis, J.A. and Economopoulos, A.P., 1968. Frequency of mating in females
    of the olive fruit fly under laboratory conditions. Journal of Economic Entomology, 61: 1309–
    1312.

Van de Pers, J.N.C., Haniotakis, G.E. and King, B.M., 1985. Electroantennogram responses from
    olfactory system in *Dacus oleae*. Entomologia Hellenica, 2: 47–53.

Vita, G., Anselmi, L. and Minelli, F., 1979. Description of a collection method and biological
    evaluation of the sexual pheromone of *Dacus oleae* (Gmelin). Com. Naz. Energ. Nucl. Rt/BIO/79/
    23, 13 pp.

Wood, D.L., 1982. The role of pheromones, kairomones and allomones in the host selection and
    colonization behavior of barke beetles. Annual Review of Entomology, 27: 411–416.

Zervas, G.A., 1982. Reproductive physiology of *Dacus oleae* (Gmel.) (Diptera: Trypetidae). Com-
    parison of a wild and artificially reared flies. Geoponica (in Greek), 282: 10–14.

Zouros, E. and Krimbas, C.B., 1970. Frequency of female bigamy in natural population of the olive
    fruit fly *Dacus oleae* as found by using enzyme polymorphism. Entomologia Experimentalis et
    Applicata, 13: 1–9.

3.4.3 *Ceratitis capitata*

O.T. JONES

1 INTRODUCTION

The availability of an efficient lure for monitoring the Mediterranean fruit fly (*Ceratitis capitata* Wiedemann) is of prime importance in programmes to control or eradicate this highly destructive pest. Although Trimedlure the parapheromone developed by Beroza et al. (1961) has fulfilled this role admirably throughout the world since the early 1960's, there have been several attempts to isolate the true pheromone of *C. capitata* in the hope of obtaining not only a more powerful attractant, but also one which would attract the females of this species. This section reviews the work done to date and underlines both the complex and often conflicting nature of the results obtained.

2 OLFACTORY STIMULI IN MATING BEHAVIOUR

The idea that olfactory stimuli are involved in the courtship behaviour of *C. capitata* was first postulated over 60 years ago (Martelli, 1910; Back and Pemberton, 1918). The first detailed laboratory study of its mating behaviour was not undertaken, however, until 40 years later. Feron (1959, 1962) described three distinct stages to the courtship process (translated here from the original French):

(i) "*Appearance of the male in Stage 1*: The male in a state of sexual excitement is immobile, resting on its legs in such a way that its entire body is raised on its legs. The abdomen is strongly retracted longitudinally at the posterior and inflated laterally. The extremity of the abdomen is raised in height and an anal ampoule appears to be inflated, shining like a droplet of liquid. The male remains in this attitude; the wings are immobile, and maintained perpendicular to the body."

(ii) "*Arrival of the female and Stage 2 of the male*: A female alights from flight some centimeters from the male. The male immediately pivots to face the female. At the same time, always resting on his legs, he lowers the anal ampoule below the abdomen, and begins a rapid vibration of the wings which are maintained laterally from the body. Approach of the female: The female faces the male and advances toward him, often very slowly. The male rests immobile, always in stage 2."

(iii) "*Stage 3 of the male and mating*: When the female is very close to the male and the heads of the two insects are about 2 or 3 mm apart, the excitement of the male is increased, characterized by a forward movement of the wings (still in vibration) with a rhythm of about 2 per second accompanied by rapid rotation of the head. The mounting of the female by the male then follows. The

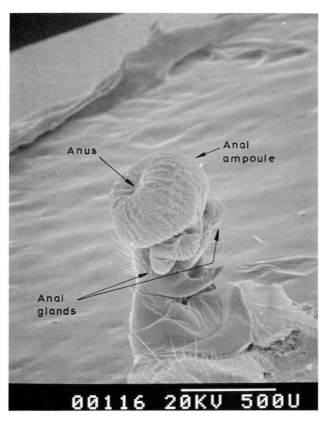

Fig. 3.4.3.1. Anal pouch (ampoule) and anal glands of a mature male *Ceratitis capitata*.

male jumps over the head of the female who raises the extremity of the abdomen with extension of the ovipositor sheath. The male turns on the back of the female, grasps the genital apparatus with his claspers and copulation is effected."

Feron (1962) conclusively demonstrated therefore that, in response to the first part of the above sequence, virgin females are highly attracted to the pheromones released from the distended tip of the anus. There has been some confusion however since the publication of this work regarding the nature of the anal ampoule. Some researchers have interpreted the ampoule as being *like* a droplet of liquid while others believe it to *be* a droplet of liquid. Careful translation of the original French plus scanning electron microscopy (SEM) studies (Fig. 3.4.3.1) and behavioural observations would indicate however that the anal ampoule is an inflated bulbous protuberance of rectal tissue.

It is quite possible that the surface of the ampoule appears moist during stage one and probably contains pheromone but males do not leave visible droplets on the surface of leaves etc. when they periodically touch the substrate upon which they are 'calling' with the tip of the abdomen. Indeed, when they 'mark' surfaces upon which they perform this behaviour, it is with quantities of pheromone which are too small to be seen with the naked eye or binocular microscope.

Olfactory stimuli therefore play an important role during the first stage of courtship but it is not known to what extent they are involved during the second and third stages. It has been shown conclusively, however, that auditory signals are involved (Rolli, 1976; Russ and Schwienbacher, 1982).

Although very useful and often rewarding in terms of the results which they give, laboratory experiments and observations can often lead to erroneous

TABLE 3.4.3.1

Chemical components of the pheromone produced by male *Ceratitis capitata*

Chemical component	Means of isolation	Identification by
Methyl-(*E*)-6-nonenoate (*E*)-6-nonenol Various fatty acids	aeration and condensation in cold trap	Jacobson et al. (1973)
(−)-*β*-fenchol (1*S-exo*-1,3,3-trimethylbicyclo-[2.2.1]heptan-2-ol)		Jacobson and Ohinata (1980)
3,4-dihydro-2*H*-pyrrole Ethyl-(*E*)-3-octenoate *E,E*-*α*-farnesene Geranyl acetate *E*-2-hexenoic acid Dihydro-3-methyl-2*H*-furanone 2-Ethyl-3,5-dimethyl-pyrazine Linalool Ethyl acetate	aeration followed by trapping in absorbent materials *or* solid sample gas chromatography- mass spectrometry	Baker et al. (1985)

conclusions simply because the insects are either not performing their full repertoir of behaviour, or are producing idiosynchratic or atypical behaviour as a consequence of being confined in small cages. The contribution of Prokopy and Hendrichs (1979) to our knowledge of *C. capitata* mating behaviour showed this to be the case. They made systematic observations of *C. capitata* mating behaviour on field-caged host trees growing in a plantation in Guatemala. It was found that, prior to the 'calling' behaviour described earlier in the literature from laboratory studies, males would first aggregate on the underside of leaves in leks from where they then attracted virgin females through calling in unisong. They also showed that an alternative positioning of males was possible for mating encounters; males sought out fruits marked with an oviposition deterring pheromone (see Chapter 3.5) deposited previously by fecund females and from where they courted any new ovipositing females.

These authors also noted that during courtship, *C. capitata* males frequently touched the leafy surface upon which they were standing with the tip of their abdomen, seemingly depositing some of their pheromone secretions and theraf-ter remained for some time within that treated area. They interpreted both this behaviour and lek formation for 'calling' as being of adaptive advantage through increasing the amount of pheromone emanating from a given locale thereby making it more attractive to the females.

3 CHEMICAL COMPONENTS OF THE PHEROMONE AND THEIR BIOLOGICAL ACTIVITY

Most of the chemical components isolated to date from the pheromone blend produced by male *C. capitata* are summarised in Table 3.4.3.1. Methyl-(*E*)-6-nonenoate, (*E*)-6-nonenol and a series of fatty acids were the first compounds to be isolated from the volatiles released by male *C. capitata*. Although the first two compounds produced an attraction response from both males and virgin females under laboratory bioassay conditions, only males were attracted to them in the field (Ohinata et al., 1977, 1979; Jacobson et al., 1973). Several papers have appeared since these early reports confirming the attraction of male *C. capitata* to methyl-(*E*)-6-nonenoate (Zumreoglu, 1982), but it is interesting to note that although over a decade has passed since its attraction to males was first demonstrated in the field, it has not displaced trimedlure as the standard attractant for monitoring this species.

Although biologically inactive, Jacobson and Ohinata (1980) recorded the presence of $(-)$-β-fenchol in male *C. capitata*, noting the fact that this was the first recorded finding of this chemical in the animal kingdom.

A recent publication by Baker et al. (1985) recorded a further nine compounds from the pheromone blend of male *C. capitata*. Preliminary laboratory bioassays indicated that the cyclic imine, 3,4-dihydro-2*H*-pyrrole, was the most active of these compounds and they tentatively concluded that this compound was the key component involved in the attraction of virgin females. In addition to the cyclic imine other components were found in large amounts; ethyl-(*E*)-3-octenoate, (*E*, *E*)-α-farnesene, geranyl acetate and (*E*)-2-hexenoic acid. The other four components, dihydro-3-methyl-furan-2-(3*H*)-one, 2-ethyl-3,-5-dimethyl-pyrazine, linalool and ethyl acetate were present as minor components. Attempts at finding any of the above compounds in aeration products of *C. capitata* females proved negative. Similarly, none of the compounds described earlier by Ohinata et al. (1977, 1979) and Jacobson et al. (1973) were found in these aeration studies.

Subsequent to the initial laboratory bioassays on the compounds described by Baker et al. (1985), more behaviourally discriminating tests were carried out in a wind tunnel followed by field evaluations (Howse and Foda, 1985) and the relative importance of the individual components in controlling the fly's field behaviour would appear to be different from that reported earlier by Baker et al. (1985). It was found that both the imine and other major components have an effect on virgin females activating them and stimulating upwind flight. However, such upwind flights were not of the zig-zag form which is usually observed in virgin females responding to a plume of odour from calling males or from males responding to a plume of trimedlure (Jones et al., 1981). Such a response was obtained, however, from both males and virgin females to one of the minor components by itself when presented to them as a plume in a wind tunnel. The insects landed near the source of the chemical and subsequently exhibited arrestment in that area. The authors of this work suggest that this component is multifunctional in nature, controlling attraction of males to other males during lek formation, attraction of virgin females to calling males, territorial behaviour of males on leaves and sexual discrimination processess on the leaf territory (Howse and Foda, 1985). These hypotheses are currently being investigated in field experiments both in Spain and Mexico. Preliminary trials in some spanish citrus orchards have produced catches of males in traps baited with one of the highly active minor components which were similar in number to those with trimedlure. In other concurrent trials, however, carried out in the same area, catches of virgin females only were obtained. It is thought that factors such as trap colour, presence of volatiles form ripe fruits as well as the synergistic or modifying effects of the other chemical components described account for these seemingly contradictory results.

Since most of our knowledge to date of pheromone mediated mating behaviour comes from studies with moths, we are probably approaching similar studies with Tephritidae with preconceived ideas. There is, for instance, no reason to believe that the chemicals isolated so far by aeration should all come from the same gland, or that a particular chemical or blend of chemicals can not produce different responses from either sex given different environmental conditions. For a clearer understanding of the chemical mediation of *C. capitata* mating behaviour, variables such as the following will have to be studied: dose response of individual components of the pheromone; additive, synergistic or even inhibitory effects of blends of these compounds; modulating effects of visual stimuli and host plant-derived odours. From the experience gained to date with this and other Tephritid species, it would seem prudent to use behavioural assays which reflect as fas as possible the field condition, backed up by field observations and trapping data.

4 ACKNOWLEDGEMENT

I thank Dr. J.L. Nation for supplying Fig. 3.4.3.1.

5 REFERENCES

Back, E.A. and Pemberton, C.E., 1918. The Mediterranean fruit fly. United States Department of Agriculture Bulletin, 640, 43 pp.

Baker, R., Herbert, R.H. and Grant, G.G., 1985. Isolation and identification of the sex pheromone of the Mediterranean fruit fly, *Ceratitis capitata* (Wied.) Journal of the Chemical Society, Chemical Communications, 824–825.

Beroza, M., Green, N., Gertler, S.I., Steiner, L.F. and Miyashita,D.H., 1961. Insect attractants. New attractants for the Mediterranean fruit fly. Journal of Agricultural Food Chemistry, 9: 361–365.

Feron, M., 1959. Attraction chimique du male de *Ceratitis capitata* Wied. (Dipt. Trypetidae) pour la femelle. Comptes Rendus des Séances de l'Académie des Sciences, Serie D: Sciences Naturelles (Paris), 248: 2403–2404.

Feron, M., 1962. L'instinct de reproduction chez la mouche Mediterraneene des fruits *Ceratitis capitata* Wied. (Dipt. Trypetidae) Comportement sexuel — comportement de ponte. Revue de Pathologie Vegetale et d'Entomologie Agricole de France, 41: 1–129.

Howse, P.E. and Foda, M.E., 1985. Pheromone communication in the Mediterranean Fruit Fly (*Ceratitis capitata* Wied). In: The NATO ASI Series, Series G, Ecological Sciences Volume 11 Pest Control, Operations and Systems Analysis in Fruit Fly Management. Springer Verlag, Berlin, p. 189.

Jacobson, M. and Ohinata, K., 1980. Unique occurrence of fenchol in the animal kingdom. Experientia, 36: 629–630.

Jacobson, M., Ohinata, K., Chambers, D.L., Jones, W.A. and Fujimoto, M.S., 1973. Insect sex attractants. 13. Isolation, identification and synthesis of sex pheromones of the male Mediterranean fruit fly. Journal of Medical Chemistry, 16: 248–251.

Jones, O.T., Lomer, R.A. and Howse, P.E., 1981. Responses of Male Mediterranean fruit flies, *Ceratitis capitata* to trimedlure in a wind tunnel of novel design. Physiological Entomology, 6: 179–181.

Martelli, G., 1910. Alcune note intorno ai costumi ed ai della mosca dele arance *Ceratitis capitata*. Bollettino del Laboratorio di Zoologia Generale e Agraria della R. Scuola Superiore d'agricoltura, Portici, 4: 120–127.

Ohinata, K., Jacobson, M., Nakagawa, S., Fujimoto, M. and Higa, H., 1977. Mediterranean fruit fly: laboratory and field evaluations of synthetic sex pheromones. Journal of Environmental Science and Health, A12: 67–78.

Ohinata, K., Jacobson, M., Nakagawa, S., Urago, T., Fujimoto, M. and Higa, H., 1979. Methyl-(*E*)-6-nonenoate: a new Mediterranean fruit fly male attractant. Journal of Economic Entomology, 72: 648–650.

Prokopy, R.J. and Hendrichs, J., 1979. Mating behaviour of *Ceratitis capitata* on a field-caged host tree. Annals of the Entomological Society of America, 72: 642–648.

Rolli, K., 1976. Die akustischen Signale von *Ceratitis capitata* Wied. und *Dacus oleae* Gmel. Zeitschrift für Angewandte Entomologie, 81: 219–223.

Russ, K. and Schwienbacher, W., 1982. Investigations on sound production of *Ceratitis capitata* L. In: Sterile Insect Technique and Radiation in Insect Control, I.A.E.A., Vienna, pp. 369–378.

Zumreoglu, A., 1982. Field cage evaluations of the male sex pheromone (Methyl-(*E*)-6-nonenoate) of the Mediterranean fruit fly (*Ceratitis capitata* Wied.). Proceedings of the CEC/IOBC International Symposium, R. Cavalloro (Editor). Fruit flies of Economic Importance, Athens, Greece, 16–19 November, 1982.

3.4.4 *Rhagoletis* spp.

B.I. KATSOYANNOS

1 INTRODUCTION

Behaviorally speaking, a sex (or mating) pheromone may be a locomotory stimulant, an arrestant, an attractant, and/or a sexual stimulant. The same chemical may elicit more than one of these responses. Certainly sex pheromone is not synonymous with sex attractant (Shorey, 1977 and references therein).

Sex pheromones, especially sex attractants, are known to play an important role in the mating behaviour of several fruit fly species. Yet, sex pheromone investigations in species of *Rhagoletis* began very recently and concern only two species, *Rhagoletis pomonella* (Walsh) and *Rhagoletis cerasi* (Linnaeus). One reason for this was the impression among *Rhagoletis* workers that such pheromones did not exist in this genus. This impression was supported by the fact that, contrary to other fruit fly species, no apparent particular behaviour could be observed indicating the release of a sex pheromone in *Rhagoletis*, such as fanning of wings or extruding a rectal ampoule to evaporate the pheromone. Also, no sex-specific glands were found in *R. pomonella* and *Rhagoletis juglandis* Cresson, the two species examined, and it has been suggested that at least in *R. pomonella*, which is known to produce a sex pheromone, the pheromone producing cells show little morphological differentiation and have been missed in the examination (Nation, 1981). Finally, various bioassay methods and different types of olfactometers failed in the past to indicate sexual attraction in *Rhagoletis*.

In the following, a brief account is presented of our knowledge of sex pheromones in *Rhagoletis* spp., without extensive reference to the mating behaviour of these flies, which is treated in more detail in Chapter 4.4.

2 SEX ATTRACTANTS

The first demonstration of a sex odour with properties of an attractant in a *Rhagoletis* species was reported from *R. pomonella* by Prokopy (1975), who found in studies conducted in large field cages that odour from virgin males was attractive to virgin females. Katsoyannos (1976), employing similar procedures as Prokopy (1975), found that mature virgin or not recently mated females of *R. cerasi* were attracted by an odour, apparently a volatile sex pheromone, released by mature males kept in groups. No attraction of males to females, males to males or females to females could be found in these studies. Based on these encouraging results, Katsoyannos (1979, 1982) elucidated further the biological and behavioral aspects of the observed attraction in *R. cerasi* in laboratory and field experiments, and tried to explain its role in the mating

---

Chapter 3.4.4 references, p. 188

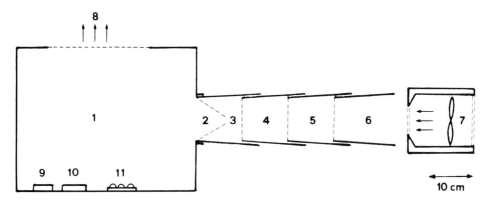

Fig. 3.4.4.1. An olfactometer unit. 1: Plexiglass test cage (40 × 30 × 30 cm) containing responding flies. 2: wire-screen funnel leading into a catch chamber (3). 4: intermediate tube. 5: bait cage containing the source of attraction (males). 6: end tube. 7: ventilator. 8: screen for airstream exit. 9, 10, 11: food, water and oviposition sites, respectively (after Katsoyannos, 1979).

behaviour of this fly. For the laboratory experiments, an olfactometer was developed, the construction and operation of which was described in detail by Katsoyannos (1979) and Katsoyannos et al. (1980). Briefly, it consists of two or more independent units (Fig. 3.4.4.1). Each unit consists of a holding cage for the virgin females (responding test flies), and a tube system through which the pheromone produced by the males (which are placed in a bait cage positioned into the tube system), is blown into the female-holding cage by an airstream generated by a ventilator. Responding females move toward the pheromone source, enter the tube through a wire screen funnel and are captured in a catch chamber where they are counted after the termination of a test. This olfactometer proved suitable for investigations of sex attractants and attractive odours of other fruit flies, such as *Ceratitis capitata* (Wiedemann) and *Dacus oleae* (Gmelin), for moths such as *Eupoecilia ambiguella* Hübner (Katsoyannos et al., 1980), and for wasps such as *Eurytoma amygdali* Enderlein (Pittara and Katsoyannos, 1985). It also proved very useful in monitoring the quality of mass-reared *C. capitata* flies (Boller et al., 1981). Field experiments consisted basically of releasing assay flies in an orchard and monitoring their response towards correctly positioned cages containing the source of attraction (Katsoyannos 1977, 1979, 1982).

Close correlation between results of the laboratory and field experiments was observed. These data confirmed the previous findings of Katsoyannos (1976) and provided additional evidence that mature *R. cerasi* males caged in groups release a volatile sex pheromone that attracts receptive (virgin and/or not recently mated), mature females. The response was positively correlated with the number of males per cage within the range tested (10–150 males). Females showed a daily rhythm of response, with a maximum between the 7th and 9th hour of an 18 h photophase in the laboratory and around noon under field conditions. Because the females loose their responsiveness for several days after mating, it was concluded, in connection with other related findings reported by Katsoyannos (1979), that *R. cerasi* is a rather oligogamous species and that high mating frequencies often observed under laboratory conditions involved forced matings.

It is beyond the scope of this brief chapter to discuss the mating behaviour of *Rhagoletis* species. However, it should be emphasized that mating success in *Rhagoletis* is probably not strongly dependent upon the male-produced sex pheromone. It is well documented that in *Rhagoletis* species mating activity occurs on host plants with host fruits being favorable meeting places for the sexes. As has been suggested (Katsoyannos 1979, 1982) for *R. cerasi* in the field, the male-produced volatile sex pheromone might serve mainly to attract receptive females to the proximity of pheromone-releasing males, which usually are concentrated in sunny host plant sites near, or on, the fruits, and that males in groups might be more effective in attracting females than individuals scattered throughout the tree.

3 MALE-ARRESTANTS

In a number of *Rhagoletis* species, and also other fruit fly genera, the female, immediately after oviposition, marks the fruit surface with a pheromone functioning as an oviposition-deterrent that discourages further egg-laying in marked fruits, thus contributing to uniformity of egg distribution among available oviposition sites (Chapter 3.5). The same pheromone, (or at least components of it, if not another deposited at the same time), plays also a role in mating behaviour by causing arriving males to spend more time on pheromone-marked than on unmarked ovipositions sites and hence functions as a male-arrestant, as has been demonstrated in *R. pomonella* (Prokopy and Bush, 1972), and *R. cerasi* (Katsoyannos, 1975, 1979). Also males of *R. pomonella*, but to a lesser extent than females, deposit a male-arrestant. The arresting effects of the female-produced pheromone were of short duration (less than 24 h) in *R. pomonella* (Prokopy and Bush, 1972), but lasted for at least 11 days in the case of *R. cerasi* under laboratory conditions (Katsoyannos, 1979).

Rhagoletis males coming in contact with the arrestant pheromone not only spend more time on the pheromone-marked oviposition sites but also appear to be sexually aroused. Significantly more homosexual activity (mating attempts among males) was observed between *R. cerasi* males caged with artificial oviposition sites treated with this pheromone than between males caged with untreated oviposition sites. This suggests that the pheromone may also act as a sexual stimulant (Katsoyannos, 1979).

In respect to the biological significance of these pheromones, it has been suggested that their deposition on host fruit by the females may increase the change of sexual encounters (Prokopy and Bush, 1972; Katsoyannos, 1975). Males arriving on a recently pheromone-marked fruit will remain there longer (arrestant effect) and will become sexually aroused (sexual stimulant effect). These males might also be stimulated to initiate the release of their volatile pheromone to attract receptive females to their vicinity (Katsoyannos, 1979).

In other *Rhagoletis* species known to deposit oviposition-deterring pheromones, research related to the above aspects has not been reported.

4 CONCLUSIONS

It is evident from the above that sex pheromone research in *Rhagoletis* is in its infancy, showing a delay in comparison to other fruit fly groups of economic importance. Even in the two *Rhagoletis* species so far investigated, i.e. *R. pomonella* and *R. cerasi*, many aspects need further elucidation. For example, the biological significance of the two kinds of phermones found in these species, i.e. the female-sex attractant and the male-arrestant and probable sexual stimulant, is not yet well understood. The range of attraction by the sex attractant and/or the amount of pheromone required to elicit response are not known with adequate precision. The possibility that this pheromone may also play a role at short distance by eliciting certain behaviours has not been investigated. Finally, nothing is known about the organs and the mechanisms of production, and little about release and perception of the pheromone. Neith-

Chapter 3.4.4 references, p. 188

er of the pheromones has been chemically identified. Similarly, the biological significance and properties of the male-arresting pheromone have been only partially investigated. However, considerable progress has been made in the research concerning the oviposition-deterring pheromone of *Rhagoletis*, which is suspected to be identical with the male-arresting one (Chapter 3.5).

Although until now the existence of sex pheromones has been demonstrated in only two *Rhagoletis* species, the similarity of behaviour among species of this genus indicates that sex pheromones probably are involved in the mating behaviour of additional species. It is anticipated that in the near future sex pheromone research in *Rhagoletis* will be intensified and expanded, providing a more solid basis for our understanding of the mating behaviour of this fruit fly group. Additional research will be needed for evaluation of possibilities to use these pheromones in control procedures.

5 ACKNOWLEDGEMENTS

Many thanks are due to Drs. M.E. Tzanakakis and J.L. Nation for critically reading the manuscript.

6 REFERENCES

Boller, E.F., Katsoyannos, B.I., Remund, U. and Chambers, D.L., 1981. Measuring, monitoring and improving the quality of mass-reared Mediterranean fruit flies, *Ceratitis capitata* Wied. 1. The RAPID quality control system for early warning. Zeitschrift für Angewandte Entomologie, 92: 67–83.

Katsoyannos, B.I., 1975. Oviposition-deterring, male-arresting, fruit-marking pheromone in *Rhagoletis cerasi*. Environmental Entomology, 4: 801–807.

Katsoyannos, B.I., 1976. Female attraction to males in *Rhagoletis cerasi*. Environmental Entomology, 5: 474–476.

Katsoyannos, B.I., 1977. Field testing of sex pheromone in *Rhagoletis cerasi*. In: E.F. Boller and D.L. Chambers (Editors), Quality Control, an Idea Book for Fruit Fly Workers. Bulletin SROP/WPRS 1977/5: pp. 85–86.

Katsoyannos, B.I., 1979. Zum Reproduktions- und Wirtswahlverhalten der Kirschenfliege, *Rhagoletis cerasi* L. (Diptera: Tephritidae). Dissertation Nr. 6409 ETH Zurich, 180 pp.

Katsoyannos, B.I., 1982. Male sex pheromone of *Rhagoletis cerasi* L. (Diptera, Tephritidae): Factors affecting release and response and its role in the mating behavior. Zeitschrift für Angewandte Entomologie, 94: 187–198.

Katsoyannos, B.I., Boller, E.F. and Remund U., 1980. A simple olfactometer for the investigation of sex pheromones and other olfactory attractants in fruit flies and moths. Zeitschrift für Angewandte Entomologie, 90: 105–112.

Nation, J.L., 1981. Sex-specific glands in tephritid fruit flies of the genera *Anastrepha*, *Ceratitis*, *Dacus* and *Rhagoletis* (Diptera: Tephritidae). International Journal of Insect Morphology and Embryology, 10: 121–129.

Pittara, I.S. and Katsoyannos, B.I., 1985. Male attraction to virgin females in the almond seed wasp, *Eurytoma amygdali* Enderlein (Hymenoptera, Eurytomidae). Entomologia Hellenica, 3: 43–46.

Prokopy, R.J., 1975. Mating behavior in *Rhagoletis pomonella* (Diptera: Tephritidae). V. Virgin female attraction to male odor. The Canadian Entomologist, 107: 905–908.

Prokopy, R.J. and Bush, G.L., 1972. Mating behavior in *Rhagoletis pomonella* (Diptera: Tephritidae). III. Male aggregation in response to an arrestant. The Canadian Entomologist, 104: 275–283.

Shorey, H.H., 1977. Interaction of insects with their chemical environment. In: H.H. Shorey and J.J. McKelvey Jr. (Editors), Chemical Control of Insect Behaviour. Theory and Application. Wiley-Interscience Publication, John Wiley & Sons, New York, Chichester, Brisbane, Toronto, pp. 1–5.

3.4.5 The Role of Pheromones in the Mating System of *Anastrepha* Fruit Flies

JAMES L. NATION

1 INTRODUCTION

The genus *Anastrepha* (Tephritidae) comprises a large group of tropical and subtropical flies. Some species lay their eggs in commercially valuable fruits such as guava, citrus, peach, mango, and loquat, and larval feeding and growth make the fruit unfit for human consumption. Fortunately, most species utilize wild or less economically important hosts. Unfortunately, the lack of an economic stimulus has contributed to a great paucity of information on all but a few *Anastrepha* species. There are published accounts of courtship and mating behavior for 3 species, and pheromone components have been identified from 2 species, but even in those the pheromone and its role in mating is incompletely known.

The general behavioral strategy used by a species to assemble the sexes for mating is commonly referred to as the mating system of the species. Characterization of the mating system includes an evaluation of (1) the number of mates acquired, (2) the manner in which mates are attracted, and (3) parental investment (Emlen and Oring, 1977). It will be the aim of this chapter to discuss the second of these, the attraction of mates, and to show how pheromones fit into the mating system of the *Anastrepha* species.

2 THE MATING SYSTEM OF *ANASTREPHA* SPECIES

Tephritid fruit flies generally have a polygenous mating system (Prokopy and Roitberg, 1984). A polygenous mating system was characterized by Emlen and Oring (1977) as one in which environmental or behavioral conditions cause clumping of females whereupon males are then able to monopolize the clumped females. They further subdivided polygenous mating systems into several categories, including resource defense polygyny and male dominance polygyny, both of which are represented among tephritids. In a resource based system, males defend resources and control mating of females that must have access to those resources. At least some species of *Rhagoletis* fruit flies have a system of resource defense polygyny, and males mate with females seeking access to these resources (Prokopy and Roitberg 1984, and references therein).

Some species in the genera *Anastrepha*, *Ceratitis* and *Dacus* have a non-resource based system that is best characterized as male dominance polygyny. In this system males do not defend resources, but they establish a dominance hierarchy among themselves, and females choose a mate based (presumably) upon male status. Male dominance polygyny often has led to evolution of communal displays by males and male competition involving male–male encounters. In a number of tephritids, including species of *Anastrepha, Ceratitis* and possibly *Dacus* it has resulted in sexual display and calling aggregations commonly called leks.

2.1 Lek behavior in fruit flies

Lek behavior in insects was documented in the Hawaiian *Drosophila* complex of flies (Spieth, 1970, 1974). Males and females of these *Drosophila* species feed communally without attempted courtship and mating. Upon feeding to repletion, males fly to twigs, small patches of shaded bark, or other nearby substrates where they immediately assume a characteristic display posture, puff pouches on the abdomen, fan the wings in brief bursts, and apparently release a sex pheromone to attract females.

Behavior similar to that of the Hawaiian *Drosophila* species has been described by Nation (1972), Perdomo (1974), Dodson (1978) and Burk (1983) for the Caribbean fruit fly, *Anastrepha suspensa* (Loew), by Morgante et al. (1983) and Malavasi et al. (1983) for *Anastrepha fraterculus* (Wiedemann), by Robacker and Hart (1985a) in *Anastrepha ludens* (Loew), and by Feron (1959), Lhoste and Roche (1960), and Prokopy and Hendrichs (1979) in *Ceratitis capitata* (Wiedemann). Although lek behavior as defined by Bradbury (1981) includes features in addition to display behavior and pheromone release, there are good reasons to believe both drosophilid and tephritid fruit flies meet the general criteria for lek behavior. The lekking behavior is one of the most powerful forces acting upon fruit fly evolution, but we also need to understand how selective forces acted on fruit flies over time to promote evolution of lek behavior.

2.2 Courtship and sexual display behavior in *Anastrepha* species

Premating behavior, including courtship and pheromone release behavior, of *A. suspensa, A. ludens* and *A. fraterculus* show many similarities. In these 3 *Anastrepha* species males attract females; there is no evidence for female produced pheromones, as in *Dacus oleae* (Gmelin) (Haniotakis, 1974), but the possibility of very close range or contact pheromones has not been thoroughly investigated in either female or male members of *Anastrepha* species. More extensive observations have been published on *A. suspensa*, and its courtship and pheromone release behavior will be summarized and such differences as are known in the other two species noted.

2.2.1 *A. suspensa* behavior

Perdomo (1974) made the first field observations of *A. suspensa* male aggregations, courtship and pheromone release behavior, and these were corroborated and extended by Burk (1983). Both workers reported male aggregations, or leks, occurring near the tips of branches of (usually) host trees, with each male on the underside of a leaf. Burk (1983) observed 7–8 males displaying and interacting within $0.25\,m^2$ of space. Males defended their leaf territories (Dodson, 1982; Burk, 1983) and fights often occurred, with peak numbers of male-male encounters occurring between 17.00 to 18.00 h.

Male behavior on the underside of leaf territories may be summarized as follows: Alert, but non-displaying males have the body raised slightly from the leaf surface, with wings at an acute angle to the long axis of the body. These males turn and face moving objects in their immediate vicinity. Sexually displaying males (Fig. 3.4.5.1) spread their wings at about 90° from the long body axis, and twist them so that the costal edge is up. Alternate wing waving occurs with first one wing moved backward through an arc of about 45° with

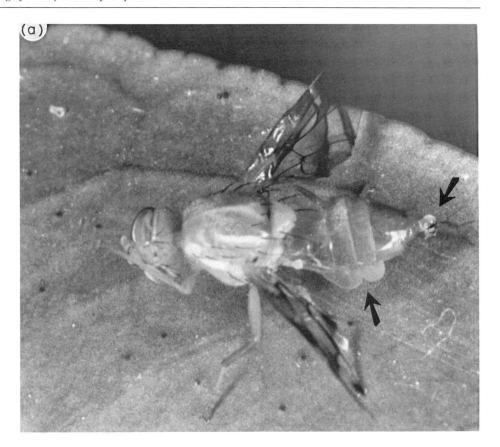

Fig. 3.4.5.1. Male display and pheromone releasing posture of *Anastrepha* species. (a). *Anastrepha ludens*. Arrows indicate lateral abdominal and anal pouches. Photograph courtesy of David Robacker, USDA, Weslaco, Texas. (b). *Anastrepha suspensa*, photography by author.

the costal edge dropped to the horizontal plane and then moved forward to a position perpendicular to the long axis of the body with the costal edge raised again to the vertical position. An observer, or another fruit fly, directly facing the displaying male sees the broad expanse of the "picture" wings moved alternatively forward and backward. Males often make short sideways runs, first in one direction and then in the other. Pheromone is apparently most actively released when the display behavior includes the puffing of lateral pouches of expandable pleural abdominal cuticle, and at intervals, eversion of

Chapter 3.4.5 references, p. 204

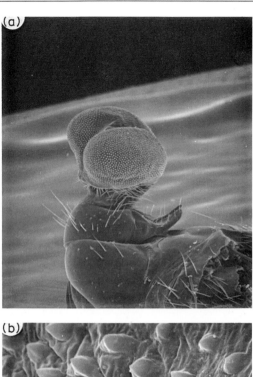

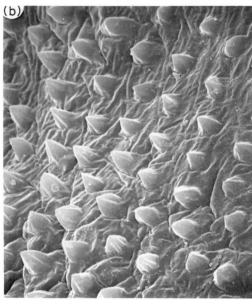

Fig. 3.4.5.2. (a). Everted anal pouch of male *Anastrepha suspensa*, 100 ×. (b). Detail of the everted pouch surface, 2000 ×, photographys by author.

anal tissue (Figs. 3.4.5.1 and 3.4.5.2) to form a pouch. The puffing behavior is punctuated by 1–3 s bursts of rapid wing fanning resulting in stereotyped tic signals (Webb et al., 1976). In addition to producing sounds possibly functioning in communication, the wing fanning directs a jet of air over the general abdominal body surface and over the everted anal pouch, and probably raises body temperature as a result of the metabolic work. Both actions aid in pheromone evaporation from the body. Frequently, after a burst of wing fanning, the leaf surface is touched by the abdominal tip, probably depositing small amounts of pheromone from the moist rectal tissue upon the leaf surface. Males often rotate the body to face a new direction and repeat the puffing, wing fanning, and dabbing of pheromone on the leaf surface. Live, sexually mature males used as bait in field traps attracted both females and males to the traps (Perdomo et al., 1975, 1976).

Perdomo (1974) and Burk (1983) reported observing male displays during morning hours, and the observed numbers of displaying males increased rapidly beginning about 16.00 h, and peaked between 17.00 and 18.00 h. The number of matings also increased in this period (Burk, 1983). Only one of the 26 matings observed in progress by Burk (1983) occurred in the morning hours, while Perdomo (1974) observed none during the morning, but saw 26 matings during the hours 13.25 to 19.55 h. Though mating typically occurred on the undersurface of leaves, Burk (1983) observed 1 mating pair on fruit, and 2 unsuccessful mating attempts there.

A female usually indirectly approached a lek of displaying males by short flights from branch to branch that culminated in arrival on the same leaf or one near a displaying male. These movements of females were reported by Perdomo (1974) to take long periods of time, often more than an hour, and frequently resulted in a female leaving the vicinity of a lek, especially if disturbed by other insects or sudden breezes. At very close range (2–3 cm) males usually ceased wing fanning and distention of pouches, spread the wings at 90° to the long body axis, and faced the female while slightly crouched as if ready to spring. Generally contact between the two, often the touching of the male by the female with her proboscis, was necessary before the male jumped above the female and, turning his body 180°, landed on top of the female facing the same direction.

2.2.2 *A. ludens* behavior

Early observations on sexual behavior of the Mexican fruit fly, *A. ludens* were recorded by Dieter Enkerlin and his students at the Instituto Technologico de Monterrey, Mexico. Enkerlin et al. (1975) showed that 4 to 9 day-old males attracted females and males in a laboratory olfactometer test. Additional work at Monterrey that generally confirmed male *A. ludens* as a source of pheromone was reviewed by Esponda–Gaxiola (1977).

Robacker and Hart (1985b) showed that underside of leaves was the characteristic site for male display and pheromone release, and for mating. Male displays included puffing of abdominal and anal pouches (Fig. 3.4.5.1A), and intermittent wing fanning. Wing waving by males was also common. In the field, mating occurred in the late afternoon and early evening (Baker et al., 1944).

2.2.3 *A. fraterculus* behavior

The behavior of *A. fraterculus* in the field and in field cages over a guava tree was described by Morgante et al. (1983) and by Malavasi et al. (1983). Male–male and male–female encounters in *A. fraterculus* occurred on leaf nodes or the under surface of leaves, but all mating pairs moved to the bottom surface of leaves. Male displays began at about 07.00 h in the summer at 18°C and 800 lux light intensity and ceased about 10.30 h. Males often displayed from a site for about 30 min before leaving. Aggregations or leks of 3–8 males within about 50 cm of each other were common. Male display behavior included puffing of lateral abdominal and anal pouches, and bursts of wing fanning. Copulation was initiated at sites other than the under surface of leaves, but a copulating female immediately walked to the under surface of a leaf.

3 CHEMICAL NATURE OF PHEROMONES IN *ANASTREPHA* SPECIES

Pheromone components have been identified from only *A. suspensa* and *A. ludens*. Two nonenols and 2 lactones with behavioral activity were isolated from *A. suspensa* male-body extracts and from aqueous washes of holding cages

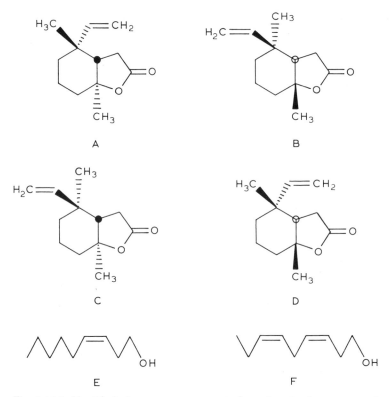

Fig. 3.4.5.3. Identified pheromone components from *Anastrepha suspensa*. Anastrephin, epiana-strephin and both nonenols also occur in *Anastrepha ludens*. A. *S,S*-(−)-anastrephin, B. *R,R*-(+)-anastrephin, C. *S,S*-(−)-epianastrephin, D. *R,R*-(+)-epianastrephin, E. (*Z*)-3-nonen-1-ol, F. (*Z,Z*)-3,6-nonadien-1-ol.

(Nation, 1975). The nonenols were identified as (*Z*)-3-nonenol and (*Z,Z*)-3,6-nonadienol (Nation, 1983) and the lactones were identified as anastrephin (*trans*-hexahydro-*trans*-4,7a-dimethyl-4-vinyl-2-(3*H*)-benzofuranone) and epi-anastrephin (*trans*-hexahydro-*cis*-4,7a-dimethyl-4-vinyl-2-(3*H*)-benzofuranone) (Fig. 3.4.5.3) (Battiste et al., 1983). Isolation of these same compounds from extracts of male Mexican fruit flies, *A. ludens*, was reported by Esponda–Gaxiola (1977), Nation (1977), Battiste et al. (1983) and Stokes et al. (1983). It is of some interest that these two flies utilize the same compounds as phero-mones, though release ratios differ. Additional unidentified compounds are present in *A. suspensa* male-volatiles and probably in those from *A. ludens*.

Anastrephin and epianastrephin are chiral lactones, and *A. suspensa* males produce an enantiomerically enriched mixture of each that approximates 55 ± 3% (−)-enantiomer to 45 ± 3% (+)-enantiomer (Battiste et al., 1983). It is not yet known if *A. ludens* males produce enantiomeric mixtures, but the females are differentially responsive to the pure enantiomers (Robacker, per-sonal communication, see also section 6). Robacker (personal communication) found these pheromone components in whole abdomen extracts of *A. ludens* in a ratio of 100 ng (*Z*)-3-nonenol, 40 ng (*Z,Z*)-3,6-nonadienol, 200 ng anastrephin, and 700 ng epianastrephin per male, and the volatiles released by males con-tained 100, 40, 60 and 300 ng, respectively, per male (Robacker and Hart, 1985b).

Although *A. suspensa* females were attracted in laboratory bioassays to individual nonenols or lactones and to various combinations (Nation, 1975), a test of a synthetic mixture (1 mg each nonenol, 0.09 mg (*S,S*)-(−)-anastrephin, 0.04 mg (*R,R*)-(+)-anastrephin, 0.3 mg (*S,S*)-(−)-epianastrephin, and 0.2 mg (*R,R*)-(+)-epianastrephin on a dental roll in Jackson traps) did not significant-ly attract flies of either sex during a 5-day field test in Homestead, Florida, in

October 1983. Insufficient quantities of the enantiomerically pure lactones prevented further exploration of possible problems such as release rate from the dental rolls and ratio of components, but this failure did lead directly to reexamination of the possibility, and discovery, of additional compounds in the pheromone blend of *A. suspensa*.

3.1 Collection of pheromone volatiles

The reexamination began with the collecting of pheromone volatiles on Tenax GC (Fig. 3.4.5.4) from displaying male *A. suspensa*, rather than preparing pheromone from whole body extracts as done earlier. The all-glass apparatus (Fig. 3.4.5.4A) was best for collecting "highly pure" samples for chemical analysis from large numbers of males (up to 500 were often used). An air inlet filter of Tenax GC scrubbed the incoming air, and all connections were made with ground glass joints. Two Tenax GC traps, each consisting of a glass tube 15 mm × 15 cm and containing 300 mg Tenax GC between silanized glass wool plugs, were connected to the water pump vacuum line. The smaller apparatus in Fig. 3.4.5.4B proved more satisfactory for routine collection of pheromone from 5 to 25 males. The rubber stopper at the outlet was covered with aluminum foil except for the exit tube opening. The water pump was adjusted to cause a flow of approximately 1 l of air per minute over the males. After a predetermined time of collection, the Tenax traps were removed, and trap 2 was eluted into trap 1 with pentane. Elution was continued until 10 ml eluate from trap 1 was collected. Preliminary analyses of the eluates from traps 1 and 2 separately showed that usually all the pheromone was found in the first trap and the second one acted as a safety trap.

3.2 Bioassay and quantitation of pheromone

Bioassays, conducted by a slightly modified method from that of Robacker and Hart (1984), showed that females of *A. suspensa* were attracted and aggregated under the paper treated with 0.09 male-hour equivalents (MHEQ) of pheromone from 5 day-old males. One MHEQ was defined as the mean quantity of pheromone released by 1 male in 1 hour; unfortunately this quantity of pheromone varies with male age and hour of collection as will be shown (see Table 3.4.5.1).

The pheromone compounds were separated by gas chromatography on a 3% OV-1 packed column, 4 mm ID × 1.8 m, and each component was quantified on the basis of an internal standard added immediately after the pheromone had

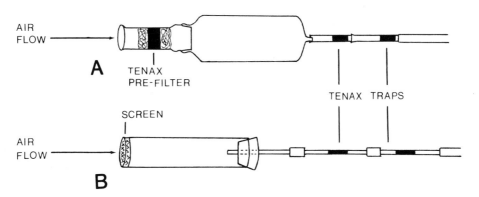

Fig. 3.4.5.4. Apparatus for collecting released volatiles from displaying male *Anastrepha suspensa*. A. All-glass apparatus with 11.5 cm × 35 cm chamber for holding large numbers of flies. B. Smaller apparatus with 4.2 cm × 25 cm glass cylinder for 5 to 25 flies.

Chapter 3.4.5 references, p. 204

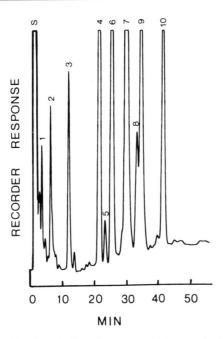

Fig. 3.4.5.5. Gas chromatographic record of separated volatiles trapped from male *Anastrepha suspensa*. Chromatographic conditions included temperature programming on 3% OV-1 column, 1.8 m × 4 mm, starting at 95°C, holding for 8 min, then 1°C/min to 125°C and holding 15 min. Nitrogen flow was 28 ml/min. S, Solvent peak, 1. monoterpene, 2. combined nonenols, 3. decanol for an internal standard, 4. tetradecane as internal standard, 5. unknown, 6. lactone B, 7. sesquiterpene, 8. anastrephin, 9. epianastrephin, 10. tridecanoic acid methyl ester as internal standard.

been eluted from the Tenax trap. A gas chromatographic record showing separation of the pheromone components and 3 internal standards added for quantitation is shown in Fig. 3.4.5.5. Collections from 25 female flies or from an empty chamber gave none of the male compounds.

Ten components were found in the male volatiles from wild *A. suspensa*. Peak 2 (Fig. 3.4.5.5) consists of (Z)-3-nonenol and (Z,Z)-3,6-nonadienol in a ratio of 1/1.5, respectively. Peak 8 represents the 2 enantiomers of anastrephin while peak 9 contains the 2 enantiomers of epianastrephin (the lactones are counted as 4 compounds in arriving at 10 components in the male volatiles). Peak 1 is a monoterpene. Peak 6 has the same molecular weight as anastrephin and epianastrephin, and the mass spectrum is similar to these two lactones; it is designated here as lactone B. Peak 5 is uncharacterized. Peak 7 is a sesquiterpene (personal communication, M.A. Battiste). Peaks 3, 4 and 10 are, respectively, decanol, tetradecane and tridecanoic acid methyl ester added as internal standards for quantitation of the pheromone components. Peaks 6 and 7 have been isolated from the male volatiles in sufficient quantity for bioassay and they aggregate females in the Robacker and Hart (1984) bioassay. Peaks 1 and 5 have not been isolated for bioassay. Data collected with this system during 1984 indicated that wild males collected as pupae from infested fruit in Homestead, Florida, produce the same volatiles and in approximately the same ratio and overall quantities as laboratory reared flies.

4 PHEROMONE GLANDS IN *ANASTREPHA* SPECIES

Beneath the lateral abdominal pouches of *A. suspensa, A. ludens, A. fraterculus,* 4 additional *Anastrepha* species, and 2 *Ceratitis* species are 2 male-specific glands, the pleural glands consisting of enlarged columnar epidermal cells,

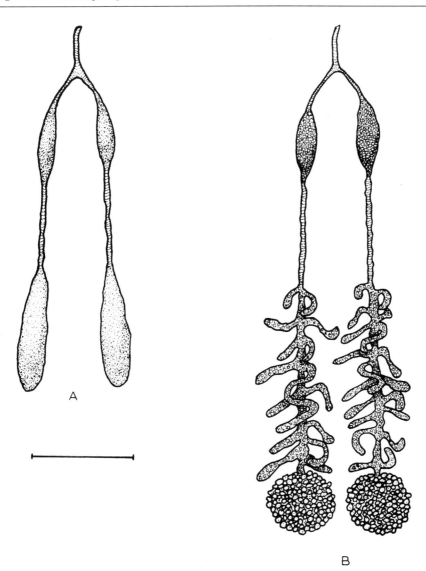

Fig. 3.4.5.6. Salivary glands from female and male *Anastrepha fraterculus*. A. Female glands, B. Male glands. Bar = 1 mm. Redrawn from an original by Humberto Martinez Arteaga of Santiago, Chile.

and the male-dimorphic salivary glands (Figs. 3.4.5.6 and 3.4.5.7) (Nation, 1981). The growth of the glands in *A. suspensa* is highly correlated with onset of sexual behavior and pheromone release (Nation, 1974). These glands may be characteristic of the genera *Anastrepha* and *Ceratitis*, but more species should be examined.

The anal pouch is everted rectal tissue that can be folded within the rectum at rest. There are no anal glands in this tissue as in *C. capitata* males or *Drosophilia grimshawi* Oldenburg. When the tissue is everted (Figs. 3.4.5.1 and 3.4.5.2) it has a shiny white appearance that might suggest a bubble of liquid. I have observed male display behavior in only *A. suspensa* and *C. capitata*, but I believe that there is no liquid bubble in either of these, and that what the observer sees is the everted anal tissue. The tissue is moist and does contain pheromone on its surface. The cuticular surface is irregular and raised into small mounds, often pointed at the apex (Fig. 3.4.5.2) that allow more surface from which the pheromone can evaporate. The epidermal cells beneath this cuticle are thin and squamous.

Hodosh et al. (1979) report that many Hawaiian *Drosophila* species, and

Chapter 3.4.5 references, p. 204

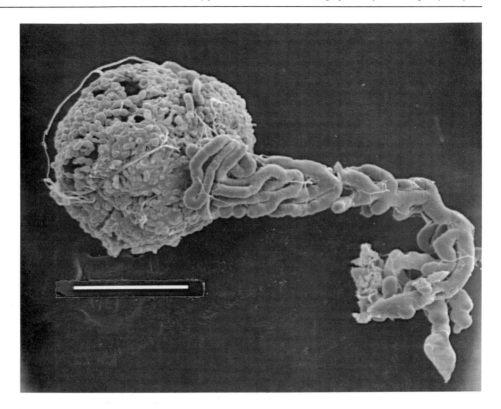

Fig. 3.4.5.7. Scanning electron micrograph of a male salivary gland from *Anastrepha suspensa*, 50 X. Scale bar = 0.5 mm. Photograph by author.

specifically *D. grimshawi*, release a droplet of liquid from paired, everted intra-anal lobes. Furthermore these lobes contain canaliculi and ducts that join to the alimentary canal just anterior to the anus. It was postulated that the droplet of liquid was produced by the epidermal cells of the pouches, secreted into the ducts, and eventually released from the anus during pheromone display behavior. While the anal pouch of *A. suspensa* superficially appears similar to that of *D. grimshawi* when everted, in *A. suspensa* there are no ducts and no droplet is released from the anus.

Male *A. suspensa* store relatively little pheromone in the body (Table 3.4.5.1), and the amount stored approximates that released in about 1 h at the peak release rate (see Table 3.4.5.3). The duration of leks in the field is unknown, but based upon the observations of Perdomo (1974) and Burk (1983) it seems likely that individual males display and release pheromone for 3–4 h or longer each afternoon. Data in Table 3.4.5.2 show that in the laboratory males are capable of synthesizing and releasing large quantities of pheromone during a 21 h continuous experiment.

The nonenols and anastrephin and epianastrephin of the pheromone are concentrated in the hindgut of males, while the sesquiterpene and lactone B are located primarily in the salivary glands and in epidermal cells, especially in the enlarged epidermal cells of the pleural glands (Table 3.4.5.1). The hindgut typically contained 70% or more of the nonenols in a whole body extract, and, on average, 58% of the anastrephin and epianastrephin of the whole body.

Characteristically, only trace quantities of the sesquiterpene and of lactone B were found in hindgut, but the 2 salivary glands/male contained about 30% of the body content of these 2 components. Salivary glands contained from traces to 7% of the nonenols and anastrephin plus epianastrephin. The lateral abdominal cuticle with the underlying epidermal and fat body cells (the pleural gland, Nation, 1974) on each side of the abdomen contained about 20% of the

TABLE 3.4.5.1

Pheromone components in a whole body extract and in various tissues of the body of laboratory reared male Caribbean fruit flies

Tissue	nanograms/male[1]			Anastrephin and Epianastrephin
	Nonenols	Lactone B	Sesquiterpene	
Salivary gland	5	86	99	20
Hindgut	76	3	3	384
Pleural gland and cuticle[2]	0[3]	52	53	trace
Whole body	69	204	332	661

[1] Mean values are given, but the means do not include cases when only trace levels below the quantitative capability of the GC method occurred. Usually salivary glands contained only traces of nonenols and anastrephin and epianastrephin, and hindgut contained only traces of sesquiterpene or lactone B in 9 experiments.
[2] The tissue taken contained the cuticle and underlying layer of epidermal and fat body cells and tergosternal muscle (Nation, 1974).
[3] The pleural glands never contained traces of nonenols and contained traces only of anastrephin and epianastrephin in 7 experiments.

body content of the sesquiterpene and lactone B, but only traces of anastrephin and epianastrephin, and no detectable nonenols.

4.1 An hypothesis of pheromone release by *A. suspensa*

Evidence from isolated gland and tissue analyses suggest that the sesquiterpene and lactone B are derived from the salivary glands and from the pleural glands. The sesquiterpene and lactone B from the salivary glands are released from the mouth, while those from the pleural glands appear to be secreted through the cuticle and allowed to evaporate from the cuticular surface. The nonenols and anastrephin and epianastrephin are concentrated in the hindgut, though the exact site of synthesis has not been located. Males release the nonenols and these two lactones in fecal deposits, and evaporate them from the anal opening and anal pouch. Males may also swallow air during the puffing behavior and expel pheromone-laden air from the gut in pulses.

5 FACTORS INFLUENCING RELEASE OF PHEROMONE BY *A. SUSPENSA*

Age of males is a major factor in pheromone release as shown previously by bioassays (Nation, 1972) and in measurements of pheromone released (Table 3.4.5.2). The earliest age that pheromone can be collected from wild males is 5 days, but it can be collected from laboratory reared males at 3 days. The peak in pheromone release by wild males occurred at or just before 14 days of age (Table 3.4.5.2) (a graphic plot of the data in Table 3.4.5.2 suggests that the maximum occurred at about 12 days of age). These results agree well with data showing that wild flies mate most actively between 12–16 days of age (Mazomenos et al., 1977; Dodson, 1982). Laboratory reared flies mature and mate earlier, at 7–9 days (Nation, 1972; Leppla et al., 1976; Mazomenos et al., 1977; Dodson, 1978; Burk and Webb, 1983), begin to release pheromone earlier than wild flies (at 3 days vs. 5 days), and achieve maximum release rate at 7 days of age.

Pheromone is released in greatest quantities during the late afternoon hours (laboratory photoperiod regime = lights on 06.00 h, off at 20.00 h) by both wild

TABLE 3.4.5.2

Pheromone released by wild Caribbean fruit fly males in laboratory tests. Collection of pheromone from 25 wild males was conducted from 12.00 h one day until 08.00 h the next day with laboratory lights continuously on during this period. Flies also had food and water

Age (days)	nanograms/male/h				
	Monoterpene	Nonenols	Lactone B	Sesquiterpene	Anastrephin and Epianastrephin
5	3.7	14.4	27.1	14.0	28.4
10	6.6	28.0	110.4	107.9	199.5
14	5.8	32.2	152.2	146.7	279.0
16	14.8	25.1	122.5	132.7	222.2
19	2.8	18.1	73.3	88.1	146.7

and laboratory reared *A. suspensa* (Table 3.4.5.3). Wild males showed a broad peak release rate between 16.00 and 20.00 hours. Laboratory reared flies were tested in their prime, at 7 days of age, and they showed a 1 h peak release rate between 16.30 and 17.30 h. The times for maximum release rate of pheromones by wild and laboratory reared males are consistent with the times observed by Perdomo (1974) and Burk (1983) for maximum male display behavior and mating in the field.

The release rate in Table 3.4.5.3 for wild males should not be interpreted as the maximum release rate of which the flies are capable, because the flies in this experiment were 21 days old, and thus past their prime age for pheromone production and release. The release rate for the laboratory reared males probably is representative of the maximum to be expected, although variation exists with different cohorts of flies, possibly related to rearing conditions, handling, and natural variation.

Pheromone release essentially ceases during the time flies are in copula, which lasts about 30 min in *A. suspensa* (Nation, 1972; Mazomenos et al., 1977).

TABLE 3.4.5.3

Daily rhythm in release of pheromone by wild and laboratory reared Caribbean fruit fly males under laboratory conditions. Photoperiod regime for flies was lights on at 06.00 h and off at 20.00 h, except during the time that pheromone was actually being collected, at which time lights were continuously on

Time	nanograms/male/h			
	Nonenols	Lactone B	Sesquiterpene	Anastrephin and Epianastrephin
21 day-old wild flies				
12.00–14.20	20.6	51.4	40.3	34.5
14.20–16.00	40.0	172.3	154.1	232.9
16.00–18.00	68.2	163.6	220.1	542.2
18.00–20.00	62.2	132.5	214.8	542.7
20.00–22.00	27.1	123.8	170.3	228.2
22.00–24.00	28.7	127.5	153.9	214.5
24.00–09.00	18.1	103.9	117.6	216.3
7 day-old laboratory reared flies[1]				
15.30–16.30		194	341	265.0
16.30–17.30		454	612	1012.0
17.30–18.30		346	468	900.0

[1] Due to certain technical difficulties, nonenols could not be accurately measured in the experiment with laboratory reared flies.

TABLE 3.4.5.4

Pheromone released by laboratory reared male Caribbean fruit flies during the first hour after mating, compared with pheromone released by a group of non-mated males from the same cohort of flies taken at random by the experimenter. Pheromone was collected during 1 h between 16.00–18.00 h in the laboratory

Rep. No.	nanograms/male/h							
	Mated males				Unmated males			
	N[1]	B	ST	L	N	B	ST	L
1	125	211	277	442	88	187	290	406
2	375	271	359	515	356	161	265	394
3	136	296	258	246	125	263	257	101
4	156	306	298	567	178	298	404	494
5	151	272	262	399	108	183	226	207

[1]N = nonenols, B = lactone B, ST = Sesquiterpene, and L = the lactones anastrephin plus epianastrephin.

The pheromone collected from 22 mating pairs contained an average of 13.3 ng nonenols, 7.1 ng lactone B, 13.8 ng sesquiterpene and 22.1 ng anastrephin plus epianastrephin per pair. Similar amounts of pheromones were collected from freshly killed (by freezing) male flies. Possibly a large percentage of that collected was already on the surface of the males' bodies when mating began.

Sivinski (1984) found that mated *A. suspensa* males are less likely to successfully re-mate within the first hour after mating, even though they are as vigorous as unmated males in acoustic advertisement. Mated males began to release pheromone immediately after mating ceased, and they produced and released during the first hour after mating at least as much as the non-mated males (Table 3.4.5.4).

The availability and quality of food that wild flies have access to is probably a factor in pheromone release. Good nutrition is important to pheromone synthesis as shown by the 40–70% reduction in pheromone released when all yeast hydrolysate was withheld from laboratory reared flies (Table 3.4.5.5).

6 ELECTROANTENNOGRAM (EAG) STUDIES

Electrophysiological recording of antennal receptor neuron responses to pheromones has been a valuable tool in identification of pheromones (Roelofs,

TABLE 3.4.5.5

Influence of nutrition upon pheromone release by laboratory reared Caribbean fruit fly males in the laboratory. Values are mean ± S.D. from 3 replicates with males of 7, 8 and 11 days of age. Flies designated "without yeast" were denied yeast hydrolysate for their entire adult life

Pheromone component	nanograms/male/h		% reduction without yeast
	With yeast	Without yeast	
Nonenols	43.1 ± 3.7	25.9 ± 11.1	39.9
B[1]	92.8 ± 12.2	38.4 ± 14.0	58.6
ST	156.6 ± 0.1	46.2 ± 20.2	70.5
Anastrephin plus epianastrephin	205.6 ± 48.7	82.2 ± 23.1	59.7

[1]B = lactone B, ST = sesquiterpene.

Chapter 3.4.5 references, p. 204

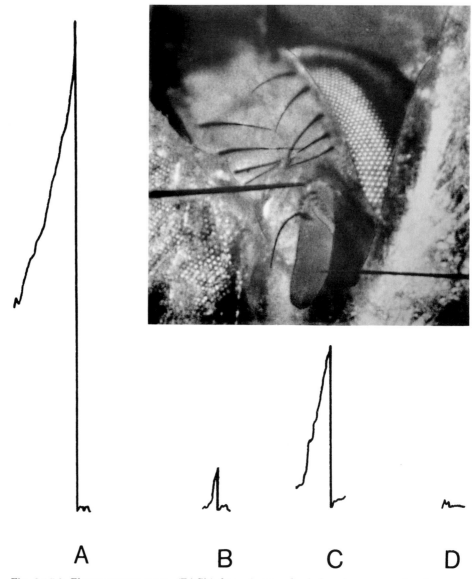

A   **B**   **C**   **D**

Fig. 3.4.5.8. Electroantennograms (EAG's) from *Anastrepha ludens* in response to pheromone extracts and some synthetic pheromone components. Inset shows placement of reference glass electrode at base of antenna with active capillary electrode contacting antennal surface near midpoint. A. Approximately 3 mV response to 1 MEQ *A. ludens* abdominal extract, B. response to 140 ng (R,R)-(+)-epianastrephin, C. response to 140 ng (S,S)-(−)-epianastrephin, D. control response to air/hexane (solvent). Photographs and EAG recordings courtesy of David Robacker, USDA, Weslaco, Texas.

1977; Moorhouse et al., 1969; Arn et al., 1975) and in study and identification of host plant attractants in several dipterans (Guerin and Städler, 1982).

Robacker and Hart (personal communication) have begun EAG studies of the response by *A. ludens* to the nonenols and lactones of its pheromone. The greatest EAG response from both male and female antennae was obtained to 1 male equivalent (1 MEQ) of abdominal extract (Fig. 3.4.5.8). Responses to single enantiomers of the lactones (tested at 1 MEQ) were in the order (S,S)-(−)-epianastrephin > (R,R)-(+)-epianastrephin > (S,S)-(−)-anastrephin > (R,R)-(+)-anastrephin. Behavioral responses in the laboratory by females were in the same order. EAG responses to 1 MEQ of (Z,Z)-3,6-nonadienol were greater than responses to 1 MEQ of (Z)-3-nonenol, and again laboratory behavioral responses followed the same pattern. Generally, tests of combinations of these alcohols and lactones tended to be additive, with the averaged EAG responses (from both females and males) to 1 MEQ of a combination of the 2 nonenols and 4 lactone enantiomers only 0.6% lower than the sum of the individual responses. This combination, however, elicited an EAG response equal to only about 60% of the response to 1 MEQ of abdominal extract. This may be indicative of additional unidentified components in the crude abdominal extract.

Preliminary EAG data from *A. suspensa* (conducted by David Robacker) in response to the 4 lactone enantiomers and 2 nonenols (each component at 0.7 nmol and at 0.35 nmol) showed mean EAG responses in the order (S,S)-$(-)$-epianastrephin $\gg$ (R,R)-$(+)$-epianastrephin and (S,S)-$(-)$-anastrephin $>$ (R,R)-$(+)$-anastrephin. The two nonenols produced nearly equal responses at 0.35 nmol each, but the response to (Z,Z)-3,6-nonadienol was greater than that to (Z)-3-nonenol at 0.7 nmol each. Although the test concentrations used with *A. suspensa* are not characteristic of the pheromone blend released by males (see Tables 3.4.5.2 and 3.4.5.3) the behavioral responses in the laboratory (conducted by the author) to the lactones and nonenols follow again the order of the EAG responses. Additional EAG studies are needed to determine the location and number of receptor sites and their specificity.

7 SUMMARY

Lek behavior, the system of male dominance polygyny, and pheromonal attraction of mates to lek sites are the salient features of behavior in the *Anastrepha* species to date, but our knowledge comes form only 3 species. The identified pheromonal components from *A. suspensa* and *A. ludens* are identical nonenols and chiral lactones, but unidentified components are present in the male-released volatiles from *A. suspensa*, and additional unidentified components are suspected in *A. ludens*.

Anastrephin, epianastrephin and nonenols are released by *A. suspensa* males from the posterior gut via the anal opening and by evaporation from the anal pouch surface. The origin of the sesquiterpene and of lactone B in *A. suspensa* seems to be the salivary glands and epidermal cells, and they are released from the mouth and by evaporation from the general cuticular body surface.

Factors known to influence the release (and by inference, synthesis) of pheromonal components by *A. suspensa* males are age, circadian rhythm, and nutritional status. Wild and laboratory-adapted *A. suspensa* released the same pheromonal components and in approximately the same proportions.

Without doubt, lek behavior and female choice have been two major forces of selection acting upon the evolution of some, if not all, *Anastrepha* species. The more than 100 species of *Anastrepha* present an extremely rich field for comparative study of mating systems, lek behavior, and pheromone chemistry. We need much additional field work to know if all members of the genus have lek behavior and what, if any, variations exist in the mating system. How did a chemical communication channel containing so many components, some of which are chiral and thus add to the complexity, evolve in *A. suspensa* and *A. ludens*? Do these species actually use the chiral information in their pheromone components? Are some pheromone receptors chiral? Do the pheromonal components play separate, but perhaps related, roles in controlling the behavior of both females and males? Perhaps through comparative studies we can learn the answers to these and other questions that will occur to those trained in behavior, ecology and chemistry.

Chapter 3.4.5 references, p. 204

8 ACKNOWLEDGEMENTS

I thank David Robacker and W.G. Hart, USDA, ARS, Weslaco, Texas for generous sharing of unpublished data and for some of the figures, and Humberto Martinez Arteaga of Santiago, Chile for the unpublished drawing in Fig. 3.4.5.6. I thank Kathy Dennis, who provided technical assistance in the laboratory and in preparation of figures; and Glinda Burnett, who typed the manuscript in a professional manner. Dr. James Lloyd and Dr. Merle Battiste, University of Florida, gave valuable criticism on early drafts of the manuscripts.

9 REFERENCES

Arn, H., Städler, E. and Rauscher S., 1975. The electroantennographic detector — a selective and sensitive tool in the gas chromatographic analysis of insect pheromones. Zeitschrift für Naturforschung, 30: 722–725.

Baker, A.C., Stone, W.E., Plummer, C.C. and McPhail, M., 1944. A review of studies on the Mexican fruitfly and related Mexican species. USDA Miscellaneous Publication No. 531, Washington, D.C.

Battiste, M.A., Strekowski, L., Vanderbilt, P., Visnick, M., King, R.W. and Nation, J.L., 1983. Anastrephin and epianastrephin, novel lactone components isolated from the sex pheromone blend of male Caribbean and Mexican fruit flies. Tetrahedron Letters, 24: 2611–2614.

Bradbury, J.W., 1981. The evolution of leks. In: R. Alexander and D. Tinkle (Editors), Natural Selection and Social Behavior, Chiron Press, Oxford, (England), pp. 138–169.

Burk, T., 1983. Behavioral ecology of mating in the Caribbean fruit fly, *Anastrepha suspensa* (Loew) (Diptera:Tephritidae). Florida Entomologist, 66: 330–344.

Burk, T. and Webb, J.C., 1983. Effect of male size on calling propensity, song parameters, and mating success in Caribbean fruit flies, *Anastrepha suspensa* (Loew) (Diptera:Tephritidae). Annals of the Entomological Society of America, 76: 678–682.

Dodson, G.N., 1978. Behavioral, anatomical, and physiological aspects of reproduction in the Caribbean fruit fly, *Anastrepha suspensa* (Loew). M.S. Thesis, University of Florida, Gainesville, FL.

Dodson, G.N., 1982. Mating and territoriality in wild *Anastrepha suspensa* (Diptera:Tephritidae) in field cages. Journal of the Georgia Entomological Society, 17: 189–200.

Emlen, S.T., and Oring, L.W., 1977. Ecology, sexual selection, and the evolution of mating systems. Science, 197: 215–223.

Enkerlin, D., Steer, D.A. and Gonzalez, E., 1975. Pheromone and radiation studies for the Mexican fruit fly *Anastrepha ludens* (Loew). Report to Insect Eradication of Pest Control Section FAO/IAEA Division, Vienna, Austria.

Esponda-Gaxiola, R.E., 1977. Contribution al estudio quimico del atrayente sexual de la mosca Mexicana de la fruta "*Anastrepha ludens* (Loew)". Thesis, Instituto Technologico y de Estudios Superiores de Monterrey, Monterrey, N.L., Mexico.

Feron, M.M., 1959. Attraction chimique du male de *Ceratitis capitata* Wied. (Dipt.:Trypetidae) pour la femelle. Comptes Rendus Hebdomadaires des Seances de l'Academie des Sciences (Paris), 248: 2403–2404.

Guerin, P.M. and Städler, E., 1982. Host odour perception in three phytophagous Diptera — a comparative study. Proceedings of the 5th International Symposium Insect-Plant Relationships, Wageningen, 1982. Pudoc, Wageningen, pp. 95–105.

Haniotakis, G.E., 1974. Sexual attraction in the olive fruit fly, *Dacus oleae* (Gmelin). Environmental Entomology, 3: 82–86.

Hodosh, R.J., Keough, E.M. and Ringo, J.M., 1979. The morphology of the sex pheromone gland in *Drosophila grimshawi*. Journal of Morphology, 161: 177–184.

Leppla, N.C., Huettel, M.D., Chambers, D.L. and Turner, W.K., 1976. Comparative life history and respiratory activity of "wild" and colonized Caribbean fruit flies. Entomophaga, 21: 353–357.

Lhoste, J. and Roche, A., 1960. Organes odoriferants des males de *Ceratitis capitata*. Bulletin de la Societie Entomologique de France, 65: 206–210.

Malavasi, A., Morgante, J.S. and Prokopy, R.J., 1983. Distribution and activities of *Anastrepha fraterculus* (Diptera:Tephritidae) flies on host and nonhost trees. Annals of the Entomological Society of America, 76: 286–292.

Mazomenos, B., Nation, J.L., Coleman, W.J., Dennis, K.C. and Esponda, R., 1977. Reproduction in Caribbean fruit flies: comparisons between a laboratory strain and a wild strain. Florida Entomologist, 60: 139–143.

Moorhouse, J.E., Yeardon, R., Beevor, P.S. and Nesbitt, B.F., 1969. Method for use in studies of insect chemical communication. Nature, London, 223: 1174–1175.

Morgante, J.S., Malavasi, A. and Prokopy, R.J., 1983. Mating behavior of wild *Anastrepha fraterculus* (Diptera:Tephritidae) on a caged host tree. Florida Entomologist, 66: 234–241.

Nation, J.L., 1972. Courtship behavior and evidence for a sex attractant in male Caribbean fruit fly, *Anastrepha suspensa*. Annals of the Entomological Society of America, 65: 1364–1367.

Nation, J.L., 1974. The structure and development of two sex specific glands in male Caribbean fruit flies. Annals of the Entomological Society of America, 67: 731–734.

Nation, J.L., 1975. The sex pheromone blend of Caribbean fruit fly males: isolation, biological activity, and partial chemical characterization. Environmental Entomology, 4: 27–30.

Nation, J.L., 1977. Pheromone research in tephritid fruit flies (Diptera:Tephritidae). Proceedings of the International Society of Citriculture, 2: 481–485.

Nation, J.L., 1981. Sex-specific glands in tephritid fruit flies of the genera *Anastrepha, Ceratitis, Dacus* and *Rhagoletis* (Diptera:Tephritidae). International Journal of Insect Morphology and Embryology, 10: 121–129.

Nation, J.L., 1983. Sex pheromone of the Caribbean fruit fly: chemistry and field ecology In: J. Miyamoto and P.C. Kearney (Editors), IUPAC Pesticide Chemistry, Human Welfare and the Environment, Vol. 2. Pergamon Press, New York, pp. 109–110.

Perdomo, A.J., 1974. Sex and aggregation pheromone bioassays and mating observations of the Caribbean fruit fly, *Anastrepha suspensa* (Loew), under field conditions. Ph.D. Dissertation, University of Florida, Gainesville, FL.

Perdomo, A.J., Baranowski, R.M. and Nation, J.L., 1975. Recapture of virgin female Caribbean fruit flies from traps baited with males. Florida Entomologist, 58: 291–295.

Perdomo, A.J., Nation, J.L. and Baranowski, R.M., 1976. Attraction of female and male Caribbean fruit flies to food-baited and male-baited traps under field conditions. Environmental Entomology, 5: 1208–1210.

Prokopy, R.J. and Hendrichs, J., 1979. Mating behavior of *Ceratitis capitata* on a field-caged host tree. Annals of the Entomological Society of America, 72: 642–648.

Prokopy, R.J. and Roitberg, B.D., 1984. Foraging behavior of true fruit flies. American Scientist, 72: 41–49.

Robacker, D.G. and Hart, W.G., 1984. A bioassay for investigation of sex pheromones of fruit flies. The Southwestern Entomologist, 9: 134–137.

Robacker, D.C. and Hart, W.G., 1985a. Courtship and territoriality of laboratory reared Mexican fruit flies, *Anastrepha ludens* (Diptera:Tephritidae), in cages containing host and nonhost trees. Annals of the Entomological Society of America, 78: 488–494.

Robacker, D.C. and Hart, W.G., 1985b. (*Z*)-3-nonenol, (*Z,Z*)-3,6-nonadienol and (*S,S*)-(−)-epianastrephin: male-produced pheromones of the Mexican fruit fly. Entomologia Experimentalis et Applicata, 39: 103–108.

Roelofs, W.L., 1977. Electroantennograms. Chemtech, 9: 222–227.

Sivinski, J., 1984. Effect of sexual experience on male mating success in a lek forming tephritid *Anastrepha suspensa* (Loew). Florida Entomologist, 67: 126–130.

Spieth, H.T., 1970. The evolutionary biology of the Hawaiian Drosophilidae. In: M.K. Hecht, W.C. Steere (Editors), Essays in Evolution and Genetics in Honor of Theodosius Dobzhansky. Appleton-Century-Crofts, New York, pp. 469–489.

Spieth, H.T., 1974. Courtship behavior in *Drosophila*. Annual Review of Entomology, 19: 385–405.

Stokes, J.B., Uebel, E.C., Warthen, Jr. J.D., Jacobson, M., Flippen-Anderson, J.L., Gilardi, R., Spishakoff, L.M. and Wilzer, K.R., 1983. Isolation and identification of novel lactones from male Mexican fruit flies. Journal of Agricultural and Food Chemistry, 31: 1162–1167.

Webb, J.C., Sharp, J.L., Chambers, D.L., McDow, J.J. and Benner, J.C., 1976. The analysis and identification of sounds produced by the male Caribbean fruit fly, *Anastrepha suspensa* (Loew). Annals of the Entomological Society of America, 69: 415–420.

Chapter 3.5 Host-Marking Pheromones

A.L. AVERILL and R.J. PROKOPY

1 INTRODUCTION

Many phytophagous insects are known to deposit host-marking pheromones following oviposition. These pheromones may serve to mediate population dispersion of individuals among available resources and decrease the probability of intraspecific encounters among immatures (Prokopy, 1981a).

By far the greatest number of phytophagous insect species known to deposit host-marking pheromones occur within the frugivorous Trypetinae (a subfamily within the family Tephritidae), but this may be due to the large number of candidate species studied. This includes: walnut husk fly, *Rhagoletis completa* Cresson (Cirio, 1972); apple maggot fly, *Rhagoletis pomonella* (Walsh) (Prokopy, 1972); European cherry fruit fly, *Rhagoletis cerasi* (Linnaeus) (Katsoyannos, 1975); black cherry fruit fly, *Rhagoletis fausta* (Osten Sacken) (Prokopy, 1975); blueberry maggot fly, *Rhagoletis mendax* Curran (Prokopy et al., 1976); eastern and western cherry fruit flies, *Rhagoletis cingulata* (Loew) and *Rhagoletis indifferens* Curran (Prokopy et al., 1976); dogwood berry flies, *Rhagoletis cornivora* Bush and *Rhagoletis tabellaria* Fitch (Prokopy et al., 1976); rose hip fruit fly, *Rhagoletis basiola* (Osten Sacken) (Averill and Prokopy, 1981); snowberry fruit fly, *Rhagoletis zephyria* Snow (Averill and Prokopy, 1982); Caribbean fruit fly, *Anastrepha suspensa* (Loew) (Prokopy et al., 1977); South American fruit fly, *Anastrepha fraterculus* (Wiedemann) (Prokopy et al., 1982a); Mediterranean fruit fly, *Ceratitis capitata* (Wiedemann) (Prokopy et al., 1978); and *Paraceratitella eurycephala* Hardy (Fitt, 1981). An unusual host-marking procedure exists in the olive fruit fly, *Dacus oleae* (Gmelin) (subfamily Dacinae): the oviposition deterrent is of plant origin, and thus, is not a pheromone. Following oviposition, a female olive fly uses her labellum to spread olive juice that exudes from the oviposition puncture over the fruit surface (Cirio, 1971; Gilolami et al., 1981). This juice acts as a deterrent to subsequent oviposition. To date, studies of other Dacinae have shown that *Dacus cucurbitae* Coquillett, *Dacus opilae* Drew and Hardy, and *Dacus cacuminatus* (Hering) do not deposit oviposition deterrents following egg-laying (Prokopy and Koyama, 1982; Fitt, 1984).

2 HISTORICAL OVERVIEW

Several early workers observed the characteristic host-marking behavior exhibited by some fruit flies. Porter (1928) was one of the first scientists to accurately describe the behavior. After observing *R. pomonella* in the field, he reported that following deposition of a single egg, the female walked rapidly around the fruit, usually with the ovipositor still extended and dragging behind. Then the female cleaned or brushed the ovipositor with the hind legs, drew the ovipositor back into place, and invariably flew or walked away. Wiesmann (1937) reported a similar behavioral scenario for *R. cerasi*, but he concluded that in this species, circling of the fruit while dragging the ovipositor occurred prior to oviposition. Although Wiesmann was a keen observer, it turns out that his description of the sequence of events in *R. cerasi* oviposition behavior was mistaken and that dragging of the ovipositor occurs primarily following successful deposition of an egg (Katsoyannos, 1975).

More than a decade later, Häfliger (1953) was the first to correctly speculate on the biological significance of this act of dragging the ovipositor. He was struck by the fact that when abundant uninfested cherries were available, they rarely contained more than one *R. cerasi* egg. In light of Wiesmann's behavioral observations, Häfliger hypothesized that this uniformity in egg dispersion was achieved by a fruit-marking procedure performed while the cherry fly dragged her ovipositor over the fruit surface. Bush (1966) also observed that in most instances where a *Rhagoletis* species infested small host fruit, the occurrence of more than one larva/fruit was rare. In agreement with Häfliger, he suggested that multiple oviposition is inhibited by a pheromonal deposition associated with egg-laying.

In his studies of *R. pomonella*, Prokopy (1972) provided the first experimental evidence that a tephritid fruit fly deposited a deterrent pheromone during the act of dragging the ovipositor following egg-laying. In field studies, Prokopy compared the proportion of females that attempted to oviposit into clean sour cherries versus cherries in which another female had recently oviposited and dragged her ovipositor. The results clearly showed the existence of a deterrent factor that was deposited during oviposition and/or dragging of the ovipositor. By picking females off fruit immediately following oviposition and allowing them to complete ovipositor dragging on another fruit, Prokopy provided strong evidence that a substance secreted by females during dragging of the ovipositor constituted the principle mechanism by which arriving females discriminate against fruits that already contain an egg.

3 CHARACTERIZATION OF HOST-MARKING PHEROMONE COMPONENTS

3.1 Pheromone collection

Diverse methods have been used to collect host-marking pheromone. In their current studies of *R. mendax*, Silk and Kuenen (personal communication) collect pheromone via methanol washes of cranberries used for oviposition. Following washing, the amount of pheromone is estimated by counting the number of eggs in each washed fruit: 1 egg = 1 dragging-bout equivalent. A similar method is used by Prokopy and coworkers in their studies of *R. pomonella* pheromone, except that *Crataegus* hawthorns or apples are used (Prokopy et al., 1982b). Other workers reported using artificial substrates for collection. In studies of *R. cerasi* pheromone, Boller and Hurter (1985) lined the walls and ceiling of plexiglas cages with glass slides, and after several weeks, scraped the faeces from the glass with razor blades. (Faeces of females are known to have high pheromone activity — see Section 5). Mumtaz and Ali-Niazee (1983b) used various solvents to remove *R. indifferens* pheromone from the surface of ceresin wax oviposition substrates.

3.2 Behavioral bioassay

As is the case for isolation and identification of any behaviorally active compound, a well-designed and accurate bioassay is an invaluable element. Unlike many other behavioral tests in which compounds are tested for elicitation of a positive response such as anemotaxis or oviposition stimulation, bioassay for host-marking pheromone activity measures a negative, or rejection, response. In general, such negative responses seem to be less specific than responses to attractants or stimulants (Boller and Hurter, 1985). Several other more specific factors may contribute to difficulty in conducting bioassays of host-marking pheromone. The candidate fractions of host-marking pheromone are applied to real or mimic host fruit, which provide strong positive stimulation, and in instances where real fruit are used, slight differences in fruit quality, which are not discernible by the human eye, may result in large differences in acceptance level by bioassay flies (Prokopy, unpublished data). Physiological state of the fly may also be critical: if the candidate female has been deprived of oviposition substrates prior to bioassay, she may not discriminate among treatments, even in cases where concentrated host-marking pheromone has been applied to fruit (Roitberg and Prokopy, 1983). Further, if a bioassay female accepts a treatment, and is not removed immediately following oviposition, considerable contamination occurs as a result of her deposition of host-marking pheromone.

In behavioral bioassays, Prokopy and co-workers attempt to control as much as possible these problems. Mature (ca. 14 days old) *R. pomonella* females are provided uninfested hawthorns for oviposition ca. 24 h prior to the start of the bioassay. On the day of bioassay, all flies are pre-tested: a candidate female is allowed to walk onto a clean fruit, followed by a pheromone-marked fruit, and then a second clean fruit. If the female attempts oviposition in each of the clean fruits, and rejects the pheromone-marked fruit, the female is used in a bioassay (Prokopy, 1981b).

For bioassay, an aliquot of a candidate pheromone fraction or control solution is swabbed onto a clean hawthorn fruit. Each fruit is attached to the end of a dissecting probe. A single mature, pre-tested *R. pomonella* female is introduced into a plexiglas-screen observation cage. From this point, 2 different bioassay procedures could be followed. In much of our previous work, a single mature *R. pomonella* female was introduced into a $15 \times 15 \times 15$ cm plexiglass-screen observation cage and an individual fruit was presented to the female in a no-choice situation. Each bioassay female was offered, at ca. 15-min intervals, a succession of 5–10 fruit. The fruit were offered in random order and blind fashion. Following the last fruit offered, all females were post-tested by offering the female a clean fruit. If she rejected it, then all data from that female were discarded, on grounds that her physiological state may not have been conducive to oviposition throughout the bioassay. If the female accepted the post-test fruit, then all data were accepted. A second bioassay procedure that we have utilized involves a choice situation in that 5–10 treated and control fruit were present in the observation cage simultaneously. Fruit were attached to dissecting probes that were placed vertically in a $30 \times 30 \times 30$ cm observation cage. A single female that had just successfully completed the pre-test was introduced into the cage. The female was observed continuously and allowed to visit fruit for up to a maximum of 2 h. If a female rejected several (ca. 5) clean fruit successively, the bioassay was terminated. Advantages of this approach are (a) an elevated level of discrimination among fruit occurs as fruit density increases (Roitberg et al., 1982), and (b) a good deal of human handling is removed, a source of considerable variation among bioassays.

For both bioassay procedures, data on acceptance (an oviposition attempt) or rejection (leaving fruit without attempting to oviposit) are recorded for each fruit. A major advantage of both bioassay procedures is that host-marking pheromone contamination of test fruit by bioassay females is minimal: when a

Chapter 3.5 references, p. 217

female did accept a test fruit, she was, immediately following egg deposition, gently transferred to a non-assay fruit, where she commenced and completed ovipositor dragging.

Several other workers have used bioassay procedures that do not involve observation of individual flies, and thus are much less labor-intensive. In ongoing studies of *R. mendax* host-marking pheromone, Kuenen and Silk (personal communication) conduct choice bioassays wherein 4 cranberries are positioned on a "tree" in the center of a cage. Five females are placed in the cage with the bioassay fruit, and after 1 h, the number of punctures in each fruit is counted. In studies of two cherry fly species, workers have used artificial oviposition substrates to bioassay fractions of host-marking pheromone. Mumtaz and AliNiazee (1983a, b) studied *R. indifferens* pheromone by using a hand atomizer to spray candidate fractions of pheromone onto 1.2 cm diam. orange domes made of ceresin wax. Usually, a bioassay was conducted by introducing 3 females (age 7–10 days) and 3 males (age 5–10 days) into a bioassay cage containing a total of 25 domes. Test and control domes were arranged in alternating rows. A test was terminated when > 50% of the domes contained an egg, but was not run longer than 48 h. Boller and Hurter (1985) also used ceresin wax domes in their studies of *R. cerasi* host-marking phero-mone. Aliquots of test compounds were applied with a cotton-tige to 2.5-cm diam. black ceresin wax domes and 10 domes were arranged in a circle within a 1 l cylindrical cage. Forty-eight hours before a bioassay, females were provided with fresh domes. At the start of a test, 5 females and 1 male were introduced to an observation cage. Eggs were counted twice a day and the bioassay terminated when ca. 25 eggs were laid per cage.

3.3 Pheromone identification and properties

Through behavioral bioassays (discussed above) conducted in conjunction with electrophysiological bioassay (recording from D-hairs, detailed in the following section), and through chemical purification and separation procedures detailed in part in Hurter et al. (1976), Boller and coworkers have recently isolated a principal component of the marking pheromone of *R. cerasi* (yet to be announced, E.F. Boller, personal communication). These workers found that the active compound(s) had a low volatility, was highly polar in solution, and had a molecular weight of < 10,000. Similar properties were described for the host-marking pheromone of *R. indifferens* (Mumtaz and AliNiazee, 1983b). Silk and Kuenen (personal communication) have partially isolated active fractions of host-marking pheromone of *R. mendax*. Despite considerable effort, Prokopy and collaborating chemists (Prokopy, 1981b) have been unable as yet to isolate an active component of the marking pheromone of *R. pomonella*.

Much of the difficulty in identification of tephritid host-marking phero-mones is due to the highly polar nature of the compound(s). Although the high solubility in water and methanol confers relative ease of collection via rinses from fruit and oviposition substrates, this solubility poses difficulty in chro-matographic separation of active and inactive components.

Regarding specific properties of tephritid host-marking pheromones, most are moderately stable under dry conditions. In an early study, conducted under laboratory conditions, Prokopy (1972) found high persistence of *R. pomonella* pheromone for at least 4 days. In a more extensive study (Averill and Prokopy, 1987b), we confirmed this finding and showed a relatively linear decline in pheromone activity both under laboratory and field conditions. There was no significant difference in rate of decline of activity between female-deposited pheromone versus an application of aqueous extract of pheromone. The half-life of the pheromone under dry conditions was 10.7 days, with some activity persisting after 3 weeks. Studies on *R. indifferens* (Mumtaz and AliNiazee, 1983b), *A. suspensa* (Prokopy et al., 1977), *C. capitata* (Prokopy et al., 1978), *R. fausta* (Prokopy, 1975), and *R. cerasi* (Katsoyannos, 1975) showed substantial persistence of host-marking pheromone for 4, 6, 6, 9, and 12 days, respectively.

In addition to those species mentioned above, the host-marking pheromones of all tephritids investigated to date are water soluble. This includes the pheromone of *A. fraterculus* (Prokopy et al., 1982a), *R. fausta* (Prokopy, 1975), *A. suspensa* (Prokopy et al., 1977), *C. capitata* (Prokopy et al., 1978), *R. cingulata*, *R. mendax*, *R. cornivora*, *R. tabelaria* (Prokopy et al., 1976), *R. basiola* (Averill and Prokopy, 1981), and *R. zephyria* (Averill and Prokopy, 1982).

4 RECEPTION OF HOST-MARKING PHEROMONE

Ablation experiments conducted by Prokopy and Spatcher (1977), wherein antennae, mouthparts, ovipositor, or pro-, meso-, or meta-thoracic tarsi were removed from *R. pomonella* females, demonstrated that only following removal of the pro-thoracic tarsi was there a significant drop in a female's ability to discriminate against pheromone-marked fruit.

Further studies of site of pheromone reception were conducted using electrophysiological techniques (Crnjar et al., 1978; Crnjar and Prokopy, 1982). Here, both the stimulating/recording and indifferent electrodes were very narrow diameter glass tubing. Heads or legs (cut off at the coxal-body joint) were impaled on the indifferent electrode that was filled with saline. Individual hairs were stimulated by a host-marking pheromone solution. Impulses were amplified and recorded. These studies demonstrated that a subgroup of chemosensory sensilla, the D-hairs on the ventral surface of the 2nd, 3rd, and 4th tarsomeres of the pro-thoracic tarsi, were stimulated by a crude extract of host-marking pheromone. Other chemosensory hairs were also highly stimulated by pheromone, such as the short hairs on the labellum and D-hairs on the meso- and meta-thoracic legs. However, in light of Prokopy and Spatcher's (1977) earlier ablation studies, which showed the pro-thoracic tarsi to be the major site of reception, Crnjar and Prokopy (1982) suggested that sensory inputs from different tarsi may be ranked at the central processing level, or further, maximum input may be received from the hairs on the pro-thoracic tarsi because they may be more likely to come into contact with the pheromone trail as the fly moves over the fruit surface. In a comprehensive study of the pro-thoracic D-hairs of *R. pomonella*, Bowdan (1984) showed that each sensillum contains several sensory cells, three of which have been identified as salt-sensitive, sucrose-sensitive, and host-marking pheromone sensitive. Compared to other stimulants, pheromone elicited a characteristic spike train. Further, cells stimulated with pheromone did not adapt as rapidly as those stimulated with salt and sucrose.

In studies of *R. cerasi*, Stadler and Boller (unpublished data) have likewise shown that the pro-thoracic tarsal D-hairs are a principal site of reception of host-marking pheromone.

5 SITE OF PRODUCTION OF HOST-MARKING PHEROMONE

To date, the only investigation of the site of production of host-marking pheromone component(s) in a fruit fly is our study on *R. pomonella* (Prokopy et al., 1982b). Extracts of male or female faeces, crop, anterior, and posterior midgut, hindgut, malphighian tubules, rectal glands, ovaries, salivary glands, bursa, and vagina were tested using behavioral and electrophysiological assay techniques. We identified the posterior half of the midgut as a principal site of production of a major pheromone component. It appears that following secretion into, and accumulation in, the gut lumen, this component is released, along with other gut contents, during deposition of the pheromone trail. Females receiving food containing amaranth dye, which is confined to the digestive tract, deposited amaranth-dyed pheromone trails, substantiating that the trial substance emanates from the digestive tract. Immature females do not produce pheromone. Pheromone is also released during defaecation by females, as faeces were active. Faeces likewise elicited a strong response in both behavioral and electrophysiological bioassays of *R. cerasi* host-marking pheromone (Boller and Hurter, 1985).

Inconsistency between behavioral and electrophysiological bioassay in our "site of pheromone" production study (Prokopy et al., 1982b) highlights a potential problem in such studies and demonstrates the advantage of pursuing a multi-pronged approach in pheromone identification. While many extracts (e.g. ovaries, male and female crop, hindgut and faeces of males) elicited significant deterrence in behavioral bioassays, they did not elicit a substantial electrophysiological response in the form of a typical pheromone-stimulated spike train. This anomaly could not be explained by salt content of these structures or male faeces because the concentration was not sufficient to elicit oviposition deterrence. However, the amount of sucrose in crop extracts was sufficient to elicit substantial feeding, and thus may account for decreased oviposition into fruit treated with these extracts (Prokopy et al., 1982b). It is possible that additional pheromone components are produced in other digestive tract structures or the ovaries.

6 RELEASE OF HOST-MARKING PHEROMONE

Many workers, studying different tephritid species, have observed considerable variation among females in time spent ovipositor dragging following egg-laying. For example, we found that following oviposition in 15 mm hawthorns, the average duration of a dragging bout by *R. pomonella* females was 30 s, but times ranged from 2–213 s (Averill and Prokopy, 1987a). Average duration of time spent dragging and range of times have also been recorded for the following species: *R. indifferens* on cherries: 51 s, range 18–86 s (Mumtaz and AliNiazee, 1983a); *A. suspensa* on 15 mm hawthorns: 36 s, range 9–74 s (Prokopy et al., 1977); *R. cerasi* on cherries: 30 s, range 2–145 s (Katsoyannos, 1975); *R. fausta* on 14 mm cherries: 17 s, range 6–37 (Prokopy, 1975); and *C. capitata* on 15 mm hawthorns: 77 s, range 4–267 s (Prokopy et al., 1978). Through behavioral bioassays of *R. pomonella*, we demonstrated a positive correlation between time spent dragging the ovipositor and activity of pheromone deposited (Averill and Prokopy, 1987c), but such a correlation has not yet been demonstrated for other tephritids.

To elucidate factors that may in part account for variability in pheromone release by *R. pomonella* females, we set up various fly and fruit treatments and studied the quantity and/or quality of pheromone trail substance deposited (Averill and Prokopy, 1987a). For some treatments (e.g. flies of different ages), we quantified the amount of pheromone deposited on a fruit by dusting newly-marked fruit with dry magnetic toner, a moisture-sensitive powder used in photocopying machines. The pheromone deposition is typically a viscous, discrete, and linear deposition and is rendered quite apparent by the powder. Trail length and width can then be measured microscopically with an ocular micrometer. Also, for some treatments, we evaluated the activity of pheromone deposited through a behavioral bioassay (see 3.2).

We found considerable variation in the amount of trail substance deposited by females, but more surprisingly, we found substantial variation in deposition among successive dragging bouts by the same female. Females fed the standard laboratory diet of sucrose and yeast hydrolysate produced pheromone equally as active as the pheromone produced by females fed a diet of natural food, insect honeydew. Older flies (28 days) or smaller flies released less or less active pheromone than younger (14 days) or larger flies. Females deposited a similar amount of trail substance on pheromone-marked fruit as on unmarked fruit. Starvation severely reduced the amount of visible trail substance deposited but resulted in a more active pheromone deposition. It is possible that reduced gut contents due to starvation may concentrate the pheromone. This inconsistency between trail measurement and behavioral bioassay results shows that, for the starvation treatment, had we not used a two-pronged approach, our results could have been misleading. Overall, we conclude that in the future, our trail substance quantification technique may be of limited utility because (1) as mentioned, results using this technique did not always parallel behavioral bioassay results, (2) the toner highlights the pheromone on only a few host fruits of *R. pomonella*, especially those that have a very smooth and waxy surface, and (3) it is labor intensive.

In other tests (Averill and Prokopy, 1987c), we found that females dragged their ovipositors for a longer time and distance on large fruit than on smaller fruit. Selective release of pheromone has likewise been demonstrated in *A. fraterculus*, also according to fruit size (Prokopy et al., 1982a). We suspect that it may be adaptively advantageous for a female, such as *R. pomonella*, to drag her ovipositor longer on larger versus small fruit because the level of deterrence elicited by comparable pheromone trails varies according to fruit size. For example, a single pheromone trail is sufficient to deter oviposition by most *R. pomonella* females on 15 mm diam fruit, whereas no deterrence is observed from a single trail on 55 mm diam fruit (Prokopy, 1972). This phenomenon is related to the observation that females may have to cross a pheromone trail a given number of times (ca. 6 for a small fruit) during pre-oviposition fruit inspection before fruit rejection is normally manifested (Prokopy, 1981b). The greater surface area of a large fruit compared with a small one results in a much less rapid increase in the ratio of pheromone-marked fruit surface area to clean fruit surface area each time a female completes a dragging circle on a large versus small fruit.

Comparative response to and/or production of host-marking pheromone by laboratory versus wild populations of fruit flies may vary. A laboratory strain of *R. pomonella*, cultured on apples for ca. 15 generations, deposited a type or amount of pheromone that was much less active than that deposited by laboratory caged wild flies (Prokopy et al., 1976). A laboratory strain of *C. capitata* flies cultured on artificial media for more than 200 generations showed minimal response to a concentration of pheromone that was highly deterrent to laboratory caged wild flies (Prokopy et al., 1978).

7 ROLE OF HOST-MARKING PHEROMONE IN MEDIATION OF LARVAL COMPETITION AND EXPLOITATION OF RESOURCES

Through a study of *R. pomonella* developing on *Crateagus mollis* hawthorns (a small, native host fruit of the fly), we substantiated Prokopy's (1972) earlier hypothesis that the amount of host-marking pheromone deposited after a single oviposition may be linked to both the amount of pheromone necessary to elicit deterrence and to the amount of food or space requirements of a developing larva (Averill and Prokopy, 1987c). First, in the field, we collected hawthorns with varying degrees of *R. pomonella* infestation and selected 3 sizes of fruit (12,

15, and 20 mm diam.), which spanned the range of naturally-occurring sizes. Each fruit was held individually. Emerging larvae were counted daily. Regardless of fruit size, we found the carrying capacity was 1 larva/fruit. Maximum larval survivorship was realized when larvae developed alone in a fruit. An increase in the initial number of eggs/fruit was reflected in declining weight of resulting individuals. Components of adult fitness, such as number of days to first egg-laying and rate of oviposition, were correlated with pupal weight. In field tests, we found that, regardless of fruit size, the amount of fruit surface marked by a female following oviposition was linked with the carrying capacity of a *C. mollis* fruit; pheromone deposited following a single egg-laying was sufficient to deter most females from further egg-laying.

In another study, we substantiated Prokopy's (1972) earlier hypothesis that, because the host-marking pheromone is both water soluble and only moderately stable, pheromone need deter oviposition only long enough to give the earliest developing larva a head start, and thus, a competitive advantage over later-developing larvae. We found that in most instances, when 2 days separated the introduction of 2 larvae into unpicked hawthorns capable of supporting only a single larva to pupation, the first-introduced larva "won"; the second larva introduced failed to complete development.

The ecological result of female recognition of already-occupied fruit via host-marking pheromone signals may be even dispersion of eggs among available host fruit. Indeed, following fruit ripening, we found that *R. pomonella* eggs were evenly dispersed among *C. mollis* fruit in nature (Averill, 1985). These results are consistent with data suggesting uniformity of eggs among apples in Quebec (LeRoux and Mukerji, 1963), but conflict with those of Reissig and Smith (1978), who found a non-random egg dispersion among *Crataegus holmsiana* hawthorns sampled in New York. There is no obvious explanation for the discrepancy among these studies.

There are a number of other tephritids in which there appears to exist a tendency towards uniformity in dispersion of eggs among available fruits. Many of these species are enumerated by Prokopy (1972, 1976) and some are known to deposit host-marking pheromones. A recent and thorough field study of *R. cerasi* egg distribution was conducted by Remund et al. (1980) and showed that, in this species, egg distribution was not random. Additional species include *R. indifferens* (AliNiazee, 1974) and *P. eurycephala* (Fitt, 1981).

8 FORAGING OF FLIES IN RELATION TO PHEROMONE-MARKED AND UNMARKED FRUIT

The outcome of a female's reaction to host fruit (whether marked or unmarked with pheromone) is determined by a summation of interacting factors. This includes availability and distribution of unmarked and marked fruit within the habitat; the physiological state of the female in regard to the number of mature eggs contained in the ovaries and time elapsed since her last oviposition; previous experience of the female with host-marking pheromone and the sequence and frequency of encounter with marked and unmarked fruit. A discussion of these topics is presented in Chapter 4.1.

We have found substantial differences in the reaction of *R. pomonella* to a given amount of host-marking pheromone under natural or semi-natural conditions versus under laboratory conditions. In earlier studies, Prokopy (1972) showed that 1 dragging bout on a 15 mm diam. cherry was sufficient to elicit complete deterrence under natural field conditions but only slight deterrence under laboratory conditions. We have conducted additional studies that confirm this pattern of decreased discrimination by females in the laboratory

(Averill and Prokopy, unpublished data). In laboratory bioassays, where we released *R. pomonella* females into observation cages containing fruit marked with various concentrations of host-marking pheromone, acceptance was 81, 65, 31, and 30% for clean (control) fruit or fruit with 1, 2, or 3 dragging bout equivalents, respectively. In contrast, in field bioassays where we released females in trees laden with fruit, acceptance was 55, 18, 18 and 15% for clean (control) fruit and fruit marked with 1, 2, or 3 dragging bout equivalents, respectively. In both series of tests we used fruit selected from the same batches. In sum, under laboratory conditions, females were less discriminating among both clean and pheromone-marked treatments. Because of this discrepancy, relating levels of female response to varying pheromone concentrations in the laboratory to a field situation is clearly inadvisable.

The host-marking pheromone of some tephritids has been shown to arrest conspecific males, apparently providing a signal to males that a female has recently been in the vicinity. This may enhance the probability of courtship encounters between males and females. Such a phenomenon has been demonstrated in *R. pomonella* (Prokopy and Bush, 1972), *R. cerasi* (Katsoyannos, 1975), and *R. indifferens* (AliNiazee and Brown, unpublished data cited in Mumtaz and AliNiazee, 1983a).

9 EVOLUTION OF HOST-MARKING

Several authors have suggested that certain ecological characteristics of a species may correlate with enhanced likelihood of frequent or severe intraspecific larval encounters. Under such conditions, evolution of host-marking may well develop. These characteristics include: (1) monophagous or oligophagous habits; (2) infestation of relatively permanent hosts, i.e. hosts that occupy habitats for at least 2 herbivore generations; (3) limited mobility by parents and/or offspring; (4) restricted feeding and/or resting sites within host plants; (5) feeding at relatively ephemeral (e.g. buds, flowers, fruits in stage susceptible to egg-laying) rather than permanent plant parts (e.g. leaves, stems); (6) exploitation of a host that is patchily distributed (Prokopy, 1981a; Thompson, 1983; Roitberg and Prokopy, 1987).

These characteristics describe not only the host exploitation patterns of many of the Trypetinae that are known to utilize host-marking pheromones, but also of many Dacinae. However, while *D. oleae* marks host fruit with olive juice that exudes from its oviposition puncture, and *D. cucurbitae*, *Dacus tryoni* (Froggatt), and *Dacus jarvisi* (Tryon) discriminate against fruit that contain larvae, none of the Dacinae produce endogenous marking pheromones (Cirio, 1972; Fitt, 1984; Prokopy and Koyama, 1982). In a comparison of *Rhagoletis* (where host-marking has proliferated) and *Dacus*, Fitt (1984) has speculated that differences in the stability of populations within patches may account for the existence or lack of a host-marking system. Whereas *Rhagoletis* typically undergo an obligate diapause and form relatively sedentary and localized populations whose levels hover near the carrying capacity of a patch, *Dacus* have no diapause, disperse soon after emergence, and form transient populations of varying and unpredictable densities. Fitt asserts that, given these factors, selection favoring host discrimination within *Rhagoletis* would be a consistent and strong pressure, whereas in *Dacus*, considerable spatial and temporal variation in degree of resource exploitation may result in too weak a selective force to favor host discrimination, especially if such a system is energetically expensive and reduces egg-laying efficiency.

As we discussed above, it appears that some sort of interference-type competition exists among *R. pomonella* larvae in hawthorn fruit, in which at least one larva survives, regardless of whether adequate fruit resource is available for multiple larvae to develop. Therefore, recognition of occupied host fruit is primarily a trait of an ovipositing female that is selectively advantageous because of interference competition among her own offspring. Secondary benefits accrue when other individuals also avoid such marked hosts. Theoretical discussion and recent models developed by Roitberg provide a clearer understanding of host-marking systems from an evolutionary and ecological perspective (Roitberg, unpublished; Roitberg and Prokopy, 1987).

Fitt (1984) has discussed the evolution of host-marking behavior in tephritids. In *Dacus*, females circle the fruit, in some cases drag the ovipositor on the fruit surface, and preen the ovipositor extensively, but deposit no pheromone. Fitt suggests that following oviposition, all of these acts serve to remove fruit debris from rasping surfaces of the ovipositor, and further suggests that in species that do deposit pheromone, marking of the fruit surface is an elaboration of this cleaning behavior. The fact that faeces contain the host-marking pheromone in *R. pomonella* and that the pheromone is produced in the midgut rather than a specialized organ, lend support to this hypothesis.

Alternatively, circling of the fruit and dragging of the ovipositor in some *Dacus* species could be a behavioral remnant of marking behavior with past but no present significance in terms of pheromone deposition (Prokopy and Koyama, 1982).

10 APPLICATIONS OF HOST-MARKING PHEROMONES IN TEPHRITID MANAGEMENT

Several researchers have suggested that if host-marking pheromones could be isolated, identified, and synthesized, then spraying host crops with pheromone might become an important new approach to tephritid management (Prokopy, 1972, 1976; Katsoyannos and Boller, 1976, 1980). Boller (1981) has suggested that application of host-marking pheromone differs from other noninsecticidal approaches. Unlike sex pheromones, where the crop is unprotected, host-marking pheromone could provide the same protection as other deterrents or repellents. Further, unlike SIT, which requires that the treatment area be isolated or restricted in size, host-marking pheromones could be used like conventional insecticides, and perhaps could be applied with conventional spray equipment. On the other hand, successful application of host-marking pheromone may be constrained by several factors (Roitberg and Prokopy, 1987). The level at which females are deterred from oviposition in treated hosts may diminish as a result of habituation following repeated exposure to the pheromone or as a result of continuous deprivation from non-pheromone treated hosts. Further, selection for females that do not respond to the pheromone may proceed at a rapid rate. For these reasons, if host-marking pheromones eventually become commercially available, this new management tool should be integrated into a multi-faceted management scheme, perhaps involving untreated trap trees and appropriate traps to capture deterred females.

By evaluating the efficacy of applications of the host-marking pheromone of *R. cerasi* in small-scale experiments under field conditions, Katsoyannos and Boller (1976, 1980) have made the most progress towards eventual application of host-marking pheromone in management of tephritids. In their initial test, Katsoyannos and Boller (1976) applied 5 sprays of an aqueous solution of partially purified pheromone (gathered from artificial oviposition substrates after ca 200,000 ovipositions and mixed with a wetting agent) to 2 entire trees or to branches of other trees. Even under conditions of relatively high fly population density and frequent rainfall, there was a 63% reduction in the total number of cherries infested. In a second test (Katsoyannos and Boller, 1980), when very little rain fell during the test period, 4 trees were sprayed once or twice with either 2 pheromone concentrations. Two sprays of the higher concentration of pheromone resulted in a 90% reduction in infestation when compared with adjacent untreated trees, and for a single spray at this concentration, infestation was reduced by 85%.

11 CONCLUSIONS

To date, more species of Tephritidae are known to deposit host-marking pheromone than species in any other family of insects. In this chapter, we have attempted to discuss in some detail biochemical, physiological, behavioral, ecological, evolutionary, and pest management facets of marking pheromones of the most thoroughly studied tephritid species. Although in total our knowledge of these facets in Tephritidae is perhaps greater than in any other family of insects, our understanding still falls far short of what is necessary for full comprehension of the true ecological significance of tephritid marking pheromones under various habitat conditions and their potential value in management of tephritids. Fuller comprehension will likely come about only after major active components of tephritid marking pheromones have been chemically synthesized, and thus available for use in experiments wherein the structure of resource patches can be manipulated in a readily quantifiable and repeatable fashion in nature. Hence, we believe the recent success of E.F. Boller and his colleagues in Switzerland in identifying a principal component of the marking pheromone of *R. cerasi* constitutes a true breakthrough in the history of tephritid marking pheromone research, and will serve as an important stepping stone toward highly productive research on tephritid marking pheromones over the next decade.

12 REFERENCES

AliNiazee, M.T., 1974. The western cherry fruit fly, *Rhagoletis indifferens* (Diptera: Tephritidae). 2. Agressive behavior. Canadian Entomologist, 106: 1201–1204.

Averill, A.L., 1985. Oviposition-deterring pheromone of *Rhagoletis pomonella*: release, residual activity and mediation of competition. Dissertation. University of Massachusetts, Amherst.

Averill, A.L., and Prokopy, R.J., 1981. Oviposition-deterring fruit marking pheromone in *Rhagoletis basiola*. Florida Entomologist, 64: 221–226.

Averill, A.L. and Prokopy, R.J., 1982. Oviposition-deterring fruit marking phermone in *Rhagoletis zephyria*. Journal of the Georgia Entomological Society, 17: 315–319.

Averill, A.L. and Prokopy, R.J., 1987a. Factors influencing the release of host-marking pheromone by *Rhagoletis pomonella* flies. Journal of Chemical Ecology, in press.

Averill, A.L. and Prokopy, R.J., 1987b. Residual activity of oviposition-deterring pheromone in *Rhagoletis pomonella* and female response to infested fruit. Journal of Chemical Ecology, 13: 167–177.

Averill, A.L. and Prokopy, R.J., 1987c. Intraspecific competition in the tephritid fruit fly *Rhagoletis pomonella*. Ecology, in press.

Boller, E.F., 1981. Oviposition-deterring pheromone of the European cherry fruit fly: status of research and potential applications. In: E.R. Mitchell (Editor), Management of Insect Pests with Semiochemicals. Plenum Press, New York, pp. 457–462.

Boller, E.F. and Hurter, J., 1985. Oviposition deterring pheromone in *Rhagoletis cerasi*: Behavioral laboratory test to measure pheromone activity. Entomologia Experimentalis et Applicata, 39: 163–169.

Bowdan, E., 1984. Electrophysiological responses of tarsal contact chemoreceptors of the apple maggot fly *Rhagoletis pomonella* to salt, sucrose and oviposition-deterrent pheromone. Journal of Comparative Physiology A, 154: 143–152.

Bush, G.L., 1966. The taxonomy, cytology, and evolution of the genus *Rhagoletis* in North America (Diptera: Tephritidae). Bulletin Harvard Museum of Comparative Zoology, 134: 431–562.

Cirio, U., 1971. Reperti sul meccanismo stimolo-risposta nell 'ovideposizione del *Dacus oleae* Gmelin (Diptera: Trypetidae). Redia, 52: 577–600.

Cirio, U., 1972. Observazioni sul comportameno di ovideposizione della *Rhagoletis completa* in laboratorio. Proceedings 9th Congress Italian Entomological Society, 99–117.

Crnjar, R.M. and Prokopy, R.J., 1982. Morphological and electrophysiological mapping of tarsal chemoreceptors of oviposition-deterring pheromone of *Rhagoletis pomonella* flies. Journal of Insect Physiology, 28: 393–400.

Crnjar, R.M., Prokopy, R.J. and Dethier, V.G., 1978. Electrophysiological identification of oviposition-deterring pheromone receptors in *Rhagoletis pomonella*. Journal of the New York Entomological Society, 86: 283–284.

Fitt, G.P., 1981. Observations on the biology and behavior of *Paraceratitella eurycephala* (Diptera: Tephritidae) in northern Australia. Journal of the Australian Entomological Society, 20: 1–7.

Fitt, G.P., 1984. Oviposition behavior of two tephritid fruit flies, *Dacus tyroni* and *Dacus jarvisi*, as influenced by the presence of larvae in the host fruit. Oecologia, 62: 37–46.

Girolami, V., Vianello, A., Strapazzon, A., Ragazzi, E. and Veronese, G., 1981. Ovipositional deterrents in *Dacus oleae*. Entomologia Experimentalis et Applicata, 29: 177–188.

Häfliger, E., 1953. Das Auswahlvermögen der Kirschenfliege bei der Eiablage (Eine statistische Studie). Mitteilungen der Schweizerischen Entomologischen Gesellschaft, 26: 258–264.

Hurter, V.J., Katsoyannos, B., Boller, E.F. and Wirz, P., 1976. Beitrag zur Anreicherung und teilweisen Reinigung des eiablageverhindernden Pheromons der Kirschenfliege, *Rhagoletis cerasi* L. (Dipt., Trypetidae). Zeitschrift für Angewandte Entomologie, 80: 50–56.

Katsoyannos, B.I., 1975. Oviposition-deterring, male-arresting, fruit-marking pheromone in *Rhagoletis cerasi*. Environmental Entomology, 4: 801–807.

Katsoyannos, B.I. and Boller, E.F., 1976. First field application of oviposition-deterring marking pheromone of European cherry fruit fly. Environmental Entomology, 5: 151–152.

Katsoyannos, B.I. and Boller, E.F., 1980. Second field application of the oviposition-deterring pheromone of the European cherry fruit fly, *Rhagoletis cerasi* L. (Diptera: Tephritidae). Zeitschrift für Angewandte Entomologie, 89: 278–281.

LeRoux, E.J. and Mukerji, M.K., 1963. Notes on the distribution of immature stages of the apple maggot, *Rhagoletis pomonella* (Walsh) (Diptera: Tephritidae) in apple in Quebec. Annals of the Entomological Society of Quebec, 8: 60–70.

Mumtaz, M.M. and AliNiazee, M.T., 1983a. The oviposition-deterring pheromone of the western cherry fruit fly, *Rhagoletis indifferens* Curran (Dipt.: Tephritidae). 1. Biological properties. Zeitschrift für Angewandte Entomologie, 96: 83–93.

Mumtaz, M.M. and AliNiazee M.T., 1983b. The oviposition-deterring pheromone of the western cherry fruit fly, *Rhagoletis indifferens* Curran (Dipt.: Tephritidae). 2. Chemical characterization and partial purification. Zeitschrift für Angewandte Entomologie, 96: 93–99.

Porter, B.A., 1928. The apple maggot. United States Department of Agriculture Technical Bulletin, 66: 1–47.

Prokopy, R.J., 1972. Evidence for a marking pheromone deterring repeated oviposition in apple maggot flies. Environmental Entomology, 1: 326–332.

Prokopy, R.J., 1975. Oviposition-deterring fruit marking pheromone in *Rhagoletis fausta*. Environmental Entomology, 4: 298–300.

Prokopy, R.J., 1976. Significance of fly marking of oviposition site (in Tephritidae). In: V.L. Delucchi (Editor). Studies in Biological Control. Cambridge University Press, Cambridge, pp. 23–27.

Prokopy, R.J., 1981a. Epideictic pheromones that influence spacing patterns of phytophagous insects. In: D.A. Nordlund, R.L. Jones, and W.J. Lewis (Editors), Semiochemicals: Their Role in Pest Control. Wiley Press, New York, pp. 181–213.

Prokopy, R.J., 1981b. Oviposition-deterring pheromone system of apple maggot flies. In: E.K. Mitchell (Editor), Management of Insect Pests with Semiochemicals. Plenum Press, New York, pp. 477–494.

Prokopy, R.J. and Bush, G.L., 1972. Mating behavior in *Rhagoletis pomonella* (Dipteria: Tephritidae). III. Male aggregation in response to an arrestant. Canadian Entomologist, 104: 275–283.

Prokopy, R.J. and Spatcher, P.J., 1977. Location of receptors for oviposition-deterring pheromone in *Rhagoletis pomonella* flies. Annals of the Entomological Society of America, 70: 960–962.

Prokopy, R.J. and Koyama, J., 1982. Oviposition site partitioning in *Dacus cucurbitae*. Entomologia Experimentalis et Applicata, 31: 428–432.

Prokopy, R.J., Reissig, W.H. and Moericke, V., 1976. Marking pheromones deterring repeated oviposition in *Rhagoletis* flies. Entomologia Experimentalis et Applicata, 20: 170–178.

Prokopy, R., Greany, P.D. and Chambers, D.L., 1977. Oviposition-deterring pheromone in *Anastrepha suspensa*. Environmental Entomology, 6: 463–465.

Prokopy, R.J., Ziegler, J.R. and Wong, T.Y., 1978. Deterrence of repeated oviposition by fruit marking pheromone in *Ceratitis capitata* (Diptera: Tephritidae). Journal of Chemical Ecology, 4: 55–63.

Prokopy, R.J., Malavasi, A., and Margante, J.S., 1982a. Oviposition-deterring pheromone in *Anastrepha fraterculus* flies. Journal of Chemical Ecology, 8: 763–771.

Prokopy, R.J., Averill, A.L., Bardinelli, C.M., Bowdan, E.S., Cooley, S.S., Crnjar, R.M., Dundulis, E.A.,Roitberg, C.A., Spatcher, P.J., Tumlinson, J.H. and Weeks, B.L., 1982b. Site of production of an oviposition-deterring pheromone component in *Rhagoletis pomonella* flies. Journal of Insect Physiology, 28: 1–10.

Reissig, W.H. and Smith, D.C., 1978. Bionomics of *Rhagoletis pomonella* in *Crataegus*. Annals of the Entomological Society of America, 71: 155–159.

Remund, U., Katsoyannos, B.I., Boller, E.F. and Berchtold, W., 1980. Zur Eiverteilung der Kirschenfliege, *Rhagoletis cerasi* L. (Dipt, Tephritidae), im Freiland. Mitteilunger der Schweizerischen Entomologischen Gesellschaft, 53: 401–405.

Roitberg, B.D. and Prokopy, R.J., 1983. Host deprivation influence on response of *Rhagoletis pomonella* to its oviposition-deterring pheromone. Physiological Entomology, 8: 69–72.

Roitberg, B.D. and Prokopy, R.J., 1987. Behavioral ecology of host marking by insect herbivores. BioScience, in press.

Roitberg, B.D., Van Lenteren, J.C., Van Alphen, J.M., Galis, F. and Prokopy, R.J., 1982. Foraging behaviour of *Rhagoletis pomonella*, a parasite of hawthorn (*Crataegus viridis*), in nature. Journal of Animal Ecology, 51: 307–325.

Thompson, J.N., 1983. Selection pressures on phytophagous insects feeding on small host plants. Oikos, 40: 438–444.

Wiesmann, R., 1937. Die Orientierung der Kirschfliege *Rhagoletis cerasi* L. Landwirtschaftliches Jahrbuch der Schweiz, 51: 1080–1109.

Chapter 3.6 Parapheromones

ROY T. CUNNINGHAM

1 INTRODUCTION

Parapheromones are one of the great mysteries of tephritid biology. The term "parapheromone" was first used by Payne et al. (1973) to describe those chemicals which are not naturally used in intraspecific communication but which do elicit responses similar to true pheromones. Since parapheromone is an artificial term to cover our ignorance, its boundaries are ill-defined. I will restrict discussion to the so-called male lures and their analogs and homologs although there are one or two compounds which have been identified as ovipositional attractants and might be called parapheromones. I will exclude the limited number of weak attractants derived from host fruits and those chemicals, such as ammonia, obviously related to food location. There are some researchers who regard the male lures of tephritid fruit flies as pseudopheromones, and there are those who argue that they are really behavior-modifying plant chemicals, kairomones. Unfortunately, we do not yet know which view, if either, is correct or whether a combination of both sides is closer to the truth. In this chapter I will present the evidence as it is now known with the hope that it will point toward some future research that could help toward our understanding of this striking phenomenon.

2 METHYLEUGENOL

Methyleugenol (Fig. 3.6.1(a)), the most powerful of the male lures and the first whose structure was identified is a constituent of many plants. It was discovered through one of those serendipitous chances that delight us all. Howlett (1912) working in Pusa, India, heard that a neighbor was bothered by a fly when he sprinkled citronella oil on his handkerchief as mosquito repellent. He discovered that all of the flies were males of *Dacus zonatus* (Saunders), an important pest species. In this very early paper well before the chemical isolation of pheromones, he laid out one of the schools of thought with regard to male lures where he says, *"If my conclusions are correct regarding the nature of the phenomena, they afford an interesting example of the imitation by artificial means of a sexual attraction probably similar in kind to that which operates in most cases of 'assembling'" (p. 418)*. In following up this lead (Howlett, 1915), demonstrated that methyleugenol was the active principle. And in this second paper he retreated from his speculation that methyleugenol might be imitating a sexual odor, toward the speculation that it was possibly a food odor and that the males have a different food requirement from the females. This same ambivalence in interpretation of the biological role of male lures is still with us today, 70 years later.

Chapter 3.6 references, p. 228

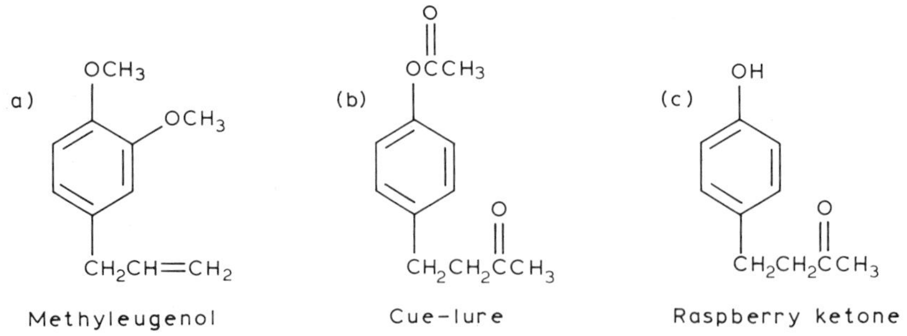

Fig. 3.6.1. Parapheromones. Male lures for *Dacus* species. (a) Methyleugenol. The powerful attractant for *Dacus dorsalis* males. A natural constituent of many plants. (b) Cue-lure. The synthetic male attractant for *Dacus cucurbitae*. (c) Raspberry ketone. Natural plant constituent attractive to male *Dacus cucurbitae*.

Methyleugenol has been found in at least 25 plants over a wide range of families. At least three of these plants were discovered to contain methyleugenol because male tephritid fruit flies were found to be attracted to them. In none of these cases were the plants breeding hosts or notable food sources (Fletcher et al., 1975; Kawano et al., 1968; Shah and Patel, 1976). On the other hand, methyleugenol has not been found in tephritid fruit flies nor have the other two chief male lures nor any of their homologs.

The strength of the attraction of methyleugenol for *Dacus dorsalis* Hendel males, the oriental fruit fly, is truly surprising. If pure liquid methyleugenol is offered, the males will drink it until they fill their crops and die. Such a powerful response must be related to similarly important phenomena in the biology of the insect. Among the Dacinae whose males have shown a strong response to methyleugenol are *D. dorsalis* (Steiner, 1951), *D. zonatus* (Howlett, 1912), *Dacus correctus* (Bezzi) (Shah and Patel, 1976), and *Dacus diversus* Coquillett (Howlett, 1915). Drew (1974) and Drew and Hooper (1981) list 39 additional species which respond to methyleugenol. There is reason to believe that this list would be greatly extended if a systematic investigation of male lure response were made of the hundreds of Malaysian species of Dacinae (D. Elmo Hardy, personal communication). Drew and Hooper (1981) investigated the response of a number of species to either methyleugenol or cue-lure (Beroza et al., 1960); the other powerful male lure for the Dacinae. Drew (1974) concluded that a species, if it responded at all, responded to one or the other, but not to both lures, and in most cases the choices followed in parallel with recognizably different taxons based on morphology. There were notable exceptions to this taxonomic division but, in general, this study reinforces the idea that the male lure response is related to a fundamentally important process embedded deep in the evolutionary development of the species.

3 CUE-LURE

Cue-lure does not occur in nature (Fig. 3.6.1(b)) but its closely related analog, the so-called raspberry ketone (Fig. 3.6.1(c)) does occur in plants (Schinz and Seidel, 1961; Gierschner and Baumann, 1967). This raspberry ketone is known as Willison's lure in Australia. About the same time that cue-lure was under development in Hawaii as a male lure for *Dacus cucurbitae* Coquillett, the melon fly. Willison's lure was found to be an attractant for *Dacus tryoni* (Froggatt), the Queensland fruit fly in Australia. Raspberry ketone was among the compounds tested in Hawaii and was found to be attractive for *D. cucurbitae* also. Correspondingly, cue-lure was found to be attractive for the *D. tryoni* males. No example has yet been found where a species is attracted to only one and not the other of these two closely related compounds. Drew (1974) and Drew and Hooper (1981) list 88 additional species responding to cue-lure. In addition, the following have been shown to respond to cue-lure: *Dacus ochrosiae* Malloch (Beroza et al., 1960), *Zeugodacus scutellatus* (Hendel) and *Zeugodacus ishigakiensis* Shiraki (Sonda, 1972). According to Drew (1974), cue-lure attraction is spread over several taxons of the tribe Dacini. It is attractive for members of the genus *Callantra* and for members of both groups of subgenera of the genus *Dacus*, that is, the *Dacus* and the *Strumeta* groups of subgenera. Methyleugenol attraction is confirmed only for the *Strumeta* group of subgenera. Thus, even more so than with methyleugenol, we would expect to find large numbers of the Malaysian fruit flies attracted to cue-lure if a systematic study were carried out.

A surprising recent development has been the finding by Hancock (1985) that methyl-*p*-hydroxybenzoate is a specific male lure for the African pest of cucurbits, *Dacus vertebratus* Bezzi. Hancock has named the compound "vert-lure". It will be extremely interesting to see if this newly discovered male lure will also attract a number of taxonomically related species.

The natural plant product, raspberry ketone, was more attractive than cue-lure in small cage tests as were several other compounds in studies done by Metcalf et al. (1983) but cue-lure was found to be superior to all other compounds in field performance by Alexander et al. (1962). Recently, T.P. McGovern and I tested raspberry ketone and other homologs and purified cue-lure in the field and again have found cue-lure to be superior contrary to the results from small cage studies. Cue-lure can degrade into raspberry ketone but the ketone is only about 10% soluble in cue-lure. The attraction of cue-lure cannot be accounted for by its degradation into raspberry ketone as has been suggested (Drew, 1974).

4 KAIROMONES

Metcalf and his colleagues in a series of studies (Metcalf et al., 1975, 1979, 1981, and 1983) of cue-lure and methyleugenol and a long series of their respective analogs and homologs have advanced the thesis that these are good examples of the co-evolution of an insect along with its associated plants. Metcalf (1979) proposed that these are kairomones that act as secondary sex attractants or rendezvous stimulants to bring the sexes together on host plants. On the face of it, this would seem to be a logical proposition. Labeyrie (1971) discussed this phenomenon in a speculative paper which reviewed the little evidence which was then available for plant kairomones acting as secondary sexual attractants. A striking phenomenon in euglossine bees has been studied by Dodson et al. (1969). The males visit orchid blossoms which are not a food source and collect the fragrant exudates on their hind tibiae. These bees establish territories to which the females are attracted, however, the females were not attracted by the pure compounds which the male bees gathered from the orchids. Among the constituents of these fragrances were eugenol and methyl cinnamate. Metcalf (1979) has speculated that *p*-hydroxycinnamic acid, a widely distributed plant compound, is the primitive precursor for both methyleugenol and raspberry ketone. A plant kairomone acting as a male lure may well be a more widely distributed phenomenon in insects than we now realize.

Further support for the idea that these are co-evolved plant kairomones comes from the fact that methyleugenol is the quintessence of a male lure for its group of species. Metcalf et al. (1975) tested a long series of related analogs and found none to be better than methyleugenol in small cage tests. Recently, T.P. McGovern and I (unpublished data) have field tested many of the same

Chapter 3.6 references, p. 228

compounds as Metcalf, plus several hundred more natural and synthetic compounds, and found none to be better than methyleugenol. In substantial agreement with the Metcalf group, we found a number of closely related analogs which were attractive, but always to a lesser degree than methyleugenol. Neither methyleugenol nor any closely related analogs have been found as a natural component in insects. If methyleugenol is merely an imitator of a male-produced pheromone as some have suggested, then we might have expected some promising leads to have been found among the thousands of other compounds that have been screened for attraction (see Beroza and Green, 1963). On the contrary, the data indicate that methyleugenol is of itself the quintessential compound and not merely a weak imitator of a true pheromone. The same is true for cue-lure and its raspberry ketone precursor. Among cue-lure responding males, Baker et al. (1982) investigated the chemical constituents of the rectal gland secretions of male *D. cucurbitae* while Bellas and Fletcher (1979) did the same for *D. tryoni* and *Dacus neohumeralis* Hardy. In no case was a cue-lure-like compound found.

On the other hand, there are some flaws in the proposition that these male lures are plant kairomones acting as secondary sexual attractants. Raspberry ketone, for example, has been reported from only 3 plants to date (Metcalf et al., 1983; Gierschner and Baumann, 1967), two of them being raspberry and the leaves of Norway maple. It is difficult to see how the large number of tropical Dacini responding to raspberry ketone or cue-lure could have co-evolved with a chemical of such restricted allopatric distribution. Methyleugenol, however, is widely distributed but not primarily in breeding host plants. It is so ubiquitous it would almost seem to be a disadvantage for the male flies to be responding to it in plants. The comments of Fletcher et al. (1975) that the *Zieria smithii* plant to which *Dacus cacuminatus* (Hering) were attracted did not seem to play any functional role in the biology of the insect as a source of adult or larval food can be applied to most of the plants so far on which males have been noticed. I have noted *D. dorsalis* on gardenia flowers and other have noted congregations on such plants as the flower of one variety of taro (P. Quentin Tomich, personal communication).

If we accept the premise that mating does not necessarily take place on breeding host plants then this lack of good correlation between the plant kairomones and breeding hosts is not a problem. The evidence for this premise is lacking or weak, but even if it is accepted there is a further difficulty relating to the time of response. Susumu Nakagawa of our laboratory (unpublished data) investigated the times of response of the males of the three pest tephritids (*D. dorsalis*, *D. cucurbitae*, and *Ceratitis capitata* (Wiedemann) in Hawaii to their respective male lures and found that peak response was in the late morning or midday hours. Fitt (1981b) did a very thorough study of the response of the males of *Dacus opiliae* (Drew and Hardy) to methyleugenol and found the peak response to be in the midmorning with a rapid abandonment of the lure at dusk. Brieze-Stegman et al. (1978) found similarly that *D. tryoni* males respond to cue-lure in the middle of the day as do *D. cacuminatus* males to methyleugenol with an abandonment of the lure at dusk. The incongruity is, of course, that many if not most of the tropical Dacini mate at dusk.

If these lures are really kairomones acting as secondary sexual attractants or if they are, as some have suggested, imitators of a male aggregation pheromone, one would logically expect their attraction to be greatest at the time of normal mating, dusk. The opposite is the case with most of the *Dacus* species.

There is a further complexity in the behaviorial repertoire of these fruit flies. In an eradication program on the Island of Rota, Steiner et al. (1965) noted that when the oriental fruit fly males had been eliminated in an area, sexually mature unmated females began to respond to the methyleugenol traps. Fitt

(1981a) studied this female response in *D. opiliae* and *Dacus tenuifascia* (May) to methyleugenol and female response in *Dacus aquilonis* (May) to cue-lure. He found that sexually mature unmated females of *D. opiliae* responded to methyleugenol around dusk which he determined is their normal mating time. Fitt found that *D. tenuifascia* on the other hand, is one of the few *Dacinae* studied to date which mates during the day. Correspondingly, the *D. tenuifascia* females responded to methyleugenol during the day but not at dusk. *D. aquilonis* has crepuscular mating habits as do most Dacinae and, the female response to cue-lure increased dramatically at dusk. With the crepuscular mating Dacinae it is difficult to reconcile the differing temporal patterns of response to the male lures by the two sexes into a logical whole. If these lures are plant kairomones acting as secondary sexual attractants used to establish lekking sites, then why isn't male response strongest at dusk for those species such as *D. dorsalis*, *D. cucurbitae*, etc. when they should be establishing their leks? The same argument can be advanced against the concept that these are pseudo male aggregation pheromones or pseudo sex attractant pheromones.

The puzzling male lure phenomenon becomes all the more striking because of its absence in most of the Tephritidae. The response is absent in some Dacinae that are fairly closely related to "male lure" species (Drew and Hooper, 1981). It is also completely absent in the temperature zone *Rhagoletis* genus. Further it has not been found in the neotropical *Anastrepha* genus. Over 8000 compounds were screened as possible attractants for the Mexican fruit fly, *Anastrepha ludens* (Loew), in the U.S. Department of Agriculture's Mexico City Laboratory (Chambers, 1977). More recently Burditt and McGovern (1979) screened 1320 compounds for attractancy to the Caribbean fruit fly, *Anastrepha suspensa* (Loew), without finding a male lure. If it were not for *Ceratitis*, we would think that the male lure phenomenon was a peculiar development of tropical Asia.

5 MEDFLY LURES

Surprisingly, the first male lure discovered was for the medfly *C. capitata*, although it was not widely recognized as such at the time. Severin and Severin (1913) give an account of its discovery and early use. As with methyleugenol, it was a serendipitous discovery. About 1907 an Australian housewife used kerosene as an ant barrier for some jam she was cooling. Large numbers of flies were attracted to the area. Her husband, Mr Devenish, recognized it as the medfly and determined that the flies were responding to the kerosene and not the jam. He caught large numbers of flies in pans of kerosene which he then installed in his orchard. Australia responded by passing regulations that enforced the use of kerosene in fruit orchards. Severin and Severin (1913) are full of scorn for the fact that most entomologists did not emphasize the fact that less than 1% of the flies caught were the damage-producing females. The lack of reduction in damage in the treated orchards soon led to an abandonment of the program.

Kerosene is a complex mixture and surprisingly little work was done in the next 30 or 40 years to try to identify the active principles contained in it. In the light of what we now know, it is likely that it contains a number of slightly attractive compounds. I have compared a few available commercial "kerosenes" in field tests and found them to be less than 1% as attractive as our modern male lure (unpublished data). The unfortunate situation is that turn-of-the-century Australian kerosene might have been a considerably different mix from what I have tested — we may never know exactly how good and what was in Mr Devenish's kerosene.

Chapter 3.6 references, p. 228

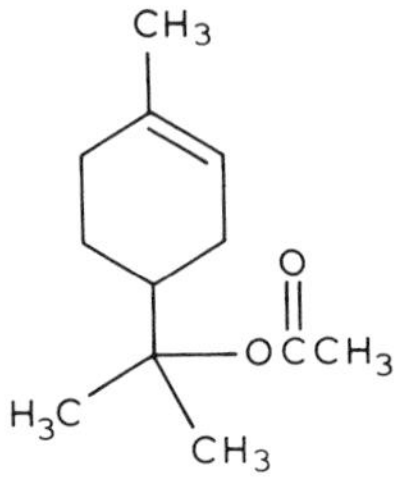

(a) Terpineol acetate

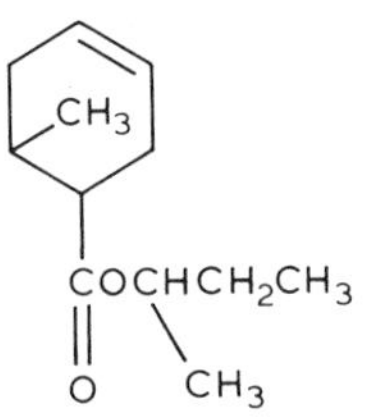

(b) Siglure

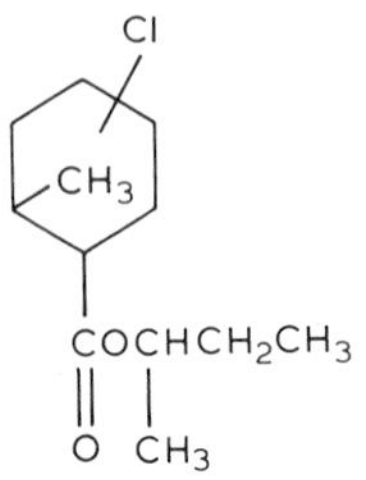

(c) Medlure

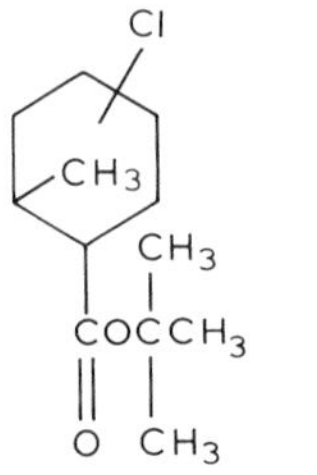

(d) Trimedlure

Fig. 3.6.2. Parapheromones. Male lures for *Ceratitis capitata*. (a) Terpineol acetate (Terpinyl acetate). A moderately attractive male lure. (b) Siglure. The first synthetic male lure for *Ceratitis capitata*. (c and d) Medlure and Trimedlure. Chloro derivatives of siglure which are superior male lures.

Riply and Hepburn (1935) working in Africa, the ancestral home of the *Ceratitis* subfamily found terpinyl acetate (Fig. 3.6.2(a)), modern terminology for this is terpineol acetate) was a male lure for the Natal fruit fly, *Ceratitis rosa* (Karsch), and 12 related species of this subfamily including the medfly, *C. capitata*. They found that response to this male lure followed recognized taxonomic lines with only species from the related genera *Ceratitis*, *Pterandrus*, *Pardalaspis*, and *Pinacochlaeta* responding. In addition to terpinyl acetate, they found male attraction to *C. rosa* by a number of botanical extracts including angelica root oil.

In the 1929 medfly eradication program in Florida, kerosene traps and sweetened food baits were used but the chief tool to delimit the infestation was examination of host fruits for infestation (Steiner et al., 1961). With the reinvasion of Florida by the medfly in 1956 the attractants screening program in the USDA laboratory in Hawaii was intensified. Angelica seed oil was found to be very attractive for the medfly males. Problems with angelica seed oil were that it was expensive, in short supply, and short lived in the field; also, the attractiveness varied greatly from one batch of angelica extract to another.

Terpinyl acetate was also found to be moderately attractive in olfactometer tests (Beroza and Green, 1963). USDA chemists in Beltsville prepared several esters of 6-methyl-3-cyclohexene-1-carboxylic acid as part of the routine screening of hundreds of compounds for medfly attraction. A number of these proved to be attractive and a long series of esters of this acid was tested. The best male lure proved to be the sec-butyl ester (Gertler et al., 1958). This was named Siglure (Fig. 3.6.2(b)) (Beroza et al., 1961) and was put into immediate use in Florida where it was found that some lots of Siglure did not perform as well as expected. Siglure can exist as a *cis* or *trans* isomer and the *trans* isomeric form was found to be far more active than the *cis* isomer (Steiner et al., 1958). While working on the isomerism problem Beroza et al. (1961) hydrohalogenated the double bond in the 6-carbon ring of Siglure and found an even more powerful and longer-lasting male lure, medlure (Fig. 3.6.2(c)) and then the even more attractive tertiary butyl analog of medlure, called trimedlure (Fig. 3.6.2(d)). Thus, we have moved from a natural compound, terpinyl acetate, to a synthetic chloro compound with greatly improved attraction, trimedlure.

The medlure and trimedlure structures introduced further complexities since they could exist in 8 different stereoisomers as opposed to 2 for Siglure (Beroza et al., 1961). The 4 isomers with the methyl and ester substituents in the *cis* configuration occur as minor components in the normal manufacture of trimedlure. McGovern and Beroza (1966) show the structure of the 4 *trans* isomers which are named in the order of their elution from a gas chromatographic column. Of the 4 *trans* isomers, B_2, which is a solid at room temperature is essentially unattractive; the other solid isomer, isomer C, is highly attractive (McGovern et al., 1966; Cunningham et al., manuscript). Of further complexity is the fact that optically active preparations of the C-isomer differ in their biological activity (Sonnet et al., 1984). As Sonnet et al. (1984) point out, the presence of this optical selectivity in the insect toward the synthetic trimedlure is further indication that a naturally occurring compound is being imitated. One would expect the "true", the archetypal pheromone or kairomone to be more powerful than this synthetic imitator.

Several workers have explored an extensive series of structures around the cyclohexene ring of Siglure (Beroza et al., 1964; Valega and Beroza, 1967; Valega et al., 1967 and Guiotto et al., 1974). No compounds were found which were much more effective than Siglure. Most modifications of the Siglure structure resulted in a diminution or complete loss of attractiveness. Recently I have bioassayed an extensive series of analogs and homologs which T.P. McGovern synthesized around the trimedlure structure (unpublished data). Our results have been similar. Many compounds show some degree of attraction but none is greatly superior to trimedlure. It would seem then that the archetypal attractant which these male lures are imitating is not to be found around the single 6-carbon ring structure. We must look elsewhere.

Good arguments can be advanced in the case of the medfly that the male lure does imitate a male sex pheromone. Unlike most Dacinae, the medfly mates during the daylight hours, primarily in the morning (Feron, 1962) and more or less coincides with the times of response to trimedlure. In the absence of males, sexually mature unmated females respond to trimedlure, medlure, and angelica seed oil (Nakagawa et al., 1970).

The case that the male lures for the medfly are really pseudopheromones would be entirely compelling except for results obtained from studies of the plant kairomone contained in angelica seed oil. Fornasiero et al. (1969) and Guiotto et al. (1972) identified the attractant principle in angelica seed oil as being α-copaene and its isomer α-ylangene plus some other unidentified sesquiterpenes of lesser attraction. They had bioassayed attraction in cage tests. Recently my colleague, M. Jacobson, and I measured the relative attraction of α-copaene and trimedlure. We found α-copaene to be 2 to 5 times more attractive for male medflies in field tests than trimedlure (unpublished data). Nakagawa et al. (1970) showed that angelica seed oil attracts mature unmated females in the absence of males just as the synthetic male lure do.

The molecular structure of α-copaene is much more complex, being a 3-ring hydrocarbon molecule (Fig. 3.6.3), than the synthetic male lures. It also differs from the other compounds in that its degree of attraction is distinctly greater. This separation is readily demonstratable in field bioassays and is not explainable simply by differences in volatility. Because of this much greater attraction one would propose that copaene is the archetypal male lure — a co-evolved plant kairomone such as Metcalf envisaged. The cyclohexane ring found in

Chapter 3.6 references, p. 228

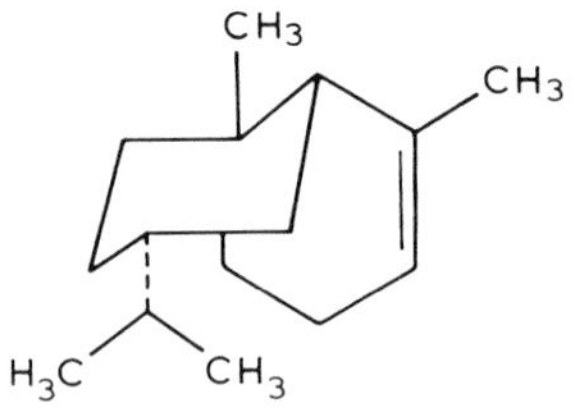

α–Copaene

Fig. 3.6.3. Parapheromone. α-copaene. A powerful plant-derived male lure for *Ceratitis capitata*.

trimedlure is also incorporated into the copaene structure and in that sense trimedlure could be considered a weaker imitator of the whole copaene molecule.

6 CONCLUSION

The examples of the male attractants for the medfly epitomizes the puzzlement and the opportunities the male lure phenomenon has given us. A completely synthetic halogen containing compound and a naturally occurring botanical compound, a possible kairomone, both elicit responses similar to that which we could expect from an insect pheromone. Nature seems to eschew the simple patterns we would like to see in her. It is likely that these male lures, these parapheromones are not simply pheromone imitators or kairomone imitators but something else having a little of both.

7 REFERENCES

Alexander, B.H., Beroza, M., Oda, T.A., Steiner, L.F., Miyashita, D.H. and Mitchell, W.C., 1962. The development of male melon fly attractants. Agricultural and Food Chemistry, 10: 270–276.

Baker, R., Herbert, R.H. and Lomer, R.A., 1982. Chemical components of the rectal gland secretions of male *Dacus cucurbitae*, the melon fly. Experentia, 38: 232–233.

Bellas, T.E. and Fletcher, B.S., 1979. Identification of the major components in the secretion from the rectal pheromone glands of the Queensland fruit flies *Dacus tryoni* and *Dacus neohumeralis* (Diptera:Tephritidae). Journal of Chemical Ecology, 5: 795–804.

Beroza, M., Alexander, B.H., Steiner, L.F., Mitchell, W.C. and Miyashita, D.H., 1960. New synthetic lures for the male melon fly. Science, 131: 1044–1045.

Beroza, M., Green, N., Gertler, S.I., Steiner, L.F. and Miyashita, D.H., 1961. New attractants for the Mediterranean fruit fly. Agricultural and Food Chemistry, 9: 361–365.

Beroza, M. and Green, N. (Editors), 1963. Materials tested as insect attractants. United States Department of Agriculture, Agricultural Handbooks No. 239, pp. 148.

Beroza, M., McGovern, T.P., Steiner, L.F. and Miyashita, D.H., 1964. t-butyl and t-pentyl esters of 6-methyl-3-cyclohexene-1-carboxylic acid as attractants for the Mediterranean fruit fly. Agricultural and Food Chemistry, 12: 258.

Brieze-Stegman, R., Rice, M.J. and Hooper, G.H.S., 1978. Daily periodicity in attraction of male tephritid fruit flies to synthetic chemical lures. Journal of the Australian Entomological Society, 17: 341–346.

Burditt, A.K. Jr and McGovern, T.P., 1979. Chemicals tested as attractants for the Caribbean fruit fly. United States Department of Agriculture Manual ARM-S-6, 47 pp.

Cunningham, R.T, Leonhardt, B.A., McGovern, T.P. and Rice, R.E., 1987. Trimedlure, the male lure for the Mediterranean fruit fly: the importance of C-isomer concentrations. manuscript in preparation.

Dodson, C.H., Dressler, R.L., Hills, H.G., Adams, R.M. and Williams, N.H., 1969. Biologically active compounds in orchid fragrances. Science, 164: 1243–1249.

Drew, R.A.I., 1974. The responses of fruit fly species (Diptera:Tephritidae) in the South Pacific area to male attractants. Journal of the Australian Entomological Society, 13: 267–270.

Drew, R.A.I. and Hooper, G.H.S., 1981. The responses of fruit fly species (Diptera:Tephritidae) in Australia to various attractants. Journal of the Australian Entomological Society, 20: 201–205.

Feron, M., 1962. L'instict de reproduction chez la mouche mediterraneenne des fruits *Ceratitits capitata* Wied. (Dipt. Trypetidae). Comportment Sexuel. Comportment de Ponte. Revue de Pathologie Vegetale et d'Entomologie Agricole de France.

Fitt, G., 1981a. Responses by female Dacinae to "male" lures, and their relationship to patterns of mating behavior and pheromone response. Entomologia Experimentalis et Applicata, 29: 87–97.

Fitt, G., 1981b. The influence of age, nutrition, and time of day on the responsiveness of male *Dacus opiliae* to the synthetic lure, methyleugenol. Entomologia Experimentalis et Applicata, 30: 83–90.

Fletcher, B.S., Bateman, M.A., Hart, N.K. and Lamberton, J.A., 1975. Identification of a fruit fly attractant in an Australian plant, *Zieria smithii*, as o-methyl eugenol. Journal of Economic Entomology, 68: 815–816.

Fornasiero, V., Guiotto, A., Caporale, G., Baccichetti, F. and Musajo, L., 1969. Identificazione della sostanza attrattiva per i maschi della *Ceratitis capitata*, contenunuto nell'olio essenziale dei semi di *Angelica archangelica*. Gazzetta Chimica Italiana, 99: 700–710.

Gertler, S.I., Steiner, L.F., Mitchell, W.C. and Barthel, W.F., 1958. Esters of 6-methyl-3-cyclohexene-1-carboxylic acid as attractants for the Mediterranean fruit fly. Agricultural and Food Chemistry, 6: 592–594.

Gierschner, K. and Baumann, G., 1967. Aromastoffe in Fruchten. In: Aroma-Geschmackstoffe, Lebensmittel Fortbildungskurs. pp. 49–89.

Guiotto, A., Fornasiero, U. and Baccichetti, F., 1972. Investigations on attractants for males of *Ceratitis capitata*. Farmaco, Edizione, Scientifica, 27: 633–670.

Guiotto, A., Rodighiero, P. and Fornasiero, U., 1974. Relationships between structure of some cyclohexene derivatives and attractiveness for the males of *Ceratitis capitata*. Farmaco, Edizione, Scientifica, 29: 95–100.

Hancock, D.L., 1985. A specific male attractant for the melon fly *Dacus vertebratus*. The Zimbabwe Science News, 19: 118–119.

Howlett, F.M., 1912. The effect of oil of citronella on two species of *Dacus*. Transaction of the Entomological Society of London. 60 Part II: 412–418.

Howlett, F.M., 1915. Chemical reactions of fruit flies. Bulletin of Entomological Research, 6: 297–305.

Kawano, Y., Mitchell, W.C. and Matsumoto, H., 1968. Identification of the male oriental fruit fly attractant in the golden shower blossom. Journal of Economic Entomology, 61: 986–988.

Labeyrie, V., 1971. Trophic relations and sex meetings in insects. Acta Phytopathologica Academiae Scientiarum Hungaricae, 6: 229–234.

McGovern, T.P. and Beroza, M., 1966. Structure of the four isomers of the insect attractant trimedlure. Journal of Organic Chemistry, 31: 1472–1477.

McGovern, T.P., Beroza, M., Ohinata, K., Miyashita, D.H. and Steiner, L.F., 1966. Volatility and attractiveness to the Mediterranean fruit fly of trimedlure and its isomers, and a comparison of its volatility with that of seven other insect attractants. Journal of Economic Entomology, 59: 1450.

Metcalf, R.L., 1979. Plants, chemicals, and insects: some aspects of co-evolution. Bulletin of the Entomological Society of America, 25: 30–35.

Metcalf, R.L., Mitchell, W.C., Fukuto, T.R. and Metcalf, E.R., 1975. Attraction of the oriental fruit fly, *Dacus dorsalis*, to methyleugenol and related olfactory stimulants. Proceedings of the National Academy of Sciences of the United States of America, 72: 2501–2505.

Metcalf, R.L., Metcalf, E.R., Mitchell, W.C. and Lee, L.W.Y., 1979. Evolution of olfactory receptor in oriental fruit fly, *Dacus dorsalis*. Proceedings of the National Academy of Sciences of the United States of America, 76: 1561–1565.

Metcalf, R.L., Metcalf, E.R. and Mitchell, W.C., 1981. Molecular parameters and olfaction in the oriental fruit fly, *Dacus dorsalis*. Proceedings of the National Academy of Sciences of the United States of America, 78: 4007–4010.

Metcalf, R.L., Mitchell, W.C. and Metcalf, E.R., 1983. Olfactory receptors in the melon fly *Dacus cucurbitae* and the oriental fruit fly *Dacus dorsalis*. Proceedings of the National Academy of Sciences of the United States of America, 80: 3143–3147.

Nakagawa, S., Farias, G.J. and Steiner, L.F., 1970. Response of female Mediterranean fruit flies to male lures in the relative absence of males. Journal of Economic Entomology, 63: 227–229.

Payne, T.L., Shorey, H.H. and Gaston, L.K., 1973. Sex pheromones of Lepidoptera. XXXVIII. Electroantennogram responses in *Autographa californica* to cis-7-dodecenyl acetate and related components. Annals of the Entomological Society of America, 66: 703–704.

Ripley, L.B. and Hepburn, G.A., 1935. Olfactory attractants for male fruit flies. Department of Agriculture, Union of South Africa. Entomology Memoir No. 9, pp. 17.

Schinz, H. and Seidel, C.F., 1961. Helvetia Chimica Acta, 44: 278.

Severin, H.P. and Severin, H.C., 1913. A historical account on the use of kerosene to trap the Mediterranean fruit fly (*Ceratitis capitata* Wied.). Journal of Economic Entomology, 6: 347–351.

Shah, A.H. and Patel, R.C., 1976. Role of tusli plant (*Ocimum sanctum*) in control of mango fruit fly, *Dacus correctus* Bezzi (Tephritidae:Diptera). Current Science, 45: 313–314.

Sonda, M., 1972. Fruit flies caught in plastic traps baited with cue-lure in Okinawa. Research

Bulletin of the Plant Protection Service, Japan No. 10, pp. 28–31.

Sonnet, P.E., McGovern, T.P. and Cunningham, R.T., 1984. Enantiomers of the biologically active components of the insect attractant trimedlure. Journal of Organic Chemistry, 49: 4639–4643.

Steiner, L.F., 1951. Lures for *Dacus dorsalis*. Hawaiian Entomological Society Proceedings, 14: 204.

Steiner, L.F., Mitchell, W.C., Green, N. and Beroza, M., 1958. Effect of *cis-trans* isomerism on the potency of an insect attractant. Journal of Economic Entomology, 6: 921–922.

Steiner, L.F., Rohwer, G.G., Ayers, E.L. and Christenson, L.D., 1961. The role of attractants in the recent Mediterranean fruit fly eradication program in Florida. Journal of Economic Entomology, 54: 30–35.

Steiner, L.F., Mitchell, W.C., Harris, E.J., Kozuma, T.T. and Fujimoto, M.S., 1965. Oriental fruit fly eradication by male annihilation. Journal of Economic Entomology, 58: 961–964.

Valega, T.M. and Beroza, M., 1967. Structure-activity relationships of some attractants of the Mediterranean fruit fly. Journal of Economic Entomology, 60: 341–347.

Valega, T.M., McGovern, T.P., Beroza, M., Miyashita, D.H. and Steiner, L.F., 1967. Candidate attractants for the control of the Mediterranean fruit fly. Journal of Economic Entomology, 60: 835–844.

Chapter 3.7 Acclimation, Activity Levels and Survival

A. MEATS

1 THE STRATEGY OF ACCLIMATION

The responses that an organism makes, in the short term, to changes in its environment can be modified, in the long term, if those changes persist. When observed in nature, these modifications or adjustments can be called *acclimatization*. The term *acclimation* has a more precise meaning. After causes and effects have been defined and quantified by laboratory experiments, a given adjustment can be referred to as *acclimation with respect to a particular cause* with respect to a particular activity, whether it occurs in experimental or natural conditions.

The ecological advantage of acclimation is that it allows levels of activity (from metabolic to behavioural) to have either, (a) a lessened response to a sustained environmental change, with the result that the organism has a measure of independence from seasonal or other factors; or (b) a more exaggerated response to a sustained change with the result that the organism has levels of activity which are markedly different according to season or other environmental circumstance.

Examples of both strategies can be seen among the types of responses that poikilothermic animals have to temperature. Such animals are particularly sensitive to changes in temperature. Some ameliorate this effect, at least on a seasonal basis, by acclimation so that (for example) enzymic and locomotory activities can occur at much the same rate in winter as they do in summer through the boosting of temperature-specific rates on sustained exposure to lower temperature. Alternatively, the response to lower temperatures may be exaggerated in winter so that a level of activity which drops at a lower temperature is further decreased on sustained exposure, leading to a seasonal state of quiescence, dormancy or even diapause according to species and circumstance. The ecological significance of both strategies is discussed, with examples in section 6.4.

Acclimation can occur with respect to temperature, illumination, salinity, osmotic pressure or oxygen tension. It can be measured in terms of adjustments made to metabolic rate, the rate of enzyme action, isozyme frequency, titre of cryoprotectants, ice nucleators, the rate of action of heart, nervous system, kidney or other organs, general motility, wingbeat frequency, mating frequency, freezing points, supercooling points, temperature thresholds for flight, walking, torpor or survival and finally thresholds for homeostasis with respect to salinity or osmotic pressure (Mutchmor and Richards, 1961; Prosser and Brown, 1961; Bowler, 1963; Nuttall, 1970, Anderson and Mutchmor, 1968, 1971; Meats and Fay, 1976; Meats, 1976b; Grice, 1976; Gordon et al., 1982; Zachariassen, 1982).

Chapter 3.7 references, p. 237

2 STEADY AND CHANGING STATES

Since acclimation is detected as the adjustment of a response with time, it can be appreciated that cases of both partial and complete adjustment can be found. Consider the adjustment of the response of metabolic rate to temperature. If an organism is reared or kept for some time at temperature A (which we may call the holding temperature) and its metabolic rate is measured for relatively short periods at a variety of other temperatures (or preferably different groups of individuals tested, with each group at a different temperature) then one may find that there is a consistent relationship between rate and test temperature. 'Consistent' here means that the relation is the same whenever it is measured in the above manner and the tested individuals are kept at holding temperature A prior to measurement. If the holding temperature is now altered to B then similar short-term tests may reveal that the relationship between metabolic rate and test temperature changes with time until a new consistent relationship is attained.

The organism can be said to be fully acclimated to A before the change in holding temperature, partially acclimated to B for some time after the change and fully acclimated to B when the new steady state is reached.

In nature, environmental conditions obviously do not change in such a simple manner. In most terrestrial environments above the ground there is a daily fluctuation of temperature with each daily regime differing from the next and with an overlying tendency for daily maxima, means and minima to rise and fall with season. In most other environments there is at least some seasonal variation which may be continuous. It would therefore seem unreasonable to expect to find in nature the existence of steady states of acclimation such as constant temperature-specific rates or thresholds of activity.

There is, however, evidence that essentially steady states can be attained. It is widely recognised that there are limits to the degree of adjustment that can be made to a threshold or temperature-specific rate (see Colhoun, 1960; Fry, 1967; Meats, 1976b; Gordon et al., 1982). The maximum degree of adjustment to cold may occur at a temperature ($x°$) well above the expected seasonal minimum, with lower temperatures causing only a similar but not greater degree of adjustment. Thus if temperatures persist below $x°$ for the best part of a season it is likely that a steady state can be reached in time for the most unfavourable part of that season, with progress towards that state being sufficient to track the demands of the environment as the season progresses from warm to cold (Brett, 1956; Meats, 1976c).

At other times of the year or in the case of other environmental factors, it is likely that the 'target' steady state, at which the acclimation process is heading, is constantly shifting with environmental conditions. The 'target' may even shift widely on a daily basis as mentioned above. The problem of predicting activity levels, thresholds etc., at any given time after any given train of events is one of the ultimate challenges of the environmental physiologist and has been met in the case of the temperature relations of the Queensland fruit fly, *Dacus tryoni* (Froggatt).

3 ACCLIMATION OF ADULT *D. tryoni* TO FLUCTUATING CONDITIONS

Meats (1976a, b, c) and Meats and Fay (1977) have shown that a given daily regime of temperature, whether constant or fluctuating, can (if sustained) produce characteristic states of acclimation with respect to cold torpor and cold-survival. Transfer to another regime produces changes towards a state typical of the new regime. The steady states expected of a fluctuating regime are predictable in terms of those attained in constant regimes. The rate of change of state at any particular time varies according to the difference between the current state (i.e. current threshold) and the steady state (threshold) to which acclimation would develop should the current regime persist. The full model for linking thermal history of any kind with torpor and survival thresholds is given by Meats (1976c).

The maximum rate and extent of cold-acclimation in *D. tryoni* is caused by temperatures below 17°C. With daily fluctuating conditions, cold acclimation occurs at nearly maximal rates even when daily maxima are as high as 20°C. Thus acclimation towards the winter condition starts in the field long before winter arrives, with the result that a maximal state of cold acclimation can be attained by the time the coldest weather occurs. Flies emerging from puparia during the pre-winter period do not have to catch up in acclimation with the older flies in the population because they attain complete acclimation to prevailing conditions during development as described below.

4 DEVELOPMENTAL ACCLIMATION IN *D. tryoni*

Maynard-Smith (1957) discovered that *Drosophila subobscura* could be acclimated to a given temperature during the immature stages. The process has been investigated in detail in the case of *D. tryoni* (Meats, 1976a, 1983).

Rapid and complete acclimation to a given temperature can occur in *D. tryoni* during two short stages (critical periods) in its life history; the stage immediately prior to the 'hopping larva' phase and the 'pharate adult' phase within the puparium. These stages correspond respectively to the last 70% and 80% of the larval and puparial stages. Acclimation takes only as long as the prevailing temperature permits these phases to be completed. In most cases this is of the order of a few days which can be contrasted with the times taken for full acclimation in adults which takes weeks or months.

Larvae therefore pupate and adults emerge from puparia fully acclimated to the prevailing conditions. The ability to acclimate appears to be absent in the earlier stages of larval or puparial development. The extent of cold acclimation to both constant and fluctuating temperatures during the critical periods is predictable with the acclimation model mentioned earlier (Meats, 1976a, b, c).

5 ACCLIMATION ABILITY OF FRUIT FLIES IN FIELD CONDITIONS

Seasonal trends in torpor threshold have been recorded in both *D. tryoni* and *Dacus oleae* (Gmelin) (Meats, 1976c; Fletcher and Zervas, 1977). In addition, Koidsumi (1936) provides evidence of seasonal changes in cold-hardiness of the immature stages of *Dacus dorsalis* Hendel.

Meats (1976a) has shown that there is a predictable link between threshold for cold torpor and the survival time of adults of *D. tryoni* at subzero temperatures and infers that ability to withstand frost also varies with season as does torpor threshold. Predictions on the extent to which flies can survive frost in the field is borne out by the observations of Fletcher (1975, 1979).

Flight threshold of *D. tryoni* is lowered on cold acclimation (Meats, 1973) but flight propensity (take-off frequency) is reduced along with metabolic rate and wingbeat frequency at any one temperature (Grice, 1976; Moses, 1982). It appears that the mating threshold of *D. tryoni* is not affected by any sort of acclimation (Meats and Fay, 1976; Fay and Meats, 1983).

Chapter 3.7 references, p. 237

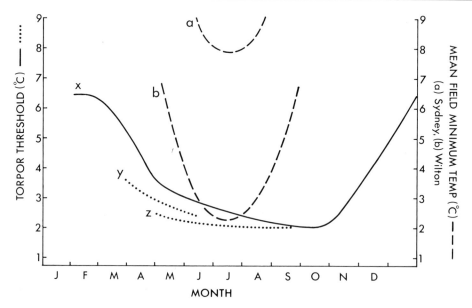

Fig. 3.7.1. A model for seasonal trends of torpor avoidance in *D. tryoni*: (x) thresholds of flies emerging in January and February; (y) emerging in mid-March; (z) emerging in mid-April. Trends in mean field minimums indicated by lines (a) for Sydney, (b) for Wilton. Lowest recorded minima for Sydney (not shown) essentially follow the trend of (b). Lowest recorded minima for Wilton shown in Fig. 3.7.2. Temperature trends (a) and (b) based on 100- and 50-year records respectively (Australian Bureau of Meteorology, Melbourne). Model based on Meats (1976c).

6 ECOLOGICAL SIGNIFICANCE OF SEASONAL ACCLIMATION IN *D. tryoni*

6.1 Significance of adjustment to torpor threshold

The torpor threshold of *D. tryoni* can be as high as 7°C in summer and drop to 2°C in winter. It has been shown that the seasonal trajectory (Fig. 3.7.1) is virtually the same in a marginally survivable habitat (Wilton) as it is in a milder one (Sydney) in which cold torpor is almost never observed (Meats, 1976c). This is mainly because (as explained earlier) the maximal rate and extent of cold acclimation can occur at any temperature below 17°C, thus the cold-acclimation trends at Wilton can occur no faster than they do at Sydney despite the fact that the mean, maximum and minimum daily temperatures at the former site tend to drop increasingly below the latter from the beginning of autumn to the middle of winter.

Developmental acclimation is much faster than post-teneral acclimation, hence flies emerging in mid and late autumn are fully acclimated to the prevailing regime as mentioned earlier (section 4). They therefore tend to be ahead (in terms of cold adaptation) of flies which emerged earlier in the season (see Fig. 3.7.1). They also comprise the majority of flies (about 94%) that successfully survive the winter (Fletcher, 1975).

Developmental acclimation is, however, of no advantage to the individuals which actually pupariate in late autumn since puparial development is arrested in winter with fatal results (Bateman and Sonleitner, 1967).

Figure 3.7.1 shows that the lowest mean monthly minimum temperature at Sydney is 8.0°C (in July) and the lowest recorded minimum temperature is 2.1°C. This indicates that the probability of flies being torpid in Sydney is extremely low. Figure 3.7.1 shows that the downward trajectory of torpor threshold in autumn makes this probably even lower than it would be if acclimation could not occur. The lowest mean monthly minimum temperature at Wilton is 2.3°C indicating that in winter temperatures will frequently go below 2°C which is the lowest that torpor threshold can go. If strong winds occur when flies are torpid they can be blown from their sheltered overwintering sites and disappear from the population (Fletcher, 1975). Temperatures that can cause torpor need not be directly fatal unless they are significantly below freezing. Frosty conditions which are likely to be directly fatal to flies are not associated with windy conditions (Foley, 1945).

6.2 Significance of adjustment to thresholds for cold-survival

Ability to survive cold is best expressed in terms of mean survival time (LE_{50}) at a given minimum, or in terms of what temperature (LT_{50}) is required to kill 50% if exposure is for a standard duration. At Wilton the nightly minimum persists for about 4 h (Meats, 1976c).

Figure 3.7.2 shows the trend in LT_{50} for 4 h estimated from the work of Meats (1976c, 1987) and Fitt (1987).

The estimation is based on the following subtle points, viz: (a) LT_{50} can vary with rate of cooling (Meats, 1987), so the rate is considered to be 1°C per hour; (b) about half the flies cooled below torpor point fall to the ground in frosty conditions (Fitt, 1987); (c) the temperature on the ground during a frost averages about 3°C below that in a Stevenson screen which in turn is the temperature flies are expected to experience in a tree (Foley, 1945; Fitt, 1987; O'Loughlin et al., 1984). These considerations lead to an estimate of LT_{50} for winter minimums of around − 4.5°C which is supported by the field observation of Fletcher (1975, 1979).

The trend illustrated in Fig. 3.7.2 uses this midwinter figure as a baseline and shows levels expected at other times of the year according to the dynamics of the model of Meats (1976c).

The indicated trend does not look very dramatic at first sight. There is only a difference in LT_{50} of 1.9°C between summer and winter. Similar, apparently trivial, differences have been noted for other invertebrates (Edney, 1964). In

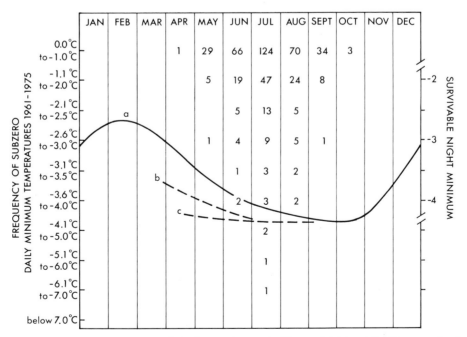

Fig. 3.7.2. A model for season trends in frost resistance in *D. tryoni*: (a) trend for flies emerging in January and February; (b) emerging in mid-March; (c) emerging in mid-April. Frost frequencies from Meats (1976c): model based on present text.

ecological terms, however, the difference is far from trivial. The lowering of LT_{50} by just under 2°C significantly improves the probability of survival at a marginally favourable site.

Figure 3.7.2 shows that there were 490 days at Wilton in the period 1961–1975 on which temperatures dropped below 0°C. If flies had not acclimated beyond their February level the number of days associated with a mortality of over 50% would have been as high as 37. With acclimation, the number of such days is reduced to no more than 6. Acclimation therefore reduces the number of potentially lethal days at Wilton to about a sixth of the number expected if flies could not acclimate. A similar answer is obtained if we compare survivable winters instead of survivable days (Meats, 1976c). The number of winters in the 15-year period which had frost below -2.65°C was 13, whereas the number which had frosts below -4.5°C was only 2. Acclimation therefore makes the Wilton climate a marginally favourable climate with respect to frost survival rather than a predominantly unfavourable one.

It should be noted, however, that acclimation has its limits and that climates which always have nightly minima in winter of -4.5°C or below are likely to be unfavourable to survival regardless of an ability to acclimate.

6.3 Ecological significance of absence of changes in mating threshold

There is apparently no acclimation with respect to mating activity in *D. tryoni* (Meats and Fay, 1976; Fay and Meats, 1985).

Perhaps there may be little or no advantage in acclimation with respect to mating frequency or threshold. *D. tryoni* mates at dusk (Tychsen and Fletcher, 1971). Meats and Khoo (1976) point out that in spring the mean dusk temperature of their study area rises by around 0.6°C per week and that it can be expected to be just above mating threshold at the time the insects are expected to become mature. Dusk temperatures will inevitably fluctuate from day to day about the mean. Unsuitably low dusk temperatures for a run of a few days can be expected to be offset by a run of temperatures which are suitably above the threshold. Thus there would seem to be a trivial advantage in adjusting to temperature if it would mean only an advantage of mating a few days earlier.

It is possible to speculate that there would be selection pressure for males to mate earlier in spring. Males evolving an ability to have a lower mating threshold would be able to mate earlier than other males and hence gain an advantage in fitness. This strategy would not work, however, if females were not also ready to mate at an earlier date, which seems unlikely as explained in the previous paragraph.

In the case of autumn conditions, the results of Fletcher (1975) indicate that mating late in the season (and hence an adaptation to extend the mating period) is largely irrelevant since subsequent weather causes resorption of oocytes and loss of sperm in spermathecae. In addition, any offspring laid before such resorption would die in the immature stage during the winter (Bateman and Sonleitner, 1967).

6.4 Ecological significance of seasonal changes in metabolic rate, flight threshold and flight propensity

D. tryoni apparently does not exhibit metabolic compensation in the cold but rather the reverse. Cold-acclimated flies have lower temperature-specific rates of metabolism that do warm-acclimated ones; their rates of consumption of sugar and protein are also extremely low (Grice, 1976; Kelly, 1981; Moses, 1982). Thresholds for forced flight are lowered on cold-acclimation but flight propensity (take off frequency) is reduced at any one temperature (Meats, 1973; Moses,

1982). All these phenomena can be linked to the fact that flies are essentially quiescent in the field during winter; they stay predominantly on the undersides of leaves in sheltered positions; submature flies stay in that condition whereas mature ones regress to a seasonally immature state (Fletcher, 1975, 1979).

This kind of response contrasts with the acclimation strategy of some other animals such as certain fish which boost temperature-specific levels of performance in winter so that activities occur at much the same rate as in summer (Fry, 1967; Gordon et al., 1982). The contrast is probably explained in terms of contrast in factors affecting survival. Fish which must avoid predators or catch prey or maintain their position relative to a current must maintain much the same abilities regardless of season. Fruit flies appear to be governed by a different combination of factors. It may be of advantage to fruit flies to lower their thresholds for torpor and forced flight in winter so that they can recover or maintain their position in the habitat after disturbance by wind or predators, but it may also be of advantage to reduce general activity in order to conserve resources in a season in which food resources are possibly low and temperatures are unfavourable for the survival of offspring in the immature stages (Bateman and Sonleitner, 1967).

7 ACCLIMATION, MASS REARING AND PEST CONTROL

Knowledge of acclimation can be of relevance to any technique of pest control that involves the release of mass-reared insects. Meats (1983), Meats and Fay (1977) and Fay and Meats (1987a, b) have identified and proved the advantages of acclimation techniques to the technique of population control by release of sterile insects. The technique works best when populations are low, localised and seasonally stressed as in spring. Unfortunately flies mass-reared at normal laboratory temperatures in the region of 25°C are not frost resistant although (as pointed out earlier) their thresholds for mating would be as adaptive as the target flies. Full developmental acclimation to cold can be achieved in mass rearing by exposing the last sixth of the pupal stage to a constant temperature of 15°C. A regime alternating between 25°C for 8 h and 12°C for 16 h per day will produce results which are almost as good and add only 4 extra days to the development period from egg to adult.

However, it is often more convenient to release the puparial rather than the adult stage. An earlier section shows that it is possible to produce cold-hardy pupae with a similar technique involving a brief period of thermal conditioning. Flies released either as puparia or adults in early spring are likely to mature at the same time as target flies (Meats and Fay, 1977 and Fay and Meats, 1986b).

8 REFERENCES

Anderson, R.L. and Mutchmor, J.A., 1968. Temperature acclimation and its influence on the electrical activity of the nervous system of three species of cockroach. Journal of Insect Physiology, 14: 243–251.

Anderson, R.L. and Mutchmor, J.A., 1971. Temperature acclimation in *Tribolium* and *Musca* at locomotory metabolic and enzyme levels. Journal of Insect Physiology, 17: 2205–2219.

Bateman, M.A. and Sonleitner, F.J., 1967. The ecology of a natural population of the Queensland fruit fly, (*Dacus* (*Strumeta*) *tryoni*). I. The parameters of pupal and adult populations during a single season. Australian Journal of Zoology, 15: 303–335.

Bowler, K., 1963. A study of the factors involved in the acclimatization to temperature and death at high temperatures in *Astacus pallipes*. I. Experiments on intact animals. Journal of Cellular and Comparative Physiology, 62: 119–132.

Brett, J.R., 1956. Some principles in the thermal requirements of fishes. Quarterly Review of Biology, 31: 75–87.

Colhoun, E.H., 1960. Acclimation to cold in insects. Entomologia Experimentalis et Applicata, 3: 27–37.

Edney, G.B., 1964. Acclimation to temperature in terrestrial isopods. I. Lethal temperatures. Physiological Zoology, 37: 364–377.

Fay, H.A.C. and Meats, A., 1983. The effect of age, ambient temperature, thermal history and mating history on mating frequency in males of the Queensland fruit fly, *Dacus tryoni*. Entomologia Experimentalis et Applicata, 35: 273–276.

Fay, H.A.C. and Meats, A., 1987a. Survival rates of the Queensland fruit fly (*Dacus tryoni*) in early spring: field cage studies with cold-acclimated wild flies and irradiated (warm- or cold-acclimated) laboratory flies. Australian Journal of Zoology, 35: 187–195.

Fay, H.A.C. and Meats, A., 1987b. The sterile insect release method and the importance of thermal conditioning before release: field cage experiments with *Dacus tryoni* in spring weather. Australian Journal of Zoology, 35: 197–204.

Fletcher, B.S., 1975. Temperature regulated changes in the ovaries of the overwintering females of the Queensland fruit fly, *Dacus tryoni*, under natural conditions. Australian Journal of Zoology, 27: 403–411.

Fletcher, B.S., 1979. The overwintering survival of adults of the Queensland fruit fly, *Dacus tryoni*, under natural conditions. Australian Journal of Zoology, 27: 403–411.

Fletcher, B.S. and Zervas, G., 1977. Acclimation of different strains of the olive fly, *Dacus oleae* to low temperatures. Journal of Insect Physiology, 231: 649–653.

Foley, J.C., 1945. Frost in the Australian Region. Commonwealth Meteorological Bureau, Bulletin No. 32, Commonwealth Meteorological Bureau, Melbourne, 144 pp + viii maps.

Fry, F.G.J., 1967. Responses of vertebrate poikilotherms to temperature. In: H.A. Ross (Editor), Thermobiology, Academic Press, London, pp. 357–409.

Gordon, M.S., Bartholemew, G.A., Grinnell, A.P., Jurgensen, C.G. and White, F.N., 1982. Animal Physiology, Principles and Applications, 4th Edition, Macmillan, New York, xvii + 635 pp.

Grice, A.C., 1976. The effects of temperature acclimation, ambient temperature and other factors on the flight ability of the Queensland fruit fly, *Dacus tryoni*. Honours thesis, University of Sydney, 121 + xxv pp.

Kelly, G., 1981. The effects of temperature, thermal history and maturity on feeding rates of the Queensland fruit fly, *Dacus tryoni*. Honours thesis, University of Sydney, iii + 98 − xlvii pp.

Koidsumi, K., 1936. On the fatal action of low temperatures upon eggs and larvae of *Chaetodacus ferrugineus dorsalis* Hendel, (in Japanese). Journal of the Society of Tropical Agriculture, Taihoku, 8: 221–236.

Maynard Smith, J., 1957. Temperature tolerance and acclimatization in *Drosophila subobscura*. Journal of Experimental Biology, 34: 85–96.

Meats, A., 1973. The abolition by low ambient temperature of tarsal inhibition of flight in certain Diptera. Search, 34: 496–497.

Meats, A., 1976a. Developmental and long-term acclimation to cold by the Queensland fruit fly (*Dacus tryoni*) at constant and fluctuating temperatures. Journal of Insect Physiology, 22: 1013–1019.

Meats, A., 1976b. Thresholds for cold-torpor and cold-survival in the Queensland fruit fly (*Dacus tryoni*) and the predictability of rates of change in survival threshold. Journal of Insect Physiology, 22: 1505–1509.

Meats, A., 1976c. Seasonal trends in acclimation to cold by the Queensland fruit fly (*Dacus tryoni*, Diptera) and their prediction by means of a physiological model fed with climatological data. Oecologia, 26: 73–87.

Meats, A., 1983. Critical periods for developmental acclimation to cold in the Queensland fruit fly, *Dacus tryoni*. Journal of Insect Physiology, 29: 943–946.

Meats, A., 1987. Survival of step and ramp changes of temperature by adults of the Queensland fruit fly (*Dacus tryoni*). Physiological Entomology, 12: in press.

Meats, A and Fay, H.A.C., 1976. The effect of acclimation on mating frequency and mating competitiveness in the Queensland fruit fly, *Dacus tryoni*, in optimal and cool mating regimes. Physiological Entomology, 1: 207–212.

Meats, A. and Fay, H.A.C., 1977. The importance of cold acclimation and stage in the release of sterile flies for population suppression in spring: a pilot caged experiment with the Queensland fruit fly, *Dacus tryoni*. Journal of Economic Entomology, 70: 681–684.

Meats, A. and Fitt, G.P., 1987. Survival of repeated frosts by the Queensland fruit fly, *Dacus tryoni*: experiments in laboratory simulated climates with either step or ramp fluctuations of temperature. Entomologica Experimentalis et Applicata, 44: in press.

Meats, A. and Khoo, K.C., 1976. The dynamics of ovarian maturation and oocyte resorption in Queensland fruit fly, *Dacus tryoni*, in daily rhythmic and constant temperature regimes. Physiological Entomology, 1: 213–221.

Moses, B., 1982. The effect of temperature acclimation on temperature-specific rates of respiration and locomotor activity in the Queensland fruit fly, *Dacus tryoni*. Honours thesis, University of Sydney, iv + 79 pp.

Mutchmor, J.A. and Richards, A.G., 1961. Low temperature tolerance in insects in relation to the influence of muscle apyrase activity. Journal of Insect Physiology, 7: 141–158.

Nuttall, R.M., 1970. The effect of acclimation upon the survival of *Ptinus tectus* and *Tenebrio molitor* when exposed to low temperatures. Entomologia Experimentalis et Applicata, 13: 217–222.

O'Loughlin, G.T., East, R.W. and Meats, A., 1984. Survival, development rates and generation times of the Queensland fruit fly, *Dacus tryoni*, in a marginally favourable climate: experiments in Victoria. Australian Journal of Zoology, 32: 353–361.

Prosser, C.L. and Brown, F.A., 1961. Comparative Animal Physiology, 2nd edition, W.B. Saunders, Philadelphia, ix + 688 pp.

Tychsen, P.H. and Fletcher, B.S., 1971. Studies on the rhythm of mating in the Queensland fruit fly, *Dacus tryoni*. Journal of Insect Physiology, 17: 2139–2156.

Zachariassen, K.E. (Editor), 1982. Special Section: Cold-Hardiness in Poikilothermic Animals. Comparative Biochemistry and Physiology, 73A: 517–639.

Chapter 3.8 Water Relations of Tephritidae

A. MEATS

1 INTRODUCTION

It can be gathered from any physiological textbook or review on water and ionic relations that organisms are essentially packets of aqueous solutions and that the components of such solutions must each be maintained within a certain range of proportions. If the insect body is considered as a single package (with some non-aqueous components) then the water content is generally found to be maintained in the region of 75% through the balancing of gains and losses that occur through the general cuticle, spiracles and the alimentary system.

Since most body fluids have a concentration in the range 300–600 mM, they would only achieve equilibrium with relative humidities in the range 99.5–99.8% (Wharton and Richards, 1978). There is therefore a tendency for water loss to air in all but virtually saturated atmospheres.

In aqueous conditions (including the environments of the endoparasites of plants and animals), the number of insects known to be adapted to hyperosmotic media is minute compared to those adapted to hypo-osmotic media. The latter have to cope with a tendency for a net influx of water by producing hypo-osmotic urine and using various mechanisms for ionic retention or acquisition. Homeostasis in such animals is never threatened except in the unusual circumstances of the environment becoming hyperosmotic or there being insufficient ions in the medium to replace urinary losses (Stobbart and Shaw, 1974).

In unsaturated air, the tendency to lose water can be countered in some xeric species by using specialized structures to reclaim it in the vapour form; despite this, the mechanism only works down to a limiting (equilibrium) humidity (see review by Machin, 1979).

Most terrestrial insects can only replace lost water by drinking or by eating food of adequate water content; quite often the food has more than enough water in it for the hydration needs of the animal (Bernays and Simpson, 1982). However, if ingestible water is not available in sufficient quantity the insect will desiccate and death will supervene if the situation is prolonged. The survival period in the absence of available water depends upon the rate of water loss (see later) and the extent of loss than can be tolerated. The proportion of body water that can be lost without immediately fatal results is termed the water reserve (Bursell, 1974b).

Chapter 3.8 references, p. 246

2 WATER RELATIONS OF TEPHRITIDS

2.1 Scope of knowledge

There has been no study of tephritids concerning the maintenance of the water reserve within viable limits. An excellent example of the type of work required is the study of grasshoppers by Sell and Houlihan (1985) who also give a review of some other investigations.

As far as tephtritids are concerned, there is a variety of information that we can use to estimate (if only approximately) the extent to which environmental moisture governs homeostasis, survival and fecundity.

2.2 Immature stages

Eggs and larvae inhabit fruit and therefore we may expect that environmental water is not a factor of immediate relevance to the survival of these stages so long as the fruit is normally hydrated. Fruit may drop prematurely or at least become hyperosmotic in a drought. Bateman (1968) reported that in drought conditions, *Dacus tryoni* (Froggatt) laid eggs into fruit that was unripe but soft and shrivelled; 89% of these died either before hatching or as larvae, whereas a figure of 40% or lower would be expected in normally hydrated fruit .

Lack of rainfall can also affect the abundance of fruit. Variation in fruit abundance (whether or not caused by variation in rainfall) naturally affects the level of the egg and larval population and hence the population dynamics in general (see Delrio and Cavalloro, 1977; Bateman, 1968; Drew and Hooper, 1983).

The puparial stage is found in the soil underneath the host tree. There are some records of puparial survival in relation to 'soil moisture' or percentage of soil water (Bateman, 1968; Trottier and Townsend, 1979; Fitt, 1981). Unfortunately, we cannot relate the drying power of a soil to its reported water content. The drying power of a given soil is related to its water tension. The relation between water content and water tension varies between soils and can only be established for a particular soil by experiment (e.g. see Meats, 1967). Soil air is virtually saturated up to 'wilting point' (1560 kPa). A given water tension (f) above wilting point can be related to relative humidity (see Meats, 1974) by $\log_{10} f = 5.492 + \log_{10}(2 - \log_{10} RH)$. The relationship between water content and relative humidity (RH) for any particular soil can be established by allowing samples of different water content to come into equilibrium with air in closed containers.

Laboratory experiments indicate that RH values of above 70, 40, 30 and 10% will permit more than 30% survival through the puparial stage in *Dacus oleae* (Gmelin), *Rhagoletis pomonella* (Walsh), *Ceratitis capitata* (Wiedemann) and *D. tryoni* respectively (Tzanakakis and Stampoulos, 1978; Neilson, 1964; Shoukry and Hafez, 1979; Meats, unpublished). Trottier and Townsend (1979) indicate that very wet soil also adversely affects survival in *R. pomonella*.

2.3 Adults

Attempts to discover the responses of adults to weather by relating trap catches of unmarked flies to weather are bound to fail. This is because adult catches are (in part) a function of the abundance and survival rate of the immature stages, which have their own relationships to moisture as outlined earlier.

Bateman (1968) reported that with *D. tryoni*, fecundity (mean eggs per female per week) was drastically reduced to about 17% of normal in drought conditions. He also suggested that dry conditions may adversely affect the chances of survival of newly emerged adults. Fletcher et al. (1978) have confirmed by experiments that low humidity is a contributing cause of ovarian regression of *D. oleae* in drought conditions.

Drew et al. (1983, 1984) suggest that environmental moisture may indirectly affect survival and fecundity by affecting food supply. They suggest that bacteria on the surface of leaves are the main source of protein for several tephritids, and that the abundance of these bacteria is reduced in dry conditions causing reduced fecundity, and possibly reduced survival rates and increased emigration rates.

Much more data are required on the relation of humidity and water supply to survival, fecundity, food supply and dispersive/aggregative behaviour.

3 A MODEL FOR ADULT TEPHRITIDS

3.1 Water reserve, water loss and rehydration

Tephritid fruit flies are small and have a very small water reserve in absolute terms. However, it follows that a tiny amount of water will suffice to rehydrate a fly that has lost most of its reserve.

A fly of 15 mg will contain about 10 mg (10 μl) of water and about 5 μl of this can be considered as water reserve (the maximum amount that can be lost without fatal results). This model accords with the facts known for *D. tryoni* (Besly, 1962). A fly like *D. tryoni* that has exhausted its water reserve needs only 5 μl of water to restore itself to a normal state of hydration. A fly can drink this quantity within minutes (Meats, unpublished) and can easily obtain it from dew (e.g. see Fletcher, 1979).

Daily formation of dew would present the fruit fly with a regular opportunity to rehydrate. Dew supply is not quite as guaranteed as this (especially in windy conditions) but is largely independent of climatic region and daytime temperature (Monteith, 1963). It is therefore instructive to predict, for a given daily, climate, the rate of loss of water reserve in terms of percentage per day. This will enable the estimation of the minimum frequency required for drinking and rehydration.

If the entire water reserve is lost in less than a day, then the insect cannot rely on dew alone. If all the reserve is lost in one day or longer, then survival on dew alone becomes a finite probability which gets higher as the number of days that the reserve can last gets higher.

3.2 Rate of water loss and survival in constant conditions

The rate of water loss at any one temperature can be related to either relative humidity (RH) or saturation deficit (SD). There was an early misconception that SD was a better measure of the drying power of air because the gradient of vapour pressure (VP) is the same between the evaporating surface and the unsaturated air at all temperatures for a given SD; the inference was that a given SD should indicate a given rate of evaporation regardless of temperature. However, the rates of diffusion, respiration, spiracular opening and possibly the permeability of the cuticle increase with temperature with the result that evaporation from many insects increases with temperature at a given SD (Bursell, 1974 a, b). The increase with temperature at a given RH is more marked because the gradient of VP also increases with temperature. The VP gradient is also dependent upon any differences in temperature (caused, for

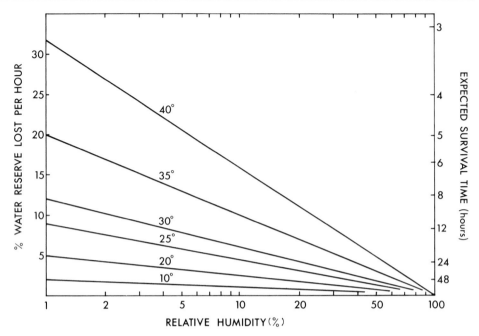

Fig. 3.8.1. A model for rate of water loss in *D. tryoni*. Rate of water loss at various temperatures is related to humidity. % water reserve lost per hour is the reciprocal of mean survival time. (Rate of water loss is constant until death point.) After Besly (1962), Meats (unpublished).

instance, by evaporative cooling) between the evaporating surface and the general air body. A wind gradient, of course, also affects evaporation rate.

There is therefore no a priori reason to prefer one measure of humidity over another. However it is important that the evaporation rate of the insect is determined by experiment at several combinations of humidity and temperature and at known (preferably several) wind speeds. The example below involves combinations of temperature and RH since the latter is more readily measured than SD.

Fig. 3.8.1 is a model for the rate of water loss (L) of *D. tryoni* in relation to temperature and RH with zero wind speed. It is based on the data of Besly (1962) and Meats (unpublished). Water loss is expressed in terms of % water reserve lost per hour. It was estimated from the reciprocal of mean survival time (in hours), which is indicated on the right hand ordinate. The relationship can be summarized by

$$L = (2 - \log_{10}\text{RH})10^{[(t - 0.03974) - 0.39279]}$$

Combinations of temperature and humidity which cause a loss of water reserve of 4.17, 8.33 and 12.5% per hour will allow survival for 1, 2 and 3 days respectively. The lowest humidity which will permit a given survival time without rehydration, varies with temperature. For instance, it is possible to survive 1 day at about 20, 40 and 54% RH at 30, 35 and 40°C respectively. This is illustrated in detail by Fig. 3.8.2 where the number of days survivable without rehydration is related to combination of temperature and humidity.

3.3 Rate of water loss and survival in daily fluctuating conditions

Temperature tends to rise each day and fall at night whereas humidity tends to do the reverse. It is possible by using the above equation to calculate the rate of loss of water reserve per hour for any combination of temperature and RH. It is therefore possible to calculate the likely percentage loss for any hour of

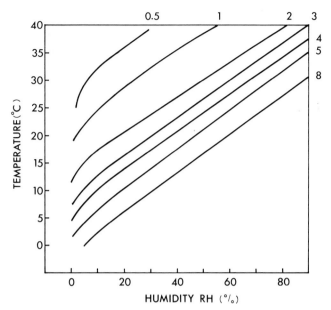

Fig. 3.8.2. Combinations of temperature and humidity that will permit *D. tryoni* to survive 0.5, 1, 2, 3, 4, 5 and 8 days without rehydration.

the day and hence the likely loss for any likely combination of daily trajectories of temperature and RH.

There is an infinite number of possible combinations of trajectories but it is apparent that the rate of water loss associated with a range of typical combinations is generally between one third and one half of that expected if the maximum temperature and minimum humidity for each combination persisted for the whole 24 h (Meats, unpublished).

The likely ecological importance of temperature and RH to desiccation can therefore be assessed in the above manner for any terrestrial insect. The evaporating power of a series of daily trajectories of temperature and RH can be calculated and related to the frequency of 'dew fall' to be associated with those trajectories. If it appears (as with *D. tryoni*) that survivable periods, even with quite adverse natural conditions, exceed dew fall frequency then we have good reason to discount water relations as being an important aspect of survival.

A more precise assessment can be made by relating dewfall frequency to chances of survival in a season in a way analogous to that given in Chapter 3.7 which relates frost frequency and frost tolerance to chances of survival in winter.

4 MICROCLIMATES AND WATER RELATIONS

As with temperature relations (Chapters 3.7 and 8.4), models based on laboratory tests should be related to microclimate since they pertain to conditions within the immediate vicinity of the insect. The foregoing section assumed (for simplicity) that microclimate was the same as macroclimate as measured in a Stevenson screen. Microclimates can be modelled in terms of macroclimates if the evaporation rate of physical models of insects placed in microclimates is compared with the evaporation rate of models in the Stevenson screen (see Wellington, 1957).

Temperature and humidity, in flowers, at leaf surfaces, in leaf canopies and in the understories of patches of trees can be less harsh than general ambient

Chapter 3.8 references, p. 246

conditions (Darby, 1933; Willmer, 1982; Meats, 1984). The surfaces (notably the undersurfaces) of mesophyllic leaves are particularly interesting and warrant further attention. Here, there is a boundary layer of saturated air. It is difficult to ascertain how thick this layer is in any given circumstance, but it appears from Willmer (1982), that it is unlikely to exceed 1 mm. This would not be sufficient for fruit flies to shelter within. However, Drew et al. (1983) suggest that fruit flies feed upon bacteria growing on the surface of leaves. The bacteria would be well within the boundary layer and therefore would be fully hydrated. It is therefore possible to suggest that the bacterial food of fruit flies could supply all the water needed. *D. tryoni* requires at least 1 mg of protein per day for maintenance and will typically consume more if given the opportunity (Kelly, 1982). It is possible that flies that are in a postition to satisfy their protein requirements get their water requirements as an incidental bonus. If wild flies obtained their protein only from leaf bacteria (80% water) it is likely that they would require at least 10 mg per day which would supply almost the equivalent of twice the water reserve per day.

5 REFERENCES

Bateman, M.A., 1968. Determinants of abundance in a population of the Queensland fruit fly. Symposium of the Royal Entomological Society of London, 4: 119–131.

Bernays, E.A. and Simpson, S.J., 1982. Control of food intake. Advances in Insect Physiology, 16: 59–118.

Besly, M.A.C., 1982. The effect of dryness upon loss of water and length of life in the Queensland fruit fly, *Dacus tryoni*. Ph.D. thesis, University of Sydney, iii + 153 pp.

Bursell, E., 1974 a. Environmental aspects — temperature. In: M. Rockstein (Editor), The Physiology of Insecta, Volume 2, Academic Press, New York, pp. 1–41.

Bursell, E., 1974 b. Environmental aspects — humidity. In: M. Rockstein (Editor), The Physiology of Insecta, Volume 2, Academic Press, New York, pp. 43–84.

Darby, H.H., 1933. Insects and microclimates. Nature, 131, 3319: p. 839.

Delrio, G. and Cavalloro, R., 1977. Reperti sul ciclo biologico e sulla dinamica di populatzione del *Dacus oleae* Gmelin in Liguria. Redia, 60: 221–253.

Drew, R.A.I. and Hooper, G.H.S., 1983. Population studies of fruit flies (Diptera: Tephritidae) in south-east Queensland. Oecologia, 56: 153–159.

Drew, R.A.I., Courtice, A.C. and Teakle, D.S., 1983. Bacteria as a natural source of food for adult fruit flies (Diptera: Tephritidae). Oecologia, 60: 279–284.

Drew, R.A.I., Zalucki, M.P. and Hooper, G.H.S., 1984. Ecological studies of eastern Australian fruit flies (Diptera: Tephritidae) in their endemic habitat. I. Temporal variation in abundance. Oecologia, 64: 267–272.

Fitt, G.P., 1981. Pupal survival of two northern Australian tephritid species and its relationship to soil conditions. Journal of the Australian Entomological Society, 20: 139–144.

Fletcher, B.S., 1979. The overwintering survival of adults of the Queensland fruit fly, *Dacus tryoni*, under natural conditions. Australian Journal of Zoology, 27: 403–411.

Fletcher, B.S., Pappas, S. and Kapatos, E., 1978. Changes in the ovaries of olive flies, *Dacus oleae*, during the summer and their relationship to temperature, humidity and fruit availability. Ecological Entomology, 3: 99–107.

Kelly, G.L., 1981. The effects of temperatue, thermal history and maturity on feeding rates of the Queensland fruit fly, *Dacus tryoni*. Honours thesis, University of Sydney, iii + 97 + xlvii pp.

Machin, J., 1979. Atmospheric water absorption in arthropods. Advances in Insect Physiology, 14: 1–48.

Meats, A., 1967. The relations between soil water tension and growth rate of larvae of *Tipula oleracea* and *T. paludosa* (Diptera) in turf. Entomologia Experimentalis et Applicata, 10: 312–320.

Meats, A., 1974. A population model for two species of *Tipula* (Diptera, Nematocera) derived from data on their physiological relations to their environment. Oecologia, 16: 119–138.

Meats, A., 1984. Thermal constraints to successful development of the Queensland fruit fly in regimes of constant and fluctuating temperature. Entomologia Experimentalis et Applicata, 36: 55–59.

Monteith, J.L., 1963. Dew: facts and fallacies. In: A.J. Rutter and F.H. Whitehead (Editors), The Water Relations of Plants. Blackwell, London, pp. 37–56.

Neilson, W.T.A., 1964. Some effects of relative humidity on the development of pupae of the apple maggot, *Rhagoletis pomonella* (Walsh). Canadian Entomologist, 96: 810–811.

Sell, D. and Houlihan, D.F., 1985. Water balance and rectal absorption in the grasshopper Oedipoda. Physiological Entomology, 10: 89–103.

Shoukry, A. and Hafez, M., 1979. Studies on the biology of the Mediterranean fruit fly, *Ceratitis capitata*. Entomologia Experimentalis et Applicata, 26: 33–39.

Stobbart, R.H. and Shaw, J., 1974. Salt and Water balance: Excretion. In: M. Rockstein (Editor), The Physiology of Insects, Volume 5, Academic Press, New York, pp. 361–446.

Trottier, R. and Townsend, J.L., 1979. Influence of soil moisture on apple maggot emergence, *Rhagoletis pomonella* (Diptera: Tephritidae). Canadian Entomologist, 111: 975–976.

Tzanakakis, M.E. and Stampoulos, D.C., 1978. Survival and egg laying ability of *Dacus oleae* (Diptera: Tephritidae) cold-stored as pupae and adults. Zeitschrift für Angewandte Entomologie, 86: 311–314.

Wellington, W.G., 1957. The synoptic approach to studies of insects and climate. Annual Review of Entomology, 2: 143–162.

Wharton, G.W. and Richards, A.G., 1978. Water exchange kinetics in insects and acarines. Annual Review of Entomology, 23: 309–328.

Willmer, P.G., 1982. Microclimate and the environmental physiology of insects. Advances in Insect Physiology, 16: 1–57.

Chapter 3.9 The Sensory Physiology of Pest Fruit Flies: Conspectus and Prospectus

M.J. RICE

1 GENERAL INTRODUCTION

Despite the serious, global pest status of many tephritid species and despite the crucial role played by sensory receptors in their feeding, mating, oviposition and other activities, very little detailed work has been done on their sensory physiology. Thus this chapter aims, not only to provide a conspectus of the information known on larval and adult tephritid sensory receptors, but to outline the many areas that are especially in need of research effort. Each of the major groups of receptors is dealt with by modality: mechanoreception, thermoreception, hygroreception, chemoreception, and photoreception. Nocireception is considered, on a number of physiological, structural and behavioural grounds to be unlikely to occur in insects (Eisemann et al., 1984). Whilst information on the sensory receptors of the Tephritidae is scanty and scattered, many useful analogies can be drawn from related Diptera. Each of the sensory modalities plays a vital role in the life of fruit flies and specific data on each would be of value to the informed pest manager. Behavioural management of pest populations offers many possibilities, particularly if sensory manipulation is based on a sound knowledge of the sensory physiology of each species.

The successful application of sensory biology to the management of pest flies, is not only impeded by the current paucity of data on sensory receptors, but also by a dated understanding of sensory physiology. More cognisance needs to be taken of the central mechanisms that interpret raw sensory input and those that determine which inputs are given attention, at any particular time. Hopefully, future experiments will be designed to investigate sensory processing and decision making and will avoid the traps of tropismic and motivational approaches. An "information processing" approach to tephritid sensory physiology will subtend advances in eco-ethological and evolutionary theory and open the door to innovative pest management procedures.

2 INTRODUCTION TO TEPHRITID SENSORY RECEPTORS

From a genetic viewpoint a fruit fly, like any living organism, is a conserver, transmuter and transmitter of hereditary information. From an ecological viewpoint it is a resource consumer and generator. From the viewpoint of sensory biology it is an information acquisition, processing and decision making system. Behaviour can be seen as the cross-roads where heredity information intereacts with resource availability and sensory information.

Thus it is the sensory physiology of a fruit fly that lends specific direction to its inherited way of life and particular method of using environmental resources for replication. In order for the insect to succeed in this, it needs not only sensory input from the external environment (ecoassessment) but also information on the positioning of its body parts and on the chemistry of its innards (autoassessment) (Rice, 1975). The life style of any insect species, or in this case the family Tephritidae, is intimately connected with the range of information that it is able to respond to and that which it has to ignore. An infinite range of information modalities and gradations is available around and within; the number and variety of receptors is limited. Each species represents a gamble, that the information its sensory receptors are able to obtain is the most useful for its way of life. In addition, a slight change in sensory information processing is probably responsible for the changes in behavioural preference that can generate sub-species and possibly, eventually species. It has been argued that multiple, multimodal sensory receptor input to the ganglia forms patterns that are compared with predicted input patterns, the discrepancy driving much of the behaviour; and that success at predicting the consequences of the motor output resultant from pattern comparisons, is what is selected for in the evolution of behaviour (Rice, 1975). If this is true it has many implications; for example the essential difference between two closely related fruit fly species may be no more than slight differences in sensory information reception, processing or decision making. Even highly divergent species may be so only because aeons ago a slight sensory physiology divergence caused their ancestors to adopt a new life style that exerted very different selection pressures. The Tephritidae appear to be an ideal group to investigate such concepts of the "sensory leading" of speciation. The genetic fixation of such leads is, of course, another question entirely.

The reason that such foundational studies in sensory biology have yet to be made lies in the magnitude of the task. In order to thoroughly characterise a part of the sensory biology of a species, at least seven components need to be considered. The current state of the art is such that only the first two of these are yielding data and for too few species. The seven major components of sensory biology are:

    i. morphological, topographical and ultrastructural typing of sensilla, their sensory neurones and accessory structures;

    ii. electrophysiological characterization of the adequate stimuli and dose/response relationships of each type of sensory neurone;

    iii. determination of the biochemical and/or biophysical bases of specificity in sensory neurones;

    iv. tracing of sensory axons into the ganglia and determination of their central connections;

    v. elucidation of the central, primary processing of multimodal sensory input;

    vi. identification of the attention system that preferentially opens the central processor to specific sensory inputs, out of the multitude of information pouring in at all times;

    vii. identification of central comparator systems that fit actual sensory input against anticipated input to make ongoing decisions.

As these different components of sensory biology yield to ultrastructural, biochemical and electrophysiological analysis, so we will be able to construct models that reveal the physiological bases of sensory behaviour. It must be admitted that currently we are very far from achieving that.

If the picture is very incomplete for the Insecta in general, it is far worse for the Tephritidae in particular. Over the whole family there have been very few papers that address any of the above seven major components in any depth at all! Losses to pest fruit fly species globally run to billions of dollars per annum. The future for insecticidal control looks gloomy because of the cost of developing and registering new compounds plus environmental concern and insect resistance. It is fully time that the sensory physiology of tephritids received a thorough treatment. This is obviously important for a number of academic reasons but, surely, for pressing commercial reasons too. The future of pest control is increasingly dependant on natural methods based on a knowledge of the ecology and behaviour of each species. As pointed out by Shorey (1977) it is absolutely essential to do the basic sensory groundwork, to provide a secure foundation for the applied pest control work.

The particular approach adopted in this chapter is intended to help open up the area to those needing information and to those who would like to become involved in further developing our understanding of tephritid senses. The physiology of the receptors that are involved in monitoring each of the five main information modalities is considered in turn: mechanoreception, thermoreception, hygroreception, chemoreception and photoreception. As the information on tephritid adults is generally slight and that on larvae non-existent, considerable reference is made to related Diptera, such as vinegar flies, house flies, tsetse flies and blowflies, which have been investigated in more detail. Each section concludes with a consideration of the most urgent research needed on tephritids and mention of the potential usefulness of such data for the development of novel control methods.

The topography of larval cephalic receptors is shown in Fig. 3.9.1 and that of the sense organs of a typical adult female tephritid in Fig. 3.9.2. The importance of *sensory context* has been argued by Rice (1975): specific inputs via one sense organ can have several different behavioural outcomes, governed by overall sensory input. Thus it is beneficial for us to keep in mind all of the sensory loci, whichever one locus we may be concentrating on.

3 TEPHRITID MECHANORECEPTOR PHYSIOLOGY

3.1 Types of mechanoreceptors

Insect mechanoreceptor neurones are divisible into two broad categories: multipolar and unipolar. The *multipolar* cells innervate soft tissues or connec-

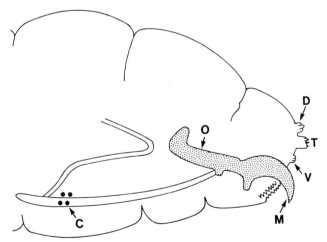

Fig. 3.9.1. Diagramatic cross-section of the head of a tephritid larva to show major sensory loci; D: the "dorsal organ", primarily olfactory. T: the "terminal organ", primarily gustatory. V: the "ventral organ", primarily gustatory. O: putative location of photoreceptors. M: mouth hooks. C: position of 4 posterior cibarial gustatory sensilla.

Chapter 3.9 references, p. 270

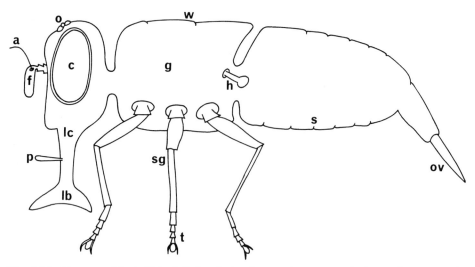

Fig. 3.9.2. Diagram of tephritid adult female to show major sensory loci. a: arista, transmits vibrations to Johnson's organ and other receptors located at pedicel/funiculus junction. (Hallberg et al., 1984a). f: funiculus, richly supplied with olfactory receptors, plus some thermo- and hygroreceptors. o: ocelli, simple photoreceptors. c: compound eye, an assembly of thousands of photoreceptors. p: palps, richly innervated with olfactory and mechanoreceptors. lc: labrum/cibarium, innervated by gustatory and mechanoreceptors. lb: labellum, richly innervated by gustatory and mechanoreceptors. sg: subgenual organs (vibration receptors of the tibia); other, diverse mechanoreceptors are found in all of the leg joints. t: tarsi richly innervated by gustatory and mechanoreceptors (probably some thermoreceptors too). w: wings, richly innervated by mechanoreceptors (probably some chemoreceptors too). h: haltares very richly innervated with mechanoreceptors (campaniform and chordotonal sensilla). n: ganglia, likely to have a range of internal gustatory and thermoreceptors. (possibly photoreceptors too). g: gut, innervated by stretch receptors. s: stretch receptors and other internal body wall receptors. ov: ovipositor, innervated with mechanoreceptors, hygroreceptors and gustatory receptors.

tive tissue and muscle. Multipolars respond to the distortion of their desheathed dendritic terminals, probably caused by bending of their many terminals (Rice, 1970). They monitor the movements of gut, body wall, joints and strand stretch receptors spanning the segments. In contrast, *unipolar* cells generally innervate cuticular structures: trichoid sensilla, campaniform sensilla and chordotonal sensilla. These three are respectively considered to be usually involved in sensing touch; cuticular stresses; movements, vibrations and sounds. These various types of mechanoreceptors are deployed throughout the body in large numbers in apparently meaningful patterns. Except in a few cases, listed by Rice (1975) and McIver (1985), the behavioural significance of the deployment pattern of these varied mechanoreceptors has not been commented on.

3.2 Data on tephritid mechanoreception

At the date of writing there does not seem to be a single paper published on the mechanoreceptors of a tephritid. Thus the results described here are fragments derived from papers written in other subjects, unpublished observations made in my laboratory or extrapolations from work on other Diptera.

Light microscope studies of larvae of *Dacus tryoni* Froggatt revealed the presence of trichoid and campaniform sensilla on every segment. The trichoid sensilla are so tiny that they are almost completely buried in their basal cups, apart from a pair on the hind end of the larva which project from raised basal domes. Inside the larvae we can anticipate a rich mechanoreceptor innervation, such as that described for a blowfly maggot (Osborne, 1963): multipolars innervating the epidermis and strand stretch receptors; unipolars innervating the trichoid and campaniform sensilla, plus some chordotonals. No doubt the viscera are innervated by multipolar neurones, as in a number of insects species (references in Rice, 1970). The electrophysiology of mechanoreceptors in tephritid larvae is unknown. Behaviourally they show responses to touch and vibrations; however these stimuli are generally complicated by simultaneous inputs through their chemoreceptors and photoreceptors (for details of which see sections 5.2 and 6.2 below). Critical sensory physiology experiments on larval tephritid mechanoreceptors remain to be done.

In the adult fly the major locus of mechanoreception is the halteres. These are the reduced second pair of wings, characteristic of Diptera, that beat at the same frequency though in antiphase to the wings, during flight. In blowflies they have 5 sets of campaniform sensilla on their surface and 2 sets of chordotonal sensilla within (Smith, 1969). These sensory organs are proven to be responsible for orientation feedback during flight. Their special mechanoreceptor innervation also suggests that they could be involved in monitoring vibrations and possibly sounds. Stridulation is common among the Tephritidae (Monroe, 1953; Webb et al., 1981), different patterns of sound being generated by vibration of the wings across large bristles on the third abdominal tergite. Characteristic sound spectra are produced during calling, courtship, aggression and flight.

The pulse train duration and pulse train interval for several species of *Dacus* are well differentiated (Fig. 3.9.3). The antenna of adult tephritids is also a locus of chordotonal receptors, inserted at the pedicel/flagellum joint. Trains of action potentials have been recorded from this region with tungsten electrodes, in response to vibrational stimuli (Simpson and Rice, unpublished).

In addition to the halteres and the antennae, the wings and legs have campaniform and chordotonal sensilla. In the female *Ceratitis capitata* Wiedemann there is an abundance of mechanoreceptor innervation of the ovipositor: campaniform sensilla on the tip, highly articulate "groove sensilla" that have both chemoreceptor and mechanoreceptor innervation, and many very short tactile sensilla along the proximal shaft (Eisemann, 1980; Marchini, 1982; Marchini and Wood, 1983; Yin and Stoffolano, in preparation). The arrangement of the sensilla is very similar to that of another piercing organ, the tsetse fly haustellum (Rice et al., 1973), an analogy that is further elaborated on by Eisemann and Rice (in preparation). The head and its appendages, thorax,

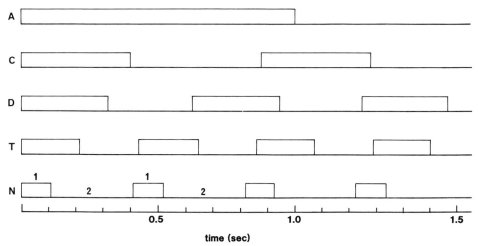

Fig. 3.9.3. Stridulation patterns of *Dacus* species. 1 = pulse; 2 = interpulse pause; (redrawn from Webb et al., 1981). A: species originally thought to be an Australian strain of *D. dorsalis*, now called *D. opiliae* Drew and Hardy. C: *Dacus cucurbitae*. D: *Dacus dorsalis* (Hawaiian strain). T: *Dacus tryoni*. N: *Dacus neohumeralis*.

Chapter 3.9 references, p. 270

wings, legs, halteres, and abdomen also have their own unique complement of trichoid mechanoreceptor sensilla. Some are robust, others delicate, some long and others relatively short. These supply information on contact with the substrate, between body parts and with other individuals; cleanliness of the surface; air current flow; and possibly are also involved in detecting vibrations, including sounds, under some circumstances. Gustatory sensilla on tarsi and labella all have mechanoreceptor neurone innervation, as well as their chemoreceptor dendrites. Information on the electrophysiology of these is needed, especially to determine whether they are phasic or tonic receptors.

The anterior food canal is richly innervated with mechanoreceptors. These occur as paired patches in the labrum and the cibarium. In *Dacus jarvisi* (Tryon) each labral patch has 10–15 trichoid sensilla of 8–12 μm length; each cibarial hair patch has 39–48 trichoid sensilla of 80–110 μm length. There is a campaniform sensillum associated with each labral sensory patch. Going on the evidence from other Diptera (Rice et al., 1973; Rice, 1973) it seems likely that each of these sensilla is innervated by a single unipolar neurone, whose axons run into the labro-cibarial sensory nerves and thence, via the tritocerebral nerves, into the tritocerebral lobes of the supra-oesophageal ganglion. The cibarial pump is also likely to be innervated by multipolar neurones, that provide feedback for the cibarial motor neurone centre to regulate cibarial pumping (Rice, 1970; David Falk cited by Dethier, 1976). Each of the hair sensilla and campaniform sensilla are likely to have a single neurone innervating the base of their hair shaft or dome, at the level of its insertion into the basal cup (Hallberg et al., 1984a). The tips of their dendrites will contain a neurotubular cytoskeleton of the tubular-body type (Rice, 1975; McIver, 1985). The tubular-body will be differentially developed in different types of sensilla and these will have different physiological properties regarding adequate stimulus, threshold, saturation, and adaptation rates. A tremendously wide range of mechanoreceptor information is thus available to the insect.

In addition, like the larva, the adult will have multipolar neurones innervating the strand stretch receptors, the inner surface of the epidermis and the viscera. The dendrites of the multipolar mechanoreceptors have no tubular bodies but do have well-developed neurotubular cytoskeletons in their receptive terminals (Rice, 1970). Thus a wealth of information is available to the fly on mechanostimulatory events inside its body, as well as outside.

3.3 Work needed on tephritid mechanoreception

All seven physiological components of auditory receptor biology need to be investigated (cf. section 2 above). This would also provide information for studies on courtship, mating, aggression, aggregation and alarm. The first step will be to locate the main auditory organ(s) in tephritids!

A possible role for leg mechanoreceptors is in the assessment of fruit size. Prokopy and Bush (1973) found that *Rhagoletis pomonella* (Walsh), *Rhagoletis zephyria* Snow, *Rhagoletis cornivora* Bush and *Rhagoletis mendax* Curran laid most of their eggs in artificial fruits of the same size as their natural host fruits. However, this could also be a visual effect (see section 6 below).

An interesting study could be made of the various mechanoreceptors on the ovipositor and how their physiology relates to the various fruits utilised by tephritid species. A start in this direction has been made by Marchini and Wood (1983) and Eisemann and Rice (in preparation).

There are species differences in labro-cibarial innervation. *Dacus cucumis* French has only 7–11 labral trichoid sensilla and 21–30 cibarial trichoid sensilla, in each of the paired sensory patches (compare with figures for *D. jarvisi* above) (Fig. 3.9.4). A comprehensive review of the labro-cibarial receptors of the Family could prove of value.

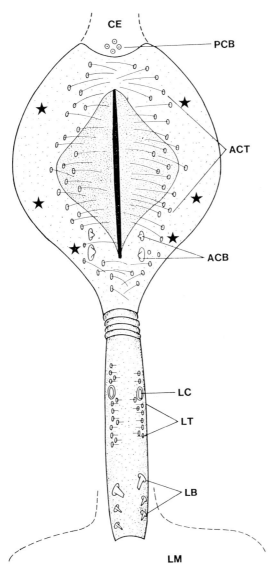

Fig. 3.9.4. Labral and cibarial receptors of *Dacus* spp. (Rice, unpublished); not to scale. PCB: 4 posterior cibarial basiconic sensilla (gustatory sensors). ACT: 40–100 anterior cibarial trichoid sensilla (tactile fluid-flow sensilla). ACB: 4 anterior cibarial basiconic sensilla (gustatory sensors). stars: 6 cibarial pump receptors (monitor cibarial pumping). LC: 2 labral campaniform sensilla (stress sensors). LT: 20–40 labral trichoid sensilla (tactile fluid-flow sensors). LB: 6 labral basiconic sensilla (gustatory sensors). LM: Labellum. OE: Oesophagus.

The mechanoreceptor physiology of larval tephritids has been entirely neglected; this needs to be remedied.

3.4 Applied uses of tephritid mechanoreception

In general, mechanoreception appears to offer less obvious opportunity for sensory manipulation than chemoreception and photoreception, Sounds of particular frequencies are sensed by auditory receptors and behavioural decisions made. Perhaps artificial sound generators, set up in orchards, could cause continual alarm reactions or disrupt mating signals. This may be worth investigating as sound energy can be generated very economically.

An increase in our understanding of the role of mechanoreceptors in oviposition may put us in the position where we can suggest surface textures that are unsuitable. This would then enable fruit growers to choose the least oviposition suitable varieties to grow and give valuable leads to plant breeders.

Chapter 3.9 references, p. 270

Further information on the role of larval mechanoreceptors could be of use to those engaged in the mass-rearing of fruit flies for sterile release control methods.

4 TEPHRITID THERMORECEPTOR AND HYGRORECEPTOR PHYSIOLOGY

4.1 Types of thermoreceptors and hygroreceptors

In recent years electrophysiological and ultrastructural work has enabled the pinpointing of the sensilla responsible for monitoring temperature and humidity. The best characterised of these are relatively small pegs (basiconic sensilla) that have dendrites running up a central canal (reviewed by Altner and Prillinger, 1980; Altner et al., 1983). The pegs have no conspicuous pores opening into the canal from the exterior, as is always found in olfactory and gustatory chemoreceptors (see section 5.1 below). There are 3 to 4 sensory neurones associated with the sensillum. One or two generally respond to temperature, two others monitor humidity. The humidity sensitive dendrites run up into the sensillum canal, the temperature sensors often terminate at or below the base. The pairs of neurones operate in a push-pull manner: an increase in stimulus will increase the rate of firing in one and decrease it in the other. Such a combination lends precision to sensor systems (Rice, 1975). Thus a four neurone thermo-hygroreceptor sensillum may contain a "wet" receptor, a "dry" receptor, a "warm" receptor and a "cold" receptor. The three neurone sensillum, commonly found in a range of insects from several orders, generally lacks the "warm" receptor neurone (Altner and Loftus, 1985).

4.2 Data on tephritid thermoreception and hygroreception

It is not known if larval tephritids have such receptors. If they do the aggregation of sensilla called the terminal organ (Chu and Axtell, 1971) is the most likely site. However, even in other Diptera electrophysiological recordings have yet to be made.

Work in my laboratory has shown that adult *D. tryoni* have both temperature and humidity receptors, the input of which synergises with the input of CO_2 receptors, to provide an "escape" reaction to vertebrate breath. Cages of *D. tryoni* were subjected to jets of pure, dry, medical air. The air was able to be moistened, warmed and/or have 5% CO_2 added to it. A jet of cold, dry, CO_2-free air evoked very little flight activity. Addition of CO_2 increased the flight activity ten fold; addition of CO_2 and warmth (37°C) increased flight activity fifteen fold; whilst addition of CO_2, warmth and saturation with water vapour caused a twenty-three fold increase in flights, in response to the jet of air (Rice and Burn, unpublished). As yet we have not recorded electrophysiologically from the receptors concerned. A suitable candidate sensillum has recently been located by Hallberg et al. (1984b) in a pit on the antenna of *Dacus oleae* (Gmelin). This is likely to be a "wet" and "dry" sensillum with a "cold" receptor. The position of these sensilla in pits that open on the outer surfaces of the funiculi of the antennae, suits them well to receive air currents both when in flight and at rest.

4.3 Work needed on tephritid thermoreception and hygroreception

The most urgent need is to obtain electrophysiological recordings so that response characteristics can be determined, especially those of the sensory neurones of the funicular pit of the adult and terminal organ of the larva. These receptors offer an excellent opportunity for work on multimodal inputs: temperature, moisture and olfactory stimuli.

4.4 Applied uses of tephritid thermoreception and hygroreception

The downward, "startle", flight response of *D. tryoni* adults when they are jetted with a stream of warm, moist air containing about 5% CO_2 suggests an application. For example, a tractor equipped with large, lateral, pyrethrin or glue-coated sheets might drive alongside fruit trees, jetting the foliage with blasts of warm, moist, CO_2-laced air. The downward flight of the fruit flies would bring them into contact with the sheets. Such a method may be adapted for surveys or even control (Burn and Rice, unpublished).

5 TEPHRITID CHEMORECEPTOR PHYSIOLOGY

5.1 Types of chemoreceptors

There are four major sources of chemical stimulation for a fruit fly: compounds dissolved in the internal body fluids; compounds in external solutions; compounds in the vapour phase in the external environment; and compounds in the solid phase in the external environment. Three broad categories of sensory receptors are involved: *internal chemoreceptors*, for example those monitoring oxygen, carbon dioxide and nutrient levels; *gustatory (taste) receptors*, for example those monitoring sugars, salts and oviposition deterring pheromone in aqueous solution, plus those monitoring solids, for example the lipids of the surface of plants or other insects, and thirdly, *olfactory (smell) receptors*, for example those responding to the kairomones of host plant volatiles, the allomones of non-host plant volatiles, the volatile pheromones of conspecifics, or the volatiles of various synthetic lures. Little is known of the internal chemoreceptors of insects in general and nothing of those in the tephritids. A good start has been made on the gustatory and olfactory receptors of several tephritid species.

Gustatory and olfactory chemoreceptors are specialised to respond to compounds presented in a solid or liquid form, or in a gaseous form, respectively. This must not be taken as implying that gustatory receptors cannot or do not monitor volatiles or that liquids cannot or do not stimulate olfactory receptors, it simply reflects the major involvement of the two classes of external chemoreceptors. The porosity of each class of receptor reflects its specialisation: gustatory receptors generally have a single pore at the tip, that allows molecules of stimulant, from a specific spatial location, ingress to the dendrites; olfactory receptors generally have many pores scattered over their surface, molecules of stimulants reaching the dendrites from any direction (cf. Dethier, 1976). There are two main types of gustatory sensilla, characterized by having one or two canals running up the setal shaft. There are four main types of olfactory sensilla in Diptera: again two types characterized by single or double canals in the seta, plus a division into those in pits and those on the surface.

Gustatory receptors have been found on all parts of the body but are especially concentrated on the labellum, labrum, cibarium, tarsi and ovipositor. Olfactory receptors are very highly concentrated on the antennae and palps.

5.2.1 Data on tephritid chemoreceptors

There does not appear to be any information published on larval chemoreceptors. *D. tryoni* larvae have been found to have a rich chemoreceptor innervation: dorsal organs (antennae), terminal organs (palps), ventral organs (labellum), and 4 dorsal cibarial papillae (Rice, unpublished). This is likely to be the situation of larval tephritids in general, it implies that more behaviour is occurring inside the fruit than has been realised. Data on adult chemoreceptors is much more extensive and is here divided into a number of sections.

5.2.2 Antennal and palpal olfactory receptors

Data on the antennal receptors is beginning to accumulate but much more is needed. The first study was that of Hallberg et al. (1984b) on *D. oleae*, who characterized the ultrastructure of five types of sensilla. More recently Giannakakis and Fletcher (1985) have studied the external morphology and topography of six types of sensilla on the antennae of male and female *D. tryoni*. There are also available data on *D. cucumis*, *Dacus cacuminatus* (Hering) and *D. jarvisi* (Gill and Rice, unpublished). In addition I have recently examined the palps of *D. tryoni*, which each have a complement of 400–500 single-walled basiconic sensilla.

Synthesising the information on the antennal receptors the following picture emerges. The major sensillum is a 15–25 μm long, trichoid, single-walled type with relatively small pores (approximately 10 nm diameter). There are about 1500 of these on each funiculus; they are innervated by two unbranched dendrites. The second most numerous type are basiconic, 8–18 μm long, single-walled, with relatively larger pores (approximately 100 nm diameter). There are about 500 of these per funiculus; they are innervated by two branching dendrites. The third type are styloconic pegs, 2–4 μm long, double-walled and heavily grooved with relatively large pores (approximately 200 nm diameter). There are approximately 300 styloconic sensilla per funiculus and they are innervated by 3 unbranched dendrites. The single funicular pit, located on the proximal abaxial surface has two types of sensilla, one sort identical with the styloconics described above, the other a typical thermo-hygroreceptor described in section 4.2 above. Preliminary results suggest that *D. cucumis* has slightly fewer receptors, *D. cacuminatus* and *D. jarvisi* slightly bigger receptors. In addition, males have slightly fewer receptors than females (Giannakakis and Fletcher, 1985). None of these differences is sufficient to explain differential responses to lures, pheromones or host odours. It is clear that the explanation for these will largely be found at the neurophysiological level. However, more morphological, topographical and ultrastructural studies are also needed on a range of tephritids, to lay sound foundations for sensory recording studies.

Very detailed work has been done on the effectiveness of lure analogues as a tool to unlock the structure of receptor molecules in the olfactory dendrites (e.g. Metcalf et al., 1975; 1979) and on fruit kairomones (e.g. Eisemann, 1980; Fein et al., 1982). The time is now ripe for electrophysiological studies, both single unit and electroantennogram. So far there is only one paper on this subject (Guerin et al., 1983). The authors found a strong degree of conformity in E.A.G. responses to compounds in a series of saturated aliphatic aldehydes, with heptanal, octanal and nonanal evoking the biggest responses in *C. capitata*, *D. oleae* and *Rhagoletis cerasi* (Linnaeus); results that were not entirely in synchrony with field lure trapping. It is likely that the E.A.G., because it represents the summed activity of all the different types of funicular receptors, will not prove precise enough for the resolution of interspecific differences. Single unit work on the electrophysiology of olfactory receptors is urgently needed.

5.2.3 Labellar, labral and cibarial gustatory receptors

The mouthpart and anterior food canal is a particularly rich locus of gustatory chemoreception, as would be expected (see section 5.2.1 above for larva). A thorough study of the labellar, labral and cibarial receptors of *D. tryoni* is being carried out in my laboratory (Njagi, Rice and Smith, unpublished data). Each labellum has approximately 300 gustatory sensilla (about 150 per labellar lobe). They are of seven different classes, ranging from the inter-pseudotracheal papillae (about 4 μm long) up to the longest fringe sensilla (about 300 μm long). The fringe sensilla all have two canals in their seta and are innervated by 4–5 neurones, one of which is a mechanoreceptor, terminating at the base. The other 3–4 neurones send unbranched dendrites up the dendritic canal, to terminate just below a pore at the tip of the seta. The inter-pseudotracheal papillae have only one canal. They are innervated by 3–4 neurones, one a mechanoreceptor, leaving 2–3 chemoreceptors to run up the canal of the peg. The tip of the peg is recurved so that the pore opens subapically, avoiding damage when the labellar lobes are brought into contact with the surfaces of plants and other insects.

In *D. cucumis*, *D. tryoni* and *Dacus bryoniae* (Tryon) there are two large and two small sensilla basiconica projecting into the distal end of the labral food canal. The large sensilla are clearly gustatory receptors (Rice, unpublished). On the distal part of the anterior wall and on the proximal part of the posterior wall of the cibarial pump there are four pit sensilla (Fig. 3.9.4), as previously described in a variety of other Diptera (Rice, 1973). These sensilla enable the fruit flies to taste the food they ingest as it enters the food canal and as it is pumped from the cibarium into the oesophagus. They are also likely to play a part in oral hygiene, checking on the cleansing of the mouthparts by salivary washing after a meal. It is interesting to note that the physiology of taste reception involves four stages: cibarial and labral gustation follows on after gustation by the inter-pseudotracheal papillae and the six size classes of labellar fringe sensilla. The significance of such sequential sensing has been touched on for mechanoreceptors by Rice (1975) but is greatly in need of more consideration, especially in terms of chemoreceptor information processing and decision making by central predictive, comparator circuits (see section 7 and Fig. 3.9.7).

There is as yet no thorough electrophysiological study of labellar, cibarial or labral gustatory receptors in a fruit fly. The best study is that of Gothilf et al. (1971) on six long labellar fringe sensilla. They recorded positive sugar receptor responses to sucrose, fructose and glucose; a lower response to mannose and L-arabinose; and no response to lactose. Inositol was found to stimulate, mannitol not. Salt receptor responses were obtained to ammonium, sodium and potassium chlorides, with less response to magnesium and calcium chlorides. Calcium chloride at 10 and 100 mM blocks the fructose evoked sugar receptor response. More recently Angioy et al. (1978) carried out some brief studies on the labellar fringe sensilla of *C. capitata* and *D. oleae*. They found that 100 mM NaCl invariably evoked polyneural firing in the labellar sensilla of *D. oleae* and *C. capitata*. At 10 mM NaCl *D. oleae* yielded trains of impulses from a single ("water") sensitive neurone, *C. capitata* still yielding the polyneural response; 10 mM CaCl$_2$ and 10 mM BaCl$_2$ solutions evoked firing in the single ("water") unit of *C. capitata*. At 100 mM these alkaline-earth chlorides evoked polyneural responses. 10 mM (NH$_4$)$_2$SO$_4$ evoked a "water" response but 10 mM NH$_4$Cl a polyneural response. The same basic pattern, with some slight differences, was observed in *D. oleae*. The usefulness of such information is limited by the absence of data on dose/response curves and by an absence of the replication and statistical analysis needed to make progress in sensory electrophysiology. Such an approach is currently being used by Njagi and Rice (unpublished data) to characterise the gustatory neurones of *D. tryoni*. Using carefully graded concentrations of LiCl, it has proved possible to recognise one "water" and two "salt" units. Replication and statistical analysis have shown up significant differences between the cells innervating different classes of sensilla. Once we have attached descriptive electrophysiological handles to the various neurones involved, progress will be swift. The work of Elizabeth Bowdan (see section below) sets the benchmark for us all.

5.2.4 Tarsal gustatory receptors

An excellent morphological and topographical study of the tarsal gustatory sensilla of *R. pomonella* is available (Crnjar and Prokopy, 1982). Using the classification proposed by Grabowski and Dethier (1954) for a blowfly, they designated various tarsal sensilla as B, C or D hairs, The B sensilla are 15–24 μm long with basal diameter about 4 μm, there are 28–29 on each of the three tarsi. The C sensilla are 38–48 μm long with basal diameter about 4.7 μm, there are 11–13 on each tarsus. The D sensilla are 48–130 μm long with basal diameter about 5.9 μm, there are 8 on each tarsus. Crnjar and Prokopy clearly demonstrated the great sensitivity of D sensilla to a material they call oviposition deterring pheromone – O.D.P. (prepared from distilled water rinses of fruit, after fruit flies had laid eggs and dragged their ovipositors on it). Stimulation with O.D.P. evoked about 8–10 action potentials in the first 0.1 second and about 46–54 in the first second after contact. They consider that this provides the main sensory input eliciting oviposition deterrence, perhaps supplemented under some circumstances by labellar chemosensilla input (Crnjar et al., 1978).

The best study of the electrophysiology of any tephritid sensory receptor is that of Bowdan (1984) who has found the responses of tarsal gustatory sensilla of *R. pomonella* to a range of concentrations of NaCl, sucrose and O.D.P. She has plotted dose/response curves, short-and long-term adaptation and impulse height and duration. This represents substantial progress towards characterising the sensory neurones concerned. A valuable outcome of this work is that each of the stimuli was shown to excite a different cell. Thus the tarsal gustatory D sensilla contain a salt-sensitive, a sucrose-sensitive and an O.D.P.-sensitive neurone.

Recent behavioural work on *D. tryoni* has identified the presence of fructose sensitive neurones (Eisemann and Rice, 1985). These neurones, whose input greatly influences oviposition, via tarsal and labellar but not ovipositor stimulation, have fructose-binding (furanose) sites. The receptor neurones concerned in oviposition behaviour apparently have no pyranose (glucose-binding) sites. Threshold is at about 4 mM and behavioural saturation at about 50 mM. Electrophysiological isolation of this neurone type is in progress. Behavioural work by Fitt (1984) suggested that *D. tryoni* and *D. jarvisi* detect the presence of larvae in fruit from the odours generated by their associated bacteria. However, in the light of the findings of Eisemann and Rice (1985) and the design of Fitt's experiments, it is also likely that tarsal and labellar gustatory receptors detect such decay products oozing through punctures in the fruit skin.

5.2.5 Ovipositor gustatory receptors

The significance of gustatory receptors on the ovipositor of tephritids was first realised by Eisemann (1980) who described four relatively large sensilla, in a groove, on each side of the tip of the ovipositor of *D. tryoni*. He conducted a carefully designed series of behavioural experiments that proved the presence of receptors sensitive to mechanical stimuli, water (liquid and vapour) and to salts such as $CaCl_2$. No behavioural response to fructose stimulation of the ovipositor was found, though molar calcium stimulated the ovipositor receptors, resulting in reduction of oviposition (Eisemann and Rice, 1985).

The behavioural roles of ovipositor gustatory receptors of *D. tryoni*, in egg-laying behaviour, have been considered by Eisemann and Rice (1985; and in preparation).

Morphological, topographical and ultrastructural information on ovipositor chemoreceptors is beginning to accumulate. Marchini (1982) and Marchini and Wood (1983) found three types of sensilla on the ovipositor of *C. capitata*: sixteen pairs of campaniform-like sensilla at the tip; five pairs of 4–5 μm long gustatory-like sensilla, subapical, four on each side in lateral grooves, and one on its own; and thirty pairs of very truncated, trichoid-like sensilla scattered over the subapical parts of the ovipositor. This seems to be the broad pattern for the ovipositor in most tephritids, though there are often just four gustatory sensilla on each side, as described by Eisemann (1980) in *D. tryoni* (Fig. 3.9.5). The four gustatory "groove sensilla" of *R. pomonella* have been sectioned by Yin and Stoffolano (personal communication) and found to have 3 to 4 sensory cells each. Two of the paired sensilla have 3 chemoreceptor and 1 mechanoreceptor dendrite, the other two pairs having 2 chemoreceptor and 1 mechanoreceptor dendrite. The grooves perfectly accommodate these articulated sensilla and they are readily folded distally by the mechanical resistance of the oviposition substrate, being protected in the grooves during penetration. The tephritid ovipositor groove sensilla are innervated by a complement of at least twenty gustatory neurones. The symmetrical arrangements of the sensilla, in lateral grooves towards the ovipositor tip, suggests that they may provide directional information on the lateral disposition of stimuli. Centrally processed, this information would then permit the fruit fly to guide the penetrating organ to the chemically "best" position in the fruit.

Ovipositor gustatory receptors are among some of the smallest known.

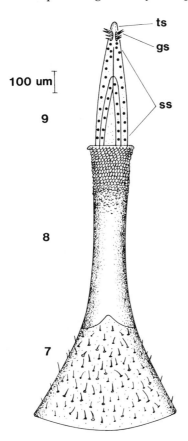

Fig. 3.9.5. Tip of ovipositor of *Dacus* spp. (after Eisemann, 1980). ts: tip sensilla (campaniform contact and stress sensors). gs: groove sensilla (gustatory and tactile sensors). ss: shaft sensilla (tactile sensors). 7,8,9: abdominal segments 7–9.

Chapter 3.9 references, p. 270

However, it is possible to record from them by means of the technique developed by Hodgson et al. (1955). For example, Rice (1977) was able to test a range of alkaline and alkaline-earth halides on a gustatory receptor of a blowfly; thresholds and dose/response relationships were determined, resulting in the characterisation of a neurone that is positively sensitive to monovalent cation concentration. A start has been made by Girolami et al. (pers. comm.) who placed a relatively large pipette over the ovipositor and obtained trains of spikes in response to solutions of glucose, fructose, NaCl, malic and quinic acids. Dose-response curves are needed of the adequate stimuli for the sensory neurones concerned. A water-sensitive oviposition response has been found, by behavioural experiments, to depend on ovipositor stimulation (Eisemann and Rice, in preparation).

5.3 Work needed on tephritid chemoreceptors

Characterisation of the morphology, topography and ultrastructure of antennal, palpal, labellar, labral, cibarial, tarsal and ovipositor chemoreceptors of the range of pest fruit flies is urgently needed. This needs to be interfaced, by detailed electrophysiology, to the chemical and evolutionary ecology of the insects (e.g. Metcalf, 1985).

5.4 Applied use of data on tephritid chemoreception

Olfactory attractants and repellents, gustatory stimulants and deterrents and chemosensory blockers and disrupters provide pest managers with some of their best options for current and future behavioural management strategies. However, as cautioned by Shorey (1977), we must avoid premature trials of techniques that are not based on a thorough knowledge of insect behaviour in a natural habitat. In addition, as argued in section 7 below, it is of great importance to have the proper theoretical foundations. It is misleading to treat the insect as a "behaviour vending" machine, delivering upon stimulation of the right chemosensory "buttons".

The central nervous system processes multimodal sensory inputs, so that (as demonstrated above in section 4.2) *combinations* of inputs are the naturally occurring, behaviour evoking factor. Context is vital. The simplest form of this is called push-pull, two types of receptors working in antiphase (Rice, 1975). This basic central mechanism is exploited in a new approach to the sensory manipulation of pest populations using kairomone and allomone chemostimuli (Rice, 1986). This "Push-Pull Strategy" could be used against fruit flies. For example, allomones derived from decay bacteria (Fitt, 1984) or from repellent plant extracts (e.g. Rice et al., 1985), or oviposition deterrent pheromone (e.g. Prokopy, 1981), could be used to "push" flies away from fruit. The frustrated tephritids could then be attracted to lethal baits laced with kairomones (Eisemann, 1980; Fein et al., 1982).

6 TEPHRITID PHOTORECEPTOR PHYSIOLOGY

6.1 Photoreceptor types

The simplest type of photoreceptor is a neurone that, by virtue of a photosensitive protein, is able to register changes in light intensity. Such neurones are found not only in simple and compound eyes but in many areas of the insect body (Truman, 1976; Arikawa and Aoki, 1982). Thus in conducting experiments on fruit fly behaviour it would be unwise to assume that the stemmata, ocelli and compound eyes are the only sites for photoreception. However, the state of the art is such that the compound eye has attracted all the attention. Larvae have no compound eyes, or indeed any externally visible photoreceptors. However, they have a marked set of photosensitive behavioural reactions. It is likely that they have lateral photoreceptors associated with the cephalo-pharyngeal skeleton as found in other fly larvae (Bolwig, 1946) (Fig. 3.9.1).

6.2 Data on tephritid photoreceptors

There are three papers on the ultrastructure of the compound eye, covering four important pest species: *Anastrepha suspensa* (Loew), *R. pomonella*, *C. capitata* and *Dacus dorsalis* Hendel (Agee et al., 1977; Davis et al. 1983; Wu et al., 1985). The tephritid compound eye is hemispherical and flattened anterio-posteriorly. It has ca. 2500–3500 ommatidia as its basic photoreceptor units. Each ommatidium has a hexagonal lens 25–30 μm wide, through which light penetrates to pass through a pseudocone and four Semper cells, before it enters the rhabdomere. The rhabdomere is made up of a central cavity, into which projects the photosensitive, pigment-ladened microvilli of eight retinula cells. These are the primary photosensitive neurones whose axons penetrate the basement membrane of the eye, carrying information on colours, patterns and movement to the processing centres of the optic lobes and brain. This arrangement is typical for the Diptera. The dorsal quadrant of the eye has been found to be the most suitable for inserting microelectrodes for electrophysiological studies, because the eye is most structurally resistant to depressions here. Studies of wild flies, laboratory reared flies and irradiated laboratory reared flies have shown that the visual pigment-containing parts of the rhabdomere are significantly reduced by laboratory rearing and greatly reduced by irradiation (Davis et al., 1983). This is obviously a very significant factor when flies are prepared for release in SIT control methods.

There are six papers published on the electrorectinogram (E.R.G.) of the fruit fly compound eye (Agee and Park, 1975; Agee and Chambers, 1980; Webb et al., 1981; Remund et al., 1981; Agee et al. 1982; Agee, 1985). So far no work has been done on single retinula cells. Because the compound eye represents a repetition of a basic unit, the E.R.G. is a reliable indicator of what is happening at the basic unit level. This is, of course, not so for the electroan-tennogram (E.A.G.) where the output of many diverse units is summed (see section 5.2.2 above). Even so, it is proper to bear in mind that the E.R.G. *is* a compound potential. Using it Agee and his co-workers have made some remark-able discoveries. Of the greatest importance is the finding that some strains of a fruit fly have ten times less light sensitivity than others, often depending on rearing diet (Fig. 3.9.6) (Agee and Park, 1975). Whilst shipment of flies, laborat-ory rearing at low density and irradiation can all have a slight depressing effect on visual sensitivity, the major factor causing depression is crowding. Flies reared at high larval densities often being a hundred times less sensitive and this is made worse by shipment and by irradiation (Agee and Chambers, 1980). Genetic selection may contribute to changes in visual sensitivity during colonisation, the sensitivity of *D. oleae* eyes to 490 nm light steadily decreased over 130 generations (Remund et al., 1981). These authors also showed that one day old flies were substantially less sensitive than ten day old flies. An attempt has been made to correlate the electrophysiological responses of the eyes of *C. capitata*, *D. oleae*, and *R. cerasi* with behavioural colour preferences. However, all three species were similar, with a broad major peak at 495–500 nm (yellow-green) and a secondary peak at 365 nm (ultraviolet) (Agee et al., 1982). For wild and laboratory populations of *R. pomonella* the E.R.G. revealed a similar shape over the spectral range 350 to 675 nm, with a

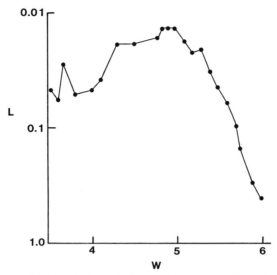

Fig. 3.9.6. Typical spectral sensitivity curve of a tephritid (from Remund et al., 1981). This is characteristic of green sensitive retinula cells, the majority type of receptor cells in the compound eye. L: light energy (μW/cm^2) needed to evoke standard E.R.G. response. W: wavelength of applied stimulus (100 nm).

broad major peak at 400 to 530 nm (blue-green to yellow) with a plateau at 600 to 625 nm (orange-red). This means that although tartar red enamel spheres are considered to provide a "super-normal" visual attractant by some workers, physiological results suggest that *R. pomonella* would "have difficulty" distinguishing that colour from black (Agee, 1985). Behavioural colour preferences are due not only to receptor sensitivity (as monitored by the E.R.G. or by single unit recording) but to the *processing* of photoreceptor input, central model comparator mechanisms, decision making circuits, or some combination of all four physiological processes (see sections 2 and 7).

6.3 Work needed on tephritid photoreceptors

The obvious gaps relate to the extra-optic receptors of larvae and, possibly, adults; photoreception in the developing pupa should also be looked at; the three ocelli of the adult must be examined. Together with these, it would be useful to have more information on the compound eyes of a greater range of species.

In relation to work on the compound eye, it is time that electrophysiologists moved from E.R.G. recording (although this should be kept for quality control testing) and began to characterise the retinula cells by single unit recording. The aim, clearly, is to determine what features of the visual environment are of behavioural significance to the fruit fly computer. It is likely to be at this level that the best shapes and colours for trapping tephritids, at different times of the day and different seasons, will be determined. The great significance of detail in visual stimuli has been shown by the work of Meats (1983), who found that contrast, grain and silhouette effects were important.

6.4 Applied uses of tephritid photoreception

As mentioned above, vision research has a long way to go before it can provide a detailed basis for the derivation of more effective field lures (Levinson and Haisch, 1983). Eventually, super-effective lures could be combined with olfactory attractants and gustatory stimulants, to act as a "sink" for fruit flies in almost any situation. There also seems to be potential for visually reducing the attractiveness of crop plants and trees, an area reviewed in detail by Prokopy and Owens (1983).

7 CONCLUSIONS

It is difficult to summarise so large an area as "fruit fly sensory physiology". Availability of data on the five senses and their ten major loci is alternately good, mediocre and poor. Very little good electrophysiology of primary receptors has been done and no electrophysiology of sensory processing. It has therefore proved a challenge to attempt a synthesis of what is known and what should be known, so that it is of use to those reading this chapter for sensory information or planning future research. There can be no doubt that this area is one of enormous applied potential and of great general biological significance. Sensory biology contributes towards our attempts to originate better fruit fly control measures and also opens a productive approach to ecoethological relations and speciation theory. Looking ahead towards the development of genetic engineering, we can anticipate the generation of some interesting new sensory phenotypes. Understanding the sensory biology of existing fruit fly types provides the background for recognition and utilisation of useful engineered types.

It is shocking to note that all the basic descriptive work has yet to be done on larval receptors. In the adult the most pressing needs are for chemoreceptor characterisation, especially olfactory receptor physiology. The sensory investigation of kairomone, allomone and pheromone reception may provide answers to several applied and basic questions. To date the blowfly is the best studied invertebrate, as regards chemoreception (Dethier, 1976) but scant information is available on the chemical senses of fructivorous Diptera (Schoonhoven, 1983a). A start needs to be made on the detailed characterisation of individual olfactory and gustatory chemosensor neurone types, by means of single cell recordings, to define the range of adequate stimuli plus their dose/response relationships. Receptor neurone input needs to be followed centrally and the behaviourally evocative circuitry identified. Once this is underway we will really have something to write about fruit fly chemosensory physiology. Work along these lines is also pressing to be done on tephritid photoreceptors. Despite some good work on E.R.G.s, we remain without information on individual retinula cell receptivity and on the processing of visual input. Work is also needed on mechanoreceptors, most especially auditory receptor neurones, and on hygroreceptors and thermoreceptors.

The classical approach to sensory physiology involves experiments that hold all inputs constant whilst one modality at a time is varied. However, this is highly unnatural, the natural situation involves simultaneous multimodal sensory input, making *sensory context* a critical factor (Rice, 1975). The only study of multimodal inputs in tephritids is the preliminary look at combinations of temperature, humidity and CO_2, of Rice and Burn (see section 4.2 above). The combination of chemical, temperature and humidity stimuli was found to have profound effects on behavioural responses. This would be an ideal area to pursue central sensory processing. Apparently the c.n.s. recognises more significance when the three inputs are provided simultaneously, presumably by matching with an internal model of reality. Work is urgently needed to elucidate the nature of such c.n.s. models. Whatever such models turn out to be in structural and physiological terms, we can reason that they must closely reflect the ecology of each species. They are likely to be a major focus of natural selection pressure. Natural selection at a particular time and

Chapter 3.9 references, p. 270

place will "favour" particular models of reality, thus shaping existing species or, perhaps, even splitting off new species. Obviously this type of work contributes to Biology as a whole, the endeavour to contribute to speciation theory ranking as one of the highest aspirations of sensory biology (cf. Paterson, 1985).

Using this sensory approach, incipient new species are considered to begin their passage towards genetic fixing with slight c.n.s. changes in the interpretation of sensory inputs. Central nervous system models of reality, thus changed, will sponsor behavioural decision differences. Because of the complex architecture of the neuropile, it is likely to be under highly polygenic control so that mutations affecting c.n.s. models of reality are certain to be common. These are probably generally deleterious or neutral; however an occasional mutation may lead to c.n.s. models for more appropriate behaviour and hence a selection advantage (Rice, 1975). Such advantageous genes would probably spread through the population, leading to a species change, such as the extension of host range seen in some tephritid species and other insects. Should reproductive isolation intervene at some point whilst the new gene(s) are spreading, then a new species might eventuate. The vital point to grasp here is that speciation is more likely to occur by c.n.s. leading, sensory receptor changes taking place subsequently. The steps involved are:

    i. occurrence of a mutation affecting the way the c.n.s. attends to or interprets sensory input;

    ii. a resultant change in behaviour, due to the c.n.s. model of reality expressing a changed model of the environment;

    iii. spread of the mutant gene and, perhaps, isolation of parent and mutant populations;

    iv. selection of sense organs better adapted to the information needs of the new model, as the mutant population adapts further;

    v. divergence of specific mate recognition systems when parent and mutant populations become separated (cf. Paterson, 1985).

To the casual reader it may appear that this is somewhat distantly related to tephritid sensory biology. However, the actual view one has of ecological relationships and speciation can greatly influence understanding of tephritid sensory physiology and behavioural management. For example, efforts to discover major differences in the primary receptor responses of various tephritid species have failed (Agee et al., 1982; Guerin et al., 1983). These workers, like Prokopy (1983) and most of us (!), have had a natural tendency to think in terms of "stimuli guiding tephritids to resources". This atttitude is no more than Loebian tropicism in modern guise (cf. Fraenkel and Gunn, 1961). Here we mistakenly treat the fly rather like a push-button vending machine: coins are inserted to provide "drive", then the sensors are stimulated (buttons pushed), resulting in a specific piece of behaviour (goods vended). Despite the extensive debunking of Loebian tropicism by the many experiments cited by Fraenkel and Gunn (1961), the philosophy of "tropisms" is widely pervasive, often in subtle ways. Tephritids and other insects are much more complex than vending machines; a truer analogy would be with a personal computer. A computer has a sophisticated, multimodal input keyboard, an internal programme that can accept or reject the input, processors that manipulate the input, make decisions and operate output devices. Note that the weight of the analogy is in the fact that different keyboards do not differ greatly; rather the internal processing of inputs by computers differ enormously. It is the emphasis, that the c.n.s. *interpretation of information* is the critical part, not so much the information reception itself, that frees us from tropismic attitudes. A structured approach is recommended to those who would research on or write about, tephritid sensory receptors (Fig. 3.9.7):

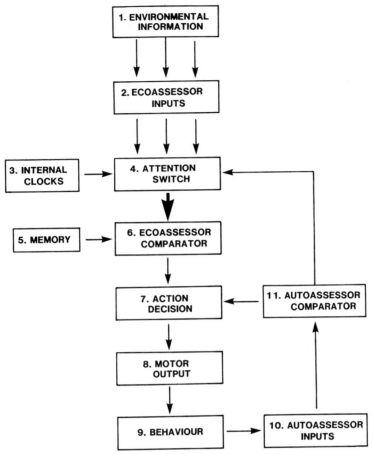

Fig. 3.9.7. An eleven-component functional model of sensory behavior (free of tropisms and "drive"). *Component 1*: represents the numerous modes and grades of information available in the external environment. *Component 2*: represents neural input on that part of the environmental information that is recognised by mechano-, thermo-, hygro-, chemo-, and photo-receptors. *Component 3*: represents internal clocks (circadian and others) that influence component 4. *Component 4*: represents the (unknown) mechanism within the c.n.s. that facilitates inputs relevant to a particular behavioural activity. How it selects which behavioural activity to pay attention to at any particular time is one of the central questions of sensory physiology. *Component 5*: represents the modifying effects of memory traces of previous experiences, both long-term and short-term. *Component 6*: represents the mechanism that compares the selected sensory input with an anticipated input relevant to the behaviour that is being attended to. *Component 7*: represents the output resulting from the comparisons of component 6. *Component 8*: represents the motor instructions that lead to muscle contractions and gland secretions, etc., that compose the observed behaviour (component 9). *Component 10*: represents the neural input of mechano- and chemoreceptors that provide information on the internal state and relationships of the insect's body parts. *Component 11*: represents the mechanism that compares the autoassessor input with an anticipated input relevant to the behaviour the insect is engaged in. Any discrepancy maintains or modifies the action decision until it achieves synchrony of autoassessor input and internal model of reality. It also produces feedback to the attention switch, able to switch ecoassessor attention appropriate to the attainment of pre-programmed behavioural objectives.

1. Multiple sensory inputs from eco- and autoreceptors are like the input from the keyboard of a computer, except that they are much more numerous and very much faster in reaction. The information selected by a species' particular sensory array from internal and external sources, is transmitted to the c.n.s. processor.

2. The primary decision of the processor is concerned with which set of inputs to pay attention to and which to ignore ($\equiv$ computer software package).

Chapter 3.9 references, p. 270

3. The characteristics of those inputs that are paid attention to are compared with an internal model of reality, an expected pattern ($\equiv$ dedicated computation).

4. The resultant c.n.s. pattern contrasts identical and discrepant information, resulting in decisions that drive motor output and hence behaviour ($\equiv$ computer output to v.d.u., chart recorder or printer).

This can be summarised: total information input $\rightarrow$ selective attention $\rightarrow$ pattern comparison $\rightarrow$ output decisions. In contrast, traditional Loebian tropicism gives: dominant information inputs $\rightarrow$ centrally summed $\rightarrow$ output decisions. Anyone familiar with tephritid ecoethological literature will immediately recognise which philosophy subtends experimental design and data interpretation: almost invariably Loebian tropicism. It is a basic thesis of this chapter that updating to the "information processing" model of sensory biology is necessary for future progress. This point cannot be emphasised too much.

The first result of such a change in philosophy will be seen in efforts to establish valid criteria for recognising differences in the c.n.s. processing of sensory input. This will benefit not only sensory physiology and neurethology but all disciplines concerned with tephritid host-range expansion, speciation, theoretical ecology and pest management. For example Prokopy and Roitberg (1984) figure sticky-coated lures to which *R. pomonella* have been attracted. They show that a spherical, apple-like lure is more attractive than other shapes. The tropismic explanation is that receptor input from the "apple" was so strong that the flies were forced to turn towards the lure and fly to it. This does not help us to understand why large numbers of flies, with identical receptors, also became trapped by cuboidal, oblong and other shaped lures. The information processing approach asks, not only what receptors are involved, but how their input is processed by the c.n.s. There are clearly differences between flies of the same species population in how their c.n.s. computers interpret the same receptor input data. It goes without saying that such differences are of interest to population ecologists, evolutionary geneticists and all those who would use sensory manipulation for the behavioural management of pests.

A number of studies on tephritids show the ability of the c.n.s. processor to control responses to particular sensory inputs. For example, male attraction to both cue-lure and methyleugenol is limited to the early part of the day (Brieze-Stegeman et al., 1978). A circadian clock has been identified as controlling the sexual responsiveness of both male and female *D. tryoni* (Tychsen, 1975). Readiness to respond to oviposition stimuli is enhanced by deprivation (Roitberg and Prokopy, 1983). The c.n.s. processor can also be conditioned by previous oviposition experiences (Prokopy, 1983). Workers in these areas often invoke the presence of "drive" to explain differences in "sensory responsiveness". It is difficult to obtain a clear understanding of the nature of drive, despite its use by many ethologists. Alcock (1979) defines drive as a "hypothetical theoretical variable". The present writer, in the absence of a better identification of drive, prefers to substitute the phenomenon of "attention" in place of "drive". Attention by the c.n.s. processor is considered to be a necessary property for manipulation of the vast amount of data input that sensory neurones transmit to the ganglia. Fruit flies have a range of behaviours such as flight, cleaning, resting, feeding, alarm, escape, courtship, mating, oviposition, etc. Each of these is made up of a number of components that generally occur in particular sequences. Like other insects, tephritids carry out one of the behavioural components at a time. It seems that the c.n.s. is a general purpose computer that can be dedicated to many different behavioural components, provided they are not required simultaneously (cf. Elsner, 1983). Thus those c.n.s. circuits which determine the sequence of behavioural components are vitally important. Orderly sequence determination seems to fail in pyrethroid poisoned insects, resulting in highly disordered sequences and unusual simultaneous expression of components. The need for circuitry to determine orderly progression of behavioural components is apparent, though as with "drive", no cellular circuitry has yet been identified. Attention circuits could *simulate* increased drive by paying more attention to sensory inputs of a particular kind, such as those related to feeding or mating. The operation of attention switching can readily be seen in fruit flies, for example the normal amount of attention paid to alarm inducing stimuli by *D. tryoni*, such as the experimenter's approaching fingers, is much reduced when flies are feeding, courting or ovipositing. Without multiplying examples, it can be discerned that "attention" and "drive" are often interchangeable concepts; attention has the great advantage, however, of relying on neural switching circuits that can be sought for with the electrode. The "attention theory" of sensory behaviour readily accommodates the modulating effects of rhythms, learning and deprivation. It is ideal for approaching dis-habituation due to novel stimuli and for understanding the wide range of activities, that a group of identically reared members of a species may engage in at any one time. The study of tephritid sensory behaviour would benefit greatly if neural "attention" mechanisms were invoked in place of the somewhat nebulous and anthropomorphic concepts of motivation, drive, action-specific energy and the like.

Thus far it has been concluded that advances in our knowledge of tephritid sensory biology will depend on:
(a) increased descriptive sensory physiology;
(b) use of c.n.s. information processor models in place of tropicisms; and
(c) use of c.n.s. attention models in place of drive.

Finally, what can be concluded about the application of sensory theory to the control of pest tephritids? To date, fruit fly behavioural control ideas have centred on two concepts: use of oviposition deterrent pheromone and use of attractant traps. Several other possibilities for sensory manipulation have been considered in sections 3.4, 4.4, 5.4, and 6.4 above. It appears that the c.n.s. often uses sensory input in a push-pull fashion, as described by Rice (1975) for mechanoreceptors and reviewed by Altner and Prillinger (1980) and Altner et al. (1983) for thermo- and hygroreceptors. Recently this information-maximizing mechanism has been developed and extended as a sensory manipulation strategy for use against pest insects (e.g. Rice, 1986). Boller and Prokopy (1976) mention a related concept, proposing that *R. pomonella* repelled by oviposition deterrent pheromone should be attracted to killing traps. Schoonhoven (1983b) also mentions that repelled females should be trapped or they will break through the oviposition inhibiting barrier (section 5.4 above). However, the Push-Pull Strategy (P.P.S.) as currently developed goes further than this, suggesting that insects deterred or repelled by allelochemicals are especially vulnerable to kairomone baited killing lures (Pyke et al., 1986, 1987). The P.P.S. involves the simultaneous action of repellent and/or deterrent sprays, applied to our crops, together with distribution of attractant and/or stimulant killing baits. It is hoped that the conspectus of tephritid senses presented here, together with the prospectus of desirable research and philosophy, will contribute to the development of such procedures for fruit fly control.

8 ACKNOWLEDGEMENTS

I would particularly like to thank Dick Drew and Gordon Hooper for, over the years, enthusiastically sharing their deep knowledge of tephritids. To my students I owe much, especially Rod Brieze-Stegeman, Don Smith, Ann Hill,

Glen Burn, Craig Eisemann and Peter Njagi. I am very grateful to Mrs Judy Attard for typing the manuscript.

9 REFERENCES

Agee, H.R., 1985. Spectral response of the compound eye of the wild and laboratory-reared apple maggot fly, *Rhagoletis pomonella*. Journal of Agricultural Entomology, 2: 147–154.

Agee, H.R. and Park, M.L., 1975. Use of the electroretinogram to measure the quality of vision of the fruit fly. Environmental Letters, 10: 171–176.

Agee, H.R. and Chambers, D.L., 1980. Fruit fly quality monitoring. Proceedings of a symposium on Fruit Fly problems. National Institute of Agricultural Sciences, Yatabe, pp. 7–15.

Agee, H.R., Phillips, W.A. and Chambers, D.L., 1977. The compound eye of the Caribbean fruit fly and the apple maggot fly. Annals of the Entomological Society of America, 70: 359–364.

Agee, H.R., Boller, E., Remund, U., Davis, J.C. and Chambers, D.L., 1982. Spectral sensitivities and visual attractant studies on the Mediterranean fruit fly, *Ceratitis capitata* (Wiedemann), olive fly, *Dacus oleae* (Gmelin) and European cherry fruit fly, *Rhagoletis cerasi* (Linnaeus) (Diptera, Tephritidae). Zeitschrift für Angewandte Entomologie, 93: 403–412.

Alcock, J., 1979. Animal Behaviour. Sinauer, Sunderland, Massachusetts.

Altner, H. and Prillinger, L., 1980. Ultrastructure of invertebrate chemo-, thermo- and hygroreceptors and its functional significance. International Review of Cytology, 67: 69–139.

Altner, H. and Loftus, R., 1985. Ultrastructure and function of insect thermo- and hygroreceptors. Annual Review of Entomology, 30: 273–295.

Altner, H., Schaller-Selzer, L., Stetter, H. and Wohlrab, I., 1983. Poreless sensilla with inflexible sockets. A comparative study of fundamental type of insect sensillum, probably comprising thermo- and hygroreceptors. Cell and Tissue Research, 234: 279–307.

Angioy, A.M., Liscia, A. and Pietra, P. 1978. Effects of salt and sugar on taste hairs of *Ceratitis capitata* Wied. and *Dacus oleae* Gmel. Bolletino Della Societa Italiana Di Biologia Sperimentale, 54: 2108–2121.

Arikawa, K. and Aoki, K., 1982. Response characteristics and occurrence of extraocular photoreceptors in Lepidopteran genitalia. Journal of Comparative Physiology, 148: 483–489.

Boller, E.F. and Prokopy, R.J., 1976. Bionomics and management of *Rhagoletis*. Annual Review of Entomology, 21: 223–246.

Bolwig, N., 1946. Senses and sense organs of the anterior end of the house fly larva. Videnskablige Meddelelser fra Dansk naturhistorik Forening; Kjobenhaun, 109: 80–217.

Bowdan, E., 1984. Electrophysiological responses of tarsal contact chemoreceptors of the apple maggot fly, *Rhagoletis pomonella* to salt, sucrose and oviposition-deterrent pheromone. Journal of Comparative Physiology, 154: 143–152.

Brieze-Stegeman, R., Rice, M.J. and Hooper, G.H.S., 1978. Daily periodicity in attraction of male tephritid fruit flies to synthetic chemical lures. Journal of the Australian Entomological Society, 17: 341–346.

Chu, I.W. and Axtel, R.C., 1971. Fine structure of the dorsal organ of the house fly larva, *Musca domestica* L. Cell and Tissue Research, 117: 17–34.

Crnjar, R.M. and Prokopy, R.J., 1982. Morphological and electrophysiological mapping of tarsal chemoreceptors of oviposition-deterring pheromone in *Rhagoletis pomonella* flies. Journal of Insect Physiology, 28: 393–400.

Crnjar, R.M., Prokopy, R.J. and Dethier, V.G., 1978. Electrophysiological identification of oviposition-deterring pheromone receptors in *Rhagoletis pomonella*. Journal of the New York Entomological Society, 86: 283–284.

Davis, J.C., Agee, H.R. and Ellis, E.A., 1983. Comparative ultrastructure of the compound eye of the wild, laboratory-reared and irradiated Mediterranean fruit fly *Ceratitis capitata* (Diptera: Tephritidae). Annals of the Entomological Society of America, 76: 322–332.

Dethier, V.G., 1976. The Hungry Fly. Harvard University Press, Cambridge, Mass.

Eisemann, C.H., 1980. An investigation of some stimuli influencing host finding and oviposition behaviours of the Queensland Fruit Fly, *Dacus (Bactrocera) tyroni* (Frogg.). Ph.D. thesis, University of Queensland, Brisbane.

Eisemann, C.H. and Rice, M.J., 1985. Oviposition behaviour of *Dacus tryoni*: the effects of some sugars and salts. Entomologia Experimentalis et Applicata, 39: 61–71.

Eisemann, C.H. and Rice, M.J., 1987. Behavioural evidence for hygro- and mechanoreception by ovipositor sensilla for *Dacus tryoni* (Froggatt) Diptera: Tephritidae. Physiological Entomology, submitted.

Eisemann, C.H., Jorgenson, W.K., Merritt, D.J., Rice, M.J., Cribb, B.W., Webb, P.D. and Zalucki, M.P., 1984. Do insects feel pain? – A biological view. Experientia, 40: 164–167.

Elsner, N., 1983. A neuroethological approach to the phylogeny of leg stridulation in gomphocerine grasshoppers. In: F. Huber and H. Markl (Editors), Neuroethology and Behavioural Physiology:

Roots and Growing Points. Springer, New York, pp. 54–68.

Fein, B.L., Reissig, W.H. and Roelofs, W.L., 1982. Identification of apple volatiles attractive to the apple maggot, *Rhagoletis pomonella*. Journal of Chemical Ecology, 8: 1473–1487.

Fitt, G.P., 1984. Oviposition behaviour of two tephritid fruit flies, *Dacus tryoni* and *Dacus jarvisi*, as influenced by the presence of larvae in the host fruit. Oecologia, 62: 37–46.

Fraenkel, G. and Gunn, D.L., 1961. The Orientation of Animals. New York; Dover Publications.

Giannakakis, A. and Fletcher, B.S., 1985. Morphology and distribution of antennal sensilla of *Dacus tryoni* (Froggatt) (Diptera: Tephritidae). Journal of the Australian Entomological Society, 24: 31–35.

Gothilf, S., Galun, R. and Bar-Zeev, M., 1971. Taste reception in the Mediterranean fruit fly: electrophysiological and behavioural studies. Journal of Insect Physiology, 17: 1371–1384.

Grabowski, C.T. and Dethier, V.G., 1954. The structure of the tarsal chemoreceptors of the blowfly, *Phormia regina* Meigen. Journal of Morphology, 94: 1–20.

Guerin, P.M., Remund, U., Boller, E.F., Katsoyannos, B. and Delrio, G., 1983. Fruit fly electroantennogram and behaviour responses to same generally occurring fruit volatiles. In: R. Cavalloro (Editor). Fruit Flies of Economic Importance. Balkema, Rotterdam, pp. 248–251.

Hallberg, E., Hogmo, O. and Nassel, D.R., 1984a. Antennal receptors in the blowfly *Calliphora erythrocephala*: II. Fine structure of the large pedicellar campaniform sensillum. Journal of Morphology, 18: 155–223.

Hallberg, E., Van der Pers, J.N.C. and Haniotakis, G.E., 1984b. Funicular sensilla of *Dacus oleae*: fine structural characteristics. Entomologia Hellenica, 2: 41–46.

Hodgson, E.S., Lettvin, J.Y. and Roeder, K.D., 1955. Physiology of a primary chemoreceptor unit. Science, 122: 417–418.

Levinson, H.Z. and Haisch, M.A., 1983. Optical and chemosensory stimuli involved in host recognition and oviposition of the cherry fruit fly *Rhagoletis cerasi* L. In: R. Cavalloro (Editor), Fruit Flies of Economic Importance. Balkema, Rotterdam, pp. 268–284.

Marchini, L., 1982. Laboratory Studies on Oviposition and on the Structure of the Ovipositor in the Mediterranean fruit fly *Ceratitis capitata* (Wied.), Ph.D. thesis, University of Manchester.

Marchini, L. and Wood, R.J., 1983. Laboratory studies on oviposition and on the structure of the ovipositor in the Mediterranean fruit fly *Ceratitis capitata* (Wied.). In: R. Cavalloro (Editor), Fruit Flies of Economic Importance. Balkema, Rotterdam, p. 113.

McIver, S.B., 1985. Mechanoreception. In: G.A. Kerkut and L.I. Gilbert (Editors), Comprehensive Insect Physiology, Biochemistry and Pharmacology, Volume 6 Nervous System: Sensory. Pergamon, Oxford, pp. 71–132.

Meats, A., 1983. The response of Queensland fruit fly, *Dacus tryoni* to tree models. In: R. Cavalloro (Editor), Fruit Flies of Economic Importance. Balkema, Rotterdam, pp. 285–289.

Metcalf, R.L., 1985. Plant kairomones and insect pest control. Illinois Natural History Survey Bulletin, 33: 175–198.

Metcalf, R.L., Mitchell, W.C., Fukoto, T.R. and Metcalf, E.R., 1975. Attraction of the oriental fruit fly, *Dacus dorsalis*, to methyl eugenol and related olfactory stimulants. Proceedings of the National Academy of Science in the U.S.A., 72: 2501–2505.

Metcalf, R.L., Metcalf, E.R., Mitchell, W.C. and Lee, L.W.Y., 1979. Evolution of olfactory receptor in Oriental fruit fly *Dacus dorsalis*. Proceedings of the National Academy of Science, U.S.A., 76: 1561–1565.

Monroe, J., 1953. Stridulation in the Queensland Fruit Fly *Dacus (Strumeta) tryoni* Frogg. Australian Journal of Science, 16: 60–62.

Osborne, M.P., 1963. The sensory neurines and sensilla in the abdomen and thorax of the blowfly larva. Quarterly Journal of Microscopical Science, 104: 227–241.

Paterson, H.E.H., 1985. The recognition concept of species. In: E.S. Vrba (Editor), Species and Speciation. Transvaal Museum, Pretoria, pp. 21–29.

Prokopy, R.J., 1981. Epideictic pheromones that influence spacing patterns of phytophagous insects. In: D.A. Nordlund, R.L. Jones and W.J. Lewis (Editors), Semiochemicals: Their Role in Pest Control. Wiley, New York, pp. 181–213.

Prokopy, R.J., 1983. Tephritid relationships with plants. In: R. Cavalloro (Editor), Fruit Flies of Economic Importance. Balkema, Rotterdam, pp. 230–239.

Prokopy, R.J. and Bush, G.L., 1973. Ovipositional responses to different sizes of artificial fruit by flies of *Rhagoletis pomonella* species group. Annals of the Entomological Society of America, 66: 927–929.

Prokopy, R.J. and Owens, E.D., 1983. Visual detection of plants by herbivorous insects. Annual Review of Entomology, 28: 337–364.

Prokopy, R.J. and Roitberg, B.D., 1984. Foraging behaviour of true fruit flies. American Scientist, 72: 41–49.

Pyke, B., Rice, M.J., Sabine, B.N.E. and Zalucki, M.P., 1986. A preliminary report on the Push-Pull Strategy against *Heliothis* spp. in cotton in Queensland. In: D. Swallow (Editor), Cotton-On: Proceedings of the Australian Cotton Conference 1986. A.C.G.R.A., Wee Waa, pp. 161–173.

Pyke, B., Rice, M.J., Sabine, B.N.E. and Zalucki, M.P., 1987. The Push-Pull Strategy – behavioural control of Heliothis. Australian Cotton Grower, 8(2): 7–9.

Remund, U., Economopoulas, A.P., Boller, E.F., Agee, H.R. and Davis, J.C., 1981. Fruit fly quality monitoring: the spectral sensitivity of field collected and laboratory-reared olive flies, *Dacus oleae* Gmel. (Dipt.: Tephritidae). Mitteilungen des Schweizerischen Entomologischen Gesell-schaft, 54: 221–227.

Rice, M.J., 1970. Cibarial stretch receptors in the tsetse fly (*Glossina austeni*) and the blowfly (*Calliphora erythrocephala*). Journal of Insect Physiology, 16: 277–289.

Rice, M.J., 1973. Cibarial sense organs of the blowfly, *Calliphora erythrocephala* (Meigen) (Diptera: Calliphoridae). International Journal of Insect Morphology and Embryology, 2: 109–116.

Rice, M.J., 1975. Insect Mechanoreceptor Mechanisms. In: R. Galun, P. Hillman, I. Parnas and R. Werman (Editors), Sensory Physiology and Behaviour. Plenum, New York, pp. 135–165.

Rice, M.J., 1977. Blowfly ovipositor receptor neurone sensitive to monovalent cation concentration. Nature: 268: 747–749.

Rice, M.J., 1986. Semiochemicals and sensory manipulation strategies for behavioural manage-ment of *Heliothis* spp. Ochsenheimer (Lepidoptera: Noctuidae). In: P.H. Twine and M.P. Zalucki (Editors), Proceedings of the Second Australian *Heliothis* Ecology Workshop. D.P.I., Brisbane, pp. 27–45.

Rice, M.J., Galun, R. and Margalit, J., 1973. Mouthpart sensilla of the tsetse fly and their function. Annals of Tropical Medicine and Parasitology, 67: 101–116.

Rice, M.J., Sexton, S. and Esmail, A.M., 1985. Antifeedant phytochemical blocks oviposition by sheep blowfly. Journal of the Australian Entomological Society, 24: 16.

Roitberg, B.D. and Prokopy, R.J., 1983. Host deprivation influence on response of *Rhagoletis pomonella* to its oviposition deterring pheromone. *Physiological Entomology*, 8: 69–72.

Schoonhoven, L.M., 1983a. The role of chemoreception in hostplant finding and oviposition in phytophagous Diptera. In: R. Cavalloro (Editor), Fruit Flies of Economic Importance. Balkema, Rotterdam, pp. 240–247.

Schoonhoven, L.M., 1983b. Report on Session 3. In: R. Cavalloro *op cit.*

Shorey, H.H., 1977. Interaction of insects with their chemical environment. In H.H. Shorey and J.J. McKelvey Jr. (Editors), Chemical Control of Insect Behaviour: Theory and Application. Wiley, New York, pp. 1–5.

Smith, D.S., 1969. The fine structure of haltare sensilla in the blowfly, *Calliphora erythrocephala* (Meig.) with scanning electron microscopic observations on the haltere surface. Tissue and Cell, 1: 443–484.

Truman, J., 1976. Extraretinal photoreceptors in insects. Photochemistry and Photobiology, 23: 215–225.

Tychsen, P.H., 1975. Circadian control of sexual drive level in *Dacus tryoni* (Diptera: Tephritidae). Behaviour, 54: 111–141.

Webb, J.C., Agee, H.R., Leppla, N.C. and Calkins, C.O., 1981. Monitoring insect quality. Transac-tions of the American Society of Agricultural Engineers, 24: 476–479.

Wu, C.Y., Chang, C.S., Tung, L.C. and Lin, J.T., 1985. Receptors in insects: I. The fine structure of the compound eye of the oriental fruit fly *Dacus dorsalis* Hendel. Bulletin of the Institute of Zoology, Academia Sinica, 24: 27–37.

10 ADDENDUM

Laboratory tests, using apple domes, have now shown that neem extracts are able to block more than 90% of egg-laying by *D. tryoni* (Permkam and Rice, in preparation). Neem, in appropriate formulation, offers a natural method of protecting fruit; the efficacy of which is likely to be much increased by the simultaneous use of kairomonal baits, in a Push-Pull Strategy.

Chapter 3.10 Temperature – Development Rate Relationships of the Immature Stages and Adults of Tephritid Fruit Flies

BRIAN S. FLETCHER

1 INTRODUCTION

It has long been recognised that temperature is the single most important factor determining development rates of the immature stages and the adult maturation rates of the majority of insects. As a result, numerous attempts have been made to determine the temperature-development rate relationships of tephritid fruit flies because of their usefulness in predicting phenological events in the field for ecological and pest management purposes, the optimization of mass rearing procedures under constant conditions and the construction of computer simulation models of population processes.

2 TEMPERATURE-DEVELOPMENT RATE MODELS USED FOR FRUIT FLIES

2.1 Linear models

The most commonly used method of predicting the development rates of individual life stages is the temperature summation model. This approach is based on the assumption that above some lower threshold for development temperature-development rate relationships are linear and, therefore, a constant number of heat units (normally expressed as day-degrees) above this threshold are needed to complete development.

To establish this relationship, the duration of individual life stages and/or total development time from egg to adult are usually determined at a series of constant temperatures in the laboratory. If the development rates (i.e. 100/development time) are then plotted against temperature, a linear regression line can be fitted through the points as shown in Fig. 3.10.1. The lower development threshold (i.e. the notional temperature at which the development rate is zero) is then determined by extrapolation of the line back to the x-axis. The thermal constant, K (i.e. the number of day-degrees above the lower threshold, t, required to complete development) can be calculated from the regression equation using the relationship $K = y(x - t)$.

When using the thermal summation model to calculate development times in fluctuating daily temperature regimes, the number of day-degrees per day ($D°$) can be determined from the formula $(T_{max} + T_{min}/2) - t$, as long as the minimum temperature (T_{min}) is above the development threshold. In other circumstances, 24 hourly readings or a method of reconstructing an appropriate daily temperature trace from T_{max} and T_{min} data is required in order to calculate day-degrees (Allen, 1976; Fletcher and Comins, 1985).

Chapter 3.10 references, p. 286

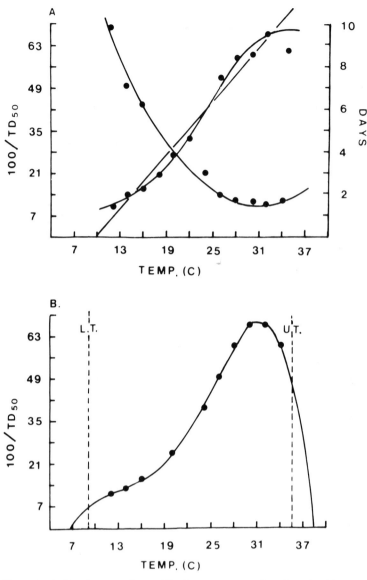

Fig. 3.10.1 A. Development times of *D. oleae* eggs at a series of constant temperatures and the linear and sigmoid curves fitted to development rates ($100/TD_{50}$) plotted against temperature B. Fourth order polynomial curve fitted to development rate data of eggs at constant temperatures, LT and UT indicate lower and upper lethal thresholds. (From Fletcher and Kapatos, 1983).

Although as first described by Davidson (1944), the temperature development rate curve obtained with constant temperatures is sigmoid rather than linear, the temperature summation model has proved to be reasonably accurate for predicting development times in the field, particularly when temperatures remain within the most favourable range for development.

Problems can arise, however, in predicting development rates at times of the year when mean temperatures tend to be around the lower temperature threshold or above the optimum level for development. There are two reasons for this, firstly the temperature development rate relationship deviates most from linearity towards the temperature extremes. Secondly, although at constant temperatures the range of temperatures at which development can proceed is limited by the upper and lower lethal thresholds, most immature life stages can survive and develop for short periods outside these thresholds (Tsitsipis, 1980; Meats, 1984).

This problem is particularly significant when predicting ovarian maturation rates of females because adults can often survive for long periods when temperatures are above or below the maturation thresholds. In such situations, ovarian maturation is not simply arrested but actually regresses due to the resorption of the developing oocytes. As a result, the thermal summation model, and indeed any other model that relates maturation rates to temperature as a straightforward additive process, is inappropriate.

2.2 Non-linear temperature development rate models

Because of the approximations inherent in linear models and the problems that arise when using them to predict development rates at times of the year when the temperature regimes are towards the upper or lower developmental limits, an increasing number of researchers have developed non-linear models to describe temperature development rate relationships and predict development or maturation times in the field (Girolami, 1979; Stark and AliNiazee, 1982; Fletcher and Kapatos, 1983a). This approach has been facilitated by the increasing availability of programmable calculators which ease the extra computational load non-linear models require.

Until recently, the most common non-linear function fitted to temperature-development rate data was the logistic equation or sigmoid curve. This, however, has the same major disadvantage as the linear models because it inadequately describes the development rate relationship towards the upper end of the temperature scale (Fig. 3.10.1). Increasingly, therefore, polynomial functions have been used to describe overall temperature-development rate relationships and used to predict development times by an algorithm which sums the calculated percent development over a series of discrete time intervals (hourly in the case of 24 hourly readings) until 100% development for the stage is reached (Fletcher and Comins, 1985; Dallwitz and Higgins, 1978).

3 TEMPERATURE-DEVELOPMENT RATE RELATIONSHIPS OF NON-DIAPAUSING IMMATURE STAGES

3.1 The immature stages of *Ceratitis capitata* (Wiedemann)

Since the pioneering work of Bodenheimer (1925), data on the duration of one or more stages of *C. capitata* at a series of constant temperatures have been published by a number of workers, e.g. McBride (1935), Messenger and Flitters (1958) and Shoukry and Hafez (1979).

Messenger and Flitters (1958) examined the effects of a series of constant temperatures on survival and duration of the egg stage in Hawaii and observed that the relationship between development rate and temperature was sigmoid rather than linear and that the rate decreased at temperatures above 32°C. By fitting a regression line to the medial values only, they calculated the lower development threshold as 11.7°C and K to be 25.74 D°. Comparison of development rates at constant and fluctuating temperatures indicated that close agreement only occurred in the medial zone. In fluctuating conditions with mean temperatures below the medial zone, development rates were accelerated and at mean temperatures above the medial zone development rates decreased, compared with the equivalent constant temperatures. The greater the diurnal temperature variation the greater the effect.

In more recent studies, Tassan et al. (1983) used Bodenheimer's and McBride's data to establish lower development thresholds and thermal constants for the egg plus larval and pupal stage, as a basis for predicting the

Chapter 3.10 references, p. 286

TABLE 3.10.1

Lower development thresholds and thermal constants for eggs of *C. capitata* calculated by Delrio et al. (1986) from their own data and that of some earlier workers

Lower threshold (°C)	Thermal constant (D°)	Temperature-development model[1]	References[2]
9.6	35.7	T	1
9.3	35.9	L	1
6.8	29.1	S	1
9.4	25.8	T	2
9.6	25.3	L	2
5.0	22.4	S	2
9.5	33.6	T	3
9.6	33.3	L	3
5.7	27.9	S	3
10.8	31.5	T	4
11.6	25.7	L	4
4.5	34.1	S	4

[1] T = Thermal summation; L = Linear regression; S = Sigmoid curve (logistic equation).
[2] Data obtained from; 1, Delrio et al. (1986); 2, Bodenheimer (1951); 3, Shoukry and Hafez (1979); 4, Messenger and Flitters (1958).

phenology of *C. capitata* in different regions of California. Egg plus larval development was calculated to take 142.8 D° above 9.7°C (using air temperatures) and the pupal stage 182.4 D° above 9.7°C (using temperatures at a soil depth of 2.5 cm).

Using the development threshold of 13.6°C determined experimentally by Messenger and Flitters (1954), Vargas et al. (1984) calculated the thermal constants for eggs, larvae and pupae to be 27.5, 117.4 and 233.9 D° respectively, based on the time taken for 50% of individuals to complete each stage at 25°C.

Delrio et al. (1986) used thermal summation, linear regression and sigmoid curves to model data on egg development times of *C. capitata* when eggs 0–1 h, or 0–12 h old, were maintained at a series of constant temperatures between 8 and 40°C. They found that newly laid eggs completed development at temperatures between 12 and 38°C. As in other studies, the temperature-development rate data were more closely described by a sigmoid curve than linear regression, although the latter predicted a temperature threshold that was closer to the observed value. Using linear regression, the thermal constants and development thresholds for eggs placed in constant conditions when 0–1 h or 0–12 h old were 31.1 D° above 11.1°C and 38.4 above 8.7°C, respectively. Egg mortality at different temperatures followed a parabolic curve with least mortality between 16.5–31.5°C and 100% mortality below 10°C and above 38°C.

When the development rate data from both sets of results were combined and re-analysed along with the data presented by Bodenheimer (1951), Messenger and Flitters (1958) and Shoukry and Hafez (1979), using thermal summation, linear regression and sigmoid curves, it was found that the lower thresholds obtained from Delrio et al.'s data were close to those using Bodenheimer's and Shoukry and Hafez's data but except for the sigmoid value, lower than those given by Messenger and Flitters' data (Table 3.10.1).

Overall, the results were in general agreement, although egg development was significantly faster in Bodenheimer's studies than in the others over most of the temperature range, and slower development at high temperatures was particularly evident in Messenger and Flitter's study. Whether these differences were due to slightly different experimental techniques or indicate actual differences in the strains of *C. capitata* used are not clear.

In a parallel study to that carried out on the egg stage, Crovetti et al. (1986) investigated the temperature-development rate relationship of *C. capitata* pupae placed in constant conditions when 0–3 h or 0–12 h old. As with eggs, it was found that the sigmoid curve gave a better overall fit to the data, but the lower development threshold predicted by linear regression was closer to the observed value. With linear regression, the thermal constants and developmental thresholds were calculated as 195.3 D° above 6.4°C for 0–3 h-old pupae and 175.8 above 9.0°C for 0–12 h-old pupae, respectively. Pupal development to emergence occurred between 12 and 33°C, and as in the case of eggs, the mortality curve was parabolic. However, the most favourable range with less than 10% mortality (21–24°C) was much narrower than for eggs.

Crovetti et al. (1986) also compared their results which came from experiments carried out at two different relative humidities (65% RH for pupae 0–3 h-old and 85% RH pupae for 0–12 h-old) with those reported by Bodenheimer (1925). Using thermal summation, linear regression and sigmoid models, the estimated lower developmental thresholds were 9.9, 9.0 and 4.3°C for pupae at 85% RH, 7.2, 6.5 and 5.0°C for pupae kept at 65% RH and 9.7, 9.0 and 4.7°C for Bodenheimer's pupae. The equivalent thermal constants were calculated to be 160.3, 175.9 and 177.7 D° (85% RH); 188.0, 198.3 and 177.0 D° (65% RH); and 182.1, 182.3 and 164.6 D° (Bodenheimer's data), respectively. Although there is some variation in these values due to the different conditions in which the pupae were held and the different lower thresholds used to calculate the thermal constants, they are not significantly different.

Pros (1983) studied duration of the egg plus larval and adult stages in oranges and other fruits under field conditions in a coastal region of Spain and fitted parabolic curves to stage duration plotted against mean temperature. He found that development times were greater in oranges than in peaches, pears, apricots, plums and figs and that overall, development rates were faster than those reported at constant temperatures by Bodenheimer (1925).

3.2 The immature stages of *Dacus cucurbitae* Coquillett

Messenger and Flitters (1958) examined the development rates of *D. cucurbitae* eggs in Hawaii under constant temperatures and found that the upper and lower limits for development were 36.4 and 11.4°C respectively, which were in close agreement with the results obtained in Formosa (now Taiwan) by Shibata (1936). Linear regression of data in the medial range gave a thermal constant of 14.54 D° above a lower threshold of 11.4°C. As with *C. capitata*, under fluctuating temperatures egg development occurred faster below optimum and slower at above optimum mean temperatures, as compared with the development rates observed in constant conditions (Messenger and Flitters, 1959).

Vargas et al. (1984) used the lower threshold for development of 14.7°C observed by Messenger and Flitters (1954) to determine the thermal constants for eggs (16.4 D°), larvae (101.5 D°) and pupae (207.2 D°) reared at 25°C under laboratory conditions in Hawaii.

Nakamori et al. (1978) investigated pupal development in a laboratory strain of *D. cucurbitae* in Okinawa at a series of constant temperatures between 12 and 34°C. No emergences occurred at the two temperature extremes and maximum survival was observed between 20 and 27°C. The thermal constant for pupae calculated from these results by linear regression was 172.7 D° above a threshold of 9.16°C.

Studies on the development rates of eggs, larvae and pupae of two different populations of *D. cucurbitae* in the Ogisawara Islands, one from Kikai Is. (28°

20′N) and the other from Ishigaki Is. (24° 25′N), at four constant temperatures between 15 and 30°C, indicated that there was no difference between them in development rates (Kawai and Yoshihara, 1981). The average thermal constants and lower thresholds were 19 D° above 10.7°C for eggs, 150.5 D° above 7.0°C for larvae, 170.5 D° above 9.7°C for male pupae and 168 D° above 9.85°C for female pupae.

In a similar study carried out by Okumura et al. (1981) thermal constants of 18.03 D° above a threshold of 10.3°C for eggs, 108.15 D° above 9.4° for larvae, and 150.35 D° above 10.6°C for pupae were obtained.

3.3 The immature stages of *Dacus dorsalis* Hendel

Messenger and Flitters (1958), studied egg development of *D. dorsalis* at a series of constant temperatures between 12.5 and 36.7°C and observed some egg hatch at 12.8 and 36.4°C, although there was increasing levels of mortality outside the range of 18.3–33.6°C. As with the other species they studied (i.e. *D. cucurbitae* and *C. capitata*), the temperature-development rate curve was sigmoid up to an upper limit above which development rates decreased rapidly. Using regression analysis of the medial points the lower developmental threshold was calculated to be 12.8°C and the thermal constant 18.1 D°. Development times in fluctuating temperatures increased at mean temperatures above the optimum range, and decreased at mean temperatures below the optimum range, as compared with development rates at constant temperatures (Messenger and Flitters, 1959).

Surprisingly, considering its economic importance and the bioclimatic studies that were carried out in Hawaii (Messenger and Flitters, 1954), only a few models describing the temperature-development rate relationships of larvae and pupae of *D. dorsalis* have been published. Using the temperature threshold of 13.3°C obtained from the earlier work of Messenger and Flitters (1954), Vargas et al. (1984) calculated the thermal constants of a laboratory strain of Hawaiian *D. dorsalis* to be 19.5 D°, 118.2 D° and 246.6 D° for eggs, larvae and pupae, respectively. Hsu and Hsu (1973) estimated the thermal constants of eggs , larvae and pupae for a Taiwanese strain to be 20.1 D° above a threshold of 11.82°C, 144.1 D° above 11.61°C and 229.1 D° above 11.1°C and Saeki et al. (1980) working with a laboratory strain in Okinawa, calculated thermal constants of 22.9 D° above 11.65°C for eggs, 85.1 D° above 11.85°C for larvae and 163.49 D° above 11.0°C for pupae, from development rate data obtained at constant temperatures between 13° and 28°C.

3.4 The immature stages of *Dacus oleae* (Gmelin)

Tsitsipis (1977, 1980) investigated the temperature-development rate relationships of eggs, larvae and pupae of *D. oleae* reared on synthetic medium under laboratory conditions at a series of constant temperatures between 7.5 and 35°C. Temperature-development rate data for larvae and pupae obtained by earlier workers (Moore, 1960; Tzanakakis et al., 1968; Manikas, 1974; Girolami, 1979) were also re-examined and shown to be similar to his own data.

No eggs hatched when maintained at 7.5 and 35°C, although some development occurred at these temperatures. As in other cases, the temperature- development rate relationship was sigmoid rather than linear, and the rate of development decreased significantly at above optimum temperatures. The thermal constant and lower developmental threshold for eggs obtained by linear regression was 41.9 D° above 7.9°C.

In studies on the larval and pupal stages, it was found that complete development could occur at temperatures between 12.5 and 35°C. The ability of pupae to survive short periods of exposure to 32.5°C increased with age. As with eggs, the larval and pupal temperature-development rate relationships were sigmoid rather than linear. Thermal constants calculated by linear regression were 186.3 D° above 7.7°C and 186.7 D° above 9.1°C for larvae and pupae, respectively.

In Italy, Girolami (1979) observed complete development of eggs at temperatures between 7.5 and 37.5°C, larvae at temperatures between 10 and 32.5°C and pupae at temperatures between 10 and 30°C, when the immature stages were reared on olives. He fitted both logistic and 3° polynomial curves to the data and found that the duration of the different stages under fluctuating temperatures were the same as those predicterd from constant temperatures.

Crovetti et al. (1982) also determined the development rates of eggs, pupae and egg-adult of *D. oleae* collected in different parts of Italy under a series of constant and fluctuating temperatures. For the egg stage, the lower thresholds and thermal constants calculated by thermal summation, linear regression and sigmoid curves were 8.2, 8.0, 5.1°C and 48.7, 52.3, 48.7 D° respectively, at constant temperatures. For the pupal stage the equivalent values were 9.5, 9.3, 5.1°C and 201.1, 196.6, 190.2 D° for constant temperatures; and 9.0, 8.7, 4.8°C and 200.1, 192.9, 213.1 D° for variable temperatures. Lower thresholds and thermal constants for the larval stage obtained by deducting the contributions of the egg and pupal stages from the overall egg-adult relationships were estimated to be 10.7, 8.5, 5.8°C and 116.6, 138.1, 128.1 D° under constant conditions and 9.1, 7.7, 4.0°C and 129.1, 144.8, 198.5 D° in fluctuating temperatures, based on the thermal summation, linear regression and sigmoid models, respectively. The thermal constants derived by this indirect method, however, are generally lower than those obtained by actual measurements of larval development rates.

Studies on the temperature-development rate relationships of eggs, larvae and pupae of *D. oleae* in conditions most closely resembling those in the field have been carried out in Crete (Neuenschwander and Michelakis, 1979) and Corfu (Fletcher and Kapatos, 1983a).

Neuenschwander and Michelakis (1979) estimated the lower thresholds and day-degree requirements of eggs and larvae by allowing females to oviposit directly into fruit in sleeve cages on trees at different times of the year, and then determining the mean duration of each stage by dissecting fruit samples. Day-degree values above several possible threshold temperatures were calculated for each experiment and the threshold that gave the smallest variance, taken over all experiments, was considered to be the true lower threshold. Based on shade temperatures recorded at the experimental site, eggs were found to require 68 D° above a threshold of 6°C, throughout the whole season. Larval requirements however, decreased from 146 D° above 10°C in October to 93 D° above 10°C in April. The apparent increase in speed of larval development from autumn to spring was attributed to the increased nutritional quality of the olives at they ripened. The three larval instars took up the same proportion of total larval development time (30, 25 and 45% respectively) at different times of the year.

Fletcher and Kapatos (1983a) calculated the lower thresholds for eggs and larvae in olives, and pupae in both fruit and in soil, at a series of constant temperatures in the laboratory and then used these thresholds to determine the number of day-degrees required to complete development when infested olives and newly formed pupae were exposed outdoors to ambient temperatures at different times of the year. The average number of day-degrees needed to complete development was found to be 46.9 D° above a threshold of 6.3°C for eggs, 209 D° above 8°C for larvae and 204 D° above 8°C for pupae. It was observed, however, that when maximum daily temperatures were higher than

Chapter 3.10 references, p. 286

30°C for any length of time, larval development slowed down and the number of day-degrees accumulated during development (around 300 D°) was considerably higher than under more optimum conditions.

3.5 The immature stages of *Dacus tryoni* (Froggatt)

Pritchard (1978) fitted a sigmoid curve to the development rates of *D. tryoni* eggs at a series of constant temperatures between 15 and 32.5°C and used it to calculated egg duration in the field based on temperature records obtained by implanting thermocouples beneath the skin of shaded fruit. The lower threshold predicted by the sigmoid model was 4.6°C but this was not verified experimentally, and no thermal constant was calculated.

No temperature-development rate models relating specifically to eggs and larvae have been published, although Meats (1981) used data on development rates from egg to adult at constant temperatures published by Bateman (1967), plus other data, to develop equations for predicting generation times at temperatures above and below 25°C (O'Loughlin et al., 1984). Meats (1984) also examined the effects of different constant and fluctuating temperatures on the survival rates of the different stages and found that the highest temperatures tolerable in fluctuating conditions were higher than those tolerable in constant conditions. Regimes near the tolerable threshold experienced during one life history stage adversely affected survival in the next stage.

3.6 Other factors influencing the development rates of the immature stages

A number of other factors, in addition to temperature, can affect the development rates of the immature stages. In the case of eggs, for example, Tsitsipis and Abatzis (1980) observed that the hatching time of eggs of *D. oleae* incubated at 20°C increased from 84.3 h when kept at 100% RH to 101.6 h when kept at 75% RH for 12 h immediately after oviposition, before transfer to 100% RH. Studies on *Dacus musae* (Tryon) indicated that the incubation time of eggs in bananas maintained at temperatures between 25–31°C varied from 3–11 days depending upon the ripeness of the fruit (Smith, 1977).

Duration of the larval stage can be affected by the type of host. In *D. dorsalis* for example (Ibrahim and Rahman, 1982), larval development was longest (11.5 days) and mortality lowest (19%) in carambola (*Averrhoa carambola*) and shortest (8.9 days) with greatest mortality (70%) in pineapple, among a range of fruit studied in Malaya. The reduction in egg plus larval stage from 11.5 days in carambola to 8.9 days in pineapple resulted in smaller pupae which also had a shorter development time (8.5 days in pineapple compared with 10.1 days in carambola). Although part of the observed difference in development times might be caused by the high mortality affecting the distribution of individuals completing the stage, with only rapid developing larvae surviving in poor quality hosts, part of the effect appears to be due to actual differences in development rate. A similar decrease in development time of larvae and pupae accompanied by increased mortality was observed when different densities of larvae were set up on 20 g pieces of papaya flesh (*Carica papaya*). Egg plus larval and pupal development times decreased from 9.0 and 12.7 days at a density of 0.5 larvae per g to 6.6 and 8.5 days, respectively, at a density of 2.5 larvae per g.

Conversely, in studies on *D. oleae* (Neuenschwander and Michelakis, 1979), larval development rates appeared to increase (based on the decrease in accumulated day-degrees) as the olive fruit ripened. In studies on *C. capitata* also (Carey, 1984), larval development times at 25°C increased from slightly more than one week in favourable hosts like mango and tomato to more than 3 weeks in quinces. The pupal stage was also extended a day or two when larvae were reared on less favourable hosts.

4 TEMPERATURE-DEVELOPMENT RATE RELATIONSHIPS OF *RHAGOLETIS* spp.

Most studies on temperature-development rate relationships in *Rhagoletis* have investigated the relationship between post-diapause pupal development and temperature, so that predictions can be made about the beginning of adult emergence for pest management purposes. The species studied include *Rhagoletis pomonella* (Walsh), *Rhagoletis cerasi* (Linnaeus), *Rhagoletis indifferens* Curran, *Rhagoletis fausta* (Osten Sacken) and *Rhagoletis cingulata* (Loew).

In all these holoartic species, the pupae enter a diapause which is terminated by exposure to a period of low temperature during the winter months. Post-diapause development then commences with the onset of warmer weather in the spring. However, diapause can be maintained by exposure to very low temperatures or extended periods of high temperature, depending upon the stage of diapause at which the treatment is received (Neilson, 1962; Haisch and Chwala, 1979; Vankirk and AliNiazee, 1982). As a result, some pupae spend two or more years in diapause.

4.1 Post-diapause development in *R. pomonella*

A number of day-degree models have been developed for the prediction of *R. pomonella* emergence in different parts of North America, including Vineland and Guelph, Ontario (Trottier, 1975; Reid and Laing, 1976; Laing and Heraty, 1984); Geneva, Highland and Wayne Country, New York State (Reissig et al., 1979) and the Saint John River Valley, New Brunswick (Maxwell and Parsons, 1969).

Using development rate data on pupae from field collected apples, that were stored at 1°C for five months and then incubated at a series of constant temperatures between 2 and 27°C in long-day (16:8, LD) conditions, Reissig et al. (1979) estimated the lower threshold for pupal development to be 6.4°C, using linear regression. This threshold was then used to determine the accumulated day-degrees (thermal units) to first emergence for a series of starting dates between 1st January and 1st June, based on daily maximum and minimum temperatures and first records of occurrence of adult captures on sticky traps over a 25-year period, in three regions of New York State.

At Geneva and in the Hudson Valley, the most satisfactory predictions were given by accumulating D° from 1st March each year. However, the use of a threshold within ± 5°C of the calculated value gave only slightly more variable predictions. The average D° accumulated between 1st March and first emergence was slightly lower in the Hudson Valley (614 ± 53) than in Geneva (641 ± 48).

The average annual deviation between observed first emergence and that predicted by using the average D° accumulations was 3.5 days for Geneva and 5.7 days for Highland. When the day-degree model developed from the Geneva data was used to predict adult emergences in 5 locations in Wayne Country, observed and predicted emergence dates varied by an average of 3.5 days and in most cases the observed date was included in the 99% confidence interval of this value (593–690 D°).

Chapter 3.10 references, p. 286

Laing and Heraty (1984) calculated the day-degrees accumulated above thresholds of 8.7 and 6.4°C after the 1st March for 1st, 10%, and 50% emergence of *R. pomonella* at Guelph, Ontario on the basis of adult captures in ground-emergence traps and sticky traps and a sine wave temperature trace reconstructed from maximum and minimum air temperatures (Allen, 1976).

Both thresholds resulted in similar levels of deviation from the mean day-degree values (equivalent on average to 5–4 days variation in prediction of emergence date). First emergence and 50% emergence occurred at 478 and 809 D° above 8.7°C, and 638 and 1021 D° above 6.4°C, respectively.

4.2 Post-diapause development in *R. indifferens*

AliNiazee (1976) developed a thermal summation model for predicting adult emergence of *R. indifferens* based on D° accumulated above an arbitrary threshold of 5°C, starting from the 1st March each year. This approach was later utilized in a computerized phenology model (AliNiazee, 1979) which predicted the start of various events in the fly's life cycle on the basis of temperature summation, including emergence (462 D°), oviposition (541 D°), egg hatch (594 D°) and pupation (795 D°). The model was developed from data collected over a 3-year period (1973–1975) in Benton County, and validated during 1976–1978 in studies carried out in two orchards in Albany, Oregon. Further assessment of the predictions relating to first emergence were carried out in a cherry growing area in Hood River Valley in a different part of Oregon. The predictions for first fly emergence over the three years were found to be within 3 days of the actual event and the actual day-degrees accumulated fell within 5% of the estimated value. Predictions of 10, 50 and 99% emergence were also reasonably accurate.

To improve the predictions of the adult emergence model, a more accurate lower temperature threshold was calculated using data on post-diapause development rates of pupae held at a series of constant temperatures between 13 and 25°C, after storage at 3°C for 200 days (Van Kirk and AliNiazee, 1981). Regression analysis using all data points gave a lower threshold of 10.2°C and when only the intermediate points were used the calculated threshold was 13.3°C. A third technique which weighted the points obtained from constant temperature treatments by their frequency of occurrence in the field gave a lower threshold of 8.3°C.

When D° summations above each of these three thresholds from 1st March up to 1st and 50% emergence of flies were compared using field data for emergences and soil temperatures collected over a five year period at Albany, the 8.3°C threshold gave the least mean deviation in accumulated D°.

More recently, Stark and AliNiazee (1982) constructed a non-linear model for predicting the post-diapause development of *R. indifferens,* by fitting an exponential curve to development rates between 13 and 22°C and a quadratic equation to development rates at temperatures between 19 and 28°C. Development rates were set at zero for temperatures below 9 and above 30°C. This model served as the basis for an algorithm which can predict the date of 1st, 10, 25 and 50% emergence on the basis of soil temperature profiles at different depths in cultivated and uncultivated ground. Comparison of the model's predictions with observed events in an orchard at Albany over a five year period (1973–1977) indicated that, on average, errors in prediction did not exceed 2 days.

4.3 Post-diapause development of *R. cerasi*

A number of temperature summation models have been proposed for *R. cerasi* by researchers in different parts of Europe including Switzerland (Boller, 1964), Austria (Muller, 1970) and Poland (Leski, 1963).

Field studies carried out in Switzerland by Boller (1964) indicated that adult emergence began when pupae had accumulated 430 D° above a threshold of 5°C starting from the 1st January, based on soil temperatures at 5 cm depth. Later studies (Boller and Remund, 1983), in which adult emergences were monitored after pupae had been buried in different cherry growing regions of Switzerland, indicated that the same model enabled accurate predictions to be made about adult emergences in different climatic zones.

After Boller and Bush (1974) reported that flies from different parts of Europe had different emergence times when kept at 25°C after storage at 2°C for 180 days, Baker and Miller (1978) studied the post diapause development of *R. cerasi* pupae from four different European countries: Italy (Rome), Czechoslavakia (Ivanka), Austria (Lackendorf) and Switzerland (Hellikon). The pupae were stored for 155 days at 4°C and then incubated in total darkness in a series of constant temperatures between 8 and 27.5°C or in four cycling temperatures between 2.1–15.3°C and 15.9–31.6°C. Using an average lower threshold of 6.8°C, which was the average of the lower thresholds (calculated by linear regression) from the development rate data obtained at different temperatures, the day-degrees accumulated to 1st, 10, 50 and 90% emergence were calculated for both males and females.

The number of day-degrees accumulated to 1st and 50% emergence by female pupae from Switzerland, Austria, Italy and Czechoslovakia were 316, 366; 322, 362; 344, 387 and 317, 350 D° respectively. The values for males were usually 10–25 D° higher. Although, the pupae from Italy were the last from which adults started to emerge at all temperatures it was not certain if this was due to differences in post-diapause development or differences in the timing of diapause termination.

Using data obtained in Switzerland, it was found that Boller's model (1964) (430 D° above 5°C) and their model (321 D° above 6.8°C) gave very similar predictions regarding adult emergences.

4.4 Post-diapause development of *R. cingulata* and *R. fausta*

An examination of field data involving mixed populations of *R. fausta* and *R. cingulata* over a 25 year period in Erie County, Pennsylvania, indicated that the number of D° above a threshold of 4.4°C accumulated from the 1st March to the start of adult emergence ranged from 918 to 1234° with a mean of 950 D° (Jubb and Cox, 1974).

The mean variation between predicted and observed dates of emergence for the 25-year observation period using this model (i.e. 950 D° above 4.4°C from 1st March) was 3.5 days. In years in which separate records of the two species were obtained, *R. fausta* emerged on average 5 days before *R. cingulata*.

4.5 Factors affecting post-diapause development

A number of factors can influence post-diapause pupal development rates of *Rhagoletis* spp. One of the most important of these is the level and duration of cold exposure, as this determines when diapause is completed (Vankirk and AliNiazee, 1982). In *R. pomonella*, differences between years in the number of D° required to complete pupal development were shown to be negatively correlated with the length of the winter cold period (i.e. weeks when air temperature was below 1°C, or when there was snow cover) indicating that the cold period had an effect either on the number of D° required to complete development or on the threshold temperature for development. The duration of the cold period required to terminate the inverse relationship with D° was found to be

Chapter 3.10 references, p. 286

21 weeks for *R. cerasi* (Baker and Miller, 1978), 28 weeks for *R. indifferens* (Brown and AliNiazee, 1977) and 30–40 weeks for *R. pomonella* (Neilson, 1962).

Other variables reported to affect emergence patterns include rainfall, amount of sunlight, soil type and the larval host (Glass, 1960; Dean and Chapman, 1973; Jubb and Cox, 1974; Neilson, 1976).

5 MATURATION RATES OF ADULT TEPHRITIDS

Despite the need to establish the temperature-maturation rate relationship of females in order to predict generation times and the yearly phenology of fruit fly populations, both in the laboratory and in the field, very few detailed studies on the relationship between egg maturation and temperature have been carried out. However, a number of workers have determined the lower thresholds and thermal constants for the female pre-oviposition periods of certain species under constant conditions.

5.1 Estimates based on the pre-oviposition period

Keck (1951) determined the pre-oviposition period of female *D. cucurbitae* in Hawaii at a series of constant temperatures and found that no eggs were deposited at or below 12.8 and at 37.8°C females did not survive long enough to reach maturity. He did not fit any mathematical model to the data but linear regression analysis of the data points between 15.6 and 32.2°C gives a thermal constant of 218 D° above 10.4°C. In studies carried out in Okinawa (Okumura et al., 1981) using constant temperatures between 13 and 28°C, the thermal constant for the pre-oviposition period of *D. cucurbitae* was estimated to be 183.02 D° above a threshold temperature of 12.08°C.

In a similar study with *D. dorsalis* (Saeki et al., 1980) the thermal constant for the pre-ovipostion period was estimated to be 148.48 D° above a threshold of 15.13°C. Based on the results published by Bodenheimer (1951), Tassan et al. (1983) used a thermal constant of 44.2 D° above a threshold of 16.6°C for the pre-oviposition of *C. capitata* in their studies associated with the 1980–1982 eradication programme in California.

On the basis of field observations, AliNiazee (1979) concluded that in *R. indifferens* the average number of day-degrees accumulated between first emergence and the start of egg-laying was 79 D° above 5°C (the arbitrary threshold chosen for studies on post-diapause development) and Laing the Heraty (1984) estimated that in *R. pomonella*, females needed to accumulate an average of 103 D° above a threshold of 8.7°C (or 113 D° above 6.4°C) to become gravid after emergence.

5.2 Estimates based on maturation of the ovaries

The only detailed studies on the relationship between temperature and ovarian maturation in tephritids have been carried out on *D. tryoni* and *D. oleae*.

Pritchard (1970) investigated the relationship between temperature and ovarian maturation in *D. tryoni* both in the laboratory and in the field. When females were kept at a series of constant temperatures between 15 and 30°C, and ovarian maturation rates determined by regular dissection of samples, mean maturation rates plotted against temperature showed the typical sigmoid relationship up to 25°C and a decrease in maturation rates at higher than optimum temperatures. Extrapolation of the development curve back to the *x*-axis indicated a lower developmental threshold of 13.5°C. Using this threshold, the number of day-degrees required for 50% of the population to mature varied between 51 D° at 15°C and 107 D° at 30°C. In the medial zone (17.5–25°C) an average of 64 D° was accumulated.

When the maturation times of females placed in large field cages at different times of the year were determined, an average of 110 D° was accumulated up to the point when 50% of the population had mature ovaries. Overall, the observed times of development in the field cages were 36% longer than predicted from the development rates observed under constant temperatures in the laboratory.

Studies on overwintering populations of *D. tryoni* south of Sydney indicated that during the winter months, when temperatures were too low for ovarian development to proceed, the developing oocytes were resorbed so that the development stage of the ovaries actually decreased as the winter progressed (Fletcher, 1975).

Fletcher and Kapatos (1983 b) studied the temperature-maturation rate relationship of *D. oleae* females by placing newly emerged groups of flies in outdoor cages at different times of the year and determining the time when 50% of the females reached maturity by dissecting samples at regular intervals.

Females provided with sugar, water, protein and olives matured faster at all times of the year than comparable groups provided with sugar, water and protein; sugar, water and olives or sugar and water only. Day-degree requirements for individual treatments were fairly uniform for the greater part of the year, with mean values of 57.9, 71.7, 91.2 and 99.6 D° above a threshold of 12°C for females fed sugar, water, protein and olives; sugar, water and protein; sugar, water and olives, and sugar and water only, respectively.

The number of day-degrees accumulated increased significantly during the summer months when higher than optimum temperatures reduced maturation rates. Because of this effect, the overall relationships between maturation rates and temperature for the different treatments were best described by polynomial regression equations. The lower developmental threshold was similar for all treatments but the upper thresholds ranged from 23°C for females fed only sugar and water, up to 29.3°C for females provided with sugar, water, protein and olives. Above and below these development thresholds, oocyte resorption and regression in the development stage of the ovaries occurred (Fletcher et al., 1978; Fletcher and Kapatos, 1983b).

5.3 Other factors affecting rates of ovarian development

Ovarian maturation rates are often more markedly affected by factors other than temperature than the development rates of the immature stages. This is because adults can live for long periods above the upper and lower temperature thresholds for ovarian development (Fletcher, 1975; Fletcher and Kapatos, 1983b) and factors other than temperature, e.g. availability of proteinaceous food or larval hosts, can have a marked effect upon development rates. In laboratory experiments, the degree of crowding and the presence of males have also been shown to influence development rates (Pritchard, 1970).

6 CONCLUSIONS

As in most other non-diapausing insects, the temperature-development time relationships for both the development of the immature stages and ovarian maturation of females are J-shaped. When the resulting development or maturation rates are plotted against temperature the relationship is sigmoid up to some upper limit above which there is a marked decrease in the rates.

Chapter 3.10 references, p. 286

Because of the non-linear shape of the overall relationship, estimation of the lower thermal threshold by linear regression is sensitive to the range of temperatures used in fitting the curve. As a result, the lower developmental thresholds and thermal constants calculated by different workers for the same species may vary over quite a wide range. In most cases however, the overall relationships are quite similar, particularly when other factors which can influence development rates are taken into account.

It appears therefore, that differences in development rates of different geographical strains of the same species are relatively minor when they occur at all. There are, however, interesting differences between species. When *C. capitata, D. cucurbitae* and *D. dorsalis* were reared in comparable conditions by Vargas et al. (1984), they observed that *D. cucurbitae* had the shortest egg stage, and longest adult pre-oviposition period. *C. capitata* had the longest egg stage and shortest adult pre-oviposition period and *D. dorsalis* was intermediate between the two. Due to the very short pre-oviposition period, *C. capitata* had the shortest generation time of the three species.

Both linear (day-degree accumulation) and non-linear models have been used relatively successfully to predict the start of adult emergence in *Rhagoletis* sp. and the yearly phenology of multivoltine species (Stark and AliNiazee, 1982; Laing and Heraty, 1984; Boller and Remund, 1983; O'Lough'in et al., 1984; Fletcher and Comins, 1985), confirming their usefulness for pest management purposes.

7 REFERENCES

AliNiazee, M.T., 1976. Thermal unit requirements for determining adult emergence of the western cherry fruit fly (Diptera: Tephritidae) in the Willamette Valley of Oregon. Environmental Entomology, 5: 397–402.

AliNiazee, M.T., 1979. A computerized phenology model for predicting biological events of *Rhagoletis indifferens* (Diptera: Tephritidae). The Canadian Entomologist, 111: 1101–1109.

Allen, J.C., 1976. A modified sine wave method for calculating degree-days. Environmental Entomology, 5: 385–396.

Baker, C.R.B. and Miller, G.W., 1978. The effect of temperature on the post-diapause development of four geographical populations of the European cherry fly (*Rhagoletis cerasi*). Entomologia Experimentalis et Applicata, 23: 1–13.

Bateman, M.A., 1967. Adaptations to temperature in geographic races of the Queensland fruit fly, *Dacus (Strumeta) tryoni* (Froggatt). Australian Journal of Zoology, 15: 1141–1161.

Bodenheimer, F.S., 1925. On predicting the development cycles of insects. I *Ceratitis capitata* Wied. Bulletin of the Societe Royale Entomologique d'Egypte, 1924: 149–157.

Bodenheimer, F.S., 1951. Citrus Entomology in the Middle East, Junk, The Hague, 663 pp.

Boller, E.F., 1964. Auftreten der Kirschenfliege (*Rhagoletis cerasi* L.) und Prognose mittels Bodentemperaturen im Jahre 1963. Schweizerische Zeitschrift fur Obst- und Weinbau, 73: 53–58.

Boller, E.F. and Bush, G.L., 1974. Evidence for genetic variation in populations of the European cherry fly, *Rhagoletis cerasi* (Diptera: Tephritidae) based on physiological parameters and hybridization experiments. Entomologia Experimentalis et Applicata, 17: 279–293.

Boller, E.F. and Remund, U., 1983. Dix annees d'utilisation des sommes de temperatures journalieres pour la prevision des vols de *Rhagoletis cerasi* et *d'Eupoecilia ambiguella* au nord de la Suisse. Bulletin of the European Organization for Plant protection, 13: 209–212.

Brown, R.D. and AliNiazee, M.T., 1977. Synchronization of adult emergence of the western cherry fly in the laboratory. Annals of the Entomological Society of America, 70: 678–680.

Carey, J.R., 1984. Host specific demographic studies of the Mediterranean fruit fly *Ceratitis capitata*. Ecological Entomology, 9: 261–270.

Crovetti, A., Quaglia, F., Loi, G., Rossi, E., Malafatti, P., Chesi, F., Conti, B., Belcari, A., Raspi, A. and Paparatti, B., 1982. Influenza di temperatura e umidita' sullo sviluppo degli stadi preimaginali di *Dacus oleae* (Gmelin). Frustula Entomologia, 5: 133–166.

Crovetti, A., Conti, B. and Delrio, G., 1986. Effect of abiotic factors on *Ceratitis capitata* (Wied.) (Diptera: Tephritidae). 2. Pupal development under constant temperatures. In: R. Cavalloro (Editor), Fruit Flies of Economic Importance 84. Balkema, Rotterdam, pp. 141–149.

Dallwitz, M.J. and Higgins, J.P., 1978. User's guide to Devar. A computer program for estimating development rate as a function of temperature. CSIRO, Division of Entomology Report No. 2, 23 pp.

Davidson, J., 1944. On the relationship between temperature and rate of development of insects at constant temperatures. Journal of Animal Ecology, 13: 26–38.

Dean, R.W. and Chapman, P.J., 1973. Bionomics of the apple maggot in eastern New York. Search, 3: 1–64.

Delrio, G., Conti, B. and Crovetti, A., 1986. Effect of abiotic factors on *Ceratitis capitata* (Wied.) (Diptera: Tephritidae). 1. Egg development under constant temperatures. In: R. Cavalloro (Editor), Fruit Flies of Economic Importance 84. Balkema, Rotterdam, pp. 133–141.

Fletcher, B.S., 1975. Temperature regulated changes in the ovaries of over-wintering females of the Queensland fruit fly, *Dacus tryoni*. Australian Journal of Zoology, 23: 91–102.

Fletcher, B.S. and Kapatos, E.T., 1983a. An evaluation of different temperature-development rate models for predicting the phenology of the olive fly, *Dacus oleae*. In: R. Cavalloro (Editor), Fruit Flies of Economic Importance. Balkema, Rotterdam, pp. 321–329.

Fletcher, B.S and Kapatos, E.T., 1983b. The influence of temperature, diet and olive fruits on the maturation rates of female olive flies at different times of the year. Entomologia Experimentalis et Applicata, 33: 244–252.

Fletcher, B.S. and Comins, H., 1985. The development and use of a computer simulation model to study the population dynamics of *Dacus oleae* and other fruit flies. Atti XIV Congresso Nazionale di Entomolgia Italiana, Palermo-Erice-Bagheria, pp. 561–575.

Fletcher, B.S., Pappas, S. and Kapatos, E., 1978. Changes in the ovaries of olive flies during the summer and their relationship to temperature, humidity and fruit availability. Ecological Entomology, 3: 99–107

Girolami, V., 1979. Studi biologici e demoecologici sul *Dacus oleae* (Gmelin) 1. Influenza dei fattori ambientali abiotici sull'adulto a sugli stadi preimmaginali. Redia, 62: 147–191.

Glass, E.H., 1960. Apple maggot fly emergence in western New York. New York State Agricultural Experiment Station Bulletin, 789, 29 pp.

Haisch, A. and Chwala, D., 1979. Über den Einfluss wechselnder Temperaturen auf den Diapause-Ablauf der Europaischen Kirschfruchtfliege, *Rhagoletis cerasi* (Diptera: Trypetidae). Entomologia Generalis, 5: 231–239.

Hsn, E., and H. Hsu, S., 1973. Biological studies on the oriental fruit fly (*Dacus dorsalis* Hendel) II. The biological effects of temperature and humidity on oriental fruit fly (*Dacus dorsalis* Hendel). Plant Protection Bulletin (Taiwan), 15: 59–86. (In Chinese with English summary).

Ibrahim, A.G. and Rahman, M.D.A., 1982. Laboratory studies on the effect of selected tropical fruit on the larvae of *Dacus dorsalis* Hendel. Pertanika, 5: 90–94.

Jubb, J.L. Jr. and Cox, J.A., 1974. Seasonal emergence of two cherry fruit fly species in Erie County, Pennsylvania: 25-year summary. Journal of Economic Entomology, 67: 613–615.

Kawai, A. and Yosnihara, T., 1981. Comparison of development and reproduction in two geographical populations of the melon fly, *Dacus cucurbitae* Coquillett. Bulletin of the Vegetable and Oriental Crops Research Station, Series C, 5; 63–73 (In Japanese with English abstract).

Keck, C.B., 1951. Effect of temperature on development and activity of the melon fly. Journal of Economic Entomology, 44: 1001–1003.

Laing, J.E. and Heraty, J.M., 1984. The use of the degree-days to predict emergence of the apple maggot, *Rhagoletis pomonella* (Diptera: Tephritidae), in Ontario. The Canadian Entomologist, 116: 1123–1129.

Leski, R., 1963. Studia na biologia: ekologica nasionnicy tresniowki *Rhagoletis cerasi* L. (Dipt., Trypetidae). Polskie Pismo Entomologiczne, B. 3-4: 193–210.

Manikas, G., 1974. A contribution to the biology and ecology of *Dacus oleae* Gmel. (Diptera: Tephritidae). Ph.D. Thesis, College of Agriculture, Athens, 75 pp. (In Greek with English summary).

Maxwell, C.W. and Parsons, E.C., 1969. Relationships between hour-degree F soil temperature summations and apple maggot adult emergence. Journal of Economic Entomology, 62: 1310–1313.

McBride, O.C., 1935. Response of the Mediterranean fruit fly to its environmental factors. Proceedings of the Hawaiian Entomological Society, 9: 99–108.

Meats, A., 1981. The bioclimatic potential of the Queensland fruit fly, *Dacus tryoni*, in Australia. Proceedings of the Ecological Society of Australia, 11: 151–161.

Meats, A., 1984. Thermal constraints to successful development of the Queensland fruit fly in regimes of constant and fluctuating temperature. Entomologia Experimentalis et Applicata, 36: 55–59.

Messenger, P.S. and Flitters, N.E., 1954. Bioclimatic studies on three species of fruit flies in Hawaii. Journal of Economic Entomology, 47: 756–765.

Messenger, P.S. and Flitters, N.E., 1958. Effect of constant temperature environments on the egg stage of three species of Hawaiian fruit flies. Annals of the Entomological Society of America, 51: 109–119.

Messenger, P.S. and Flitters, N.E., 1959. Effect of variable temperature environments on egg development of three species of fruit flies. Annals of the Entomological Society of America, 52: 191–204.

Moore, I., 1960. A contribution to the ecology of the olive fly, *Dacus oleae* (Gmel.). Israel Agricultural Research Station Special Bulletin, 26: 53 pp.

Muller, W., 1970. Agrarmeteorologische Untersuchungen über des Erstauftreten der Kirschfliege (*Rhagoletis cerasi* L.) in Österreich, Pflanzenschutz–Berichte, 41: 193–210.

Nakamori, H., Soemori, H. and Kakinohana, H., 1978. Effect of temperature on pupal development of the melon fly, *Dacus cucurbitae* Coq. and a method to control timing of adult emergence. Japanese Journal of Applied Entomology and Zoology, 22: 56–59.

Neilson, W.T.A., 1962. Effects of temperature on development of overwintering pupae of the apple maggot, *Rhagoletis pomonella* (Walsh). The Canadian Entomologist, 94: 924–928.

Neilson, W.T.A., 1976. The apple maggot (Diptera: Tephritidae) in Nova Scotia. The Canadian Entomologist, 108: 885–892.

Neuenschwander, P. and Michelakis, S., 1979. Determination of the lower thermal thresholds and day-degree requirements for eggs and larvae of *Dacus oleae* (Gmel.) (Diptera: Tephritidae) under field conditions in Crete, Greece. Bulletin de la Société Entomologique Suisse, 52: 57–74.

Okumura, M., Ide, T. and Takagi, S., 1981. Studies of the effect of temperature on the development of the melon fly, *Dacus cucurbitae* Coquillett. Research Bulletin of the Plant Protection Service of Japan, 17: 51–56. (In Japanese with English abstract).

O'Lough'in, G.T., East, R.A. and Meats, A., 1984. Survival, development rates and generation times of the Queensland fruit fly, *Dacus tryoni*, in a marginally favourable climate: experiments in Victoria. Australian Journal of Zoology, 32: 353–361.

Pritchard, G., 1970. The ecology of a natural population of the Queensland fruit fly, *Dacus tryoni*. III. The maturation of female flies in relation to temperature. Australian Journal of Zoology, 18: 77–89.

Pritchard, G., 1978. The estimation of natality in a fruit-infesting insect (Diptera: Tephritidae). Canadian Journal of Zoology, 56: 75–79.

Pros, R.J., 1983. Importance of ecological studies for application S.I.T. against *Ceratitis capitata* Wied. In: R. Cavalloro (Editor), Fruit Flies of Economic Importance. Balkema, Rotterdam, pp. 68–73.

Reid, J.A.K. and Laing, J.E., 1976. Developmental threshold and degree-days to adult emergence of the apple maggot *Rhagoletis pomonella* (Walsh) collected in Ontario. Proceedings of the Entomological Society of Ontario, 107: 19–22.

Reissig, W.H., Barnard, J., Weires, R.W., Glass, E.H. and Dean, R.W., 1979. Prediction of apple maggot fly emergence from thermal unit accumulation. Environmental Entomology, 8: 51–54.

Saeki, S., Katayama, M. and Okumura, M., 1980. Effect of temperature upon the development of the oriental fruit fly and its possible distribution in the mainland of Japan. Research Bulletin of the Plant Protection Service of Japan, 16: 73–76. (In Japanese with English abstract).

Shibata, K., 1936. Notes on the autoecology of some fruit flies. III. On the velocity and threshold of development of eggs of *Chaetodacus cucurbitae* Coquillet (melon fly). Journal of the Society of Tropical Agriculture of Taihoku Imperial University, 8: 373–380.

Shoukry, A. and Hafez, M., 1979. Studies on the biology of the Mediterranean fruit fly *Ceratitis capitata*. Entomologia Experimentalis et Applicata, 26: 33–39.

Smith, E.S.C., 1977. Studies on the biology and commodity control of the banana fly, *Dacus musae* (Tryon) in Papua New Guinea. Papua New Guinea Agricultural Journal, 28: 47–56.

Stark, S.B. and AliNiazee, M.T., 1982. Model of post-diapause development in the western cherry fruit fly. Environmental Entomology, 11: 471–474.

Tassan, R.L., Hagen, K.S., Cheng, A., Palmer, T.K., Feliciano, G. and Blough, T.L., 1983. Mediterranean fruit fly life cycle estimations for the California eradication program. In: R. Cavalloro (Editor), Fruit Flies of Economic Importance. Balkema, Rotterdam, pp. 564–570.

Trottier, R.J., 1975. A warning system for pests in apple orchards. Canadian Agriculture, 95: 1154–1159.

Tsitsipis, J.A., 1977. Effect of constant temperatures on the eggs of the olive fly, *Dacus oleae* (Diptera: Tephritidae). Annales de Zoologie — Ecologie Animale, 9: 133–139.

Tsitsipis, J.A., 1980. Effect of constant temperatures on larval and pupal development of olive fruit flies reared on artificial diet. Environmental Entomology, 9: 764–768.

Tsitsipis, J.A. and Abatzis, C., 1980. Relative humidity effects at 20°C, on eggs of the olive fly, *Dacus oleae* (Diptera: Tephritidae), reared on artificial diet. Entomologia Experimentalis et Applicata, 28: 92–99.

Tzanakis, M.E., Economopoulos, A.P. and Tsitsipis, J.A., 1968. Artificial rearing of the olive fly. Progress report. In: Radiation, Radioisotopes and Rearing Methods in the Control of Insect Pests. International Atomic Energy Agency, Vienna, pp. 123–130.

Vankirk, J.R. and AliNiazee, M.T., 1981. Determining low-temperature threshold for pupal development of the western cherry fruit fly for use in phenology models. Environmental Entomology, 10: 969–971.

Vankirk, J.R. and AliNiazee, M.T., 1982. Diapause development in the western cherry fruit fly, *Rhagoletis indifferens* Curran (Diptera: Tephritidae). Zeitschrift für Angewandte Entomologie, 93: 440–445.

Vargas, R.I., Miyashita, P. and Nishida, T., 1984. Life history and demographic parameters of three laboratory-reared tephritids. Annals of the Entomological Society of America, 7: 651–656.

PART 4

BEHAVIOUR

Chapter 4.1 Fruit Fly Foraging Behavior

R.J. PROKOPY and B.D. ROITBERG

1 INTRODUCTION

Over the past decade, there has been growing emphasis on detailed quantitative analysis of the behavioral repertoire of insects within the framework of behavioral ecology. This framework is particularly useful because it calls attention to the manner in which ecological factors affect the contribution of behavior to insect survival and reproductive success. One aspect of insect behavioral ecology presently of widespread interest is resource foraging behavior. A central issue in foraging behavior, as proposed in Hassel and Southwood (1978), Kamil and Sargent (1981), Krebs and Davies (1984), Pyke (1984), and elsewhere, concerns how an individual adjusts its foraging activities in response to the characteristics and distribution of its potential resources. Changes in behavior as a function of the spatial and temporal distribution of resources may affect foraging efficiency, and ultimately fitness. A thorough understanding of foraging behavior requires integration of mechanistic approaches to behavior analysis which accentuate proximal causation with evolutionary-ecological approaches that accentuate adaptive significance.

Until rather recently, most insect foraging behavior investigations have been focused on assessing which resource patch types a forager visits, which particular resources within a patch are utilized, how long a forager remains in a patch, and how a forager moves efficiently between patches (Pyke et al., 1977). Currently, three additional questions posed by foraging behaviorists are receiving increased attention: (a) what is the process by which a forager samples resources to arrive at an average value for previous and present locales, (b) how does a forager go about resolving tradeoffs in attempting to satisfy all types of resource requirements, and (c) should the forager aim at maximizing efficient use of time and energy in acquiring the benefits of a resource, should it aim at minimizing variation in amount or quality of acquired resource, or should it aim at reducing risks during foraging by minimizing exposure to such potential hazards as predators or harmful abiotic factors within or between resource patches?

Providing meaningful answers to these and other questions arising out of foraging behavior analysis has proven challenging. Much of the challenge lies in gaining sufficient background information on the physiology, behavior, and ecology of the insect under study. At least six elements are desirable, if not required, in this regard: (a) precise identification of each essential resource type utilized by the insect in nature, (b) an estimate of the quantity, quality, and distribution of each resource type in the 'normal' habitat, (c) familiarity with the activity pattern of the insect in space and time in nature, (d) characterization of the resource finding behavior of the insect in nature, (e) some

---

Chapter 4.1 references, p. 304

knowledge of the sorts of competitors, natural enemies, and abiotic factors that do or could affect the insects' foraging behavior in nature, and (f) information on the roles that genetic background and physiological state might play in the insects' behavior and whether the insect is capable of learning and remembering.

It is also very helpful when studying insect foraging behavior to recognize that the insect might perceive resource-containing environments at several hierarchical levels: the habitat, the patch, and the resource item itself (Hassel and Southwood, 1978). A habitat (e.g., the margin of a forest) can be considered a collection of patches (e.g., scattered groups of trees). A patch may be identified as an aggregation of resource items (e.g., individual trees, limbs, leaves, fruit clusters, fruit). The 'rules' that govern foraging behavior at one hierarchical level may be quite different from those that govern behavior at a different hierarchical level. In some habitats stimuli from non-resource items may mask or otherwise interfere with ready detectability of resources, or even elicit repulsion from resources. In other habitats (especially agricultural croplands), non-resource items may be less abundant, thereby facilitating ready detectability of resource stimuli. Whichever, one should be aware that over evolutionary time, the structure of the resource environment has undoubtedly played a major role in shaping the foraging behavior of the insect and the manner in which the insect responds to environmental stimuli.

Some degree of qualitative insight into insect foraging behavior can be gained from investigating the distribution and movement patterns of insect populations in time and space. However, quantitative analysis of foraging behavior requires detailed tracking of the movement of individuals over time in a variety of settings that vary in quantity, quality, or distribution of resources. Conducting such analyses under the heterogeneous conditions typical of totally natural settings poses substantial difficulties. On the other hand, much more than has been done to date could be done toward this end through creating, under semi-natural conditions, prescribed resource-containing patches that mimic natural patches in structure and quality, and tracking the movements of individuals of prescribed (so far as possible) physiological state.

We believe the foraging behavior of tephritid flies can be best approached and best understood within the context of these introductory remarks. In many respects, frugivorous tephritids (the focus of this chapter) would seem highly suitable, if not ideal organisms for quantitative analysis of foraging behavior. The economic import of several tephritid species has given rise to a comparative wealth of basic knowledge of tephritid physiology, behavior, and ecology. To date, however, this knowledge has been gained almost exclusively from studies of populations of tephritids. Only in the case of *Rhagoletis pomonella* (Walsh) has a concerted attempt been made to analyze quantitatively the foraging behavior of individuals, and even here, only with respect to intra-tree foraging of females for egglaying sites.

To review all the background information acquired to date on the physiology, behavior, and ecology of populations of tephritids foraging for each essential resource type (food, water, mates, egglaying sites, shelter) would not only require a vastly greater amount of space than is available, but would also partially duplicate a recent review by Prokopy and Roitberg (1984) and would overlap too greatly with numerous other chapters of this book, particularly reviews on nutrition (Chapter 3.1), mating systems (Chapter 4.4), and dispersal and migration (Chapter 8.2, Vol. 3B). Hence, we will limit this chapter to a discussion primarily of the foraging of frugivorous females for egglaying sites. The first and largest section (2) will deal with qualitative aspects of host foraging in a variety of frugivorous tephritid species. The last and smaller section (3) will focus on quantitative aspects of intratree host fruit foraging behavior of *R. pomonella* as a case history.

2 QUALITATIVE ASPECTS OF HOST FORAGING BEHAVIOR

2.1 Identification of host plants

If one were to investigate the food foraging behavior of tephritid adults, one would be faced immediately with considerable difficulty, given current state of knowledge, in identifying precisely the types of substances serving as adult food resources in nature. These substances may involve insect honeydew, floral nectar, extra-floral nectarious secretions, sap oozing from wounds to plant structures, bird dung, insect frass, bacteria, yeasts, and very likely other food sources (Chapter 3.1). Moreover, many tephritid adults, particularly honey-dew-feeders, have been observed to lower the proboscis and apparently commence feeding at sites where, to the human eye, no food item could be identified.

Fortunately, precise identification of resources sought by females of frugivorous species for egglaying is more readily accomplished. This is particularly true of pest species, where extensive arrays of fruit have been evaluated in nature and the laboratory for ability to support larval development. Even so, reports are rather numerous of cases where substantial numbers of eggs are laid by females in fruit that are unable to support any larvae to maturity (e.g., Glasgow, 1933; Baker et al., 1944; Neilson, 1967; Greany et al., 1983; Carey, 1984). In such cases, identification of a particular plant species as not being a larval host may have little bearing on the individual's foraging behavior should the individual seek that plant species for oviposition anyway.

2.2 Determination of host plant quantity, quality, and distribution

After identification of the potential host plants of a tephritid, the next factor to be considered is the quantity and quality of hosts within a habitat and their distribution over space and time with respect to non-host plants and plants furnishing other sorts of tephritid resources. Lack of knowledge of this sort would call into question the value of even qualitative-type conclusions on foraging behavior patterns.

It is time-consuming but not otherwise difficult to map the location of each host and non-host plant in a habitat and map the location of each fruit on a host plant. What does pose inherent difficulty, however, is measuring the quality of host fruit, particularly so because what might appear to humans as fruit of equal quality at a given moment (e.g., fruit of same size, color, softness) might actually be of very unequal quality to tephritids as egglaying or larval development sites. To illustrate, Papaj and Prokopy (unpub. data) assessed the propensity of *R. pomenella* females to attempt oviposition into con-varietal apples picked from adjacent trees over successive time intervals. Even among apples picked on the same day from the same branch and having a virtually identical physical appearance to each other, substantial differences in acceptability were found. This constraint is unlikely to ease over the near future.

To our knowledge, there exists no study of tephritid fly host foraging behavior in nature that has incorporated a thorough documentation of host plant and host fruit availability in space and time. There do exist, however, several studies aimed primarily at taking a census of the distribution of populations of

---

Chapter 4.1 references, p. 304

released and/or native tephritid males and females over space and time in a habitat containing host plants that do incorporate some form of mapping the distribution of host plants and fruit in relation to non-host plants. These include investigations on: *R. pomonella* by Maxwell and Parsons (1968), Neilson (1971), and Johnson (1983); *Rhagoletis cerasi (L.)* by Boller et al. (1971); *Rhagoletis fausta* (Osten Sacken) by Prokopy (1976); *Rhagoletis mendax* Curran by Smith and Prokopy (1981); *Ceratitis capitata* (Wiedemann) by Ripley et al. (1940), Cirio and de Murtas (1974), and Vargas et al. (1983a); *Anastrepha fraterculus* (Wiedemann) by Malavasi et al. (1983); *Dacus cucurbitae* Coquillett by Nishida and Bess (1957); *Dacus tryoni* (Froggatt) by Sonleitner and Bateman (1963) and Fletcher (1973); and *Dacus dorsalis* Hendel by Vargas et al. (1983b). Researchers studying tephritid fly foraging behavior in the future would do well to pay much more careful attention to this crucial factor of foraging behavior analysis.

2.3 Activity patterns in space and time in nature

Several studies have been conducted in which odor and/or visual traps have been placed in host plants bearing fruit in a stage suitable for oviposition as well as in host plants without fruit and non-host plants. The general intent was to monitor patterns of movement of released or native flies through periodic counts of fly captures in the traps. Most studies of this sort have involved use of traps baited with trimedlure, methyleugenol, or cuelure, which capture largely or exclusively males of *Ceratitis* and *Dacus* species. The behavior of host-foraging females cannot be inferred accurately from the behavior of males, however.

From the more limited number of studies that employed traps that captured females, one principal conclusion relevant to female foraging behavior can be drawn. Females with mature ovaries tend to remain on or very near fruiting host plants so long as the fruit is acceptable for egglaying. If the plants are non-hosts or hosts with low quality fruit, mature females either arrive in low numbers and/or emigrate rather rapidly, and in some species, may fly considerable distances before finding host plants with acceptable fruit (Neilson, 1971; Boller et al., 1971; Fletcher and Kapatos, 1981; Michelakis and Neuenschwander, 1981; Drew and Hooper, 1983). In *R. pomonella* (Johnson, 1983), and probably also most tephritids, such intertree flights by females in search of fruit are diurnal.

In other investigations, insect nets rather than traps have been used to collect each sex of fly. From their analyses of net captures of *D. tryoni*, Sonleitner and Bateman (1963), Bateman and Sonleitner (1967), and Fletcher (1973) concluded, as above, that gravid females are prone to depart rapidly from host trees that no longer bear acceptable fruit and are apt to move long distances (possibly several tens of kilometers) in search of new hosts (see also Chapman, 1982). From their insect-net-capture analysis of movements of *D. cucurbitae*, Nishida and Bess (1957) determined that gravid females move from shelter sites in neighboring non-host vegetation into fields containing fruiting host plants in the morning, oviposit during the day, and return to neighboring non-host vegetation in the early evening before dark.

The most extensive efforts to census tephritid fly foraging activities in space and time in nature have involved systematic direct observations of (a) the varying location over time (the course of a single day and/or entire season) of individual flies on host and non-host plants coupled with (b) the entire repertoire of activities of randomly selected individuals over periods of several minutes during a day or season. To date, three studies of this sort have been published: on *R. fausta* by Prokopy (1976); on *R. mendax* by Smith and Prokopy

(1981); and on *A. fraterculus* by Malavasi et al. (1983). In seven other studies in nature, quantitative data of this sort have been taken on individual flies on fruiting host plants but not on flies on non-fruiting hosts or non-hosts. These include investigations on *R. pomonella* (Prokopy et al., 1972), *Rhagoletis cingulata* (Loew) (Smith, 1984), *Rhagoletis conversa* (Brèthes) (Frias et al., 1984), *Rhagoletis tabellaria* Fitch (Smith, 1985a), *Rhagoletis cornivora* Bush (Smith, 1985b), *Anastrepha suspensa* (Loew) (Burk, 1983), and *Toxotrypana curvicauda* Gerstaecker (Landolt and Hendrichs, 1983).

As one might imagine, there exists considerable diversity among these ten species in location and periodicity of foraging behavior activities. For example, oviposition occurred predominantly in the morning in *A. suspensa*, mostly in the afternoon in *R. pomonella* and *R. mendax*, and somewhat evenly throughout the day in *R. conversa*, *A. fraterculus*, and *T. curvicauda*. In some species (*R. mendax*, *R. pomonella*, *R. conversa*), there was considerable overlap in time of day of substantial egglaying and time of day of acquisition of two other essential resources: food and mates. In other species (*A. fraterculus*, *A. suspensa*, *T. curvicauda*), these activities appeared more segregated over time. Segregation of feeding, mating, and oviposition according to time of day may be more common in sub-tropical or tropical polyphagous tephritids (possibly owing to more frequent presence of other tephritid species on the same host, to greater numbers of non-tephritid competitors, or to more constraining abiotic conditions) than in temperate monophagous or oligophagous species.

Despite such diversity in foraging activity pattern among these ten species, most if not all of them do exhibit some elements in common. Foremost is the tendency of mature females as well as males present on fruiting host plants to move toward the upper parts of such plants in late afternoon or near dusk, and either to spend the night at the tops of host plants under shelter on the undersides of leaves or to seek shelter under larger-size leaves at the tops of (taller) nearby non-host plants (recall also *D. cucurbitae*). The next day, when light and temperature conditions again became favorable for foraging activity, gravid females either returned to host plants, or, if already there, fanned out from the tops of hosts to various lower sections of the plants. Such a rhythmic type of movement pattern should facilitate fly discovery of many fruit borne by the host. Three other elements also appear common: (a) except in those species where feeding occurs mainly on wounded fruit, females make relatively few visits to fruit until they are gravid; (b) oviposition attempts are far more frequent in what appear to humans as unripe or slightly ripe fruit than in ripe fruit, and (c) gravid females spend very little time, other than in oviposition-related behavior, on fruit that are acceptable for egglaying.

2.4 Host finding behavior

2.4.1 Host stimuli

Probably the major long-range stimuli guiding mature frugivorous tephritids to host plants are volatile components of ripening fruit. Fruit volatiles have been found to elicit positive responses of gravid females in *Pterandrus rosa* (Karsch) (Ripley and Hepburn, 1929), *C. capitata* (Sanders, 1968; Guerin et al., 1983), *D. dorsalis* (Tanaka, 1965), *D. cucurbitae* (Tanaka, 1965), *D. tryoni* (Pritchard, 1969), *Dacus oleae* (Gmelin) (Fiestas Ros de Ursinos et al., 1972; Girolami et al., 1983; Guerin et al., 1983); *R. pomonella* (Prokopy et al., 1973), and *R. cerasi* (Haisch and Levinson, 1980). In *D. oleae* (Orphanidis and Kalmoukas, 1970) and *R. cerasi* (Haisch and Levinson, 1980), volatiles from certain non-hosts appear repellent to gravid females. It is unknown whether volatiles from non-fruiting parts of hosts (e.g., foliage, woody tissue) affect the behavior of frugivorous tephritids, although a study by Keiser et al., (1975) suggests they

---

might. In only one species, *R. pomonella*, has the entire complex of volatiles in the 'head space' around host fruit been collected at sequential intervals during the fruit ripening process, isolated, and analyzed systematically for fly response. In an elegant study, Fein et al. (1982) (also Averill, Roelofs, and Reissig, unpub. data) employed electroantennogram, laboratory olfactometer, and field bioassay techniques to show that seven volatile esters of apple fruit elicited upwind movement of gravid females (as well as males) toward the source. None of the esters stimulated as much response when presented alone as when in the complete mixture, a phenomenon that will likely hold true for most or all other tephritids. Visual characters of host plants (e.g., color, form, size) may also play a role in fly detection of hosts from a distance (Chapter 4.2).

After female arrival on a plant, the architectural structure of the plant and chemical and physical properties of individual leaves may provide cues to the identity of the plant (potential host or non-host) and may affect the amount of time a female is willing to spend foraging on the plant in an attempt to find a fruit (Diehl et al., 1986; Prokopy et al., 1986).

After using visual and/or olfactory cues to locate an individual fruit (Chapter 4.2), gravid females then employ a variety of fruit characters (shape, size, color, surface structure, chemistry) to assess whether the fruit is acceptable for egglaying or not (Chapter 4.2; references in Prokopy, 1977, 1983; Schoonhoven, 1983; Girolami et al., 1983; Vita, Correnti, Minnelli, and Pucci, unpub. data; Seo et al., 1983; Fitt, 1983; McDonald and McIniss, 1985).

2.4.2 Distance of detection of host stimuli

While studies over the past two decades or so have given rise to at least some degree of information on the nature of plant cues affecting host finding and acceptance behavior in several tephritids, we unfortunately remain virtually totally ignorant of the distance over which plant odor and visual stimuli are first detectable by fruit foraging females. To date, the only published information on this aspect concerns *R. pomonella* and suggests that most fruit foraging females have difficulty detecting ca. 2 m tall trees at distances greater than ca. 3 m (Roitberg and Prokopy, 1982) and in detecting clusters of small fruit (ca. 1 cm diameter) at distances greater than ca. 40 cm (Roitberg, 1985a).

2.4.3 Search paths of foragers

As with distance detection of host stimuli, there exists almost no information on the nature of search paths of females foraging for hosts. This is particularly true at the interplant level of foraging behavior. Thus, we have essentially no knowledge of whether female response to plants disseminating host fruit odor involves true taxis (i.e., directed orientation to the wind that bears molecules of odor, or to the source of odor itself), or whether it involves kinetic processes (i.e., flies accumulating at the odor source purely as a result of changes in speed of locomotion or rate of turning in response to intensity of odor stimulation). Nor do we know how a female might alter its search path according to level of resource distribution (habitat, patch, plant), or according to concentration of odor or apparency of visual stimuli. The only published information that does exist on female search paths for hosts concerns intra-tree fruit foraging patterns in *R. pomonella* (see section 3).

2.5 Effects of competitors, natural enemies, and abiotic factors on foraging behavior

Several cases of possible intraspecific competition among tephritids for food, mates, or shelter sites have been reported (Prokopy, 1977; Chapter 4.4). But the most intense form of intraspecific competition, with greatest consequences to individual fitness, is that for oviposition sites. As documented in Chapter 3.5, there exist (a) a number of examples among tephritids where an overload of eggs (hence larvae) in a fruit can be highly detrimental to larval survival, and (b) a number of species in which females are able to assess the presence of conspecific eggs through detection of host marking pheromones or plant exudates deposited on the fruit surface after oviposition, or assess the presence of conspecific larvae themselves. After determining that a fruit is occupied by a conspecific egg or larva, a foraging female is likely to leave that fruit rather rapidly in search of another (see Chapter 3.5 for factors affecting female readiness to depart from fruit occupied by conspecifics). Occasionally, two or more females may occupy a fruit simultaneously (e.g., Pritchard, 1969), but this is unlikely to be an important factor affecting female foraging behavior under anything but very low fruit density or very high fly density conditions. Only in *R. pomonella* has the effect of contact with occupied fruit on female foraging behavior been examined in a somewhat natural setting (see section 3). Except for a study by Fitt (1984) showing that *D. tryoni* and *Dacus jarvisi* (Tryon) females discriminate against fruit occupied by larvae of either species, little is known about possible effects of interspecific competitors on the host foraging behavior of tephritids.

The eggs, larvae and pupae of tephritids are susceptible to a broad range of natural enemies (Chapter 8.3). Adults in nature (except for newly eclosed ones), however, appear largely unaffected by natural enemies (but see Landolt and Hendrichs, 1983). This is undoubtedly due in part to their alert avoidance, under diurnal conditions, of larger approaching objects (e.g., see Monteith, 1972) and possibly also to their strong, mimicry-like resemblance in morphology and movement pattern to jumping spiders. Eisner (1984) indicates jumping spiders are quick, hard to catch, and have poisonous mouthparts that can injure predators. He suggests jumping spider morphology is an aposematic character that serves to warn predators of potential difficulty in capturing and keeping the spider. Eisner postulates that some tephritid species may gain a similar advantage through mimicking jumping spiders. It is also possible that resemblance of jumping spiders to tephritid flies allows such spiders to approach flies very closely and consume them (Burk, 1982). This has been observed in *Ceratitis* and *Anastrepha* flies confined in field cages (Prokopy and Hendrichs, unpub. data), but was not found in an extensive study in nature by Monteith (1972) of interactions between jumping spiders and *R. pomonella* flies. Until it is firmly demonstrated that predation constitutes a significant mortality factor to tephritid adults, analysis of the effects of potential predators on female fruit foraging behaviour will be of little value. Although not predators in the sense of inflicting injury on females, males of some tephritid species may, when at high density in relation to females, cause fruit foraging females to sacrifice valuable foraging time through repeated forced-mating (Opp and Prokopy, unpub. data).

In addition to competitors and natural enemies, abiotic factors such as light, temperature, humidity, and wind might affect female host foraging behavior. As far as we know, tephritid females do not forage for fruit during darkness. We have already indicated that female movement from lower to upper parts of host plants toward evening in response to diminishing light, and re-infestation of lower plant parts the following morning in response to increasing light, is probably adaptively advantageous through maximizing fly exposure to fruit throughout the canopy, even though the principal advantage might be acquisition of better shelter at night. Changes in light intensity can thus be considered a strong organizing force in female fruit foraging behavior. When temperature is high, most tephritids seem to forage for fruit primarily on the more shaded parts of plants. When temperature is low, they tend to forage on

Chapter 4.1 references, p. 304

the more sun-lit parts, or, if also cloudy, may cease to forage altogether (Boller, 1965). In shade, temperature has been found to have a significant impact on the rate at which *R. pomonella* females search for fruit (section 3). Humidity may have comparatively little effect on fruit foraging behavior except when extremely high humidity causes oviposition to cease (Boller, 1965), or when very low humidity, coupled with strong wind and high temperature, elicits emigration of mature females from fruiting host plants to more protected sites (Ripley et al., 1940; Fletcher, personal, communication; Prokopy, unpub. data). Even under favorable temperature and humidity conditions, wind of only moderate strength (e.g. 7 km/h) can reduce the foraging efficiency of females (Aluja and Prokopy, unpub. data). After arrival on fruit, tephritid females seem more likely to oviposit on the shaded and/or leeward side of fruit than on the sunlit and/or windward side (Feron, 1957; Boller, 1965; Pritchard, 1969).

2.6 Influence of genetic background, physiological state, and previous experience on foraging behavior

Only very recently has evidence arisen in *R. pomonella* (Prokopy et al., 1982a; Diehl, 1984), and in *C. capitata* (Prokopy et al., 1984) that suggests existence of genetically based differences in host acceptance propensity among different conspecific populations of tephritid females. How such interpopulational genetic differences might affect traits of foraging other than acceptance of fruit after arrival remains to be determined.

There have been numerous anecdotal observations of tephritid females attempting oviposition in non-host fruit in nature when host fruit were unavailable. But evidently only two studies, one on *R. pomonella* (Roitberg and Prokopy, 1983) and the other on four Australian *Dacus* species (Fitt, 1983), have examined in a quantitative fashion the effect of fruit deprivation on female propensity to oviposit in normally unacceptable fruit. These studies, each under laboratory conditions, suggest females of *R. pomonella* and *D. tryoni* more readily accept normally unacceptable fruit with increasing time of fruit deprivation. Neither study, however, examined the effect of deprivation on the process of foraging behavior. Interestingly, in *D. oleae* (Fletcher et al., 1978), females resorb their ovaries and cease foraging for fruit when abiotic conditions become unfavourable. They resume ovarian development and fruit foraging when abiotic conditions again are favorable, but then only after being primed by exposure to host fruit volatiles.

There exists no evidence that prior experience of a tephritid female as a larva affects any aspect of its fruit foraging behavior. In both *R. pomonella* (Prokopy et al., 1982b, 1986) and *C. capitata* (Cooley et al., 1986), however, females that oviposit in a given fruit type several times are highly prone to reject other (normally equally acceptable) fruit types when offered. In both species, the process of conditioning to hosts has proven reversible, with memory of ovipositional experience on a given fruit type lasting several hours or days. The influence of such conditioning on traits of female foraging for hosts in a semi-natural setting is currently being examined in *R. pomonella* (Papaj and Prokopy, unpub. data).

3 QUANTITATIVE ASPECTS OF INTRATREE HOST FORAGING BEHAVIOR IN *RHAGOLETIS POMONELLA*

As noted in the Introduction, a key question most animals must address while foraging for resources is how long to remain in a patch. A precise answer to this question, based on a premise of maximization of reproductive fitness, is

likely to be too complex for most animals, including man, to cope with analytic-
ally. There exist, however, many simple rules of behavior that, if followed, can
promote highly efficient foraging. Evidence that insects and other animals
employ such rules is substantial (Krebs and Davies, 1984). Thus, a more reasonable question to ask might be 'what, if any, rules do tephritids employ to
'decide' how long to remain in a host tree while searching for fruit and what
are the consequences of such decisions?' As pointed out earlier, only in the case
of *R. pomonella* has there been an attempt to answer this question. Thus, we
will concentrate on this species as a model for the general case while acknow-
ledging major differences in foraging might exist among different species.

As do most other animals, *R. pomonella* forages for hosts that are generally
patchily distributed in both space and time. While it may pay females to invest
much time (energy) in searching within heavily fruiting, uninfested trees, it
might be very costly to continue searching in trees harboring few or poor
quality fruit. *R. pomonella* females apparently solve this potential problem
through employment of several endogenous clocks that influence their propensity for emigration from host trees.

Following arrival in trees, *R. pomonella* females apparently set an internal
clock that determines how much time (energy) will be allotted to searching for
host fruit. If no hosts are located within that time, emigration ensues. Under
semi-natural conditions, wild-type, lab-maintained flies allot ca. 4 min for
search within small semi-dwarf apple trees, devoid of fruit, before emigrating
(Roitberg et al., 1982). If, however, fruit are discovered before the allocated time
has elapsed, several outcomes are possible. If fruit are highly unacceptable on
the basis of specific biochemical attributes, flies are likely to emigrate quickly,
as evidenced by the rapid emigration of flies from apple trees that were em-
bellished with clusters of buckthorn *Rhamnus* (non-host) fruit (Roitberg et al.,
1982). If, on the other hand, a female locates a host fruit suitable for oviposition, she then resets her search allocation clock to some new higher value
following oviposition. As before, following oviposition, the female is likely to
continue searching for new hosts so long as the clock has not expired (i.e. a
Giving Up Time (GUT) clock). The exact value at which the GUT is set is
dependent upon the time that has elapsed since the previous oviposition. Short
inter-oviposition intervals cause a high setting of the GUT clock, and vice-
versa (Roitberg and Prokopy, 1984). Further, if a female encounters an egg-
infested, pheromone-marked fruit and finds that fruit unacceptable for oviposition, she then resets her GUT clock at a new, lower value. The exact value that
the clock setting is decremented is again dependent upon an elapsed-time
interval, in this case the interval being the time since the fly's previous visit to
a marked fruit. In such a manner, a female will continue to search for and visit
fruit until search time expenditure exceeds the most recent search allocation
setting. The proposed tephritid GUT model is shown in Fig. 4.1.1. Finally, there
is some circumstantial evidence females actually measure search allocation in
terms of energy expenditure, in that GUT are best estimated via the time
females spend searching actively for hosts or by the number of visits to leaves
(Roitberg and Prokopy, 1984).

An important feature of the GUT clock system is that it provides a simple
means by which flies may sensitively track changes in availability of both
suitable and unsuitable hosts. Further, under such behavioral control, flies are
likely to remain for lengthy periods in trees harboring good quality fruit and
quickly exit from trees harboring a high proportion of already-infested fruit.
An important point is that this model does not attempt to predict what a fly
'ought to do' under particular ecological conditions (i.e. predict optimal behavior). Rather, it is a mechanistic model that describes how flies in reality do
behave under particular conditions. We have found that accurate predictions

Chapter 4.1 references, p. 304

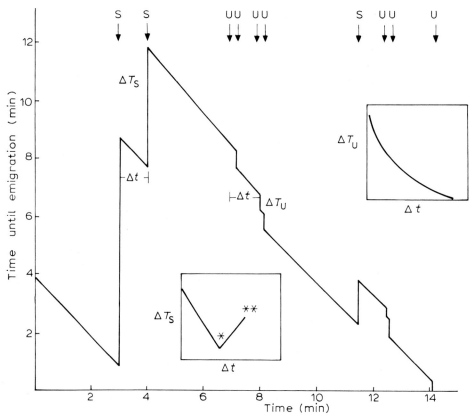

Fig. 4.1.1. Effect of visits by *R. pomonella* flies to suitable (s) and unsuitable (u) fruit on the amount of time flies allot to active search until emigration. Insert at lower left shows the effect of inter-oviposition interval on the increment to search time. * = point at which time-until-emigration is fixed at 3.8 min; ** = point at which emigration occurs. Insert at upper right shows the effect of interval between visits to unsuitable fruit on the decrement to search time. This figure is taken from Roitberg and Prokopy (1984).

of fly foraging parameters (e.g. number of hosts visited, number of ovipositions, total time spent searching, etc.) are in fact possible through employment of simulation models that incorporate (1) fruit density and distribution, (2) fly movement patterns, (3) fly visual capability and (4) endogenous search allocation clocks (Elkinton et al., in manuscript).

Having examined the probable means by which *R. pomonella* females allot search time within trees, we now turn to other features of the habitat that may cause flies to modify search allocation. First, a common prediction from optimal foraging models is that foragers should allocate more search time to patches when such patches are distantly spaced from other patches then when they are near (Charnov, 1976). We recently demonstrated that *R. pomonella* females do in fact alter their search allocation as predicted by the models: flies remained longer and visited more leaves before exiting trees when such trees were spaced 1.6 m apart than when separated by 3.2 m (Roitberg and Prokopy, 1982). Second, temperature acts at least indirectly on search allocation by directly influencing the speed at which flies search branches. Since flies apparently measure energy expended rather than time expended, flies foraging at different temperature (i.e. different speeds) may allot different search times when measured against a real-time clock, even though energy expenditures may be similar. Added to this complexity are differences (a) in the number and length of rests initiated by flies foraging at different temperatures on host plants, and (b) in patterns of rest initiation on host versus non-host plants (Diehl and Prokopy, unpub. data). Thus, it appears that the foraging performance of females can be influenced dramatically by both biotic and abiotic components of the habitat.

Finally, contemporary foraging theory assumes implicitly that the traits that determine how an animal forages are heritable in a quantitative fashion, such that small changes in GUT clock settings, for example, can occur across generations (Pyke, 1984). There is little evidence to support this assumption, though few experiments have been designed to test it. We have obtained preliminary evidence, however, that there are consistent differences in search energy allocation among individual flies even when those flies have been reared under identical laboratory conditions (Roitberg et al., unpub. data). Whether such differences are heritable remains to be seen.

In summary, we have described, in a quantitative fashion, some aspects of the intra-tree foraging behavior of *R. pomonella* as a model for the general case. The concepts we have presented should be of interest at several levels. For example, we have discussed how females are more likely to emigrate from trees the longer the time period between ovipositions. We have also shown, however, that should females locate hosts, they are more likely to attempt oviposition in such hosts as time between oviposition increases. The rates at which these two variables change in value may have dramatic effects on any pest management program that employs tactics that utilize behavior modification techniques (e.g. treatment of fruit with marking pheromone). Toward this end, we have employed many of the concepts and parameters described above in simulation models to demonstrate how synthetic marking pheromone and fruit-mimicking traps might someday be employed effectively in novel, environmentally sound *R. pomonella* management programs (Roitberg, 1985b). Whether such programs would be effective against other tephritid species is not yet clear.

4 CONCLUSION

In this chapter we attempted to provide a brief overview of the sorts of questions currently being addressed by investigators examining the foraging behavior of an array of invertebrate and vertebrate animals. We also discussed a variety of factors that researchers might consider when analyzing animal foraging behavior. We took this approach because it is our belief that analysis of the foraging behavior of tephritid flies for food, mates, egglaying sites, or refugia is best viewed within a conceptual framework developed over the past decade or two by zoologically-oriented behavioral ecologists. We further believe that tephritid flies, by virtue of their economic importance and hence the availability of a considerable amount of basic biological information, are potentially highly suitable organisms for addressing and even advancing current concepts of animal foraging behavior. To date, however, little of this potential has been realized, primarily because tephritid foraging behavior has been approached almost exclusively from qualitative analysis of the behavior of populations. Until there is greater research emphasis on quantitative analysis of the foraging behavior of tephritid individuals over time in a variety of semi-natural or natural settings that vary in quantity, quality, or distribution of resources, we will remain largely ignorant of numerous facets of foraging behavior that could contribute greatly to new and better approaches to management.

Chapter 4.1 references, p. 304

5 REFERENCES

Baker, A.C., Stone, W.E., Plummer, C.C. and McPhail, M., 1944. A review of studies on the Mexican fruit fly and related Mexican species. United States Department Agriculture Miscellaneous Publications, 531: 1–155.

Bateman, M.A. and Sonleitner, F.J., 1967. The ecology of a natural population of the Queensland fruit fly, *Dacus tryoni*. I. The parameters of the pupal and adult populations during a single season. Australian Journal of Zoology, 15: 303–335.

Boller, E.F., 1965. Beitrag zur Kenntnis der Eiablage und Fertilität der Kirschenfliege *Rhagoletis cerasi*. Mitteilungen Schweizerischen Entomologischen Gesellschaft, 38: 193–202.

Boller, E.F., Haisch, A. and Prokopy, R.J., 1971. Sterile insect release method against *Rhagoletis cerasi*: preparatory ecological and behavioral studies. In: International Atomic Energy Agency Symposium, Athens, 1970, Sterility Principle for Insect Control or Eradication, pp. 77–86.

Burk, T., 1982. Evolutionary significance of predation on sexually signalling males. Florida Entomologist, 65: 90–103.

Burk, T., 1983. Behavioral ecology of mating in the Caribbean fruit fly, *Anastrepha suspensa*. Florida Entomologist, 66: 330–344.

Carey, J.R., 1984. Host-specific demographic studies of the Mediterranean fruit fly *Ceratitis capitata*. Ecological Entomology, 9: 261–270.

Chapman, M.G., 1982. Experimental analysis of the pattern of tethered flight in the Queensland fruit fly, *Dacus tryoni*. Physiological Entomology, 7: 143–150.

Charnov, E.L., 1976. Optimal foraging: the marginal value theorem. Theoretical Population Biology, 9: 129–136.

Cirio, U. and De Murtas, I., 1974. Status of Mediterranean fruit fly control by the sterile-male technique on the island of Procida. In: International Atomic Energy Agency Symposium, Vienna, 1972, The Sterile-Insect Technique and its Field Application, pp. 5–16.

Cooley, S.S., Prokopy, R.J., McDonald, P.T. and Wong, T.T.Y., 1986. Learning in oviposition site selection by *Ceratitis capitata* flies. Entomologia Experimentalis Applicata, 40: 47–51.

Diehl, S.R., 1984. The role of host plant shifts in the ecology and speciation of *Rhagoletis* flies. Ph.D. thesis. University of Texas at Austin, 256 pp.

Diehl, S.R., Prokopy, R.J. and Henderson, S., 1986. The role of stimuli associated with branches and foliage in host selection by *Rhagoletis pomonella*. In: R. Cavalloro (Editor), Economically Important Fruit Flies. Balkema, Rotterdam, pp. 191–196.

Drew, R.A.I. and Hooper, G.H.S., 1983. Population studies of fruit flies in south-east Queensland. Oecologia, 56: 153–159.

Eisner, T., 1984. Consumer fraud. Natural History, 93(11): 112.

Elkinton, J.S., Roitberg, B.D. and Prokopy, R.J., 1988. Simulation models of intra-tree fruit foraging by *Rhagoletis pomonella* females, in press.

Fein, B.L., Reissig, W.H. and Roelofs, W.L., 1982. Identification of apple volatiles attractive to the apple maggot, *Rhagoletis pomonella*. Journal of Chemical Ecology, 8: 1473–1486.

Feron, M., 1957. Le comportement de ponte de *Ceratitis capitata*. Influence de la lumiere. Revue Pathologie Vegetale Entomologie Agricole France, 36: 127–143.

Fiestas Ros de Ursinos, J.A., Constante, E.G., Duran, R.M. and Roncero, A.V., 1972. Etude d'un attractif naturel pour *Dacus oleae*. Annals Entomological Society France, 8: 179–188.

Fitt, G.P., 1983. Factors limiting the host range of tephritid fruit flies, with particular emphasis on the influence of *Dacus tryoni* on the distribution and abundance of *Dacus jarvisi*. Ph.D. thesis. University of Sydney, Australia, 291 pp.

Fitt, G.P., 1984. Oviposition behavior of two tephritid fruit flies, *Dacus tryoni* and *Dacus jarvisi*, as influenced by the presence of larvae in the host fruit. Oecologia, 62: 37–46.

Fletcher, B.S., 1973. The ecology of a natural population of the Queensland fruit fly, *Dacus tryoni*. IV. Immigration and emigration of adults. Australian Journal of Zoology, 21: 541–565.

Fletcher, B.S. and Kapatos, E., 1981. Dispersal of the olive fly, *Dacus oleae*, during the summer period on Corfu. Entomologia Experimentalis Applicata, 29: 1–8.

Fletcher, B.S., Papas, S. and Kapatos, E., 1978. Changes in the ovaries of olive flies *Dacus oleae* during the summer, and their relationship to temperature, humidity, and fruit availability. Ecological Entomology, 3: 99–107.

Frias, D., Malavasi, A., and Morgante, J.S., 1984. Field observations of distribution and activities of *Rhagoletis conversa* on two hosts in nature. Annals of the Entomological Society of America, 77: 548–551.

Girolami, V., Strapazzon, A. and De Gerloni, P.F., 1983. Insect/plant relationships in olive flies: general aspects and new findings. In: R. Cavalloro (Editor), Fruit Flies of Economic Importance. Balkema, Rotterdam, pp. 258–267.

Glasgow, H., 1933. The host relations of our cherry fruit flies. Journal of Economic Entomology, 26: 431–438.

Greany, P.D., Styer, S.C., Davis, P.L., Shaw, P.E. and Chambers, D.L., 1983. Biochemical resistance of citrus to fruit flies. Entomologia Experimentalis Applicata, 34: 40–50.

Guerin, P.M., Remund, U., Boller, E.F., Katsoyannos, B.I. and Delrio, G., 1983. Fruit fly electroantennogram and behavior responses to some generally occurring fruit volatiles. In: R. Cavalloro (Editor), Fruit Flies of Economic Importance. Balkema, Rotterdam, pp. 248–251.

Haisch, A. and Levinson, H.Z., 1980. Influences of fruit volatiles and coloration on oviposition of the cherry fruit fly. Naturwissenschaften, 67: 44–45.

Hassel, M.P. and Southwood, T.R.E., 1978. Foraging strategies of insects. Annual Review Ecology Systematics, 9: 75–98.

Johnson, P.C., 1983. Response of adult apple maggots to pheromone traps and red spheres in a non-orchard habitat. Journal of Economic Entomology, 76: 1279–1284.

Kamil, A.C. and Sargent, T.D., 1981. Foraging Behavior: Ecological, Ethological, and Psychological Approaches. Garland, New York, 534 pp.

Keiser, I., Harris, E.J., Miyashita, D.H., Jacobsen, M. and Perdue, R.E., 1975. Attraction of ethyl ether extracts of 232 botanicals to Oriental fruit flies, melon flies, and Mediterranean fruit flies. Lloydia, 38: 141–152.

Krebs, J.R. and Davies, N.B., 1984. Behavioral Ecology: An Evolutionary Approach, 2nd Ed. Sinauer, Sunderland, MA, 493 pp.

Landolt, P.J. and Hendrichs, J., 1983. Reproductive behavior of the papaya fruit fly, *Toxotrypana curvicauda*. Annals Entomological Society America, 76: 413–417.

Malavasi, A., Morgante, J.S. and Prokopy, R.J., 1983. Distribution and activities of *Anastrepha fraterculus* flies on host and nonhost trees. Annals Entomological Society America, 76: 286–292.

Maxwell, C.W. and Parsons, E.C., 1968. The recapture of marked apple maggot adults in several orchards from a release point. Journal of Economic Entomology, 61: 1157–1159.

McDonald, P.T. and McIniss, D.O., 1985. Effect of host fruit size on number of eggs per oviposition in *Ceratitis capitata*. Entomologia Experimentalis Applicata, 37: 207–211.

Michelakis, S. and Neuenschwander, P., 1981. Etude des deplacements de la population imaginale de *Dacus oleae* en Crete, Greece. Acta Oecologia, 2: 127–137.

Monteith, L.G., 1972. Status of adult predators of the apple maggot, *Rhagoletis pomonella*, in Ontario. Canadian Entomologist, 104: 257–262.

Neilson, W.T.A., 1967. Development and mortality of the apple maggot, *Rhagoletis pomonella*, in crab apples. Canadian Entomologist, 99: 217–219.

Neilson, W.T.A., 1971. Dispersal studies of a natural population of apple maggot adults. Journal of Economic Entomology, 64: 648–653.

Nishida, T. and Bess, H.A., 1957. Studies on the ecology and control of the melon fly *Dacus cucurbitae*. Hawaii Agricultural Experiment Station Technical Bulletin, 34: 1–44.

Orphanidis, P.S. and Kalmoukas, P.E., 1970. Negative chemotropism of *Dacus oleae* adults against essential oils. Annals Institut Phytopathologie Benaki, 9: 288–294.

Pritchard, G., 1969. The ecology of a natural population of Queensland fruit fly, *Dacus tryoni*. II. The Distribution of eggs and its relation to behavior. Australian Journal of Zoology, 17: 293–311.

Prokopy, R.J., 1976. Feeding, mating, and oviposition activities of *Rhagoletis fausta* flies in nature. Annals Entomological Society America, 69: 899–904.

Prokopy, R.J., 1977. Stimuli influencing trophic relations in Tephritidae. Colloques Internationaux Centre National Recherche Scientifique, France, 265: 305–336.

Prokopy, R.J., 1983. Tephritid relationships with plants. In: R. Cavalloro (Editor), Fruit Flies of Economic Importance. Balkema, Rotterdam, pp. 230–239.

Prokopy, R.J. and Roitberg, B.D., 1984. Foraging behavior of true fruit flies. American Scientist, 72: 41–49.

Prokopy, R.J., Bennett, E.W. and Bush, G.L., 1972. Mating behavior in *Rhagoletis pomonella*. II. Temporal organization. Canadian Entomologist, 104: 97–104.

Prokopy, R.J., Moericke, V. and Bush, G.L., 1973. Attraction of apple maggot flies to odor of apples. Environmental Entomology, 2: 743–749.

Prokopy, R.J., Averill, A.L., Cooley, S.S., Roitberg, C.A. and Kallet, C., 1982a. Variation in host acceptance pattern in apple maggot flies. Proceedings 5th International Symposium Insect-Plant Relationships, Wageningen, pp. 123–129.

Prokopy, R.J., Averill, A.L., Cooley, S.S. and Roitberg, C.A., 1982b. Associative learning in egglaying site selection by apple maggot flies. Science (Washington) 218: 76–77.

Prokopy, R.J., McDonald, P.T. and Wong, T.T.Y., 1984. Inter-population variation among *Ceratitis capitata* flies in host acceptance pattern. Entomologia Experimentalis Applicata, 35: 65–69.

Prokopy, R.J., Papaj, D.R., Cooley, S.S. and Kallet, C., 1986. On the nature of learning in oviposition site acceptance by apple maggot flies. Animal Behavior, 34: 98–107.

Prokopy, R.J., Papaj, D.R. and Wong, T.T.Y., 1986. Fruit-foraging behavior of Mediterranean fruit fly females on host and non-host plants. Florida Entomologist, 69: 651–657.

Pyke, G.H., Pulliam, H.R. and Charnov, E.L., 1977. Optimal foraging: a selective review of theory and tests. Quarterly Review Biology, 52: 137–154.

Pyke, G.H., 1984. Optimal foraging theory: a critical review. Annual Review Ecology Systematics, 15: 523–575.

Ripley, L.B. and Hepburn, G.A., 1929. Olfactory and visual reactions of the natal fruit fly, *Pterandrus rosa*, as applied to control. South African Journal Science, 24: 449–458.

Ripley, L.B., Hepburn, G.A. and Anderssen, E.E., 1940. Fruit fly migration in the Kat River valley. South Africa Department Agriculture Forestry, Plant Industry Series, 49: 1–17.

Roitberg, B.D., 1985a. Search dynamics of fruit parasites. Journal of Insect Physiology, 31: 865–872.

Roitberg, B.D., 1985b. Foraging theory and pest management: An ecological fantasy. In: Symposium on 'Recent biological and ecological considerations in the regulation of fruit fly populations by total population management'. Annual Meeting Pacific Branch Entomological Society of America, June 1985.

Roitberg, B.D. and Prokopy, R.J., 1982. Influence of intertree distance on foraging behavior of *Rhagoletis pomonella* in the field. Ecological Entomology, 7: 437–442.

Roitberg, B.D. and Prokopy, R.J., 1983. Host deprivation influence on response of *Rhagoletis pomonella* to its oviposition deterring pheromone. Physiological Entomology, 8: 69–72.

Roitberg, B.D. and Prokopy, R.J., 1984. Host visitation sequence as a determinant of search persistence by fruit-parasitic tephritid flies. Oecologia, 62: 7–12.

Roitberg, B.D., Van Lenteren, J.C., Van Alphen, J.J.M., Galis, F. and Prokopy, R.J., 1982. Foraging behavior of *Rhagoletis ponmonella*, a parasite of hawthorn, *Crataegus*, in nature. Journal of Animal Ecology, 51: 307–325.

Sanders, W., 1968. Die Eiablagehandlung der Mittelmeerfruchtfliege *Ceratitis capitata*. Ihre Abhangigkeit von Farbe und Gliederung des Umfeldes. Zeitschrift für Tierpsychologie, 25: 588–607.

Schoonhoven, L.M., 1983. The role of chemoreception in hostplant finding and oviposition in phytophagous Diptera. In: R. Cavalloro (Editor), Fruit Flies of Economic Importance. Balkema, Rotterdam, pp. 240–247.

Seo, S.T., Tang, C.S., Sanidad, S. and Takenaka, T.H., 1983. Hawaiian fruit flies: variation of index of infestation with benzl isothiocyanate concentration and color of maturing papaya. Journal of Economic Entomology, 76: 535–538.

Smith, D.C., 1984. Feeding, mating, and oviposition by *Rhagoletis cingulata* flies in nature. Annals Entomological Society America, 77: 702–704.

Smith, D.C., 1985a. General activity and reproductive behavior of *Rhagoletis tabellaria* flies in nature. Journal Kansas Entomological Society, 58: 737–739.

Smith, D.C., 1985b. Feeding, mating, and oviposition by *Rhagoletis cornivora* flies in nature. Journal New York Entomological Society, 93: 1052–1056.

Smith, D.C. and Prokopy, R.J., 1981. Seasonal and diurnal activity of *Rhagoletis mendax* flies in nature. Annals Entomological Society America, 74: 462–466.

Sonleitner, F.J. and Bateman, M.A., 1963. Mark recapture analysis of a population of Queensland fruit fly, *Dacus tryoni*, in an orchard. Journal of Animal Ecology, 32: 259–269.

Tanaka, N., 1965. Artifical egging receptacles for three species of tephritid flies. Journal of Economic Entomology, 58: 177–178.

Vargas, R.I., Harris, E.J. and Nishida, T., 1983a. Distribution and seasonal recurrence of *Ceratitis capitata* on the island of Kauai in the Hawaiian Islands. Environmental Entomology, 12: 303–310.

Vargas, R.I., Nishida, T. and Beardsley, J.W., 1983b. Distribution and abundance of *Dacus dorsalis* in native and exotic forest areas on Kauai. Environmental Entomology, 12: 1185–1189.

Chapter 4.2 Response to Shape, Size and Color

B.I. KATSOYANNOS

1 INTRODUCTION

The general aspects of the process of host selection by fruit flies have been recently reviewed by Prokopy (1977a, 1983) and Zwölfer (1983). It is recognized that host finding in tephritids involves a combination of visual and chemical (attractant and/or repellent) olfactory responses. The sequence in which olfactory and visual stimuli operate in tephritid host detection is largely unknown (Prokopy, 1977a).

Prokopy (1983), divided the principal factors guiding tephritids to essential plant associated resources into those perceptible from a distance (beyond or within the plant canopy) and those perceptible upon direct contact. He listed as many as seven principal kinds of plant-associated stimuli perceptible from a distance, among which the visual properties of foliar or oviposition site color, silhouette (contrast), pattern, and size.

General aspects of insect vision and its role in the host plant detection and selection by herbivorous insects with extensive references to tephritids, have been recently reviewed by Prokopy and Owens (1983). According to them, there are three principal properties of individual plants and plant structures that may serve as visual cues for foraging insects, including fruit flies: spectral quality, dimensions and pattern. Background characteristics had a substantial influence on the level of apple maggot, *Rhagoletis pomonella* (Walsh) discrimination between colors as found in a laboratory flight chamber, and field experiments (Owens and Prokopy, 1984). Background characteristics affected fruit fly behavior also in other cases (e.g. Haisch and Levinson, 1980), apparently modifying the detectability of an object by increasing or decreasing its contrast against the background. In fact, the detectability of an object by an insect is a function of object dimensions, contrast against the background, optical properties of the medium, and intensity of illumination. Contrast may arise from intensity or spectral differences, pattern differences, or motion differences, all of which may provide stimuli to the insect (Prokopy and Owens, 1983 and references therein).

Behavioral responses to visual stimuli constitute a part of the visual ecology and can be best treated in terms of insect/plant relationships. After a brief account of the fruit fly visual system, an overview will be presented of the present knowledge of fruit fly response to visual stimuli of shape, size and color as they affect host plant detection, and within it, location and selection of essential plant-associated resources. Although visual stimuli, in particular shape and pattern of the bracts are important for the host recognition by tephritids infecting flower buds of Compositae (Zwölfer, 1983 and references therein), and also by a few other tephritid groups, (see Prokopy, 1977a), this

Chapter 4.2 references, p. 321

chapter deals only with frugivorous fruit flies such as species of *Dacus, Anastrepha, Ceratitis* and *Rhagoletis.*

2 VISION IN FRUIT FLIES

In considering the fruit fly responses to visual stimuli, especially to objects of different shape, size and color, it should be kept in mind that although we have a rather concrete idea of how insects may see, their visual world is surely quite different from our own.

The insect visual receptors include compound eyes, ocelli and stemmata. True image formation is a property solely of the compound eye which is a unique structure with enormous flexibility of selective adaption (Prokopy and Owens, 1983).

Only limited information exists concerning the ultrastructural morphology of visual receptors in fruit flies. The compound eye of the apple maggot *R. pomonella* and the Caribbean fruit fly, *Anastrepha suspensa* (Loew) consists of ca. 2600 and ca. 3600 ommatidia respectively, and has been described in detail by Agee et al. (1977). Similarly, the compound eye of the Mediterranean fruit fly, *Ceratitis capitata* (Wiedemann) consists of ca. 3000 ommatidia (Davis et al., 1983).

Fruit flies, like most insects, respond to a bandwidth of the spectrum extending from near-ultraviolet to red (360–650 nm) and in general they respond positively to colors reflecting most of their energy between 500 and 600 nm and relatively little energy below 500 nm (for ref. see Prokopy, 1977a). In some species the visual sensitivity of the compound eye has been measured using the electroretinogram technique which provides a physiological measurement of visual sensitivity at the receptor level (Agee et al., 1982; Remund et al., 1981). The spectral sensitivity of *C. capitata*, the olive fruit fly, *Dacus oleae* (Gmelin) and the European cherry fruit fly, *Rhagoletis cerasi* (Linnaeus), the three species investigated, was basically similar, with a broad major peak of response at 485–500 nm and a second peak at 365 nm. As indicated by Agee et al. (1982), differences in color preference behavior can be a function of sensitivity at the photoreceptor level but a specific color to which a species is most attracted need not correspond with the wavelength to which the photoreceptors of that species are most sensitive, i.e., attraction may be controlled by higher order of neural input than at the first-order receptor level. Furthermore, measurements such as the electroretinogram can be very useful for assessing the quality of laboratory reared insects and have been used for evaluating the quality of several fruit fly species (Agee and Chambers, 1980 and references therein).

3 FIELD RESPONSES TO TREE MODELS FROM A DISTANCE

The first step in the host selection process by phytophagous insects, including fruit flies, classically consists of host habitat location, and within it, of the location and selection of host plants. Fruit flies under certain conditions may perform those two orientation phases in order to arrive on host plants bearing suitable resources. In many cases however, depending on the species and the particular habitat, fruit flies may spend almost all their life on or among their host plants and thus do not need to locate them from a distance.

The odor of host fruit is a specific stimulus known to attract a number of fruit flies to host plants bearing fruits in a stage suitable for oviposition, while odor of non-host plants may be repellent. Within a habitat, general visual characteristics of plants are of principal importance for the location of vegetation (Prokopy, 1977a and references therein).

Only a limited amount of information exists with respect to the visual location of host plants by tephritids and is mostly derived from field experiments in which sticky-coated colored boards of different shapes and sizes were used, posted in clearings near or between host and non-host trees to mimic tree models. The responses to these models of natural populations, or released flies, has been assessed in *R. pomonella* and the black cherry fruit fly *Rhagoletis fausta* (Osten Sacken) (Moericke et al., 1975), in *D. oleae* (Prokopy and Haniotakis, 1975), and the Queensland fruit fly, *Dacus tryoni* (Froggatt), (Meats, 1983). Some related information can be derived also from laboratory experiments with *C. capitata* (Sanders, 1968b). These investigations showed that within a habitat, fruit flies utilize visual cues to locate vegetation. At least for those fruit flies whose host plants are trees, the shape and size of the host tree plays a role. Color of foliage and darkness of silhouette are also of prime importance.

As pointed out by Prokopy (1977a), within the insect visible spectrum of ca. 300–650 nm, green leaves reflect maximally between 500–600 nm. This band of energy seems important in guiding tephritids to vegetation. Conversely, reflected or transmitted energy below 500 nm may be repulsive to alighting tephritids. Thus, visual location may involve the dual processes of attraction to foliar reflected light and repulsion by non-foliar reflected or transmitted light. Although visual stimuli are important in the vegetation location process, none of them are specific to the hosts in any of the species investigated, including the color of olive leaves which differs to a certain degree from the general leaf-green color (Prokopy and Haniotakis, 1975).

Despite the great importance of the above findings to our understanding of host selection mechanisms, much more information is needed in order to form a more solid idea of how fruit flies locate vegetation from a distance or from nearby as well as about the sequence in which the visual and chemical cues operate. For detailed and interesting discussions on these aspects see Moericke et al. (1975) and Prokopy and Owens (1983).

4 FIELD RESPONSES WITHIN TREE CANOPY

Upon arrival on host plants, fruit flies locate plant structures primarily, if not exclusively, by vision. The shape, size and color of leaves and fruits are of great significance. Inanimate objects mimicking visual characteristics of these structures, when suspended on host plants, attract fruit flies. Consequently, such objects have proved suitable for trapping purposes.

Visual responses in the field to objects of different shape, size and color were and continue to be a very popular field of intensive research, not only by behavioral ecologists, but mostly, by applied entomologists intent on creating new effective traps or improving trap efficiency and selectivity. As a consequence, most of the related studies have been conducted using objects suitable for trapping purposes, i.e. cheap, practical and easy to handle, such as sticky coated small rectangles and/or spheres. Usually the final outcome is measured in terms of the number of flies captured on these objects over a certain time period. Studies reporting more detailed behavioral observations are rather limited.

4.1 Within tree response to different shapes

The first desirable step in assessing visual responses in the field is to test fruit fly behavior toward different shapes. This was accomplished in some species where the attractiveness of 2-dimensional or 3-dimensional sticky coated objects of different colors and/or sizes was compared. This includes

Chapter 4.2 references, p. 321

studies of the apple maggot fly, *R. pomonella* (Prokopy, 1968, 1975; Prokopy and Owens, 1978), the European cherry fruit fly, *R. cerasi* (Boller, 1969; Prokopy, 1969), the black cherry fruit fly, *R. fausta*, the cherry fruit fly, *Rhagoletis cingulata* (Loew) and the western cherry fruit fly, *Rhagoletis indifferens* Curran (Prokopy, 1975; Reissig, 1976), the olive fruit fly, *D. oleae* (Prokopy and Haniotakis, 1976), the Queensland fruit fly, *D. tryoni* and other Australian *Dacus* spp. (Hill and Hooper, 1984) and the Mediterranean fruit fly, *C. capitata* (Nakagawa et al., 1978).

In general, when the response to different shapes such as spheres, cylinders, cones, cubes, rectangles or cross-rectangles was compared, spherical models were more attractive for both sexes than other shapes. As suggested by Boller (1969), spheres may be more attractive because they can be seen by the flies from all directions. However, significant preference of discs over rectangles of the same surface suggests that shape *per se* is an important cue and that flies may detect and prefer spheres over the other shapes partly, if not solely, on the basis of their circular outline (Prokopy, 1969). The biological basis of the stronger attraction to spherical objects over objects of other shapes is probably related to the close similarity of this shape with the shape of the host fruit. Although shape *per se* of the objects tested, in conjunction with their size and color, is an important factor affecting fruit fly response, it has been found in a number of cases, that the modus of the object's exposure within host plant canopy e.g. panel orientated in 45° compared to a vertical one, affects strongly the responses of the flies. Usually a particular behavioral pattern of a given species such as the usual modus of the within plant movement, is responsible for the observed differences in fruit fly responses to differently exposed, otherwise equivalent shapes (Prokopy, 1975; Reissig, 1976; Prokopy and Coli, 1978; Davis et al., 1984; Hill and Hooper, 1984; Neilson et al., 1984).

4.2 Within tree responses to small and medium size colored rectangles

A considerable amount of research deals with field responses to small and medium size (4 × 5 to 30 × 40, usually 15 × 20 cm) rectangles or crossed (x-shaped, wing) rectangles, with the aim of developing effective traps for monitoring and control of fruit flies (see Economopoulos, Vol. 3B, Chapter 9.2). Similar to many other herbivorous insects (Prokopy and Owens, 1983 and ref. therein), yellow was the most attractive color for almost all fruit flies tested (Table 4.2.1), usually followed by orange and green colors, which were also strongly attractive. In some cases these two colors were as attractive as yellow (*D. oleae*, Prokopy et al. (1975)) or orange was the most attractive color (*A. suspensa*, Greany et al. (1977)). All available information supports the conclusion, that fruit fly response to flat surfaces of a certain color such as yellow or orange, is primarily a positive response to the hue and not to the total amount of energy reflected by that color (Prokopy, 1968, 1969, 1972; Prokopy and Boller, 1971b; Prokopy et al., 1975; Prokopy and Economopoulos, 1976; Prokopy and Coli, 1978; Greany et al., 1977; Cytrynowicz et al., 1982; Hill and Hooper, 1984).

Interpretation of fruit fly responses to colored objects with the help of light reflectance curves has been attempted in a number of investigations. In general fruit flies were strongly attracted to those small or medium size flat surfaces reflecting most of the energy between 500–600 nm and little energy below 500 nm. Within the insect visible spectrum of ca. 300–650 nm, green leaves reflect maximally between 500–600 nm with a peak at ca. 550 nm. Thus, it has been suggested, that most fruit flies might be responding to yellow, which reflects maximally, and proportionally more energy than the other colors,

TABLE 4.2.1

Fruit fly species for which yellow is found to be the most attractive color in small to medium size
(4 × 5 to 30 × 40 cm), panels or X-shaped rectangles. References with an asterisk indicate that
fluorescent colors were tested

Fruit fly species	Selected references
Rhagoletis cerasi Linnaeus	Boller, 1969; Prokopy, 1969; Prokopy and Boller, 1971b★
Rhagoletis pomonella (Walsh)	Still, 1960; Oatman, 1964; Prokopy, 1968, 1972★
Rhagoletis cingulata (Loew)	Frick et al., 1954; Still, 1960; Prokopy, 1975; Reissig, 1976
Rhagoletis fausta (Osten Sacken)	Still, 1960; Reissig, 1976
Rhagoletis completa Cresson	Gibson and Kearby, 1978★; Riedl and Hoying, 1980, 1981★
Rhagoletis suavis (Loew)	Gibson and Kearby, 1978★
Rhagoletis mendax Curran	Prokopy and Coli, 1978★; Neilson et al., 1984
Dacus oleae (Gmelin)	Girolami and Cavalloro, 1973; Prokopy et al. 1975★
Dacus tryoni (Froggatt)	Bateman, 1976★; Hill and Hooper, 1984★
Dacus cucurbitae Coquillett	Bateman, 1976★
Dacus dorsalis Hendel	Bateman, 1976★
Dacus neohumeralis Hardy	Hill and Hooper, 1984★
Dacus cacuminatus (Hering)	Hill and Hooper, 1984★
Anastrepha ludens (Loew)	Baker et al., 1944
Anastrepha suspensa (Loew)	Greany et al., 1977[1], 1978★[1]
Anastrepha fraterculus (Wiedemann)	Cytrynowicz et al., 1982
Ceratitis capitata (Wiedemann)	Prokopy and Economopoulos, 1976★; Bateman, 1976★; Cytrynowicz et al., 1982

[1] In these studies, orange color was more attractive than yellow.

between 500–600 nm, as if to foliage on which to find food (Prokopy, 1968, 1969,
1972; Prokopy and Boller, 1971b; Prokopy et al., 1975; Prokopy and Coli, 1978).
On the other hand, strong attraction of *A. suspensa* to enamel or fluorescent
orange colored panels was considered to indicate fruit-seeking for oviposition
rather than foliage-seeking behavior (Greany et al., 1977, 1978).

Several works (Prokopy, 1972; Prokopy et al., 1975; Greany et al., 1977, 1978;
Cytrynowicz et al., 1982; Hill and Hooper, 1984) suggest, that hue purity, i.e.
high reflectance within a narrow spectral region, is more important for fruit fly
attraction than the quantity of energy reflected within the larger spectral
region corresponding to the attractive hue. In fact, fluorescent colors reflecting
maximally within a narrow spectral region were highly attractive for several
fruit flies (see Table 4.2.1). Prokopy (1972) proposed, that daylight fluorescent
Saturn Yellow color is so highly attractive to *R. pomonella* flies because it
constitutes a '*super-normal*' foliage-type stimulus eliciting food-seeking and/or
host plant-seeking behavior. Since then, this suggestion has been used to
explain attraction to yellow by almost all fruit flies subsequently investigated.

Very low responses of fruit flies to aluminium foil covered rectangles indicate, that radiation in the ultra-violet portion of the spectrum (300–390 nm)
where this foil reflects a large amount of energy, does not usually attract fruit
flies, at least when in combination with other wavelengths in the spectrum (e.g.
Prokopy, 1972; Prokopy et al., 1975; Prokopy and Coli, 1978).

4.3 Field response to colored spheres of different sizes

Field response to spheres is of particular interest because of the similarity
of these objects to the host fruits of fruit flies, which serve as oviposition place
and in some species, also as rendezvous for mating. (The most relevant reports

Chapter 4.2 references, p. 321

are listed in Table 4.2.2) As will be apparent from the following more detailed consideration of the main findings, generalizations on fruit fly preference to spheres of certain sizes and/or colors as also on the biological significance of the observed responses, are rather difficult. In most cases sufficient detailed behavioral field observations which could provide the basis for the explanations needed, are not available to confirm conclusively the different suggestions made.

4.3.1 *Rhagoletis* spp.

Prokopy (1968) reported that *R. pomonella* flies of both sexes in the field showed a significant preference for 7.5 cm diameter spheres of dark color (such as red, blue and black) over light colored spheres (such as yellow, green and white) of the same size. As however the diameter of a sphere increased from 7.5 to 45 cm, there was an orderly and significant decrease in the attractiveness of those that were dark colored (red) but a progressive increase in the attractiveness of those that were yellow. A similar size dependent color preference for spheres has been found also in *R. cerasi* (Prokopy, 1969). According to Prokopy (1968, 1969), these flies were attracted to small, dark-colored spheres because they react to such spheres as they react to fruits which serve as oviposition site and also as a rendezvous for mating. On the other hand, more *R. cingulata* and *R. fausta* flies were captured on yellow, 7.5 cm diameter spheres than on darker spheres of the same size and a large (13 cm diameter) red sphere caught twice as many flies as the smaller red spheres (7.5 cm and 5.0 cm diameter) which captured similar numbers of flies (Reissig, 1976). Prokopy, (1977b) for these two *Rhagoletis* species, and Prokopy and Coli (1978) for *Rhagoletis mendax* Curran, reported that the flies were more attracted to 7.5 cm red spheres (the only color tested), than to smaller ones approximating the size of the flies' native host fruits and suggested that these 7.5 cm spheres represent a *super-normal fruit-type* stimulus for these flies. In respect to the detection mechanisms of colored spheres by the flies in the field, Prokopy (1968, 1969) suggested for *R. pomonella* and *R. cerasi* that dark colored small spheres are preferred over those of lighter colors on the basis that they stand out in stronger contrast against the background of foliage and other reflected or transmitted light and hence are more readily detectable than the latter by the flies, and not on the basis of true color discrimination, i.e. of color hue *per se*. However, Prokopy (1973) reported that *R. pomonella* flies were just as attracted to red enamel 7.5 cm diameter spheres

TABLE 4.2.2

References concerning fruit fly field responses to spheres of different size and/or color

Fruit fly species	Selected references
Rhagoletis cerasi Linnaeus	Boller, 1969; Prokopy, 1969
Rhagoletis pomonella (Walsh)	Oatman, 1964; Prokopy, 1968, 1973, 1977b; Prokopy and Owens, 1978
Rhagoletis cingulata (Loew)	Reissig, 1976; Prokopy, 1977b
Rhagoletis fausta (Osten Sacken)	Reissig, 1976; Prokopy, 1977b
Rhagoletis completa (Cresson)	Riedl and Hoying, 1981
Rhagoletis mendax Curran	Prokopy and Coli, 1978; Neilson et al., 1984
Dacus oleae (Gmelin)	Prokopy and Haniotakis, 1976
Dacus tryoni (Froggatt)	Hill and Hooper, 1984
Dacus neohumeralis Hardy	Hill and Hooper, 1984
Anastrepha fraterculus (Wiedemann)	Cytrynowicz et al., 1982
Ceratitis capitata (Wiedemann)	Nakagawa et al., 1978; Cytrynowicz et al., 1982; Katsoyannos, 1987a,b,c,d

as to daylight fluorescent red ones which reflect considerably more energy and to the human eye are more readily detectable (on the basis of contrast against the background) than enamel spheres in apple trees. He suggested that possibly darkness *per se* of a sphere is a positive factor in the flies' ability to detect it and that some flies may have flown to the spheres in response to stimuli other than that of sphere itself.

4.3.2 *Dacus* spp.

Prokopy and Haniotakis (1976) assessed the response of *D. oleae* females to fruit mimics and found that they were about equally attracted to black, red and yellow wooden, 3.1 × 2.0 cm sticky-coated olive models, slightly less attracted to green and orange models and least attracted to blue, white and aluminium foil-covered models. Of four different sizes of black spherical models the larger (7.5 cm diam.) were more attractive than the smaller ones. They concluded that color *per se* of olives and/or the degree of contrast against the background is a cue utilised by the flies in locating olives and that olive fruit flies can more readily locate larger olives than smaller ones. Also, that a black sphere of 7.5 cm in diameter is a super-normal olive to the flies. Katsoyannos (1984, unpubl.) working with the same fly in the field in Chios, Greece, also found, that among sticky-coated spheres of three sizes and seven colors suspended on olive trees, the larger (7.0 cm diam.) were the most attractive for both sexes. For all sizes, yellow and orange spheres were the most attractive, followed by black, red, green and blue ones which captured also large numbers of flies, while only a limited number of flies was captured on white spheres. The response was similar during the early season when almost all females were immature and later, when the majority of females were mature. A large proportion (more than 50%) of the flies were captured during late afternoon and at dusk, when in this species mating activity predominates; but host fruit in this, as well as in other *Dacus* species, do not seem to serve as a rendezvous for mating. It has been generally concluded from these experiments, that colored spheres, perceived probably on the basis of color hue as well as of brightness and contrast against the background, do not represent, at least for a large part of the responding *D. oleae* flies, a fruit mimic substituting an oviposition site, but something else, not yet identified. Similar results (and interpretations) have been recently reported for *D. tryoni* (Hill and Hooper, 1984).

4.3.3 *Ceratitis capitata*

Nakagawa et al. (1978) reported, that *C. capitata* flies of both sexes were attracted to 7.5 cm diameter spheres, black and yellow being the most attractive of eight colors tested. Similarly for *R. pomonella* and *R. cerasi* when an array of sphere sizes was tested, the attraction to flies increased as the size of yellow spheres increased from 1.8 to 18 cm diameter. Similar results are reported by Cytrynowicz et al. (1982). Katsoyannos (1987a, b) found that among spheres of six sizes and seven colors suspended on fig and citrus trees in Chios, Greece, wild *C. capitata* flies, mostly mature females, were strongest attracted to yellow, 7.0 cm diam. spheres. Orange, green, red and black were less attractive than yellow, while blue and white were the least attractive. Nakagawa et al. (1978), suggested that *C. capitata* flies were highly attracted to 7.5 cm black spheres because these stood out in stronger contrast against the background. Cytrynowicz et al. (1982), observed that the maximum captures of *C. capitata* females occurred on those spheres which reflected proportionally less light below 530 nm. Because green leaves usually reflect maximally close to this wavelength, they suggested, that ready-to-oviposit females would be attracted to ripening fruit because this fruit reflects proportionally less ca. 530 nm light than does unripe (green) fruit and surrounding foliage. Finally, Katsoyannos

Chapter 4.2 references, p. 321

(1987a) reported that the response to the spheres was strongest during late afternoon and was correlated with observed feeding activity on ripe fruits. He concluded that the flies' response to colored spheres, especially the yellow ones, was a feeding response to super-normal ripe-fruit-type stimuli and not a response to oviposition site-type stimuli. Analysis of those results in respect to the color properties of the spheres (Katsoyannos, 1987c), showed that the response depended primarily on the hue of the colors tested while their brightness and the degree the spheres contrasted to the background had a minor effect.

4.4 Factors affecting field responses

The spectrum and quantity of natural illumination falling upon an insect at any given moment and available to it for visual discrimination of plants and plant structures may be extremely variable according to a multitude of environmental factors and the insect's precise position within a habitat (Prokopy and Owens, 1983). This variability no doubt, may affect considerably fruit fly field responses to objects of different shape, size and color. Some additional environmental factors, besides those of a meteorological character such as temperature, wind or rain, which affect fruit fly responses in the field include the object's position near or within the tree canopy (radius, compass direction, height above ground, vicinity to leaves and/or fruits), fruit density, distribution and maturity, infestation level, foliage density and physiological stage, host status and variety of the tree or plant on which the object is suspended and its position in relation to the other vegetation in the area, frequency of removal from the objects (when sticky coated), of the insects captured, and others (Prokopy, 1968; Reissig, 1975; Johnson, 1983; Drummond et al., 1984; Hill and Hooper, 1984; Katsoyannos, 1987b).

Except for environmental factors, fruit fly response to visual stimuli also depends strongly on the physiological state of the individuals of the population which in turn affects their 'motivation' to express a certain behavior. This state varies with the season, the proportion of immature flies in a population being usually greater in the beginning of the growing season. Thus, seasonal differences in response of *R. pomonella* flies to spheres of different sizes might be due to differences in the response between immature and mature flies (Prokopy, 1968, 1977b). Furthermore, it has been suggested, that the frequency of fruit fly encounters with a certain type of objects, e.g. apple fruit, may alter *R. pomonella* behavior through *learning* to prefer a certain type of object matching the size of host fruit (Prokopy, 1977b). Finally, the presence or absence of seriochemicals such as oviposition-deterring pheromones on fruits, by affecting the foraging behavior of fruit flies (see Chapter 4.1, p. 293), may have a considerable influence on fruit fly field responses toward fruit mimics or other objects.

From the above it is obvious, that interpretation in the behavioral context of field capture data gained under such a large variety of uncontrolled factors, is extremely difficult and may sometimes lead to wrong conclusions.

5 LABORATORY RESPONSES TO OVIPOSITION SITE MIMICS

5.1 Generalities

With a few exceptions, where general field responses to shape, size and color were verified in the laboratory (Prokopy, 1972; Cytrynowicz et al., 1982; Owens and Prokopy, 1984), the vast majority of laboratory experiments related to fruit fly responses to these physical stimuli, concern the oviposition behavior of the flies. On the other hand, only a few occasional field observations exist concerning the effect of these stimuli on the oviposition behavior. For this reason, the following deals only with laboratory responses of gravid females to objects mimicking oviposition sites.

As pointed out by Prokopy (1977a), once the fruit fly has arrived on a host plant, physical characteristics of host fruits perceptible by vision, are of principal importance for the detection and selection of the suitable oviposition site. In some cases, chemical short range olfactory stimuli were also found to affect the selection of oviposition sites (Féron, 1962; Sanders, 1962; Tanaka, 1965; Haisch and Levinson, 1980).

After arrival upon a potential oviposition site, its acceptance for oviposition depends on its surface structure and condition, and fruit chemistry (for fruits) (Prokopy, 1977a and references therein). The presence or absence of oviposition-deterring pheromones or other analogous substances is decisive for acceptance after arrival (Averill and Prokopy, Chapter 3.5, p. 207). It has been suggested, that shape, size and color play also a role in this phase of the host selection process. However, there are certain serious doubts as to whether this is the case, because such conclusions are based either on observations on groups of flies (Prokopy and Boller, 1971a; Prokopy and Bush, 1973) which might have affected the outcome of these observations (see section 5.6), or on inadequate observations. For example, Wiesmann (1937), claimed, that a *R. cerasi* female, upon arrival on a cherry to oviposit, begins to circle rapidly several times, occasionally dragging the ovipositor over the fruit surface. He concluded that this behavior is a 'cinesthetical orientation', through which the female determines the curvature of the object upon which it has landed and checks the physical suitability of the object as an oviposition place, i.e. if it is a round object of a certain size. As shown by Katsoyannos (1975) however, fast circling with dragging of the ovipositor occurs only after oviposition and indicates the deposition of an oviposition-deterring pheromone. Thus Wiesmann (1937) possibly described the very characteristic post-ovipositional behavior as if it occurred before oviposition after arrival on a potential oviposition site, and explained it as a cinesthetical orientation. Other workers, probably under the influence of Wiesmann's fascinating theory, gave similar explanations to their results, for example Féron (1962), who indicates, that a 'posture reflex' determines fruit size acceptance by *C. capitata* females. For these reasons the effects of shape, size and color on the oviposition behavior, will be examined assuming, that they affect visually the selection of the oviposition site before arrival, from a short distance, as demonstrated in the most cases.

5.2 Development of artificial oviposition devices, based on shape, size and color

In frugivorous fruit flies the stimuli eliciting oviposition are not specific to the host plants suitable for larval development, because inanimate objects devoid of host plant material can elicit a high degree of egg laying in the laboratory (Prokopy, 1977a). This attribute of fruit flies has led several workers to develop artificial oviposition devices for the needs of rearing, based on host fruit physical characteristics. Thus an orange-colored hollow hemispheric dome, of citrus fruit size, made of cheese cloth, paraffin and petrolatum was found to be a highly satisfactory oviposition device for the Mexican fruit fly, *Anastrepha ludens* (Loew), with color and shape being important characteristics (McPhail and Guiza, 1956). Likewise, an inverted waxed-paper cup was reported to be an effective oviposition device for *D. tryoni* (Friend, 1957). Colorless paraffin wax domes of 'hen's egg' size were used as oviposition sites for *D. oleae* by Hagen et al. (1963). Artificial oviposition substrates, based on shape, size and color for *C. capitata* have been reported by Féron et al. (1958), Féron (1962), Sanders (1962, 1968a,b) and Tanaka (1965), who reported a lemon-shaped, orange-colored plastic juice dispenser as an effective egging receptacle for this fly, and by adding in its interior fruit juice, making it effective also for the oriental fruit fly, *Dacus dorsalis* Hendel and the melon fruit fly, *Dacus cucurbitae* Coquillett.

Taking into account certain characteristics of cherries reported by Wiesmann (1937) to be important in eliciting oviposition behavior in *R. cerasi*, Boller (1968) developed small gelatin spheres and Haisch (1968) developed small foam balls wrapped with parafilm as artificial egging devices for this insect. Two years later, Prokopy and Boller (1970, 1971a), using the technique of Hagen et al. (1963), formed hollow wax devices of various characteristics and found, that domes made of soft ceresin wax Type 1577 (Deutsche Erdoel AG, Hamburg 13, W. Germany), of appropriate size and color, elicited a high degree of oviposition in *R. cerasi*. Since that time, ceresin wax domes were used also for other fruit flies such as *C. capitata* (Boller and Katsoyannos, 1978, unpubl; Katsoyannos et al., 1986) and *D. oleae* (Katsoyannos and Pittara, 1983; Katsoyannos et al., 1985), to which they also elicit a high degree of oviposition. This success of ceresin wax domes is, to a high degree, attributable to the texture of this material as shown for *R. cerasi* (Prokopy and Boller, 1971a).

5.2.1 Effect of shape

All available information strongly suggests, that shape of a potential oviposition site is of predominant importance for its selection, followed by size. Consequently, responses of gravid females to different shapes are quite clear-cut and variability within populations is rather low. Convex surfaces such as spheres, hemispheres, ellipsoids, and in some cases the convex surface of cylinders or cones, elicit by far greater egg-laying behavior than objects having only flat surfaces such as cubes. Because of the preference for convex surfaces, most investigators omit to test effects of non-convex objects. As a consequence, the number of reports comparing different shapes is rather limited and concerns *R. pomonella*, (Prokopy, 1966), *R. cerasi*, (Wiesmann, 1937; Prokopy and Boller, 1971a), *Rhagoletis completa* (Cresson) (Cirio, 1972), *C. capitata* (Féron, 1962; Sanders, 1962) and *A. suspensa* (Greany and Szentesi, 1979).

5.2.2 Effect of size

Although the information concerning effects of size derived from laboratory experiments is rather limited, the size of natural or artificial oviposition sites seems to be of great importance for its acceptance for oviposition, ranking second in importance, after shape. In most cases, a close similarity was found between the size of natural host fruit and the preferred size of artificial substrates. For example, in *R. cerasi*, females usually accepted for oviposition, sizes between 4 and 24 mm diameter (Wiesmann, 1937) and most preferred were ceresin wax hollow hemispheres (domes) of 10–20 mm diameter (Prokopy and Boller, 1971a; Katsoyannos, 1979). The usual size of host fruits of this species is 5–30 mm diameter. Similarly, Prokopy and Bush (1973) reported a close correspondence between the size of artificial fruits preferred and the size of natural host fruits of four *Rhagoletis* species belonging to the *R. pomonella* species group, i.e. *R. pomonella, R. mendax, R. zephyria* Snow and *R. cornivora* Bush. They suggested that fruit size *per se* may play an important role in the process of host selection by these species. On the other hand, it has been reported, that *R. completa* females preferred for oviposition artificial fruits larger than the size of their natural host fruits (Cirio, 1972). Similarly, when *D. oleae* females were given a choice, they preferred to oviposit in ceresin wax domes larger than the diameter of the largest-fruit olive varieties or the length of the fruit of most varieties, which suggests that these domes may have represented a super-normal fruit-type stimulus for the females (Katsoyannos and Pittara, 1983). Finally, in *C. capitata* the larger artificial fruits were usually preferred over smaller ones, at least when not very large sizes were included in the tests (Féron, 1962; Sanders, 1968a; Katsoyannos and Boller, 1978, unpubl.). A strong inter-population variation in fruit-size acceptance by the females has been recently reported using natural host fruits (Prokopy et al., 1984). A similar variation in artificial fruit size preference between wild and artificially reared *C. capitata* females of different origins was observed by Katsoyannos and Boller (1978, unpubl.).

5.2.3 Effect of color

A rather limited number of reports deal with the effect of color on the selection of oviposition site by fruit flies. Most of them concern comparisons between a few selected colors. Others are limited to comparisons between black and white (e.g. Greany and Szentesi, 1979 for *A. suspensa*). In some investigations the color of the background and/or the degree of substrate's contrast with the background also has been considered. In general, the color properties of the oviposition substrates are important factors for the selection of the oviposition site. These factors however, seem to be less important than shape and size because they affect strongly the females' responses only in choice situations whereas in no choice tests the females usually oviposit well, although eventually with a certain delay, to substrates of any color, as long as these substrates are of suitable shape and size.

5.2.3.1 *Rhagoletis* spp.

Working in the laboratory with ceresin wax domes, Prokopy and Boller (1971a), found that the attractiveness of these domes for gravid *R. cerasi* females was reduced in the absence of color (white domes) or when the color was very light such as yellow or light grey. Domes of darker colors such as red, dark orange, light orange, blue, dark green, dark grey and black were equally preferred for oviposition. They concluded, that *R. cerasi*, like *R. pomonella* (Prokopy, 1966), prefers dark-colored over light-colored (yellow) oviposition sites against a white background and they suggested that the flies locate fruits in the field on the basis of their shape, size and also color-contrast against the background. Color-contrast effects have been also reported in this species by Haisch and Levinson (1980). On the other hand, Katsoyannos (1979) found that black domes on a grey background were preferred over yellow or green ones, but were less preferred than red and orange colored domes. These results, as well as the results of Prokopy and Boller (1971a), suggest that hue discrimination may also be involved in the selection of oviposition site by *R. cerasi*. Finally, Cirio (1972), reported that *R. completa* females preferred green and yellow artificial oviposition sites, matching the color of the host fruit of this fly, in preference to black ones.

5.2.3.2 *Dacus oleae*.

The response of gravid *D. oleae* females to 18 mm diameter ceresin wax domes of different colors has been recently studied in detail in the laboratory (Katsoyannos et al., 1985). The females showed a strong preference for oviposition in yellow and orange colored domes followed by green ones. Red, blue, black and white domes received only a limited number of eggs, when compared in choice tests with a preferred color. It was found that the observed preference for certain colors primarily depended on the color hue and not on the intensity of the total reflected light or the degree the colored domes contrasted with the background. The females strongly preferred hues reflecting maximally between 650 and 610 nm with an optimum at ca. 580 nm. Because yellow or orange colored olives of hues similar to those used in their experiments do not occur in the field, Katsoyannos et al. (1985), suggested that these hues may constitute a super-normal oviposition site stimulus for the females.

5.2.3.3 *Ceratitis capitata*. Effects of color on the oviposition behavior of *C. capitata* have been reported by several workers with mostly conflicting results. This is not surprising however, because most of this work has been conducted with flies reared artificially for an unknown number of generations and usually of unknown origin. But it is well known that this species possesses a great genetic variability and genetic plasticity, which is reflected also in its behavior. Thus, according to Féron (1962) differences in light intensity and quality may elicit variable response and the 'brilliancy' rather than the color of oviposition sites favours more or less egg laying. On the other hand, Sanders (1968b) reported that among differently colored spheres against a monochromatic background the females preferred those with sharper contrasts. However, among equal fruit models or real fruits placed on both sides of a background divided in two halves (black and white or black and grey), the females preferred those substrates placed in the darker part of the arena, irrespective of the degree of contrast with the background. He concluded, that the color of the background is more important than the color-contrast degree. Both the aforementioned works were conducted with artificially reared flies. Boller and Katsoyannos (1978, unpubl.), working with artificially reared and wild *C. capitata* flies of different origins, found considerable differences in the color preference for oviposition sites between different strains. Black domes in a grey background were preferred over white ones by all strains tested. But while females from pupae field-collected in Costa Rica preferred orange-colored domes over black ones, the reverse was true for F_1 flies of the same origin, reared artificially. A similar dependence of color preference on larval food has been indicated by Féron (1962) and has been clearly demonstrated by Katsoyannos and Papadopoulou (1985, unpubl.), who found that F_1 to F_3 flies reared on an artificial larval diet, showed a different color preference for oviposition sites from the original wild flies. When however the insects were reared again on fruits for one generation, they expressed the same color preference as their wild ancestors. In another recent study Marchini and Wood (1983), assessed color preference of *C. capitata* females using spherical fruit mimics. Colors were compared in two halves (upper and lower) of the same artificial fruit. Unfortunately essential details such as the flies' origin and holding conditions are not given in this brief report but it is indicated, that yellow, green and orange were particularly favoured, the most attractive being yellow. Black was much less favoured and red was the least attractive color of all. However, in a recent detailed study Katsoyannos et al. (1986), found that irrespective of the background color (black, white or grey), wild *C. capitata* females originating from figs collected in Chios, Greece, showed strong preference for oviposition in black, blue and red colored ceresin wax domes of 18 mm diameter. Orange and green domes received fewer eggs, while yellow and white domes received only a very limited number of eggs. These results in combination with others obtained using neutrally colored domes, suggest that these flies select the oviposition site not only on the basis of shape and size, but also on the basis of color properties; color hue and brightness were found to be of great importance while the degree the oviposition site contrasts to the background was of minor importance.

5.3 Experimental approaches and factors affecting female response to oviposition sites in the laboratory

In most laboratory experiments on factors affecting oviposition behavior, two kinds of measurements are usually made, separately or in combination: (a) More frequently counts of eggs deposited in the different treatments after a certain time, i.e. the end product of the behavioral chain leading to oviposition. (b) Detailed continuous observations of various behavioral manifestations of the flies such as sequence, frequency, distribution and duration of visits to the oviposition substrates and other activities, and/or momentary observations of the flies' activities and distribution at regular intervals. Direct behavioral observations on individual females give very valuable information about the mechanisms by which the flies locate and select the substrates and the real preference for certain substrates. Such data, when represented and analysed as flow diagrams, which illustrate the sequence of the different behavioral elements (for example of such diagrams see Katsoyannos, 1975, 1979, Katsoyannos and Pittara, 1983), are much more informative than tabulated data. In some instances they might even lead to conclusions quite different from those made on the basis of tabulated data alone (for a related example see Fig. 2 in Katsoyannos and Pittara, 1983 and its interpretation therein). Flow behavioral diagrams also give valuable information concerning the ovarian pressure (or oviposition drive) of the females, a factor of considerable importance for the acceptance of oviposition sites, which is often neglected.

Experiments concerning oviposition behavior are usually of the no-choice type (only one treatment per time is offered), or of the choice type (two or more treatments at the same time). Experiments of the no-choice type provide information concerning mostly the acceptance limits by the flies for the treatments offered (e.g. maximum or minimum size of oviposition site into which eggs can be laid), as well as on the persistence of the non-acceptance under high ovarian pressure. For example domes of a too small size which otherwise are not acceptable for oviposition by *R. cerasi* or *D. oleae*, may elicit oviposition of a small part of the population when the gravid females have no other choice and are caged with them for a few to several days (Katsoyannos, 1976 and 1984, unpubl.). Experiments of the choice type provide information concerning mostly differences in the preference between treatments which are within the acceptable limits. For example, while in no-choice tests no differences were found between domes of different sizes or different colors, *D. oleae* females showed a strong preference for certain of these sizes or colors when they were compared in choice tests (Katsoyannos and Pittara, 1983; Katsoyannos, 1984, unpubl.). Finally, some experiments have been conducted using groups of females or groups of both sexes (e.g. Prokopy and Boller, 1971a; Prokopy and Bush, 1973). This may create difficulties in conducting direct continuous observations and may sometimes lead to misinterpretations resulting from undesirable interactions among flies, such as those due to territorial and aggressive behavior. These negative effects can be avoided, when single females instead of groups are used. In this way the performance of the individuals can also be measured and the intra-population variability of response can be quantified (Katsoyannos and Boller, 1977). It is obvious from the above, that some details of the experimental design (e.g. groups of flies vs. individuals, choice vs. no-choice tests) may directly affect the behavior of the flies and subsequently the measurements made and the conclusions drawn.

A serious, but usually neglected factor affecting oviposition preference is the oviposition-deterring pheromones (ODP) deposited after oviposition on the substrate by several fruit fly species (see Averill and Prokopy, Chapter 3.5, p. 207). The presence of such pheromones or similarly acting chemicals may affect the results of direct observations by increasing considerably the female mobility and decreasing the visit-duration on pheromone-marked substrates (Katsoyannos, 1979). As a result, number of visits on preferred and non-preferred substrates may be equalised, leading to the wrong conclusion, that the

Chapter 4.2 references, p. 321

preferred substrates were discriminated and selected by the females only after arrival on them. In experiments, where only the number of eggs laid is counted, the real preference of the female may also be masked as an artifact of the concentration-dependence of the deterring effects of ODP. For example, when two different sizes are compared and the same number of eggs is initially deposited in each (i.e. equal preference), the concentration of ODP per surface unit will be higher in the smaller substrates. This leads to a shift of the preference in favor of the larger substrates, as has been shown in *R. cerasi* (Katsoyannos, 1979). To avoid this negative effect, Katsoyannos and Boller (1977), proposed a distinction in such experiments between initial preference, based on the first eggs deposited and the persistence of preference, based on eggs deposited after a certain time. Using more than one substrate per treatment can reduce to a certain degree the negative effects of this factor. Some additional factors identified as affecting fruit fly response to oviposition sites in the laboratory include: Fruit fly origin and genetical make up, larval food, and learning through adult experience (Katsoyannos, 1979; Prokopy et al., 1982, 1984).

6 CONCLUDING REMARKS

I have tried to review our current knowledge on fruit fly responses to objects of different shape, size and color, emphasizing the main findings and their interpretations. Some questions and/or doubts arising in respect to the experimental approaches employed and the interpretations of the results have been discussed.

It is apparent, that considerable progress has been achieved in this field of research, especially during the last two decades. However, our knowledge on fruit fly responses to some physical stimuli remains rather meagre, essentially restricted to registering and classifying some simple behavioral responses in a limited number of species. The underlying neurophysiological and genetical mechanisms remain largely unknown.

Certain aspects of fruit fly responses such as field response to spheres and to small rectangles of different colors, have been investigated rather extensively in a few species of economic importance, because of the significance of the related findings for trapping purposes. For most species however, even these aspects have been either inadequately investigated or not at all. Factors, which may affect the responses in the field or in the laboratory, are either not yet identified, or not adequately considered. Specially designed laboratory experiments, which could provide the explanations needed, are rather limited and restricted to almost only the ovipositional responses of a few economically important species. In this respect, it is very fortunate, that in most tephritids a high degree of oviposition can be elicited into inanimate objects. This attribute facilitates considerably the conduct of studies on oviposition behavior of fruit flies, offering excellent research possibilities. Still other aspects have only recently begun to be the object of more intensive research such as the influence of the genetic make-up and genetic variability on the insect's behavior, the fruit fly responses in relation to their foraging behavior, or the effects due to certain neurophysiological processes such as learning.

It is obvious, that although we have a certain idea of how some fruit flies visually detect and respond to some objects of certain optical characteristics, much more is needed in order to obtain an improved understanding concerning the biological significance of the observed responses and the underlying neurophysiological and genetic mechanisms.

7 ACKNOWLEDGEMENT

I sincerely thank Prof. M.E. Tzanakakis for critically reading the manuscript.

8 REFERENCES

Agee, H.R. and Chambers, D.L., 1980. Fruit fly quality monitoring. In: Proceedings of a Symposium on Fruit Fly Problems, Kyoto and Naha, August 9–12, 1980. National Institute of Agricultural Sciences, Yataba, Ibaraki 305, Japan: 7–15.

Agee, H.R., Phillis, W.A. and Chambers, D.L., 1977. The compound eye of the Caribbean fruit fly and the apple maggot fly. Annals of the Entomological Society of America, 70: 359–364.

Agee, H.R., Boller, E., Remund, U., Davis, J.C. and Chambers, D.L., 1982. Spectral sensitivities and visual attractant studies in the Mediterranean fruit fly, *Ceratitis capitata* (Wiedemann), olive fly, *Dacus oleae* (Gmelin), and the European cherry fruit fly, *Rhagoletis cerasi* (L.) (Diptera, Tephritidae). Zeitschrift für Angewandte Entomologie, 93: 403–412.

Baker, A.C., Stone, W.E., Plummer, C.C. and McPhail, M., 1944. A review of studies on the mexican fruit fly and related mexican species. United States Department of Agriculture, Miscellaneous Publications No. 531, 155 pp.

Bateman, M.A., 1976. Fruit flies. In: V.L. Delucchi (Editor), Studies in Biological Control. Cambridge University Press, pp. 11–49.

Boller, E.F., 1968. An artificial oviposition device for the European cherry fruit fly, *Rhagoletis cerasi*. Journal of Economic Entomology, 61: 850–851.

Boller, E., 1969. Neues über die Kirschenfliege: Freilandversuche im Jahre 1969. Schweizerische Zeitschrift für Obst- und Weinbau, 105: 566–572.

Cirio, U., 1972. Osservazioni sul comportamento di ovideposizione della *Rhagoletis completa* Cresson (Diptera, Trypetidae) in laboratorio. In: Atti del IX Congresso Nazionale Italiano di Entomologia. Siena 21-25 Giugno 1972: 99–117.

Cytrynowicz, M., Morgante, J.S. and De Souza, H.M.L., 1982. Visual responses of South American fruit flies, *Anastrepha fraterculus*, and Mediterranean fruit flies, *Ceratitis capitata*, to colored rectangles and spheres. Environmental Entomology, 11: 1202–1210.

Davis, J.C., Agee, H.R. and Ellis, E.A., 1983. Comparative ultra-structure of the compound eye of the wild, laboratory-reared, and irradiated Mediterranean fruit fly, *Ceratitis capitata* (Diptera: Tephritidae). Annals of the Entomological Society of America, 76: 322–332.

Davis, J.C., Agee, H.R. and Chambers, D.L., 1984. Trap features that promote capture of the Caribbean fruit fly. Journal of Agricultural Entomology, 1: 236–248.

Drummond, F., Groden, E. and Prokopy, R.J., 1984. Comparative efficacy and optimal positioning of traps for monitoring apple maggot flies (Diptera: Tephritidae). Environmental Entomology, 13: 232–235.

Féron, M., 1962. L'instinct de reproduction chez la mouche méditerranéenne des fruit *Ceratitis capitata* Wied. (Dipt., Trypetidae). Comportement sexuel. Comportement de ponte. Revue de Pathologie Végétale et d'Entomologie Agricole de France, 41: 1–129.

Féron, M., Delanoue, P. and Soria, F., 1958. L'élevage massif artificiel de *Ceratitis capitata* Wied. Entomophaga, 3: 45–53.

Frick, K.E., Simkover, H.G. and Telford, H.S., 1954. Bionomics of the cherry fruit flies in eastern Washington. Washington Agricultural Experimental Station, Technical Bulletin, 13: 1–66.

Friend, A.H., 1957. Artificial infestation of oranges with the Queensland fruit fly. Journal of Australian Institute of Agricultural Sciences, 23(1): 77–80.

Gibson, K.E. and Kearby, W.H., 1978. Seasonal life history of the walnut husk fly and husk maggot in Missouri. Environmental Entomology, 7: 81–87.

Girolami, V. and Cavalloro, R., 1973. Metodi cromotropici per indagini di popolazione degli adulti di *Dacus oleae* Gmelin. Note ed Appunti Sperimentali di Entomologia Agraria, 14: 13–29.

Greany, P.D. and Szentesi, A., 1979. Oviposition behavior of laboratory-reared and wild Caribbean fruit flies (*Anastrepha suspensa*: Diptera: Tephritidae): II. Selected physical influences. Entomologia Experimentalis et Applicata, 26: 239–244.

Greany, P.D., Agee, H.R., Burditt, A.K. and Chambers, D.L., 1977. Field studies on color preferences of the Caribbean fruit fly, *Anastrepha suspensa* (Diptera: Tephritidae). Entomologia Experimentalis et Applicata, 21: 63–70.

Greany, P.D., Burditt Jr., A.K., Agee, H.R. and Chambers, D.L., 1978. Increasing effectiveness of visual traps for the Caribbean fruit fly, *Anastrepha suspensa* (Diptera: Tephritidae), by use of fluorescent colors. Entomologia Experimentalis et Applicata, 23: 20–25.

Hagen, K.S., Santas, L. and Tsekouras, A., 1963. A technique of culturing the olive fly, *Dacus oleae* Gmelin on synthetic media under xenic conditions. In: Radiation and Radioisotopes Applied to

Insects of Agricultural Importance. Proceedings Symposium Athens 22–26 April 1963, International Atomic Energy Agency, Vienna, pp. 333–356.

Haisch, A., 1968. Preliminary results in rearing the cherry fruit fly (*Rhagoletis cerasi* L.) on a semi-synthetic medium. In: Radiation, Radioisotopes and Rearing Methods in the Control of Insect Pests. Proceedings Symposium Tel Aviv 1966, International Atomic Energy Agency, Vienna, pp. 69–78.

Haisch, A. and Levinson, H.Z., 1980. Influences of fruit volatiles and coloration on oviposition of the cherry fruit fly. Naturwissenschaften, 67: 44–45.

Hill, A.R. and Hooper, G.H.S., 1984. Attractiveness of various colours to Australian tephritid fruit flies in the field. Entomologia Experimentalis et Applicata, 35: 119–128.

Johnson, P.C., 1983. Response of adult apple maggot (Diptera: Tephritidae) to Pherocon A.M. traps and red spheres in a non-orchard habitat. Journal of Economic Entomology, 76: 1279–1284.

Katsoyannos, B.I., 1975. Oviposition-deterring, male-arresting, fruit-marking pheromone in *Rhagoletis cerasi*. Environmental Entomology, 4: 801–807.

Katsoyannos, B.I., 1979. Zum Reproduktions- und Wirtswahlverhalten der Kirschenfliege, *Rhagoletis cerasi* L. (Diptera: Tephritidae). Dissertation Nr. 6409 ETH Zurich, 180 pp.

Katsoyannos, B.I., 1987a. Field responses of Mediterranean fruit flies to colored spheres suspended on fig, citrus and olive trees. Proceedings of the 6th International Symposium on Insect-Plant Relationships, Pau, France, 1–5 July 1986. Dr. W. Junk Publishers, Dordrecht, in press.

Katsoyannos, B.I., 1987b. Some factors affecting field responses of Mediterranean fruit flies to colored spheres of different sizes. Proceedings of 2nd International Symposium on Fruit Flies. Colymbari, Crete, Greece, 16–21 September 1986. Elsevier Science Publishers, Amsterdam, in press.

Katsoyannos, B.I., 1987c. Effect of color properties of spheres on their attractiveness for *Ceratitis capitata* flies in the field. Zeitschrift für Angewandte Entomologie, in press.

Katsoyannos, B.I., 1987d. Field responses of Mediterranean fruit flies to spheres of different color patterns and to yellow crossed panels. Proceedings of CEC/IOBC International Symposium on Fruit Flies of Economic Importance. Rome, Italy, 7–10 April 1987. A.A. Balkema, Rotterdam, in press.

Katsoyannos, B.I. and Boller, E.F., 1977. Testing *Rhagoletis cerasi* for discrimination of fruit size. In: E.F. Boller and D.L. Chambers (Editors), Quality Control, an Idea Book for Fruit Fly Workers, Bulletin SROP/WPRS 1977/5, pp. 64–65.

Katsoyannos, B.I. and Pittara, I.S., 1983. Effect of size of artificial oviposition substrates and presence of natural host fruits on the selection of oviposition site by *Dacus oleae*. Entomologia Experimentalis et Applicata, 34: 326–332.

Katsoyannos, B.I., Patsouras, G. and Vrekoussi, M., 1985. Effect of color hue and brightness of artificial oviposition substrates on the selection of oviposition site by *Dacus oleae*. Entomologia Experimentalis et Applicata, 38: 205–214.

Katsoyannos, B.I., Panagiotidou, K. and Kechagia, I., 1986. Effect of color properties on the selection of oviposition site by *Ceratitis capitata*. Entomologia Experimentalis et Applicata, 42: 187–193.

Marchini, L. and Wood, R.J., 1983. Laboratory studies on oviposition and on the structure of the ovipositor in the Mediterranean fruit fly *Ceratitis capitata* (Wied.). In: R. Cavalloro (Editor), Fruit Flies of Economic Importance. Proceedings of the CEC/IOBC International Symposium, Athens, Greece, 16–19 November 1982, p. 113.

McPhail, M. and Guiza, F.E., 1956. An oviposition medium for the Mexican fruit fly. Journal of Economic Entomology, 49: 570.

Meats, A., 1983. The response of the Queensland fruit fly, *Dacus tryoni*, to tree models. In: R. Cavalloro (Editor), Fruit Flies of Economic Importance. Proceedings of the CEC/IOBC International Symposium, Athens, Greece, 16–19 November 1982, pp. 285–289.

Moericke, V., Prokopy, R.J., Berlocher, S. and Bush, G.L., 1975. Visual stimuli eliciting attraction of *Rhagoletis pomonella* (Diptera: Tephritidae) flies to trees. Entomologia Experimentalis et Applicata, 18: 497–507.

Nakagawa, S., Prokopy, R.J., Wong, T.T.Y., Ziegler, J.R., Mitchell, S.M., Urago, T. and Harris, E.J., 1978. Visual orientation of *Ceratitis capitata* flies to fruit models. Entomologia Experimentalis et Applicata, 24: 193–198.

Neilson, W.T.A., Knowlton, A.D. and Fuller, M., 1984. Capture of blueberry maggot adults, *Rhagoletis mendax* (Diptera:Tephritidae), on Pherocon® AM traps and on tartar red dark sticky spheres in lowbush blueberry fields. The Canadian Entomologist, 116: 113–118.

Oatman, E.R., 1964. Apple maggot trap and attractant studies. Journal of Economic Entomology, 57: 529–531.

Owens, E.D. and Prokopy, R.J., 1984. Habitat background characteristics influencing *Rhagoletis pomonella* (Walsh) (Dipt., Tephritidae) fly response to foliar and fruit mimic traps. Zeitschrift für Angewandte Entomologie, 98: 98–103.

Prokopy, R.J., 1966. Artificial oviposition devices for apple maggot. Journal of Economic Entomology, 59: 231–232.

Prokopy, R.J., 1968. Visual responses of apple maggot flies, *Rhagoletis pomonella* (Diptera: Teph-
    ritidae): Orchard studies. Entomologia Experimentalis et Applicata, 11: 403–422.

Prokopy, R.J., 1969. Visual responses of European cherry fruit flies - *Rhagoletis cerasi* L. (Diptera,
    Trypetidae). Polskie Pismo Entomologiczne, 39: 539–566.

Prokopy, R.J., 1972. Response of apple maggot flies to rectangles of different colors and shades.
    Environmental Entomology, 1: 720–726.

Prokopy, R.J., 1973. Dark enamel spheres capture as many apple maggot flies as fluorescent
    spheres. Environmental Entomology, 2: 953–954.

Prokopy, R.J., 1975. Selective new trap for *Rhagoletis cingulata* and *R. pomonella* flies. Environ-
    mental Entomology, 4: 420–424.

Prokopy, R.J., 1977a. Stimuli influencing trophic relations in Tephritidae. Colloques Internation-
    aux du C.N.R.S., Comportment des Insectes et Millieu Trophique, 265: 305–336.

Prokopy, R.J., 1977b. Attraction of *Rhagoletis* flies (Diptera: Tephritidae) to red spheres of different
    sizes. The Canadian Entomologist, 109: 593–596.

Prokopy, R.J., 1983. Tephritid relationships with plants. In: R. Cavalloro (Editor), Fruit Flies of
    Economic Importance. Proceedings of the CEC/IOBC International Symposium, Athens,
    Greece, 16–19 November 1982, pp. 230–239.

Prokopy, R.J. and Boller, E.F., 1970. Artificial egging system for the European cherry fruit fly.
    Journal of Economic Entomology, 63: 1413–1417.

Prokopy, R.J. and Boller, E.F., 1971a. Stimuli eliciting oviposition of European cherry fruit flies,
    Rhagoletis cerasi (Diptera: Tephritidae), into inanimate objects. Entomologia Experimentalis et
    Applicata, 14: 1–14.

Prokopy, R.J. and Boller, E.F., 1971b. Response of European cherry fruit flies to colored rectangles.
    Journal of Economic Entomology, 64: 1444–1447.

Prokopy, R.J. and Bush, G.L., 1973. Ovipositional responses to different sizes of artificial fruit by
    flies of *Rhagoletis pomonella* species group. Annals of the Entomological Society of America, 66:
    927–929.

Prokopy, R.J. and Coli, W.M., 1978. Selective traps for monitoring *Rhagoletis mendax* flies. Protec-
    tion Ecology, 1: 45–53.

Prokopy, R.J. and Economopoulos, A.P., 1976. Color responses of *Ceratitis capitata* flies. Zeitschrift
    für Angewandte Entomologie, 80: 434–437.

Prokopy, R.J. and Haniotakis, G.E., 1975. Responses of wild and laboratory-cultured *Dacus oleae*
    to host plant color. Annals of the Entomological Society of America, 68: 73–77.

Prokopy, R.J. and Haniotakis, G.E., 1976. Host detection by wild and lab-cultured olive flies. In:
    T. Jermy (Editor), The Host-Plant in Relation to Insect Behaviour and Reproduction. Symp.
    Biol. Hungary, no. 16: 209–214.

Prokopy, R.J. and Owens, E.D., 1978. Visual generalist-visual specialist phytophagous insects:
    Host selection behaviour and application to management. Entomologia Experimentalis et
    Applicata, 24: 409–420.

Prokopy, R.J. and Owens, E.D., 1983. Visual detection of plants by herbivorous insects. Annual
    Review of Entomology, 28: 337–364.

Prokopy, R.J., Economopoulos, A.P. and McFadden, M.W., 1975. Attraction of wild and laboratory-
    cultured *Dacus oleae* flies to small rectangles of different hues, shades, and tints. Entomologia
    Experimentalis et Applicata, 18: 141–152.

Prokopy, R.J., McDonald, P.T. and Wong, T.T.Y., 1984. Inter-population variation among *Ceratitis
    capitata* flies in host acceptance pattern. Entomologia Experimentalis et Applicata, 35: 65–69.

Prokopy, R.J., Averill, A.C., Cooley, S.S., Roitberg, C.A. and Kallet, C., 1982. Variation in host
    acceptance pattern in apple maggot flies. In: J.H. Visser and A.K. Minks (Editors), Proceedings
    of the 5th International Symposium on Insect-Plant Relationships. PUDOC, The Netherlands:
    123–129.

Reissig, W.H., 1975. Performance of apple maggot traps in various apple tree canopy positions.
    Journal of Economic Entomology, 68: 534–538.

Reissig, W.H., 1976. Comparison of traps and lures for *Rhagoletis fausta* and *R. cingulata*. Journal
    of Economic Entomology, 69: 639–643.

Riedl, H. and Hoying, S.A., 1980. Seasonal patterns of emergence, flight activity and oviposition
    of the walnut husk fly in northern California. Environmental Entomology, 9: 567–571.

Riedl, H. and Hoying, S.A., 1981. Evaluation of trap designs and attractants for monitoring the
    walnut husk fly, *Rhagoletis completa* Cresson (Diptera: Tephritidae). Zeitschrift für Ange-
    wandte Entomologie, 91: 510–520.

Remund, U., Economopoulos, A.P., Boller, E.F., Agee, H.R. and Davis, J.C., 1981. Fruit fly quality
    monitoring: The spectral sensitivity of field-collected and laboratory-reared olive flies, *Dacus
    oleae* Gmelin (Dipt., Tephritidae). Mitteilungen der Schweizerischen Entomologischen
    Gesellschaft, 54: 221–227.

Sanders, W., 1962. Das Verhalten der Mittelmeerfruchtfliege *Ceratitis capitata* Wied. bei der
    Eiablage. Zeitschrift für Tierpsychologie, 19: 1–28.

Sanders, W., 1968a. Die Eiablagehandlung der Mittelmeerfruchtfliege *Ceratitis capitata* Wied. Ihre

Abhängigkeit von Grösse und Dichte der Früchte. Zeitschrift für Tierpsychologie, 25: 1–21.

Sanders, W., 1968b. Die Eiablagehandlung der Mittelmeerfruchtfliege *Ceratitis capitata* Wied. Ihre Abhängigkeit von Farbe und Gliederung des Umfeldes. Zeitschrift für Tierpsychologie, 25: 588–607.

Still, G.W., 1960. An improved trap for deciduous tree fruit flies. Journal of Economic Entomology, 53: 967.

Tanaka, N., 1965. Artificial egging receptacles for three species of tephritid flies. Journal of Economic Entomology, 58: 177–178.

Wiesmann, R., 1937. Die Orientierung der Kirschfliege, *Rhagoletis cerasi* L., bei der Eiablage. (Eine sinnesphysiologische Untersuchung). Landwirtschaftliches Jahresbuch der Schweiz, 51: 1080–1109.

Zwölfer, H., 1983. Life systems and strategies of resource exploitation in tephritids. In: R. Cavalloro (Editor), Fruit Flies of Economic Importance. Proceedings of the CEC/IOBC International Symposium, Athens, Greece, 16–19 November 1982, pp. 16–30.

Chapter 4.3 Behavioural Partitioning of the Day and Circadian Rhythmicity

P.H. SMITH[1]

1 INTRODUCTION

Many types of behaviour, all necessary for survival and reproduction, compete with each other for expression. Switching between different behaviours occurs frequently, in response to changes in stimulus conditions and levels of motivation and due to interactions between behaviours such as inhibition and facilitation. In addition, there is an overriding level of organisation which causes some behaviours to be expressed more frequently at particular times of day independent of stimulus conditions. In this case central control through a circadian mechanism is usually involved. In tsetse flies, for example, a number of motivational systems show similar diel patterning of overt behaviour and of responsiveness, a bimodal distribution with peaks following dawn and dusk. Circadian control of most of these responses has been demonstrated (Brady, 1975). By contrast, in milkweed bugs of the genus *Oncopeltus*, oviposition, mating and feeding behaviours, each under circadian control, occur at distinctly different times of day (Alden et al., 1983; Caldwell and Dingle, 1967; Caldwell and Rankin, 1974).

In many tephritids diel patterning of behaviour has been developed to such a degree that the day is to an extent partitioned by the major motivational systems so that conflict between them is minimised.

2 DIEL PARTITIONING IN THE BEHAVIOUR OF *Dacus tryoni*

Time partitioning in behaviour is very distinct in the Queensland fruit fly, *Dacus tryoni* (Froggatt). At constant temperature under a laboratory light/dark (LD) cycle with hour-long constant twilights (10 lux) *D. tryoni* males and virgin females show the daily pattern of spontaneous flight activity in Fig. 4.3.1. The flies are inactive at night with activity ceasing rapidly as darkness begins. There is an intense period of activity during the dusk twilight and a lesser peak during the dawn twilight with a low level of activity through the middle of the photophase.

Mated females display a radically different pattern of flight activity (Fig. 4.3.1). Activity during the photophase is greatly enhanced. The dawn peak is missing and activity does not increase during the dusk twilight above the level that preceded it under high light intensity. Mating by males has no effect on their activity pattern.

Patterns of flight activity are aggregate effects of the number of motivation-

---

[1]The author was previously known as Peter H. Tychsen.

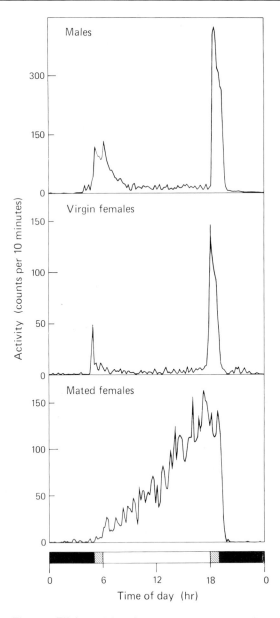

Fig. 4.3.1 Flight activity of males, virgin females and mated females of *Dacus tryoni* in a laboratory light/dark cycle with hour-long constant light intensity (10 lux) dawn and dusk. Each actograph contained 10 flies and operated as in Smith (1983). Flight activity was monitored in 10 min blocks for a 10-day period and the data aggregated (P.H. Smith, unpublished, 1985).

al systems that flight serves, predominantly those controlling sexual behaviour, oviposition and feeding. The changing patterns occur because the intensities of these behaviours vary according to the time of day and stage of the life cycle.

Mating in *D. tryoni* is entirely limited to the dusk twilight under such an LD cycle, and is accompanied by high levels of flight activity. In nature sexual activity is limited to a period of about 30 min as the light intensity falls at dusk (Fig. 4.3.2) (Tychsen and Fletcher, 1971; Tychsen, 1977). No mating occurs at dawn though, as Fig. 4.3.1 demonstrates, an increase in flight activity in males and virgin females occurs at this time.

Daily cycles in the responsiveness associated with a number of aspects of sexual behaviour have been demonstrated. To measure responsiveness, flies must be presented with a standard stimulus situation and the intensity of

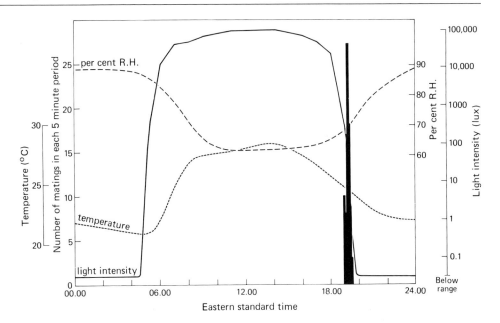

Fig. 4.3.2 Histogram of number of matings of *Dacus tryoni* occurring in each 5 min period during a day under field conditions. The changes in relative humidity, temperature, and light intensity are also shown (from Tychsen and Fletcher, 1971).

behaviour recorded. Light intensity within a certain range is a dominant stimulus requirement for many aspects of sexual behaviour in *D. tryoni*. Mating activity falls off rapidly at light intensities above and below the optimal value of 10 lux (Smith, 1979) and similar effects of light intensity have been demonstrated for the female's response to a male-produced pheromone (Fletcher and Giannakakis, 1973) and for stridulation by males (Giannakakis, 1976) (Fig. 4.3.3). Therefore for assaying responsiveness in aspects of sexual behaviour samples of flies are taken from a population and exposed to periods at 10 lux, either after a step-down from 10,000 lux at all times of day or from the lighting conditions appropriate for the time of day. The behaviour is then

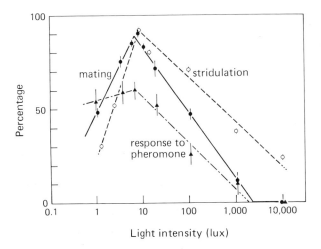

Fig. 4.3.3 Percentage mating, stridulation and response to pheromone of *Dacus tryoni* during exposure to different light intensities at the normal time of dusk in a laboratory light/dark cycle. *Mating*: 50 females and 100 males in a 25 × 25 × 38 cm cage, 30 min observation, 10 replicates (Smith, 1979). *Stridulation*: 5 males in a 2.5 × 10 cm glass tube, 15 min observation, 10 replicates (Giannakakis, 1976). *Response to pheromone*: 4 male equivalents of pheromone presented to 5 females in a 25 × 25 × 38 cm cage, 15 min observation, 10 replicates (Fletcher and Giannakakis, 1973). The vertical lines indicate ± S.E.

Chapter 4.3 references, p. 338

recorded for 30 min or less. Closely correlated diel cycles of the responsiveness associated with mating (similar in both sexes, see Tychsen and Fletcher, 1971), male stridulation (which is closely correlated with pheromone release) and female response to pheromone have been demonstrated in this way (Fig. 4.3.4). Under the photoperiod chosen (14 h light, 1 h dusk at 10 lux, 9 h darkness), no behaviour is observed in the assay when it is applied during the first 8 h of the photophase. After this time the intensity of the behaviour rises to a maximum at the normal time of dusk then declines again. The rate of decline after dusk in an LD cycle is difficult to measure unequivocally. The rate shown in Fig. 4.3.4A occurs when a strictly standardized assay of responsiveness is used but during a day in which the photophase is extended up till the time of each assay. The true rate of decline in the intact light/dark cycle is more rapid, more like that in Fig. 4.3.4B but this is also not a true estimate since the mating responsiveness assay applied after dark in this case lacks a rapid decline in light intensity which may be an important component of the stimulus.

A similar diel rhythm in the response of virgin females to the 'male' lure, cue-lure, is reported by Fletcher (1969) though this is better documented for other dacines (see below).

After they have mated once, females show greatly reduced sexual activity, becoming unreceptive to mating attempts (Tychsen, 1972) and ceasing to respond to pheromone (Fletcher and Giannakakis, 1973), for many (> 14) days. Associated with this, their flight activity does not increase during dusk. The flight activity of mated females is higher during the photophase because oviposition occurs then. Watson (1971) has shown that responsiveness to oviposition stimuli is maximal near the centre of the photophase (Fig. 4.3.5). Feeding is maximal during the early morning hours, at least under a LD 8:16 photoperiod (Barton Browne, 1956).

Male responsiveness to the synthetic lure, cue-lure, is maximal in the morn-

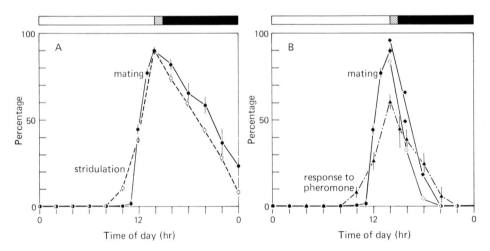

Fig. 4.3.4 Percentage mating, stridulation and response to pheromone of *Dacus tryoni* in assays at constant 10 lux applied at different times in a laboratory light/dark cycle. A. All assays involved transfer from daytime light intensity (10,000 lux) to 10 lux. For assays after the normal time of dusk the daytime light intensity was extended until the time of the assay. B. Assays involved transfer from the lighting conditions prevailing at the time (either daytime light intensity or darkness) to 10 lux.

Mating: 50 females and 100 males in a 25 × 25 × 38 cm cage, 30 min observation, 10 replicates except after dusk in B where two populations were tested repeatedly during a single night. Two samples of flies from each population were assayed at each time and the means of the two values plotted (Tychsen and Fletcher, 1971; Tychsen, 1978). *Stridulation*: 5 males in a 2.5 × 10 cm glass tube, 15 min observation, 10 replicates (Giannakakis, 1976). *Response to pheromone*: 4 male equivalents of pheromone presented to 5 females in a 25 × 25 × 38 cm cage, 15 min observation, 10 replicates (Fletcher and Giannakakis, 1973). The vertical lines indicate ± S.E.

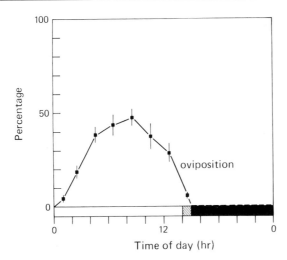

Fig. 4.3.5 Percentage of *Dacus tryoni* females ovipositing in an assay applied at different times during a laboratory light/dark cycle. Samples of 20 gravid females were taken from a large population and exposed (in a 25 × 25 × 38 cm cage) to a fresh apple hollow for 15 min. The vertical lines indicate ± S.E. (Watson, 1971).

ing and reduces towards the late afternoon (Fletcher, personal communication, 1985) which parallels the pattern of catches in cue-lure baited traps reported by Brieze-Stegeman et al. (1978).

3 TIME PARTITIONING AMONG OTHER ADULT TEPHRITIDS

3.1 Sexual behaviour

Other Tephritids also show partitioning of aspects of adult behaviour by time of day. In many species mating is limited to the late afternoon and dusk. In fact, this is the common situation among the Dacini, having been reported in *D. tryoni* (Tychsen and Fletcher, 1971), *Dacus dorsalis* Hendel (Roan et al., 1954; Bess and Haramoto, 1961), *Dacus oleae* (Gmelin) (Causse et al., 1966), *Dacus cucurbitae* Coquillett (Back and Pemberton, 1917; Kuba et al., 1984), *Dacus cacuminatus* (Hering) (Myers, 1952a), *Dacus passiflorae* (Froggatt) (Simmonds, 1936), *Dacus halfordiae* (Tryon) (Barton Browne, 1957), *Dacus bryoniae* (Tryon) (Barton Browne, 1957), *Dacus ciliatus* Loew (Syed, 1969), *Dacus hageni* De Meijere (Syed, 1970), *Dacus scutellaris* (Bezzi) (Syed, 1970), *Dacus diversus* Coquillett (Syed, 1970), *Dacus zonatus* (Saunders) (Syed et al., 1970), *Dacus kraussi* Hardy (Fletcher, personal communication, 1972), *Dacus cucumis* French (Hooper, 1975), *Dacus jarvisi* (Tryon) (Fitt, 1981b), *Dacus opiliae* Drew and Hardy, *Dacus aquilonis* (May) (Fitt, 1981a), *Dacus decurtans* (May), *Dacus endiandrae* (Perkins and May), *Dacus musae* (Tryon), and *Dacus aglaiae* Hardy (G.P. Fitt, personal communication, 1985).

Among these dusk mating species there is variation in the time when mating begins under natural lighting conditions. *D. tryoni*, *D. cucurbitae* and *D. dorsalis* begin mating at light intensities similar to that at sunset under a clear sky (Tychsen and Fletcher, 1971; Suzuki and Koyama, 1980; Arakaki et al., 1984). By contrast, *D. opiliae* begins mating about 40 min before sunset (Fitt, 1981a) and *D. cucumis* 120 min before sunset (Hooper, 1975).

There are exceptions among Dacini to the dusk mating rule. *D. neohumeralis* Hardy mates only during the middle of the photophase under high light intensity (Tychsen, 1977; Smith, 1979). This species is sympatric with *D. tryoni* and very closely related to it. In fact, the time of mating is the only known barrier

Chapter 4.3 references, p. 338

to hybridisation between them (Lewontin and Birch, 1966; Vogt, 1977). This case is an example of partitioning of the day between species as a sexual isolating mechanism. *Dacus tsuneonis* Miyake and *Dacus tenuifascia* (May) are also exceptional. Both initiate copulation at any time during the photophase (Miyake, 1919; Fitt, 1981a).

Among tephritids that do not belong to the tribe Dacini mating at dusk is unusual. Mating normally occurs at any time of the photophase when the temperature is in a favourable range and the light intensity is sufficient to assure adequate vision. This has been recorded for *Rhagoletis pomonella* (Walsh) (Prokopy et al., 1972), *Rhagoletis completa* (Cresson) (Boyce, 1934), *Rhagoletis conversa* (Brèthes) (Frías et al., 1984), *Rhagoletis mendax* Curran (Smith and Prokopy, 1981), *Ceratitis capitata* (Wiedemann) (Myrburgh, 1962; Prokopy and Hendrichs, 1979), *Toxotrypana curvicauda* Gerstaecker (Landolt and Hendrichs, 1983), *Trupanea bisetosa* (Coquillett) (Cavender and Goeden, 1982), *Eurosta solidaginis* (Fitch) (Uhler, 1951), *Epochra canadensis* (Loew) (Severin, 1917), *Zonosemata electa* (Say) (Peterson, 1923), *Icterica seriata* (Loew) (Foote, 1967), *Icterica circinata* (Loew) (Foote, 1967), *Procecidochares sp.* (Silverman and Goeden, 1980) and *Adrama biseta* Malloch. (G.P. Fitt, personal communication, 1985).

Reports of observations of mating behaviour without specific reference to time of day imply that daytime mating is also the case in *Urophora sirunaseva* Hering (Zwölfer, 1969), *Urophora cardui* (Linnaeus) (Zwölfer et al., 1970), *Tephritis stigmatica* (Coquillett) (Tauber and Toschi, 1965a), *Aciurina ferruginea* (Doane) (Tauber and Tauber 1967), *Paroxyna albiceps* (Loew) (Novak and Foote, 1968), *Zonosemata vittigera* (Coquillett) (Cazier, 1962), and *Procecidochares utilis* Stone (Haseler, 1965).

Some of these species concentrate their mating during a part of the photophase. For example *T. bisetosa* and *C. capitata* both mate dominantly during the morning; *T. curvicaudata* more frequently during the afternoon.

Dusk mating occurs in a few non-dacines. Mating is concentrated in the late afternoon and early evening in *Anastrepha ludens* (Loew), *Anastrepha oblicua* (Macquart), *Anastrepha suspensa* (Loew) (McPhail and Bliss, 1933; Baker et al., 1944; Perdomo et al., 1975; Aluja et al., 1983) and *Euleia fratria* (Loew) (Tauber and Toschi, 1965b). Most mating occurs during the dusk period in *Pterandrus rosa* (Karsch) (Myrburgh, 1962) and *Dirioxa pornia* (Walker) (Pritchard, 1967).

Anastrepha fraterculus (Wiedemann) has a diel patterning of sexual behaviour in sharp contrast to all other tephritids reported to date. Sexual behaviour in this species occurs exclusively from 1 h after dawn till mid-morning (Malavasi et al., 1983).

The response to pheromone by virgin females parallels the daily mating rhythm in *D. opiliae, D. aquilonis* (Fitt, 1981a), *D. jarvisi* (Fitt, 1981b), *D. neohumeralis* (own observations), and *D. tryoni* (Fletcher and Giannakakis, 1973). Response to 'male' lures by virgin females similarly parallels the mating rhythm in *D. opiliae, D. tenuifascia* and *D. aquilonis* (Fitt, 1981a). Male response to the lure is during the photophase in *D. opiliae*. The flies feed avidly on the lure (Fitt, 1981a).

3.2 Oviposition

All reports indicate that tephritids oviposit during the photophase. For many species no further details are provided e.g. *A. ludens* (McPhail and Bliss, 1933), *D. cucurbitae* (Back and Pemberton, 1917), *U. siruna-seva* (Zwölfer, 1969), *U. cardui* (Zwölfer et al., 1970), *T. bisetosa* (Cavender and Goeden, 1982), *E. solidaginis* (Uhler, 1951) and *Procecidochares sp.* (Silverman and Goeden, 1980). However, from cases where more detail is given it is obvious that a wide variety of distributions of ovipositing activity occur. Oviposition peaks at midday in *D. tryoni* (Watson, 1971), *D. dorsalis* (Arai, 1975), *C. capitata* (Prokopy and Hendrichs, 1979) and *A. fraterculus* (Malavasi et al., 1983). It peaks in the late afternoon in *Dacus umbrosus* (Fabricius) (Cendaña and Namocale, 1953), *D. cucurbitae* (Nishida and Bess, 1957), *R. completa* (Boyce, 1934), *T. curvicauda* (Landholt and Hendrichs, 1983), *I. circinata* (Foote, 1967) and *I. seriata* (Foote, 1967). It is uniform throughout the photophase in *R. conversa* (Frías et al., 1984).

3.3 Feeding

The time of feeding of tephritids has rarely been reported. Christensen and Foote (1960) comment that fruit flies may feed at any time of day. Laboratory observations on *D. tryoni, D. cucurbitae* and *D. dorsalis* suggest that feeding is more prevalent in the morning (Barton Browne, 1956; Suzuki and Koyama, 1980; Arakaki et al., 1984). Field observations on *R. mendax* indicate fairly uniform feeding throughout the photophase (Smith and Prokopy, 1981).

3.4 The daily routine

To approach the question of how a fly's daily routine is organised it is necessary to carry out simultaneous observations on a number of behaviours. Locomotion, grooming, feeding, stridulation and physical encounters in groups of 10 males or 10 females have been recorded simultaneously in the laboratory in *D. cucurbitae* (Suzuki and Koyama, 1980) and *D. dorsalis* (Arakaki et al., 1984). For a more realistic indication, however, these observations need to be carried out in the field. Here many aspects differ from the laboratory circumstance, Temperature, light, humidity, wind and physical aspects of the environment like the distances between resources and difficulty in locating them all affect the daily organisation of activities. Recently an exciting start has been made through observations in field cases large enough to contain host trees. Much detail has been added to the general timing information as described above. In addition, much more refined partitioning of time has sometimes been demonstrated through these observations. For example, Causse and Féron (1967) working with *C. capitata* showed that, in laboratory cages, mating was uniformly high for about the first 7 h of the photophase, then declined to be very low for the remainder. In flies confined in a field cage surrounding a coffee tree Prokopy and Hendrichs (1979) found a similar period of mating activity but that two distinct mating strategies partitioned the time. In the mid-morning and late afternoon males were predominantly on the fruit where they approached ovipositing females and attempted to mate with them. In the late morning and early afternoon males released pheromone and fanned their wings on the bottom of leaves waiting for females to approach them. At this time males often aggregated into leks.

4 DIEL PATTERNING OF BEHAVIOUR DURING DEVELOPMENT

Two behaviours during development have been observed to show diel rhythmicity; larval jumping and eclosion. Where it has been looked for, an egg hatch rhythm has not been found (Arai, 1975).

Larval jumping occurs when larvae, which have completed their feeding, jump from the fruit or rearing medium onto the substrate where they burrow before pupariating. This behaviour begins before dawn and peaks at first light in *C. capitata* (Causse, 1974), *P. rosa* (Myrburgh, 1963), *R. completa* (Boyce, 1934), *A. ludens* (McPhail and Bliss, 1933), *D. ciliatus* (Malan and Giliomee,

1968), *D. dorsalis* (Arai, 1975, 1976a), *D. oleae* (Laudého et al., 1978) and *D. tryoni* (Bower, 1976). Rain can disrupt the pattern since it causes larvae to leave the fruit (McPhail and Bliss, 1933; Bower, 1976).

Eclosion of the adult from the puparium usually occurs underground and is followed by burrowing upward to the surface of the soil where the newly emerged adult hardens its wings before take-off. As in many Diptera (Saunders, 1982) this occurs close to dawn among Tephritidae including *P. rosa* (Myrburgh, 1963), *C. capitata* (Myrburgh, 1963), *R. completa* (Boyce, 1934), *R. mendax* (Lathrop and Nickels, 1932), *A. ludens* (McPhail and Bliss, 1933), *T. bisetosa* (Cavender and Goeden, 1982), *D. cacuminatus* (Myers, 1952b), *D. cucurbitae* (Back and Pemberton, 1917), *D. dorsalis* (Arai 1975, 1976b) and *D. oleae* (Laudého et al., 1978).

In *D. tryoni* at constant temperature with LD, 8:16 h the emergence peak is centred 4.75 h after lights-on (Bateman, 1955). However under a more typical light cycle (LD, 12:12 h) this species also emerges near dawn (own data).

5 PHYSIOLOGICAL MECHANISMS IN RHYTHMIC BEHAVIOUR

Though it has frequently been found that a central circadian mechanism plays a major role at the level of behavioural organisation operating in partitioning the diel, occurrence of a behaviour at a particular time of day does not demonstrate that an endogenous or 'circadian' rhythm is involved (Saunders, 1982; Brady, 1974). Behaviours can be limited by stimulus conditions e.g. the light intensity at dusk, or rhythmicity observed in a behaviour may be produced by circadian control applied at some earlier stage of the life cycle. For example, in the mosquito, *Aedes taeniorhynchus* (Wiedemann), pupation is under circadian control. Eclosion is not under circadian control but, since it occurs at a certain time after pupation dependent on the temperature, rhythmicity is impressed upon it (Provost and Lum, 1967; Nayar, 1968).

Some cases of rhythmic behaviour in tephritids have been subjected to the rigorous experimental procedures required to prove they are under circadian control and in a few cases details of the interaction between the circadian mechanism and its zeitgebers and the mechanism and the observed behaviour have been obtained. None of these studies has yet progressed so far as to provide such important basic data in the characterisation of a circadian rhythm as a phase response curve.

5.1 Investigations on behaviours during development

Causse (1974) has studied the role of light in controlling the rhythm of larval jumping in *C. capitata* in some detail. Under natural lighting conditions and artificial LD cycles of similar photoperiod, larval jumping occurs within a narrow peak 3–4 h wide reaching a maximum just before dawn. There is a distinct larval jumping rhythm across the range of photoperiods LD 23:1 h to LD 3:21 h with the peak some hours before lights-on in short days, some hours after in long days, such as occurs in many circadian rhythms (Tychsen, 1978; Aschoff, 1965). A secondary peak of eclosion is evident at lights-on in these cycles since switching on the light directly stimulates jumping behaviour when it occurs within the 'gate' set by the rhythm.

The rhythm is under circadian control, it persists in constant darkness with a period indistinguishable from 24 h for at least 8 days. Therefore, when cultures remain in constant darkness from the egg stage, a perceptible, though low amplitude, rhythm of jumping is observed. Rhythmicity damps out rapidly in constant light becoming arrhythmic within 2 or 3 days (Fig. 4.3.6).

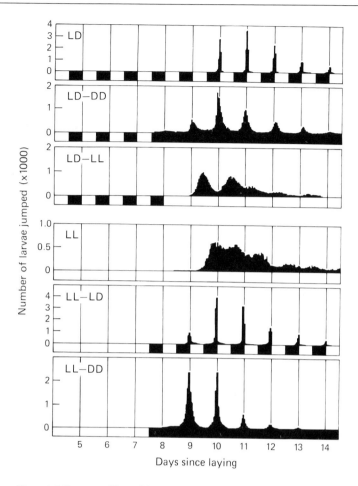

Fig. 4.3.6 Pattern of larval jumping from *Ceratitis capitata* cultures exposed from laying to light/ dark cycles (LD), or a transition from light/dark to constant darkness (LD–DD) or to constant light (LD–LL) on 8th day after laying. Similarly from cultures exposed to constant light (LL) since laying or a transition from constant light to light/dark cycles (LL–LD) or constant darkness (LL–DD) on 8th day after laying (Causse, 1974).

No rhythmicity persists in cultures left in constant light from laying. However, rhythmicity can be readily initiated in such arrhythmic cultures. LD cycles immediately initiate and entrain the rhythm even if commenced when the 5 day jumping period has already begun (Fig. 4.3.6). The rhythm is also readily initiated by a transition from constant light to constant darkness. Such a transition is effective at any time during development from egg-laying onwards though the amplitude of the rhythm is lesser the earlier in development the transition is applied (Fig. 4.3.7).

The weak rhythmicity that remains in cultures maintained in constant darkness from egg laying is made much more precise by a pulse of light applied just before jumping begins. Pulses of 12 h or more are equivalent to the constant light to constant darkness transition, producing precise rhythmicity and the same phase relationship. Shorter pulses produce progressively less precise rhythmicity down to 15 min which has no discernible effect on the phase or precision of the rhythm.

This rhythm behaves, therefore, as far as has been determined as would that of *Drosophila pseudoobscura* Frolova eclosion (Saunders, 1982, Chapter 3). All larval stages are equally sensitive to the initiating and phase shifting effects of light changes.

Arai (1976a) found similar properties in the effects of light on the jumping rhythm of *D. dorsalis*. This rhythm was however rather less precise and no

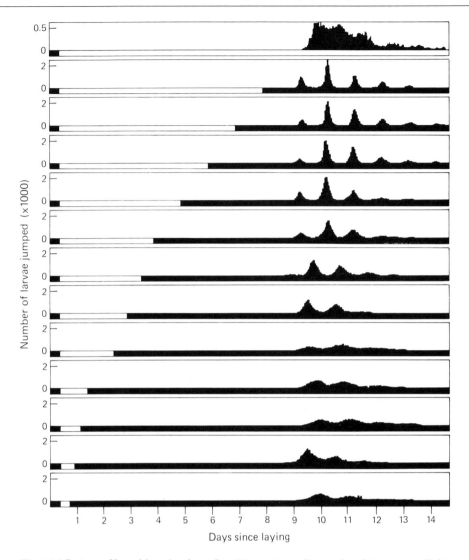

Fig. 4.3.7 Pattern of larval jumping from *Ceratitis capitata* cultures placed in constant light on the day of laying then transferred to constant darkness at different times during development (Causse, 1974).

rhythmicity at all could be demonstrated under constant light. He also investigated the role of temperature cycles in controlling the rhythm. In either constant light or constant darkness with thermoperiods (25–20°C) from 16:8 to 8:16 h, rhythmic jumping was observed with a peak corresponding to the drop in temperature. This response would therefore be 180° out of phase with the response to the light component in a natural day/night cycle. After being exposed to thermoperiods the rhythm persists for approximately 3 cycles at 20°C, but no persistence could be detected at 25°C (Fig. 4.3.8).

As in *C. capitata* lights-on strongly stimulates jumping activity if it occurs within the circadian 'gate' set by the light cycle. At other times of day it has no effect. Under a thermoperiod the same is true of the 25–20°C transition though the circadian 'gate' set by thermoperiod is rather wider. The effect of a lights-on transition applied during a thermoperiod is unexpected. Lights-on only causes jumping during the 20°C phase of a 25–20°C thermoperiod.

The eclosion rhythms of *D. tryoni* and *D. dorsalis* have been studied. Bateman (1955) studied the rhythm of eclosion in *D. tryoni*. He reports that peak eclosion occurs about 4.75 h after lights-on in a LD 8:16 h light cycle. The rhythm persists in constant darkness with a period very close to 24 h for at least

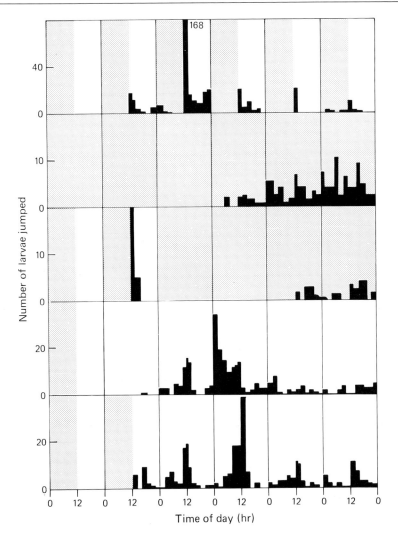

Fig. 4.3.8 Pattern of larval jumping from cultures of *Dacus dorsalis* in constant light exposed to thermoperiods. Temperatures 25°C (stipple), 20°C (white). The treatments were: continuous thermoperiods, transfer from thermoperiods to constant 25°C on days 8 and 10 after laying, and transfer from thermoperiods to constant 20°C on days 8 and 9 after laying (Arai, 1976a).

the time from egg laying to emergence, about 17 days in these experiments. Light cycles applied during larval life phase shift and entrain the rhythm but after pupariation the rhythm is unresponsive to light, neither LD cycles nor constant light affect the degree or phase of rhythmicity.

The *D. tryoni* eclosion rhythm, on this evidence, would therefore share with *Sarcophaga argyrostoma* Robineau-Desvoidy, *Cochliomyia hominivorax* (Coquerel) and *Lucilia cuprina* (Wiedemann) (see Smith, 1985) the characteristic of being very unresponsive to light changes during pupal life. This is in strong contrast to the eclosion rhythms of *D. pseudoobscura* and *Drosophila melanogaster* Meigen both of which show uniformly high response to light throughout larval and pupal life. Recent experiments of my own (P.H. Smith, unpublished) do not confirm Bateman's conclusions. In the latter experiments pupae of *D. tryoni* were shown to be sensitive to both LD cycles and to transitions from constant darkness to constant light.

Bateman also concluded that the LD cycles applied to the female parent determine the phase of the eclosion rhythm. This was regarded as an exceptional case but some data suggesting a similar transfer of eclosion rhythmicity from the parent has been recently obtained by Saunders (1979) in *Sarcophaga argy-*

rostoma. It is, however, premature to agree with Bateman until these experiments are repeated.

Bateman's data differ substantially from those collected by Arai (1976b) on *D. dorsalis.* Arai studied the response of the eclosion rhythm to light and temperature cycles during the pupal stage. In this species both photoperiods and thermoperiods applied to pupae entrain the eclosion rhythm and after such entrainment persistence of the rhythm can be demonstrated for 4 or more days under constant light and constant darkness following photoperiods, and at constant 25 and 20°C following thermoperiods. Persistence of the rhythm could not be demonstrated at 15°C following thermoperiods.

In the eclosion rhythm of *D. dorsalis* the effect of light cycles on the pupa is overridden by that of temperature cycles. Thermoperiods (25–20°C) 12:12 h cause eclosion to peak at the rise in temperature. Photoperiods LD 12:12 result in eclosion peaking at lights-on. When the two cycles are combined with phase differences of 0, 6, 12, and 18 h eclosion always peaks at the rise in temperature (Fig. 4.3.9). It seems likely that temperature fluctuations alone are maintaining in the pupa entrainment of the circadian rhythm controlling eclosion, which had been set up by light and temperature cycles experienced by the larvae. This is not surprising since the light intensity experienced by the buried pupae is very low whereas temperature fluctuations within the first few centimetres of the soil are not less than those in air temperature.

5.2 Investigations of mating behaviour

Detailed studies of mating rhythms have been carried out in *D. tryoni* and *C. capitata.* The narrow daily period of mating at dusk in *D. tryoni* results from an interaction between a response to light intensity and a daily fluctuation in mating responsiveness ('readiness to mate') as described in section 2. The form of the daily fluctuation of mating responsiveness is affected by photoperiod (Tychsen, 1978), its phase showing a similar phase relationship to each LD

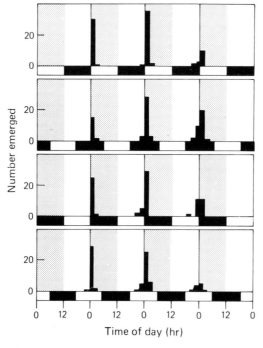

Fig. 4.3.9 Interaction between thermoperiods and photoperiods applied throughout pupal life on the timing of eclosion in *Dacus dorsalis*. Temperatures are 25°C (stipple) and 20°C (white) (Arai, 1976b).

cycle to that in other circadian rhythms. The amplitude of the fluctuation is however markedly lower for the photoperiods distant from the normal (Tychsen, 1978). The rhythm of mating responsiveness has been shown to be under circadian control. It persists under constant low light intensity (10 lux) for at least 4 days with a period of about 28 h. Under high light intensity, however, it is considerably disturbed. After reaching a normal maximum during the first cycle under constant light, it declines slowly and never returns to zero. By the third cycle all fluctuations are completely damped out and mating responsiveness remains constant at a fairly high level (Fig. 4.3.10).

The input from the circadian mechanism to sexual behaviour in males has been studied further by Tychsen (1975). The course of the daily change in mating responsiveness of males (termed 'sexual drive level' in Tychsen, 1975) is unaffected by mating or other expression of sexual behaviour. Its level seems to be continuously determined by the circadian mechanism. Treatments which cause the circadian clock to recommence running after being 'stopped' (in the terminology of Pittendrigh (1966)) by continuous light, result in immediate changes in the mating responsiveness consistent with their effect on the phase of the circadian rhythm.

Causse and Féron (1967) studied the rhythm of sexual activity in *C. capitata*. In this species, in both natural and artificial light cycles, the mating responsiveness is maximal soon after lights-on, stays maximal for about 7 h then declines to zero where it remains for 3–5 h. Both males and females show a similar rhythm of mating responsiveness though females show intervals of zero responsiveness whereas males always show some level of sexual activity in the assays. The rhythm of sexual responsiveness is affected by photoperiod in a way similar to that described for circadian rhythms in many species (Tychsen, 1978, Aschoff, 1965) (Fig. 4.3.11).

If the photophase is lengthened past about 12 h the mating responsiveness recovers to maximal level and remains maximal subsequently. Because of this lack of persistence in constant light, Causse and Féron (1967) concluded that the rhythm is not endogenous. However destruction of behavioural rhythms in constant light is a common occurrence. It occurs in the mating rhythm of *D. tryoni* (see above) and Truman (1972) regarded it as a general property of circadian behavioural rhythms in insects.

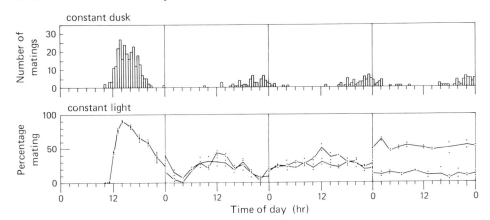

Fig. 4.3.10 Mating rhythm of *Dacus tryoni* in constant dusk (10 lux) and constant light. In both cases the flies entered constant conditions at 0000 hours on day 1 of the plot which corresponds to lights-on in the preceding LD, 14:1:9 h cycle. Mating in constant dusk was scored in a population of 100 males and 1000 virgin females. Mated females were removed and replaced with virgins maintained under similar conditions. In constant light samples of 50 virgin females and 100 males were taken from the population and exposed to a standard dusk. The results on day 1 are as in Fig. 4.3.4A. For subsequent days two populations were sampled repeatedly on each day, two replicate assays being carried out at each time (Tychsen and Fletcher, 1971).

Chapter 4.3 references, p. 338

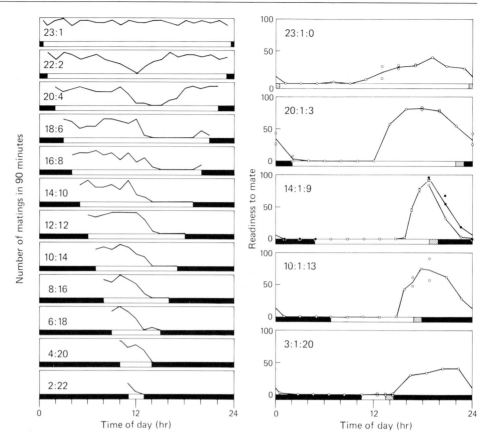

Fig. 4.3.11 Effect of photoperiod on the rhythm of mating in *Ceratitis capitata* (left) and *Dacus tryoni* (right). Assays of mating responsiveness were carried out throughout the photophase in *Ceratitis* and throughout the diel in *Dacus*. Assays: *Ceratitis capitata*: 10 males, 10 virgin females, matings in 90 min counted (Causse and Féron, 1967). *Dacus tryoni*: 100 males, 50 virgin females, transferred from prevailing lighting conditions into constant 10 lux, matings in 30 min counted (Tychsen, 1978).

The probable role of circadian rhythms in many of the behaviours partitioning the diel in tephritids, suggest them as a productive topic of study in considering behavioural organisation by the circadian clock.

6 REFERENCES

Alden, B., Dingle, H. and Possidente, B., 1983. Diel organisation of behaviour in milkweed bugs, *Oncopeltus* spp. Physiological Entomology, 8: 223–230.

Aluja, M., Hendrichs, J. and Cabrera, M., 1983. Behavior and interactions between *Anastrepha ludens* Linnaeus and *A. oblicua* on a field caged mango tree. 1. Lekking behavior and male territoriality. In: R. Cavalloro (Editor), Fruit Flies of Economic Importance. A.A. Balkema, Rotterdam, pp. 122–133.

Arai, T., 1975. Diel activity rhythms in the life history of the oriental fruit fly, *Dacus dorsalis* Hendel (Diptera: Trypetidae). Japanese Journal of Applied Entomology and Zoology, 19: 253–259. (in Japanese, with English abstract)

Arai, T., 1976a. Effects of light and temperature on the diel cyclicity of the larval jumping behaviour of the oriental fruit fly, *Dacus dorsalis* Hendel (Diptera: Trypetidae). Japanese Journal of Applied Entomology and Zoology, 20: 9–14. (in Japanese, with English abstract)

Arai, T., 1976b. Effects of temperature and light-dark cycles on the diel rhythm of emergence in the oriental fruit fly, *Dacus dorsalis* Hendel (Diptera: Trypetidae). Japanese Journal of Applied Entomology and Zoology, 20: 69–76. (in Japanese, with English abstract)

Arakaki, N., Kuba, H. and Soemori, H., 1984. Mating behaviour of the oriental fruit fly, *Dacus dorsalis* Hendel (Diptera: Tephritidae). Applied Entomology and Zoology, 19: 42–51.

Aschoff, J., 1965. The phase-angle difference in circadian periodicity. In: J. Aschoff (Editor), Circadian Clocks. North-Holland, Amsterdam, pp. 262–276.

Back, E.A. and Pemberton, C.E., 1917. The melon fly in Hawaii. United States Department of Agriculture Bulletin, 191: 64 pp.

Baker, A.C., Stone, W.E., Plummer, C.C. and McPhail, M., 1944. A review of studies on the Mexican fruit fly and related Mexican species. United States Department of Agriculture, Miscellaneous Publications, 531, 155 pp.

Barton Browne, L., 1956. The effect of light on the fecundity of the Queensland fruit fly, *Strumeta tryoni* (Frogg.). Australian Journal of Zoology, 4: 125–145.

Barton Browne, L., 1957. The effect of light on the mating behaviour of the Queensland fruit fly *Strumeta tryoni* (Frogg.). Australian Journal of Zoology, 5: 145–158.

Bateman, M.A., 1955. The effect of light and temperature on the rhythm of pupal ecdysis in the Queensland fruit-fly, *Dacus (Strumeta) tryoni* (Frogg.). Australian Journal of Zoology, 3: 22–33.

Bess, H.A. and Haramoto, F.H., 1961. Contributions to the biology and ecology of the Oriental fruit fly, *Dacus dorsalis* in Hawaii. Hawaii Agricultural Experiment Station Technical Bulletin, 44: 1–30.

Bower, C.C., 1976. The ecology and behaviour of larvae of the Queensland fruit fly *Dacus tryoni* (Frogg.). PhD Thesis, University of Sydney Library.

Boyce, A.M., 1934. Bionomics of the walnut husk fly, *Rhagoletis completa*. Hilgardia, 8: 363–579.

Brady, J., 1974. The physiology of insect circadian rhythms. Advances in Insect Physiology, 10: 1–115.

Brady, J., 1975. Circadian changes in central excitability – the origins of behavioural rhythms in tsetse flies and other animals? Journal of Entomology (A), 50: 79–95.

Brieze-Stegeman, R., Rice, M.J. and Hooper, G.H.S., 1978. Daily periodicity in attraction of male tephritid fruit flies to synthetic chemical lures. Journal of the Australian Entomological Society, 17: 341–346.

Caldwell, R.L. and Dingle, H., 1967. The regulation of cyclic reproductive and feeding activity in the milkweed bug, *Oncopeltus*, by temperature and photoperiod. Biological Bulletin, 133: 510–525.

Caldwell, R.L. and Rankin, M.A., 1974. Separation of migratory from feeding and reproductive behaviour in *Oncopeltus fasciatus*. Journal of Comparative Physiology, 88: 383–394.

Causse, R., 1974. Étude d'un rythme circadien du comportement de prénymphose chez *Ceratitis capitata* Wiedemann (Diptère Trypetidae). Annales de Zoologie – Écologie Animale, 6: 475–498.

Causse, R. and Féron, M., 1967. Influence du rythme photopériodique sur l'activité sexuelle de la mouche méditerranéene des fruits: *Ceratitis capitata* Wiedemann (Diptère Trypetidae). Annales des Épiphyties, 18: 175–192.

Causse, R., Féron, M. and Serment, M.-M., 1966. Rythmes nycthéméraux d'activité sexuelle inverses l'un de l'autre deux Diptères Trypetidae, *Dacus oleae* Gmelin et *Ceratitis capitata* Wiedemann. Compte Rendu de l'Académie des Sciences, Paris, 262: 1558–1560.

Cavender, G.L. and Goeden, R.D., 1982. Life History of *Trupanea bisetosa* (Diptera: Tephritidae) on wild sunflower in southern California. Annals of the Entomological Society of America, 75: 400–406.

Cazier, M.A., 1962. Notes on the bionomics of *Zonosemata vittigera* (Coquillett), a fruit fly on solanum. Pan-Pacific Entomologist, 38: 181–186.

Cendaña, S.M. and Namocale, P.P., 1953. A biological study of *Bactrocera umbrosa* (Fabricius) (Trypetidae, Diptera). Eight Pacific Science Congress, Entomology Section, 1347–1370.

Christenson, L.D. and Foote, R.H., 1960. Biology of fruit flies. Annual Review of Entomology, 5: 171–192.

Fitt, G.P., 1981a. Responses of female dacinae to 'male' lures and their relationship to patterns of mating behaviour and pheromone response. Entomologia Experimentalis et Applicata, 29: 87–97.

Fitt, G.P., 1981b. Inter- and Intraspecific responses to sex pheromones in laboratory bioassays by females of three species of Tephritid fruit flies from northern Australia. Entomologia Experimentalis et Applicata, 30: 40–44.

Fletcher, B.S., 1969. The structure and function of the sex pheromone glands of the male Queensland fruit fly, *Dacus tryoni*. Journal of Insect Physiology, 15: 1309–1322.

Fletcher, B.S. and Giannakakis, A., 1973. Factors limiting the response of females of the Queensland fruit fly, *Dacus tryoni*, to the sex pheromone of the male. Journal of Insect Physiology, 19: 1147–1155.

Foote, B.A., 1967. Biology and immature stages of fruit flies: The genus *Icterica* (Diptera: Tephritidae). Annals of the Entomological Society of America, 60: 1295–1305.

Frías, D.L., Malavasi, A. and Morgante, J.S., 1984. Field observations of distribution and activities of *Rhagoletis conversa* (Diptera: Tephritidae) on two hosts in nature. Annals of the Entomological Society of America, 77: 548–551.

Giannakakis, A.-M., 1976. Behavioural and physiological studies on the Queensland fruit fly, *Dacus tryoni*, in relation to the male sex pheromone. PhD thesis, University of Sydney Library.

Haseler, W.H., 1965. Life-history and behaviour of the crofton weed gall fly *Procecidochares utilis* Stone (Diptera: Trypetidae). Journal of the Entomological Society of Queensland, 4: 27–32.

Hooper, G.H.S., 1975. Sterilization of *Dacus cucumis* French (Diptera: Tephritidae) by gamma radiation. 1. Effect of dose on fertility, survival and competitiveness. Journal of the Australian Entomological Society, 14: 81–87.

Kuba, H., Koyama, J. and Prokopy, R.J., 1984. Mating behaviour of wild melon flies, *Dacus cucurbitae* Coquillett (Diptera: Tephritidae) in a field cage: Distribution and behaviour of flies. Applied Entomology and Zoology, 19: 367–373.

Landolt, P.J. and Hendrichs, J., 1983. Reproductive behaviour of the papaya fruit fly, *Toxotrypana curvicauda* Gerstaecker (Diptera: Tephritidae). Annals of the Entomological Society of America, 76: 413–417.

Lathrop, F.H. and Nickels, C.B., 1932. The biology and control of the blueberry maggot in Washington County, Me. United States Department of Agriculture Technical Bulletin No. 275, 76 pp.

Laudého, Y., Liaropoulos, C. and Canard, M., 1978. Etude, pendant la période automnale, du rythme de sortie hors des fruits des larves due derniere âge de la mouche de l'olive *Dacus oleae* (Gmel.) (Diptera, Trypetidae). Annales de Zoologie – Écologie Animale, 10: 37–50.

Lewontin, R.C. and Birch, L.C., 1966. Hybridisation as a source of variation for adaptation to new environments. Evolution, 20: 315–336.

McPhail, M. and Bliss, C.I., 1933. Observations on the Mexican fruit fly and some related species in Cuernavaca, Mexico, in 1928 and 1929. United States Department of Agriculture Circular No. 255, 24 pp.

Malan, E.M. and Giliomee, J.H., 1968. Aspeckte van die bionomie van *Dacus ciliatus* Loew (Diptera: Trypetidae). Journal of the Entomological Society of South Africa, 31: 373–389.

Malavasi, A., Morgante, J.S. and Prokopy, R.J., 1983. Distribution and activities of *Anastrepha fraterculus* (Diptera: Tephritidae) flies on host and nonhost trees. Annals of the Entomological Society of America, 76: 286–292.

Miyake, T., 1919. Studies on the fruit flies of Japan. Contribution 1. — Japanese Orange-Fly. Bulletin of the Imperial Central Agricultural Experiment Station in Japan, 2: 85–165.

Myers, K., 1952a. Oviposition and mating behaviour of the Queensland fruit-fly (*Dacus (Strumeta) tryoni* (Frogg.)) and the solanum fruit-fly (*Dacus (Strumeta) cacuminatus* (Hering)). Australian Journal of Scientific Research, Series B, Biological Sciences, 5: 264–281.

Myers, K., 1952b. Rhythms in emergence and other aspects of behaviour of the Queensland fruit-fly (*Dacus (Strumeta) tryoni*, Frogg.) and the solanum fruit-fly (*Dacus (Strumeta) cacuminatus*, Hering). Australian Journal of Science, 15: 101–102.

Myrburgh, A.C., 1962. Mating habits of the fruit flies *Ceratitis capitata* (Wied.) and *Pterandrus rosa* (Ksh.). South African Journal of Agricultural Science, 5: 457–464.

Myrburgh, A.C., 1963. Diurnal rhythms in emergence of mature larvae from fruit and eclosion of adult *Pterandrus rosa* and *Ceratitis capitata*. South African Journal of Agricultural Science, 6: 41–46.

Nayar, J.K., 1968. The pupation rhythm in *Aedes taeniorhynchus*. IV. Further studies of the endogenous diurnal (circadian) rhythm of pupation. Annals of the Entomological Society of America, 61: 1408–1417.

Nishida, T. and Bess, H.A., 1957. Studies on the ecology and control of the melon fly *Dacus (Strumeta) cucurbitae*. Hawaii Agricultural and Experimental Station Technical Bulletin, 34, 44 pp.

Novak, J.A. and Foote, B.A., 1968. Biology and immature stages of fruit flies: *Paroxyna albiceps* (Diptera: Tephritidae). Journal of Kansas Entomological Society, 41: 108–119.

Perdomo, A.J., Baranowski, R.M. and Nation, J.L., 1975. Recaptures of virgin female Caribbean fruit flies from traps baited with males. Florida Entomologist, 58: 291–295.

Peterson, A., 1923. The pepper maggot, a new pest of peppers and eggplants. New Jersey Agricultural Experiment Station Bulletin, 373: 1–23.

Pittendrigh, C.S., 1966. The circadian oscillation in *Drosophila pseudoobscura* pupae: a model for the photoperiodic clock. Zeitschrift für Pflanzenphysiologie, 54: 275–307.

Pritchard, G., 1967. Laboratory observations on the mating behaviour of the island fruit fly *Rioxa pornia* (Diptera: Tephritidae). Journal of the Australian Entomological Society, 6: 127–132.

Prokopy, R.J. and Hendrichs, J., 1979. Mating behaviour of *Ceratitis capitata* on a field-caged host tree. Annals of the Entomological Society of America, 72: 642–648.

Prokopy, R.J., Bennett, E.W. and Bush, G.L., 1972. Mating behaviour in *Rhagoletis pomonella*. II. Temporal organization. Canadian Entomologist, 104: 97–104.

Provost, M.W. and Lum, P.T.M., 1967. The pupation rhythm in *Aedes taeniorhynchus* (Diptera: Culicidae). I. Introduction. Annals of the Entomological Society of America, 60: 138–149.

Roan, C.C., Flitters, N.E. and Davis, C.J., 1954. Light intensity and temperature as factors limiting the mating of the oriental fruit fly. Annals of the Entomological Society of America, 47: 593–594.

Saunders, D.S., 1979. The circadian eclosion rhythm in *Sarcophaga argyrostoma*: delineation of the responsive period for entrainment. Physiological Entomology, 4: 263–274.

Saunders, D.S., 1982. Insect clocks, 2nd edition. Pergamon Press, Oxford.

Severin, H.H.P., 1917. The currant fruit fly. Maine Agricultural Experiment Station Bulletin, 264: 177–247.

Silverman, J. and Goeden, R.D., 1980. Life history of a fruit fly, *Procecidochares* sp., on the ragweed, *Ambrosia dumosa* (Gray) Payne, in southern California. Pan-Pacific Entomologist, 4: 283–288.

Simmonds, H.W., 1936. Fruit fly investigations 1935. Bulletin, Department of Agriculture Fiji 19, 18 pp.

Smith, D.C. and Prokopy, R.J. 1981. Seasonal and diurnal activity of *Rhagoletis mendax* flies in nature. Annals of the Entomological Society of America, 69: 462–466.

Smith, P.H., 1979. Genetic manipulation of the circadian clock's timing of sexual behaviour in the Queensland fruit flies, *Dacus tryoni* and *Dacus neohumeralis*. Physiological Entomology, 4: 71–78.

Smith, P.H., 1983. Circadian control of spontaneous flight activity in the blowfly, *Lucilia cuprina*. Physiological Entomology, 8: 73–82.

Smith, P.H., 1985. Responsiveness to light of the circadian clock controlling eclosion in the blowfly, *Lucilia cuprina*. Physiological Entomology, 10: 323–336.

Suzuki, Y. and Koyama, J., 1980. Temporal aspects of mating behaviour of the melon fly, *Dacus cucurbitae* Coquillett (Diptera: Tephritidae): A comparison between laboratory and wild strains. Applied Entomology and Zoology, 15: 215–224.

Syed, R.A., 1969. Studies on the ecology of some important species of fruit flies and their natural enemies in West Pakistan. Pakistan Commonwealth Institute of Biological Control Station Report, 12 pp.

Syed, R.A., 1970. Studies on Trypetids and their natural enemies in West Pakistan. *Dacus* species of lesser importance. Pakistan Journal of Zoology, 2: 17–24.

Syed, R.A., Ghant, M.A. and Murtaza, M., 1970. Studies on Trypetids and their natural enemies in West Pakistan. III. *Dacus (Strumeta) zonatus* (Saunders). Commonwealth Institute of Biological Control, Technical Bulletin, 13: 1–16.

Tauber, M.J. and Tauber, C.A., 1967. Reproductive behaviour of the gall-former *Aciurina ferruginea* (Doane) (Diptera: Tephritidae). Canadian Journal of Zoology, 45: 907–913.

Tauber, M.J. and Toschi, C.A., 1965a. Life history and mating behaviour of *Tephritis stigmatica* (Coquillett). Pan-Pacific Entomologist, 41: 73–79.

Tauber, M.J. and Toschi, C.A., 1965b. Bionomics of *Euleia fratria* (Loew) (Diptera: Tephritidae). 1. Life history and mating behaviour. Canadian Journal of Zoology, 43: 369–379.

Truman, J.W., 1972. Circadian rhythms and physiology with special reference to neuroendocrine processes in insects. In: Circadian Rhythmicity, Proceedings of the International Symposium on Circadian Rhythmicity, Wageningen, 1971, Pudoc, Wageningen, pp. 111–135.

Tychsen, P.H., 1972. Mating behaviour and the control of sexual responsiveness in the Queensland fruit fly, *Dacus tryoni* (Frogg.). PhD Thesis, University of Sydney Library.

Tychsen, P.H., 1975. Circadian control of sexual drive level in *Dacus tryoni* (Diptera: Tephritidae). Behaviour, 54: 111–141.

Tychsen, P.H., 1977. Mating behaviour of the Queensland fruit fly, *Dacus tryoni* (Diptera: Tephritidae), in field cages. Journal of the Australian Entomological Society, 16: 459–465.

Tychsen, P.H., 1978. The effect of photoperiod on the circadian rhythm of mating responsiveness in the fruit fly, *Dacus tryoni*. Physiological Entomology, 3: 65–69.

Tychsen, P.H. and Fletcher, B.S., 1971. Studies on the rhythm of mating in the Queensland fruit fly, *Dacus tryoni*. Journal of Insect Physiology, 17: 2139–2156.

Uhler, L.D., 1951. Biology and ecology of the goldenrod gall fly, *Eurosta solidaginis* (Fitch). Cornell University Agricultural Experiment Station Memoir, 300: 51 pp.

Vogt, W.G., 1977. A re-evaluation of introgression between *Dacus tryoni* and *Dacus neohumeralis* (Diptera: Tephritidae). Australian Journal of Zoology, 25: 59–69.

Watson, C., 1971. The role of olfactory and visual stimuli in the ovipositional behaviour of female *Dacus tryoni* (Froggatt). Honours Thesis, University of Sydney Library.

Zwölfer, H., 1969. *Urophora siruna-seva* (Hg.) (Dipt.: Trypetidae), a potential insect for the biological control of *Centaurea solstitialis* L. in California. Commonwealth Institute of Biological Control Technical Bulletin, 11: 105–155.

Zwölfer, H., Englert, W. and Pattullo, W., 1970. Investigations on the biology, population ecology and the distribution of *Urophora cardui* L. Commonwealth Institute of Biological Control European Station Delémont — Switzerland Report, Weed Projects for Canada, Progress Report No. XXVII, 17 pp.

Chapter 4.4 Reproductive and Mating Behaviour

J. SIVINSKI[1] and T. BURK

1 INTRODUCTION

The sexual behaviour of tephritid fruit flies is interesting for several reasons. As far back as 1862, entomologists spoke of tephritids' 'graceful behaviour' (Loew, quoted in Zwölfer, 1974). More recently, entomologists such as Prokopy (1980) and Burk and Calkins (1983) have suggested that an increased comprehension of tephritid sexual behaviour could lead to more effective tephritid control. This upturn in interest in the behaviour of fruit flies has coincided with the growth of the branch of biology known variously as behavioural ecology or animal sociobiology, and there has been a cross-fertilization of ideas between entomologists and behavioural ecologists that has resulted in satisfying increases in our knowledge. In this chapter, our aim is to present an ecological framework for understanding tephritid sexual behaviour, to review some recent studies in the context of that framework, and to illustrate the depth of understanding that is potentially available by referring in some detail to results from a series of studies of one tephritid species. We will concentrate on studies not covered in the recent reviews by Prokopy (1980) and Burk (1981); for our detailed example, we have chosen the Caribbean fruit fly, *Anastrepha suspensa* (Loew).

2 CONCEPTUAL BACKGROUND

The general ecological framework for discussing the variety of tephritid sexual behaviour was provided by Bateman (1972). He divided the tephritids into r-selected and K-selected species; this distinction cut across the gall-former vs. fruit-infester spectrum, and was based mainly on climatic consideration (e.g., tropical vs. temperate species), especially as climatic and other ecological factors affected the distribution and availability of hosts in time and space. Bateman was able to correlate a variety of factors, such as dispersal abilities, with a species' r or K classification. Prokopy (1980) and Burk (1981) independently applied Bateman's model to explain the variety of mating encounter sites and behaviour in tephritids. Each suggested that, in tropical species, widely dispersed hosts would fruit asynchronously, leading to the evolution of fruit fly species that were multivoltine, polyphagous and good dispersers. In temperate species, short, synchronous host seasonality would

[1] Postdoctoral fellow employed through a cooperative agreement between the Insect Attractants, Behavior, and Basic Biology Research Laboratory, and the Department of Entomology and Nematology, University of Florida, Gainesville, FL 31611.

Chapter 4.4 references, p. 350

lead to the evolution of univoltine, monophagous species. Both authors predicted that monophagous species, especially those whose hosts were somewhat patchily distributed, would mate directly on the oviposition substrate. In polyphagous species, decreased likelihood of male-female encounters at the oviposition site would allow the evolution of non-resource-bound mating systems, especially lek mating systems (leks are male communal display aggregations that are visited by receptive females).

Burk (1981) went on to attribute differences in the type and complexity of sexual signals in tephritids to these differences in their mating systems. He argued that species that mated directly on oviposition substrates, at least when hosts were patchily distributed, would have simple signalling systems. Such species would lack long-distance attractants, would exhibit male territoriality centered on the oviposition site, and their courtship signals would be simple, usually species-identifying wing waving. Males successful at establishing territories on such female-required resources would be assured of a steady supply of arriving females, and would be able to force copulations on arriving females in return for access to the oviposition substrate. However, in many monophagous or oligophagous species, the host plant is extremely abundant and evenly distributed, as is sometimes the case for gall-forming species. In such cases, females can be considered to have a choice of flying to a host with or without a resident male. In these situations, males would evolve more complex signalling systems to attract females and induce them to mate. In many such species, males supplement close-range wing waving with long-distance sex pheromone attractants. Females may still be willing to swap a copulation for access to a host plant, if response to a male pheromone allows her to more quickly and efficiently locate a suitable oviposition site. A further incentive to females is provided in the relatively small number of species in which the male offers a salivary secretion as a nutritional offering to a female prior to or during mating. Most of the species that are currently known to produce such nutritional offerings are gall formers whose hosts are abundant and widespread (see review in Dodson, 1978).

The most complex signalling systems in tephritids are found in the polyphagous fruit-infesting species (mainly tropical or subtropical in distribution, including such important pests as *Ceratitis capitata* (Wiedemann) and several *Dacus* and *Anastrepha* species). In such species, the females are liberated to a free choice of mate(s), and are expected from Darwinian sexual selection theory to have evolved highly discriminating mate choices. Sexual selection theory suggests that females should choose mates on the basis of material benefits offered or, failing any such benefits, on the basis of male vigour indicative of superior survival characteristics or the possession of characteristics that would make a female's sons attractive in turn to the next generation of females; see review of sexual selection theory as applied to insects in Thornhill (1980) and Thornhill et al. (1983). Males in these species will often aggregate in leks, and evolve elaborate sexual signalling systems for at least three reasons: (1) Males have to attract females to the communal display site, and they compete with each other to attract the most females; (2) Males also may compete aggressively to establish display territories in leks (perhaps in particularly favorable locations — see below for *A. suspensa*), leading to the evolution of male weapons and threat ornamentation and displays; (3) Females making discriminating mate decisions will become involved in a coevolutionary 'arms race' with displaying males (Dawkins and Krebs, 1979), leading to the open-ended evolution of ever-more-complicated courtship signalling interactions. In most lekking polyphagous species, we see the evolution of acoustic calling, courtship, and aggressive songs (Monro, 1953; Webb et al., 1976; Webb, unpublished data), which coexist in behaviour repertoires with long-distance, male-produced sex pheromones and elaborate visual displays involving wing waving and dancing.

Burk (1981) produced the preceding generalizations based on a review of the pre-1981 literature and on preliminary observations of *A. suspensa*. With a few enlightening exceptions considered below, the scheme has been successful in predicting subsequent findings, some of which we now review.

3 MATING BEHAVIOUR

3.1 Tephritids in General

Steck (1984) reported on the sexual behaviour of *Chaetostomella undosa* (Coquillett), which forms galls in seed heads of the thistle, *Cirsium cymosum*. Males alternately patrolled and sat in wait on the crown and upper leaves of the host. Males approached conspecifics and attempted to mount them. Males subsequently fought, while copulation ensued if the conspecific was a female. Steck reported, 'Precopulatory behavior was devoid of stylized wing movements or mouthpart contact . . . .' Dodson (1986a) observed two gall-formers associated with rabbitbrush (*Chrysothamnus* spp.), *Aciurina bigeloviae* (Cockerell) and *Valentibulla dodsoni* Foote and Blanc (see also Thornhill et al., 1983). Males established positions on host plants, and scanned for arriving conspecifics. If females were spotted, males walked along leaves and stems to intercept them. Mountings followed, with males using greatly enlarged front femora to gain a 'leglock'. Virgin females waved wings to attract the attention of males, while mated females seemed to resist male advances, so that only about 65% of interactions resulted in copulations. Courtship interactions involved only wing waving, although Dodson reports that this was often elaborate.

The papaya fruit fly, *Toxotrypana curvicauda* Gerstaecker, studied by Landolt and Hendrichs (1983), may offer some exceptions to the scheme given above. It is a tropical species, but nevertheless monophagous on papayas. At first, its sexual behaviour seemed to correspond to predictions based on its host specificity. Males were found to establish territories on papaya trees via overt aggressive interactions. Having done so, they positioned themselves on papaya fruit and 'puffed', expanding pleural membranes to release sex pheromone. Males approached arriving females from behind, and mounted them, often as they attempted to oviposit. However, Landolt et al. (1985) have since found a more complex courtship in the laboratory that takes place at a time flies are not normally on their host. They argue that as yet undiscovered lek sites exist (see Sivinski and Webb (1985a) for a description of the acoustic courtship signals). Hopefully, further field work will determine how well the papaya fruit fly fits into the Burk-Prokopy model.

Examples of recent studies of sexual behaviour in polyphagous lekking tephritids include those of melon flies (*Dacus cucurbitae* Coquillett) by Kuba et al. (1984) and of Mexican fruit flies (*Anastrepha ludens* (Loew)) by Robacker and Hart (1985). In *D. cucurbitae*, as in polyphagous *Dacus* species previously studied, males gather at specific times of day in aggregations or leks. Males do not reside on fruit, but set up single-leaf territories on leaves, from which they emit sex pheromone and produce calling songs. Females fly to such male aggregations to mate. In *A. ludens*, males establish single-leaf territories underneath leaves during a sexual activity period in the late afternoon. Male aggression is extensive, and the winning territory holders 'puff' to produce sex pheromone and also produce calling songs via rapid wing vibrations. Leks usually form in host trees, but away from the host fruit. Females approached territory-holding males, and precopulatory orientation, approach, and wing-waving behaviour on the part of both sexes has been observed. Males mount females from a face-to-face orientation by flying over them and turning around to mount. As in *A. suspensa*, after mounting, males produce 'precopulatory' or courtship songs. Similar observations have been made for other *Anastrepha* species (*A. suspensa* — see below; *Anastrepha fracterculus* (Wiedemann), Morgante et al., 1983; Malavasi et al., 1983): we are now at the exciting point with this genus (the same applies to *Dacus*) where we can begin to perform careful comparative studies. Especially valuable would be studies of the monophagous species in the two very large genera, *Anastrepha* and *Dacus*.

Other recent studies of tephritid sexual behaviour have pointed out shortcomings of the generalizations mentioned above; but these exceptions are readily explainable from ecological or sexual selection considerations. For example, Dodson (1986b) has observed lekking-type aggregations in a monophagous gall-former, an unnamed *Procecidochares* species in the *minuta* species group. Males of this species aggregate on conspicuous nonhost saltbush plants in north central New Mexico. Males are alert and react quickly to incoming males or females, although signalling is simple (only wing-flicking was observed). The male aggregations on conspicuous objects are not uncommon in species whose population densities are low and whose adults are widely dispersed (this phenomenon is called 'hilltopping' by students of butterflies, wasps, and botflies). This situation seems to apply to the *Procecidochares* species, as Dodson points out.

A more general shortcoming of the simple scheme outlined above is that it underestimates the extent and importance of adaptive intraspecific variation in sexual behaviour. Behavioural ecologists in the last decade have extensively modeled the possibility of alternative male and female mating tactics, and many examples have been discovered in insects and other animals (see good reviews by Rubenstein, 1980; Cade, 1980). Even in 1980 and 1981, several examples of alternative sexual behaviour tactics were known in tephritids. Smith and Prokopy (1980) had discovered that early season matings in the apple maggot (*Rhagoletis pomonella* (Walsh)) took place in lek-type aggregations in host plant foliage and were female-initiated, while later in the season males defended host fruits and attempted to force copulations on arriving females. In the Mediterranean fruit fly, *C. capitata*, Prokopy and Hendrichs (1979) observed a within-day switch in location of sexual activities. Wing waving males chased and courted females on fruit in the morning and afternoon, but gathered in leks to display via sex pheromones and calling songs at midday. Similar patterns have been found in the most recent studies. Smith and Prokopy (1982) found an early-season late-season switch in *Rhagoletis mendax* Curran exactly like that in *R. pomonella*. Burk (1983) found that *A. suspensa* males engaged in on-fruit courtships early in the morning (the main female oviposition period), but formed leks in the late afternoon. However, the simple rule, 'monophagous species mate on fruit, polyphagous species mate on foliage in leks' perhaps can be defended even in such cases, when the relative success of various male tactics is examined. In *Rhagoletis*, male-initiated matings on fruit are usually successful, resulting in multiple matings by most females; so most matings probably do take place on fruit. In *Ceratitis* and *Anastrepha*, most courtships on fruit do not lead to copulation, and most matings therefore actually take place in the female-initiated interactions in leks. It seems likely (and some unpublished evidence by Hendrichs for *A. suspensa* supports the possibility) that on-fruit courtships in the polyphagous species are far less effective as a tactic than obtaining a display territory in a lek, and are probably performed by subordinate males unable to establish a lek territory. In Dawkins' (1980) terminology, these subordinate males are 'making the best of a bad job'.

So far, exploration of alternative tactics and their reproductive consequences has focused on male tephritids, but there are likely to be similar adaptive intraspecific female differences in mating behaviour (Thornhill, 1984). Students of tephritids, like other students of animal behaviour, need to pay much more detailed attention to female sexual behaviour.

3.2 *Anastrepha suspensa*

The generalizations and brief reviews given so far fail to give a real flavour of the types of questions being posed and answers being obtained in recent studies of tephritid sexual behaviour. To do that, we need to look in somewhat more detail at studies of a particular species. The one that has been examined most closely so far is the Caribbean fruit fly, *A. suspensa*.

As noted, the Caribbean fruit fly is a polyphagous species that infests over a hundred hosts. In the morning, males are often found on fruit where they attempt, with relatively poor success, to mate with ovipositing/feeding females (Burk, 1983). In the afternoon, particularly late afternoon, males aggregate on host foliage and it is here that the complex sexual repertoire associated with polyphagy is performed. Males defend leaf territories with a mixture of postures, sounds, proboscis extention, and butting. Large size and prior residency are the major components of a successful defense (Burk, 1984). Females fly into leks and sometimes visit several leaf territories before mating or leaving the vicinity altogether. In the laboratory, females show a ca. 3/1 preference for larger males (Burk and Webb, 1983). The advantages of having a large mate are as yet unknown, but could include acquiring superior genes for offspring or obtaining greater material benefits from a more substantial ejaculate (see Sivinski, 1984). Besides mating more often, large males mate longer, perhaps passing more sperm, which could in turn lessen the possibility of remating (Webb et al., 1984; see Farrias et al., 1972 discussion of sperm transfer and mating duration in the Mediterranean fruit fly, Tzanakakis et al., 1968 for a positive relationship between coital and subsequent refractory period length in *Dacus oleae* (Gmelin)). Preliminary evidence in the field suggests that bigger males occupy more advantageous positions in leks. These sexual advantages to being large should be borne in mind, since the most attractive versions of some of the following displays are associated with greater size. We will argue that females use these displays not so much to determine males of the correct species as to find the best available mates within her species.

A number of signals and what appear to be signals are emitted by males on their leaves. The pleural abdominal glands are puffed and 'anal' pouch membranes everted to release a pheromone (discussed by Nation, Chapter 3.4.5). Initial data suggest that the pheromone plays a role in both long and short range communication with females and males. Certain pheromone components occur in the male head and it is possible that these are also used to communicate as the male intermittently dabs the female head with his mouthparts during mating (Sivinski, unpublished data). Wing waving, alternating movements of patterned wings reminiscent of semaphoring, is performed and its production increases in the presence of females (Sivinski and Webb, 1985b). These graceful wing motions accompany arcing sideways movements, the whole constituting a complex visual display. Periodically during the afternoon, puffing males produce a 'calling song' made up of repeated ca. 0.5 s bursts of sound at ca. 0.5 s intervals (bursts = pulse trains = PT; intervals = PTI, fundamental frequency = 149 Hz). The other principal part of the acoustic repertoire is the relatively continuous precopulatory song performed just after the male mounts the female for an average of ca. 150 s and at a slightly higher frequency ($\bar{\chi} = 167$ Hz) (Webb et al., 1976; Webb et al., 1984). These sounds are

Chapter 4.4 references, p. 350

generated by beating sexually dimorphic wings. Males have a more oval, surface area-boosting shape, as do males in the acoustically signalling tephritids *T. curvicauda* and *C. capitata*. The 'mute' *R. pomonella* has no such dimorphism (Sivinski and Webb, 1985c).

We have found that the relative form of acoustic displays influences the reproductive fate of their sender; that is, not all males sound alike, not all males have equal sexual success, and females in part choose males on the basis of the song they sing. Subsequently they offer a potential means for tephritid breeders to quantify the sexual competitiveness of their stock. Consider first the calling song. This sound is produced in the absence of another fly, although as we will see, it continues in a modified form in the presence of a conspecific (close neighbors also will sing earlier than isolated individuals; Burk, 1984). Under semi-natural conditions in field cages over guava host-trees, females are more likely to be caught in traps baited with recorded calling song than in silent controls (Webb et al., 1983). It is not clear how far calling song travels, but the acoustic morphology of the pulse train gives further evidence for a range beyond the immediate vicinity of the sender. *A. suspensa* pulse trains have two distinct frequencies, unlike the monotonic repeated pulse train songs of papaya fruit fly, *T. curvicauda*, and Mediterranean fruit fly, *C. capitata*, which are directed only at nearby individuals (Sivinski and Webb, 1985a). Frequency bimodality has been suggested to be a means of escaping filtering by vegetation in vibrational signals (Michelson et al., 1982), or to provide locational cues through the differential attenuation of the two frequencies (Morris et al., 1975). Either function is consistent with broadcasting toward a relatively distant receiver. Flies irradiated 48 h before eclosion (5 kR) lose much of their bimodality and may be less effective signallers (Webb et al., 1985).

In the laboratory, virgin females, but not mated females or males, increase their activity (flying and walking) during the broadcast of a recorded calling song (Sivinski et al., 1984). This increase is a plausible reaction for a mate-searching female to make toward a 'long range' attractant. The superficially similar song of the papaya fruit fly, *T. curvicauda*, results in great quiescence of its females, but here the song is sung only at close range and may hold a choosing female in place until she can further ascertain the quality of the singer (Sivinski and Webb, 1985a). Virgin female caribflies do not respond to either the song of *Dacus neohumeralis* Hardy or a caribfly song conspicuous for its long pulse train interval (Sivinski et al., 1984). Thus, sexually receptive females distinguish, and discriminate among conspecific signals. The stimulating character of short pulse train interval is correlated with the attractive quality of large male size (Burk and Webb, 1983, see however the weaker relation in Webb et al., 1984). Males tend to decrease their pulse train interval in the presence of females corroborating a female preference for short interval songs (Sivinski and Webb, 1985b). It is tempting to conclude that females try to judge male size/vigor by interval length and that males expend more energy in front of a female audience in order to appear as attractive as possible. If so, breeders should note the positive relationship between radiation dosage and pulse train interval (Webb et al., 1985).

Besides its role in female attraction and courtship, calling song plays a part in male-male encounters. Songs sung near other males have a higher frequency and longer pulse trains. Broadcasts of such modified songs have the same quieting effect on males as the clearly agonistic aggression song (Sivinski and Webb, 1985b).

Precopulatory song is sung by males as they mount the back of a potential mate. Often, but not always, the sound lasts until the aedeagus is completely threaded through the ovipositor. Similar sounds are produced if a female becomes restless during copulation (Webb et al., 1984). The context of singing argues that the proximate role of the song is to quiet the female and maintain her cooperation in the insemination. We propose that its ultimate function is as a last and crucial display of fitness, i.e., the song will gain further cooperation if it demonstrates the quality of its singer. The power of precopulatory song, 3.3X that of the calling song, reflects the effort put into its production. A correlate of this power, sound pressure level (intensity of the sound) is an important attribute of a song. Dealated, essentially mute, males placed with females are more likely to copulate if a precopulatory song is broadcast at 90 dB. The same song at 52 dB has no significant effect (Sivinski et al., 1984). This demonstrates the causality behind the finding that males who mate tend to have louder songs than those who fail, i.e., females prefer loud songs and are not choosing mates by criteria only incidentally correlated to volume (see Burk and Webb, 1983; Webb et al., 1984). Parenthetically, the lower sound pressure level of failed songs is not an artifact produced by the females' vigorous movements during the rejection. Males that both fail and succeed show no difference in the sound pressure level of their successful and unsuccessful efforts (Webb et al., 1984).

Sound pressure level is correlated to male size and so could be used by females searching for large mates (Webb et al., 1984). Interestingly, while size is inversely related to calling song fundamental frequency, there is no such correlation in precopulatory song. Perhaps females 'prefer' a particularly narrow range of sound and males have been strongly selected to perform it.

In addition to intensity, the form of the precopulatory song is important as well. Inappropriate calling song broadcast at 90 dB to females and dealated males has no effect on rejection rate. There is considerable variance in the length of the precopulatory song, which can last from a second to 25 min. However, the longer the song, generally, the shorter the copulation. This inverse relationship between song length and mating duration may be due to a less attractive male having more difficulty eliciting female cooperation so that it takes him longer to insert his genitals and leaves him less able to maintain the coupling in the face of female 'restlessness' (Webb et al., 1984). Heavily irradiated males (10 kR) tend to have longer songs sung at a lower frequency (Webb et al., 1985). A few flies (3% of one large sample) mate without producing a song at all. These, likewise, have significantly shorter copulations (Sivinski et al., 1984).

4 CONCLUSION

We hope that we have outlined something of the surprising breadth of fruit fly sexual behaviour. Males communicate to females in many 'media' and from a number of stages. Some of these advertisements appear to be both self-revealing and complicated by the possibility of deceit. Tephritid displays are all the more remarkable when one recalls the alternative behavioural suites males employ at different times and under different regimes of competition. Finally, it is satisfying to note an organizing generalization: that the richness of a species' courtship can be predicted from the distribution of its oviposition sites.

5 ACKNOWLEDGMENTS

Tim Forrest, Peter Landolt and J.C. Webb made helpful comments on the manuscript, which was professionally prepared by Elaine Turner.

Chapter 4.4 references, p. 350

6 REFERENCES

Bateman, M.A., 1972. The ecology of fruit flies. Annual Review of Entomology, 17: 493–518.

Burk, T., 1981. Signaling and sex in acalyptrate flies. Florida Entomologist, 64: 30–43.

Burk, T., 1983. Behavioral ecology of mating in the Caribbean fruit fly, *Anastrepha suspensa* (Loew). Florida Entomologist, 66: 330–344.

Burk, T., 1984. Male–male interactions in Caribbean fruit flies, *Anastrepha suspensa* (Loew) (Diptera: Tephritidae): territorial fights and signalling stimulation. Florida Entomologist, 67: 542–548.

Burk, T. and Calkins, C.O., 1983. Medfly mating behavior and control strategies. Florida Entomologist, 66: 3–18.

Burk, T. and Webb, J.C., 1983. Effect of male size on calling propensity, song parameters, and mating success in Caribbean fruit flies (*Anastrepha suspensa* (Loew). Annals of the Entomological Society of America, 76: 678–682.

Cade, W., 1980. Alternative male reproductive behaviors. Florida Entomologist, 63: 30–45.

Dawkins, R., 1980. Good strategy or evolutionarily stable strategy? In: G.W. Barlow and J. Silverberg (Editors), Socibiology Beyond Nature/Nurture. Westview Press, Boulder, Colorado, pp. 331–367.

Dawkins, R. and Krebs, J.R., 1979. Arms races between and within species. Proceedings of the Royal Society of London, Series B, 205: 489–511.

Dodson, G.N., 1978. Behavioral, anatomical, and physiological aspects of reproduction in the Caribbean fruit fly, *Anastrepha suspensa* (Loew). M.S. Thesis, University of Florida, Gainesville, 68 pp.

Dodson, G., 1986a. The significance of sexual dimorphism in the mating system of two species of tephritid flies (*Aciurina bigeloviae* and *Velantibulla dodsoni*) (Diptera: Tephritidae). Canadian Journal of Zoology, 65: 194–198.

Dodson, G., 1986b. Lek mating system and large male aggressive advantage in a gall-forming tephritid fly (Diptera: Tephritidae). Ethology, 72: 99–108.

Farrias, G.T., Cunningham, R.T. and Nakagawa, S., 1972. Reproduction in the Mediterranean fruit fly: abundance of stored sperm affected by duration of copulation, and affecting egg hatch. Journal of Economic Entomology, 65: 914–915.

Kuba, H., Kogama, J. and Prokopy, R.J., 1984. Mating behavior of wild melon flies, *Dacus cucurbitae* Coquillett (Diptera: Tephritidae) in a field cage: distribution and behavior of flies. Applied Entomology and Zoology, 19: 367–373.

Landolt, P.J. and Hendrichs, J., 1983. Reproductive behavior of the papaya fruit fly, *Toxotrypana curvicauda* Gerstaecker (Diptera: Tephritidae). Annals of the Entomological Society of America, 76: 413–417.

Landolt, P.J., Heath, R.R. and King, J.R., 1985. Behavioral response of female papaya fruit flies, *Toxotrypana curvicauda* Gerstaecker (Diptera: Tephritidae) to male produced sex pheromone. Annals of the Entomological Society of America, 78: 751–755.

Malavasi, A., Morgante, J.S. and Prokopy, R.J., 1983. Distribution and activities of *Anastrepha fraterculus* (Diptera: Tephritidae) flies on host and non-host trees. Annals of the Entomological Society of America, 76: 286–292.

Michelsen, A.F., Fink, F., Gogala, M. and Trave, D., 1982. Plants as transmission channels for insect vibrational songs. Behavioral Ecology and Sociobiology, 11: 269–281.

Monro, J., 1953. Stridulation in the Queensland fruit fly, *Dacus (strumeta) tryoni* (Frogg.). Australian Journal of Science, 16: 60–62.

Morgante, J.S., Malavasi, A. and Prokopy, R.J., 1983. Mating behavior of wild *Anastrepha fraterculus* (Diptera: Tephritidae) on a caged host tree. Florida Entomologist, 66: 234.

Morris, G.K., Kerr, G.E. and Gwynne, D.T., 1975. Calling song function in the bog katydid *Metreoptera sphagnoeum* (F. Wilkes) (Orthoptera): (Tettigoniidae): female phonotaxis to normal and altered song. Zeitschrift für Tierpsychologie, 37: 502–514.

Prokopy, R.J., 1980. Mating behavior of frugivorous tephritidae in nature. Proceedings, Symposium on Fruit Fly Problems, XVI, International Congress Entomology, Kyoto, pp. 37–46.

Prokopy, R.J. and Hendrichs, J., 1979. Mating behavior of *Ceratitis capitata* on a field caged host tree. Annals of the Entomological Society of America, 72: 642–648.

Robacher, D.C. and Hart, W.G., 1985. Microhabitat choice and courtship behavior of laboratory-reared Mexican fruit flies *Anastrepha ludens* (Diptera: Tephritidae) in cages containing host and non-host trees. Annals of the Entomological Society of America, in press.

Rubenstein, D.I., 1980. On the evolution of alternative mating strategies. In: J.E.R. Staddon (Editor), Limits to Action: the Allocation of Individual Behavior. Academic Press, New York, pp. 65–100.

Sivinski, J., 1984. Effect of sexual experience on male mating success in a lek forming tephritid *Anastrepha suspensa* (Loew). Florida Entomologist, 67: 126–130.

Sivinski, J. and Webb, J.C., 1985a. The form and function of acoustic courtship signals of the papaya fruit fly *Toxotrypana curvicauda*. Florida Entomologist, 68: 634–641.

Sivinski, J. and Webb, J.C., 1985b. Changes in Caribbean fruit fly acoustic signal with social situation (*Anastrepha suspensa*: Diptera: Tephritidae). Annals of the Entomological Society of America, 79: 146–149.

Sivinski, J. and Webb, J.C., 1985c. Sound production and reception in the caribfly *Anastrepha suspensa*. Florida Entomologist, 68: 273–278.

Sivinski, J., Burk, T. and Webb, J.C., 1984. Acoustic courtship signals in the caribfly *Anastrepha suspensa*. Animal Behavior, 32: 1011–1016.

Smith, D.C. and Prokopy, R.J., 1980. Mating behavior of *Rhagoletis pomonella* (Diptera: Tephritidae). VI. Site of early-season encounters. Canadian Entomologist, 112: 585–590.

Smith, D.C. and Prokopy, R.J., 1982. Mating behavior of *Rhagoletis mendax* (Diptera: Tephritidae) flies in nature. Annals of the Entomological Society of America, 75: 388–392.

Steck, G.J., 1984. *Chaetostomella undosa* (Diptera: Tephritidae) biology, ecology, and larval description. Annals of the Entomological Society of America, 77: 669–678.

Thornhill, R., 1980. Competitive, charming males, and choosy females: was Darwin correct? Florida Entomologist, 63: 5–30.

Thornhill, R., 1984. Alternative female choice tactics in the scorpionfly *Hyobittacus apicalis* (Mecoptera) and their implications. American Zoologist, 24: 367–383.

Thornhill, R., Dodson, G. and Marshall, L., 1983. Sexual selection and insect mating behavior. The American Biology Teacher, 45: 310–319.

Tzanakakis, M.E., Tsitsipis, J.A. and Economopoulos, A.P., 1968. Frequency of mating in females of the olive fruit fly under laboratory conditions. Journal of Economic Entomology, 61: 1309–1312.

Webb, J.C., Sharp, J.L., Chambers, D.L., McDow, J.J. and Benner, J.C., 1976. The analysis and identification of sounds produced by the male Caribbean fruit fly, *Anastrepha suspensa* (Loew). Annals of the Entomological Society of America, 69: 415–420.

Webb, J.C., Burk, T. and Sivinski, J., 1983. Attraction of female Caribbean fruit flies (*Anastrepha suspensa* (Loew)) (Diptera: Tephritidae) to males and male-produced stimuli in field cages. Annals of the Entomological Society of America, 76: 996–998.

Webb, J.C., Sivinski, J. and Litzkow, C., 1984. Acoustical behavior and sexual success in the Caribbean fruit fly, *Anastrepha suspensa* (Loew) (Diptera: Tephritidae). Environmental Entomology, 13: 650–656.

Webb, J.C., Sivinski, J. and Smittle, B., 1985. Acoustical courtship signals and sexual success in irradiated caribfly *Anastrepha suspensa* (Loew) (Diptera: Tephritidae). Florida Entomologist, 70: 103–109.

Zwölfer, H., 1974. Innerartliche kommunikationsstrategie und isolationsmechanismus bei Bohrfliegen (Diptera: Trypetidae). Entomologica Germanica, 1: 11–20.

Chapter 4.5 Host Plant Resistance to Tephritids: an Under-exploited Control Strategy

PATRICK D. GREANY

1 INTRODUCTION

In his discussion of tephritid-host plant relations, Schoonhoven (1982) noted that "in contrast to insect control programmes for other crops, very little attention has been given to the possibilities of breeding resistant varieties". While little plant breeding for fruit fly resistance has been performed, scattered accounts of innate host plant resistance to tephritids do exist. Unfortunately, most of these studies did not inquire into the mechanisms affording resistance, and stopped short at demonstration. Some are simply anecdotal observations. Tephritid-host plant relations seem to have escaped notice in most general reviews on insect-plant relations and host plant resistance (e.g., Beck, 1965; Chapman, 1974; Chapman and Bernays, 1978; Hedin, 1977, 1983; Jermy, 1976; Maxwell, 1976; Maxwell et al., 1972; Maxwell and Jennings, 1980). Moreover, most attention has been given to the host plant relations of tephritid adults rather than the larvae, as is apparent from the principal reviews of the ecology and trophic relations of fruit flies (Bateman, 1972; Christenson and Foote, 1960; Prokopy, 1977, 1982). This is true despite the fact that it is the larval stage that accounts for the principal damage caused by fruit flies. Much of the information that is available pertains to the role of symbiotic microorganisms in host plant relations, rather than the influence of plant allelochemics upon fruit fly larvae. Although the role of symbionts is reviewed elsewhere in this book, it will receive some attention here as it pertains to the ability of fruit fly larvae to successfully develop in host plants.

Many opportunities exist for improved control of tephritids through manipulation of plant resistance. These include: (1) use of plant growth regulators to control plant senescence and the expression of genes for resistance factors, (2) employing conventional breeding schemes and genetic engineering approaches to improve plant resistance, (3) interfering with symbionts, and (4) direct utilization of plant allelochemics as post-harvest fumigants. These possibilities will be discussed below.

2 EXAMPLES OF PLANT RESISTANCE TO TEPHRITIDS

Review of the literature yielded numerous examples of plants that are resistant to fruit flies. The following is not an exhaustive compilation of all known instances. Instead, emphasis is placed on studies that provided information on the bases for resistance. Many reports indicated that the susceptibility of the plants increased with senescence; special reference is made to these reports as it may be possible to manipulate this phenomenon through use of plant growth regulators.

Chapter 4.5 references, p. 360

2.1 Citrus

One of the best documented examples of plant resistance to tephritid attack is that of citrus fruit (reviewed by Greany et al., 1983, 1985). Citrus susceptibility varies according to: (1) the species of fruit fly, (2) the degree of peel senescence, and (3) the type of fruit.

Three determinants appear to be involved: the oviposition behavior of the fly, the oil content of the peel, and the softness of the peel. As will be discussed below, these factors together determine the outcome of a given pest-fruit interaction. Variations in the chemical composition of the peel oil related to senescence and fruit type, as well as the internal properties of the fruit (such as the pH of the juice) appear to have little effect on fruit susceptibility.

The Japanese orange fly, *Dacus tsuneonis* Miyake, is relatively well adapted to citrus, perhaps because it lays its eggs below the flavedo (Miyake, 1919), the region of the peel where the oils occur. Similarly, the competence of the medfly, *Ceratitis capitata* (Wiedemann), to attack citrus fruit may be attributable to its habit of depositing a large number of eggs at one site (Back, 1915; Bodenheimer, 1951).

Further, medfly females may repeatedly attack a previously-visited site, so that the resistance of the fruit is gradually broken down (Back and Pemberton, 1918). The presence of many larvae in one cavity may confer mutual benefit through co-operativeness, which helps to insure that at least some larvae penetrate the oily flavedo region.

Species that are less severe pests of citrus, such as the Caribbean fruit fly, *Anastrepha suspensa* (Loew), often deposit only one or a few eggs at each site (Greany et al., 1983).

Several studies have shown that citrus peel oils are toxic to *C. capitata* eggs and larvae (Back, 1915; Back and Pemberton, 1918; Bodenheimer, 1951; Ortu, 1978). This also proved true for *A. suspensa* (Greany et al., 1983; Styer and Greany, 1983). In fact, *C. capitata* and *A. suspensa* eggs proved equally susceptible to citrus peel oils (Greany and Ohinata, unpublished), so that the difference in ability of the two flies to attack citrus may be due to the differences in oviposition behavior rather than to differences in tolerance to the plant allelochemics.

In general, senescent fruit are markedly more susceptible than early season fruit; however, lemons are so nearly immune that even over-ripe fruit are usually refractory (Spitler et al., 1984; Greany et al., 1983 and references therein). Oranges are somewhat more resistant than grapefruit to most species of fruit flies, but they do become susceptible as the peel becomes senescent. Senescent Marsh grapefruit proved ca. 5 × more susceptible than early-season fruit to laboratory infestation by *A. suspensa* (Greany et al., 1985). McDonald et al. (1987) correlated the senescence of Marsh grapefruit with reduced peel oil content and increased peel softness.

As mentioned above, composition of the peel oil appears to have less of an effect than oil abundance upon fly susceptibility. Changes in the composition of the peel oil throughout the season did not correlate with increased susceptibility of grapefruit to *A. suspensa* (Greany et al., 1985). Similarly, while lemon peel oil contains relatively more of the compounds found by Styer and Greany (1983) to be most toxic to *A. suspensa* eggs and larvae (particularly the oxygenated terpenoids such as citral), lemon oil proved to be no more toxic to this species than grapefruit or orange oil (Styer and Greany, 1983). Lemon peel possesses ca. 2 × more oil (per gram of flavedo) than grapefruit peel and this may be the more important determinant of resistance of lemons than oil composition (Greany et al., 1983).

Increased peel softness of senescent citrus fruit is a well recognized phenomenon (Coggins and Lewis, 1965; Coggins et al., 1969). Bodenheimer (1951) considered peel softness to be a key factor in the susceptibility of citrus fruit to attack by the medfly. A similtaneous decline in peel oil content and peel firmness would be expected to result in a marked increase in larval survival because of the reduced amount of oil in the flavedo and because of reduced difficulty for the larvae to migrate through this inhospitable region to the spongy albedo and into the pulp, where no deleterious allelochemics have been found (Greany et al., 1983). This explanation appears to fit the observed facts associated with senescence-related susceptibility.

Another possible factor in the resistance of citrus fruit is the presence of limonoids and naringin in the peel. Naringin has been found to be capable of deterring oviposition by *A. suspensa* (Szentesi et al., 1979), and limonin was shown by Klocke and Kubo (1982) to be an antifeedant against two non-citrus pests. These agents may be candidates for the manipulation of oviposition and feeding by fruit flies in citrus.

Means by which the resistance of citrus might be increase and extended in time will be discussed below.

2.2 Apples

Both the Queensland fruit fly, *Dacus tryoni* (Froggatt), and the walnut husk fly, *Rhagoletis completa* (Cresson), are variably successful in attacking different apple varieties (*Malus* spp.), with the outcome depending upon the maturity of the fruit (Bower, 1977; Dean and Chapman, 1973, Goonewardene et al., 1975; Pree, 1977; Reissig, 1979). Larval mortality for both species was significantly greater in unpicked than picked fruit, for reasons that were not determined, but which were suspected to be due to chemical rather than physical differences (Bower, 1977; Reissig, 1979). While medfly females readily oviposit into 'Golden Delicious' apples (McDonald, 1986), medfly larvae were found to be poorly adapted to apples (Rivnay, 1950).

Crabapples (*Malus toringoides* Hughes) proved quite resistant to the apple maggot fly, *Rhagoletis pomonella* (Walsh) (Neilson, 1967; Pree, 1977). Pree correlated resistance with total phenol content and found that gallic, tannic, and o-coumaric acids, quercetin, naringenin, and d-catechin at 1000 ppm in an artificial medium prevented larval development. The desire was expressed by Pree (1977) and Goonewardene et al. (1975) to find an alternative to conventional broad spectrum insecticides for control of *R. pomonella* through incorporation of resistance genes. Pree (1977) indicated that while many of the phenolic acids shown to be toxic in bioassays occcur in commercial apple varieties, they occur at much lower levels than required for activity. He hoped that analysis of resistant apple varieties for individual phenol compounds might reveal a unique compound toxic in small amounts to *R. pomonella* larvae, but no further work appears to have been done toward this end.

2.3 Papayas

Papayas are infested with varying success by the Oriental fruit fly, *Dacus dorsalis* Hendel, depending upon fruit ripeness (Seo et al., 1982). The medfly is unable to attack immature papayas because of a "milky juice" that exudes from immature fruit; only fruit that are too ripe to be desirable for human consumption are suitable for this fly (Back and Pemberton, 1918). While papayas have been recorded as a host for the Caribbean fruit fly (Swanson and Baranowski, 1972), it is a rarity for papayas to be successfully attacked by this species. The Mexican fruit fly, *Anastrepha ludens* (Loew), was not reported to attack papayas (Baker et al., 1944).

Seo et al. (1983) related resistance of papayas to the Oriental fruit fly to two compounds that are released from damaged, unripened papayas: linalool and benzyl isothiocyanate (BITC). The latex produced by green papayas in response to probing by *D. dorsalis* females is an emulsion of benzylglucosinolate, a precursor of BITC (Seo et al., 1983). Release of BITC occurred within 5 min of mixing the substrate benzylglucosinolate in the latex with thioglucosidase from idioblasts ruptured upon injury of green papaya tissues (Seo et al., 1983). They indicated that the concentration of BITC could have a direct impact on the susceptibility of papayas because of its toxicity and deterrence to oviposition. Oviposition deterrence in papayas was less for the Oriental fruit fly than for the medfly and the melon fly, *Dacus cucurbitae* Coquillett, which is in agreement with the relative amount of damage these 3 species cause in papayas.

Seo et al. (1983), did not evaluate the potential effect of linalool on the flies, but this compound proved quite toxic to eggs and larvae of *A. suspensa* and *R. pomonella* (Greany, unpublished data; Pree, personal communication, respectively). Additional studies on the role of linalool in the resistance of papayas to fruit flies would be warranted.

2.4 Other fruits

An assortment of other fruits have been found to be resistant to fruit fly attack, but these reports generally did not include detailed studies on the mechanisms affording resistance. For example, Joel (1978, 1980) found that mango fruits (*Mangifera indica* L.) produce a secretion from ducts that is deleterious to *C. capitata*, and assumed that the secretion is of a defensive nature. Armstrong (1983) found that three banana cultivars were resistant to attack by *D. cucurbitae*, *D. dorsalis*, and *C. capitata* in Hawaii, and related this to the exudation of latex from unripe bananas. Much earlier, Back and Pemberton (1918) suggested that the immunity of commercial varieties of bananas to *C. capitata* was related to the presence of tannins in the peel and the flow of sap from egg punctures in unripe bananas.

Bush (1957) found that some strains of avocados were relatively resistant to *A. ludens*, and suggested that resistance to this fly should be taken into account in plant breeding efforts. Similarly, Armstrong et al. (1983) found that 'Sharwil' avocados were resistant to *C. capitata*, *D. cucurbitae*, and *D. dorsalis* at the mature green harvest stage of ripeness. Chan and Tam (1985) showed that the alkaloid α-tomatine was toxic to medfly larvae and suggested that this could account for the poor success of this species in green tomatoes. This agent disappears almost completely from tomatoes when the fruit are most acceptable to attack by the medfly, according to these authors.

Howard and Kenney (1987) found that some cultivars of carambollas were relatively resistant to attack by *A. suspensa*, and that immature fruit were more resistant than ripe fruit. Khandelwal and Nath (1978) found that some strains of watermelon were resistant to *D. cucurbitae*. Chelliah and Sambandam (1974a, b) found that *D. cucurbitae* was better able to attack some varieties of the melon *Cucumis callosus* (Rottl.) than others, and determined that this was due to a combination of non-preference and antibiosis, perhaps due to phenols and cucurbitacins.

Armstrong et al. (1979) and Armstrong and Vargas (1982) found that several varieties of pineapple were resistant to *D. dorsalis*, *D. cucurbitae*, and *C. capitata*. Macion et al. (1968) correlated resistance of pineapples to *D. dorsalis* with the liquid content of the fruit. Fruit acidity had no effect on larval survival. They suggested that the liquid content was a heritable trait, and could be a basis for selection for fruit fly resistance. Fullaway (1949) noted, however, that although it was uncommon for pineapples to be attacked by *D. dorsalis*, this species was successfully reared from overripe pineapples in Hawaii and Formosa. This is in keeping with the general observation that it is the senescent fruit that are most susceptible to fruit fly attack. This also was recognized by Boyce (1934), who regarded the hardness of the husk of walnuts to be the principal determinant of walnut resistance to the walnut husk fly, *Rhagoletis completa* (Cresson). He found that hardness was a function of senescence and that it varied with different types of walnuts, with a corresponding difference in their resistance to this fly.

Numerous host records exist for the fruit flies of major economic importance (e.g., Swanson and Baranowski (1972), Von Windeguth et al. (1973), Kapoor and Agarwal (1982)). These tend to be of value primarily in the quarantine sense, by enumerating the fruits that might require fumigation prior to shipment. It should not be inferred that all the listed fruit are equivalent in susceptibility to attack, and the potential effect of fruit senescence on susceptibility should be kept in mind when reviewing these host lists.

3 ROLE OF SYMBIONTS

One of the long-accepted tenets on the biology of tephritid fruit flies is that they use microorganisms as symbionts (reviewed by Buchner, 1965; Courtice et al., 1984; Hagen and Tassan, 1972; Rossiter et al., 1982; Fitt, 1983 and Howard et al., 1985). While this topic will be fully reviewed elsewhere (Chapter 3.1), some discussion is needed here in relation to the potential roles played by symbionts in tephritid host-plant relations. Possible functional roles that have been suggested for bacterial symbionts include: (1) pectinolytic activity, (2) nitrogen fixation, (3) nutrient provision, (4) detoxification of host fruit allelochemicals, and (5) protection against pathogens (Howard et al., 1985).

The bacterial species considered symbionts typically also are plant pathogens (e.g., *Klebsiella oxytoca* and *Enterobacter cloacae*, which cause rot in apples — Rossiter et al., 1982). While no symbiotic relationship was claimed, fruit flies were shown to vector fungal and bacterial plant pathogens in at least two instances (Ito et al., 1979; Lozano and Bellotti, 1978). In some cases (e.g., *A. suspensa*, as studied by Boush et al., 1972), no particular bacterial species could be consistently isolated from the fruit fly species in question, as would indicate an obligate association. However, this does not preclude facultative use by fruit flies of fortuitously present microorganisms that are capable of serving their needs in a given host plant.

Bacteria associated with the apple maggot fly can degrade a variety of insecticides (Boush and Matsumura, 1967), at least some of which by the action of mixed-function oxidases (MFOs). These bacterially-produced enzymes also could play a major role in protecting herbivores against chemical stress from secondary plant substances, many of which can be detoxified by MFOs (Brattsten et al., 1977). Whether the enzymes are produced by symbionts or through de *de novo* biosynthesis by the fruit fly larvae themselves, the MFOs could play an important role in allowing the flies to survive in a given host plant.

As mentioned above, one of the potential benefits symbionts may provide is the degradation of pectin, which could facilitate their entry into the fruit and escape from allelochemicals in the peel. The ability of the aphid *Schizaphis graminum* Rondani to hydrolyze pectic substances proved to be a key issue in determining the resistance of different sorghum varieties (Campbell and Dreyer, 1985). Rossiter et al. (1982) indicated that pectin degradation was a possible role for the putative bacterial symbiotes of the apple maggot fly.

Chapter 4.5 references, p. 360

However, recent studies by Howard et al. (1985) showed that not all strains of
K. oxytoca isolated from apple maggot flies produced pectinolytic enzymes, and,
along with other data, their studies have thrown the entire issue of fruit fly
microorganism symbiosis into question. In fact, this group now believes that
the bacteria associated with apple maggot flies may be more deleterious than
beneficial (Guy L. Bush, personal communication).

This is a dramatic change from conventional beliefs about fruit fly symbionts: it brings into question the relevance of the remarkable morphological
adaptations exhibited by fruit flies, such as the esophageal bulb in the head of
adult flies, gastric caecae in the gut of the larvae, and crypts in the ovipositor,
all of which become filled with bacteria (reviewed by Buchner, 1965). Resolution of the fruit fly-microorganism symbiosis issue will prove quite interesting,
as well as important, given the potential for controlling fruit flies by depriving
them of their associated microorganisms, as has been suggested by Hagen and
Tassan (1972) and Tsiropoulos (1981).

4 MANIPULATION OF RESISTANCE

It would seem from the many examples cited above that tephritids are not
generally able to attack fruit indiscriminantly without suffering the consequences. Candidate host fruit often possess innate resistance barriers, either of
a physical or a biochemical nature, that fruit flies are unable to penetrate, even
with the aid of putative bacterial symbionts. In some cases it has been suggested that varieties of fruit exhibiting resistance be bred with varieties possessing other horticultural virtues in the hope of producing fruit displaying the
combined features. It appears that very little attention has been given to this
proposition, perhaps because of the long maturation period for most fruit crops
and the commensurate long-term commitment required.

Another possible means of manipulating resistance of fruit to fruit fly attack
is through the use of plant growth regulators which cause the plant to produce
more of a given resistance agent or which sustain the innate resistance properties of the plant. Some effort has already been made toward this end for a
variety of insect crop-plant systems (see Tingey and Singh, 1980; Campbell et
al., 1984; Campbell, in press; Hedin, 1985).

One notable success in the use of bioregulators for improved plant resistance
involves the use of gibberellic acid (GA$_3$) to delay senescence of citrus fruit peel
and thereby reduce fruit susceptibility to plant pathogens (Coggins, 1973). This
practice is now being used routinely throughout citrus-growing areas. Fortuitously, gibberellic acid treatment does not inhibit internal maturation, so
that the fruit ripen properly despite possessing a non-senescent, resistant peel.

Since fruit flies, like plant pathogens, are better able to attack senescent
fruit, this author initiated cooperative research in Florida and Israel on use of
GA$_3$ to manipulate the resistance of citrus fruit to *A. suspensa* and *C. capitata*,
respectively. These studies indicated that GA$_3$ has considerable promise for
reducing the susceptibility of citrus fruit to tephritids. Oranges were easier to
protect than grapefuit, and it was easier to defend the fruit against *A. suspensa*
than *C. capitata* (reflecting the greater competence of the latter species to
attack citrus). Results of tests in Florida against *A. suspensa* are described by
Greany et al. (1987); a summary of tests in Israel on *C. capitata* is in preparation.

It appears that the principal benefit from use of GA$_3$ is related to its effect
on the physical properties of the fruit peel: the rate of softening of the peel was
reduced as a function of GA$_3$ treatment. However, the rate of decline in peel oil
content was not affected (McDonald et al., 1987). We believe that because of
retained peel integrity, the fly larvae were exposed to the peel oil for a longer

time in treated fruit, and were more likely to be intoxicated than in the softer control fruit (Greany et al., 1987).

We are now investigating the possibility of synergizing the effect of GA$_3$ by use of another agent, DCPTA (2-diethylamino- ethyl-3,4-dichlorophenylether). This agent has been reported to induce citrus fruit to produce more peel oil, possibly through de-repression of genes for key enzymes involved in the biosynthesis of isoprenoids (Yokoyama et al., 1984). The desired outcome would be an internally ripe fruit with a non-senescent, resistant peel. An advantage of this approach, as compared to conventional bait-spray programs for control of adult fruit flies, is that it should not be injurious to beneficial insects that are important in controlling other citrus pests.

5 CONCLUSIONS

It appears that there is much opportunity for control of tephritids through manipulation of host plant resistance properties, either through conventional breeding strategies, or through bioregulation of resistance properties many plants exhibit prior to onset of senescence.

Most pre-harvest control programs for fruit flies have been directed toward the adults (e.g., the bait spray programs) rather than the eggs or larvae. By manipulating the plant's resistance features, these critical and injurious life stages may be attacked. In addition, by elucidating the allelochemicals responsible for biochemical resistance, it may prove possible to use these natural products (or synthetic analogs) instead of conventional fumigants for postharvest disinfestation. This possibility was demonstrated by Davis et al. (1976), who used naturally-occurring fruit volatiles to kill *A. suspensa* larvae in guavas.

As suggested by Hagen and Tassan (1972), it may also be possible to derange fruit fly/host plant relations by eliminating their symbionts. This may depend upon finding a means that does not pose environmental hazards, which is the objection often raised to the use of antibiotics in the field. It is conceivable that instead of using antibiotics to kill the symbionts, it may be possible to use pathogenic microorganisms in bait sprays to kill the flies directly, or to use genetically-engineered bacteria in the bait which will outcompete the natural flora and render the flies incompetent to attack their host fruit. This of course presupposes a need by fruit flies for symbiotic microorganism to assist them in their host plant relations.

Finally, genetic engineering of plants holds great promise for improving their resistance to plant pests of all types, including fruit flies. It seems logical to assume that this effort will be expedited by studies to reveal the key mechanisms used by plants to resist fruit fly attack. Once these are understood, it may be possible either to alter the expression of genes of interest that are already present, or to selectively introduce new genes that will confer the desired degree of resistance without causing undesirable side effects.

6 ACKNOWLEDGEMENTS

I would like to express my appreciation to the following individuals for their excellent cooperation in the research on use of gibberellic acid to reduce the susceptibility of citrus to fruit flies: Yoram Rössler (Israel), and Carrol Calkins, Tim Hatton, Roy McDonald, Gordon Rasmussen, Bill Schroeder, and Phil Shaw (USA). I would also like to thank the reviewers of this manuscript, Drs. Calkins, McDonald, and Schroeder, plus Dr. L.M. Schoonhoven, for their many helpful suggestions.

Chapter 4.5 references, p. 360

7 REFERENCES

Armstrong, J.W., 1983. Infestation biology of three fruit fly (Diptera: Tephritidae) species on 'Brazilian,' 'Valery,' and 'Williams's' cultivars of banana in Hawaii. Journal of Economic Entomology, 76: 539–543.

Armstrong, J.W. and Vargas, R.I., 1982. Resistance of pineapple variety '59–656' to field populations of Oriental fruit flies and melon flies (Diptera: Tephritidae). Journal of Economic Entomology, 75: 781–782.

Armstrong, J.W., Vriesenga, J.D. and Lew, C.Y.L., 1979. Resistance of pineapple varieties D-10 and D-20 to field populations of Oriental fruit flies and melon flies. Journal of Economic Entomology, 72: 6–7.

Armstrong, J.W., Mitchell, W.C. and Farias, G.J., 1983. Resistance of 'Sharwil' avocados at harvest maturity to infestation by three fruit fly species (Diptera: Tephritidae) in Hawaii. Journal of Economic Entomology, 76: 119–121.

Back, E.A., 1915. Susceptibility of citrus fruits to the attack of the Mediterranean fruit fly. Journal of Agricultural Research, 3: 311–330.

Back, E.A. and Pemberton, C.E., 1918. The Mediterranean fruit fly. USDA Tech. Bull. #640. 43 pp.

Baker, A.C., Stone, W.E., Plummer, C.C. and McPhail, M., 1944. A review of studies on the Mexican fruit fly and related Mexican species. USDA Misc. Pub. 531, 155 pp.

Bateman, M.A., 1972. The ecology of fruit flies. Annual Review of Entomology, 17: 493–518.

Beck, S.D., 1965. Resistance of plants to insects. Annual Review of Entomology, 10: 207–232.

Bodenheimer, F.S., 1951. Citrus Entomology in the Middle East. W. Junk Pub Co., The Hague, 661 pp.

Boush, G.M. and Matsumura, F., 1967. Insecticidal degradation by *Pseudomonas melophthora*, the bacterial symbiote of the apple maggot. Journal of Economic Entomology, 60: 918–920.

Boush, G.M., Saleh, S.M. and Baranowski, R.M., 1972. Bacteria associated with the Caribbean fruit fly. Environmental Entomology, 1: 30–33.

Bower, C.C., 1977. Inhibition of larval growth of the Queensland fruit fly, *Dacus tryoni* (Diptera: Tephritidae) in apples. Annals of the Entomological Society of America, 70: 97–100.

Boyce, A.M., 1934. Bionomics of the walnut husk fly, *Rhagoletis completa*. Hilgardia, 8: 363–579.

Brattsten, L.B., Wilkinson, C.F. and Eisner, T., 1977. Herbivore-plant interactions: mixed-function oxidases and secondary plant substances. Science, 196: 1349–1352.

Buchner, P., 1965. Endosymbiosis of animals with plant microorganisms. Interscience, New York, 887 pp.

Bush, G.L., Jr., 1957. Some notes on the susceptibility of avocados in Mexico to attack by the Mexican fruit fly. Journal of the Rio Grande Valley Horticultural Society, 11: 75–78.

Campbell, B.C., 1987. The effects of plant growth regulation and herbicides on host plant quality to insects. In: B. Heinrichs (Editor), Plant Stress Interactions. John Wiley & Sons, New York, (in press).

Campbell, B.C. and Dreyer, D.L., 1985. Host-plant resistance of sorghum: differential hydrolysis of sorghum pectic substances by polysaccharases of greenbug biotypes (*Schizaphis graminum*, Homoptera: Aphididae). Archives Insect Biochemistry and Physiology, 2: 203–215.

Campbell, B.C., Chan, B.G., Creasy, L.L., Dreyer, D.L., Rabin, L.B., and Waiss, A.C., Jr., 1984. Bioregulation of host plant resistance to insects. In: R.L. Ory and F.R. Rittig (Editors), Bioregulators: Chemistry and Uses. ACS Symposium Series 257, Am. Chem. Soc., Washington, D.C., pp. 193–203.

Chan, H.T., Jr. and Tam, S.Y.T., 1985. Toxicity of α-tomatine to larvae of the Mediterranean fruit fly (Diptera: Tephritidae). Journal of Economic Entomology, 78: 305–307.

Chapman, R.F., 1974. The chemical inhibition of feeding by phytophagous insects: a review. Bulletin of the Entomological Res., 64: 339–363.

Chapman, R.F. and Bernays, E.A. (Editors), 1978. Insect and Host plant. Entomologia Experimentalis et Applicata, 24: 965 pp.

Chelliah, S. and Sambandam, C.N., 1974a. Mechanism of resistance in *Cucumis callosus* (Rottl.) Cogn. to the fruit fly, *Dacus cucurbitae* Coq. (Diptera: Tephritidae) I. Non-preference. Indian Journal of Entomology, 36: 98–102.

Chelliah, S. and Sambandam, C.N. 1974b. Mechanism of resistance in *Cucumis callosus* (Rottl.) Cogn. to the fruit fly, *Dacus cucurbitae* Coq. (Diptera: Tephritidae). II. Antibiosis, 36: 290–296.

Christenson, L.D. and Foote, R.H., 1960. Biology of fruit flies. Annual Review of Entomology, 5: 171–192.

Coggins, C.W., Jr., 1973. Use of growth regulators to delay maturity and prolong shelf life of Citrus. Acta Horticulturae, 34: 469–472.

Coggins, C.W., Jr. and Lewis, L.N., 1965. Some physical properties of the navel orange rind as related to ripening and to gibberellic acid treatments. Proceedings of the American Society for Horticultural Science, 86: 272–279.

Coggins, C.W., Jr., Scora, R.W., Lewis, L.N. and Knapp, J.C.F. 1969. Gibberellin-delayed senescence and essential oil changes in the navel orange rind. Journal of Agricultural and Food Chemistry, 17: 807–809.

Courtice, A.C. and Drew, R.A.I., 1984. Bacterial regulation of abundance in tropical fruit flies (Diptera: Tephritidae). Australian Zoologist, 21: 251–268.

Davis, P.L., Munroe, K.A. and Selheime, A.G., 1976. Laboratory bioassay of volatile naturally occurring compounds against the Caribbean fruit fly. Proceedings of the Florida State Horticultural Society, 89: 174–175.

Dean, R.W. and Chapman, P.J., 1973. Bionomics of the apple maggot in eastern New York. Search Agriculture (Geneva New York), 3: 1–62.

Fitt, G.P., 1983. Factors limiting the host range of tephritid fruit flies with particular reference to the effect of *Dacus tryoni* on the distribution and abundance of *Dacus jarvisi*. Ph.D. Thesis, Univ. Sydney.

Fullaway, D.T., 1949. *Dacus dorsalis* Hendel, in Hawaii. Proceedings of the Hawaiian Entomological Society, 13: 351–355.

Goonewardene, H.F., Kwolek, W.F., Dolphin, R.E. and Williams, E.B., 1975. Evaluating resistance of apple fruits to four insect pests. HortScience, 10: 393–394.

Greany, P.D., Styer, S.C., Davis, P.L., Shaw, P.E. and Chambers, D.L., 1983. Biochemical resistance of citrus to fruit flies: Demonstration and elucidation of resistance to the Caribbean fruit fly, *Anastrepha suspensa*. Entomologia Experimentalis et Applicata, 34: 40–50.

Greany, P.D., Shaw, P.E., Davis, P.L. and Hatton, T.T., 1985. Senescence-related susceptibility of Marsh grapefruit to laboratory infestation by *Anastrepha suspensa* (Diptera: Tephritidae). Florida Entomologist, 68: 144–150.

Greany, P.C., McDonald, R.E., Shaw, P.E., Schroeder, W.J., Howard, D.F., Hatton, T.H., Davis, P.L. and Rasmussen, G.K., 1987. Use of gibberellic acid to reduce grapefruit susceptibility to attack by the Carribbean fruit fly, *Anastrepha suspensa* (Diptera: Tephritidae). Tropical Science, 27: (in press).

Hagen, K.S. and Tassan, R.L., 1972. Exploring nutritional roles of extracellular symbiotes on the reproduction of honeydew feeding adult chrysopids and tephritids. In: J.G. Rodriguez (Editor), Insect and Mite Nutrition. Elsevier, Amsterdam, pp. 323–351.

Hedin, P.A. (Editor), 1977. Host Plant Resistance to Pests. Am. Chem. Soc. Symp. Ser. 62, ACS, Washington, D.C., 286 pp.

Hedin, P.A. (Editor), 1983. Plant Resistance to Insects. Am. Chem. Soc. Symp. Series 208, ACS, Washington, D.C., 375 pp.

Hedin, P.A. (Editor), 1985. Bioregulators for Pest Control. Am. Chem. Soc. Symp. Ser. 276, ACS, Washington, D.C., 540 pp.

Howard, D.F. and Kenney, P., 1987. Infestation of carambolas by laboratory-reared Caribbean fruit flies (Diptera: Tephritidae): Effects of fruit ripeness and cultivar. Journal of Economic Entomology, (in press).

Howard, D.J., Bush, G.L. and Breznak, J.A., 1985. The evolutionary significance of bacteria associated with *Rhagoletis*. Evolution, 39: 405–417.

Ito, P.J., Kunimoto, R. and Ko, W.H., 1979. Transmission of Mucor rot of guava fruits by three species of fruit flies. Tropical Agriculture, 56: 49–52.

Jermy, T. (Editor), 1976. The Host-Plant in Relation to Insect Behaviour and Reproduction. Plenum Press, New York, 321 pp.

Joel, D.M., 1978. The secretory ducts of mango fruits: a defense system effective against the Mediterranean fruit fly. Israel Journal of Botany, 27: 44–45.

Joel, D.M., 1980. Resin ducts in the mango fruit: a defence system. Journal of Experimental Botany, 31: 1707–1718.

Kapoor, V.C. and Agarwal, M.L., 1982. Fruit flies and their increasing host plants in India. In: R. Cavalloro (Editor), Fruit Flies of Economic Importance. A.A. Balkema, Rotterdam, The Netherlands, pp. 252–257.

Khandelwal, R.C. and Nath, P., 1978. Inheritance of resistance to fruit fly in watermelon. Canadian Journal of Genetics and Cytology, 20: 31–34.

Klocke, J.A. and Kubo, I., 1982. Citrus limonoid by-products as insect control agents. Entomologia Experimentalis et Applicata, 32: 299–301.

Lozano, J.C. and Bellotti, A., 1978. *Erwinia carotovora* var. *carotovora*, causal agent of bacterial stem rot of cassava: etiology, epidemiology and control. PANS, 24: 467–479.

Macion, E.A., Mitchell, W.C. and Smith, J.B., 1968. Biophysical and biochemical studies on the nature of resistance of pineapples to the Oriental fruit fly. Journal of Economic Entomology, 61: 910–916.

Maxwell, F.G., 1976. Host-plant resistance to insects–chemical relationships. In: H.H. Shorey and J.J. McKelvey (Editors), Chemical Control of Insect Behavior. Wiley, New York, pp. 299–304.

Maxwell, F.G. and Jennings, P.R., 1980. Breeding Plants Resistant to Insects. Wiley, New York, 683 pp.

Maxwell, F.G., Jenkins, J.N. and Parrott, W.L., 1972. Resistance of plants to insects. In: Advances in Agronomy, 24: 187–265.

McDonald, P.T., 1986. Larger egg clutches following host deprivation in colonized *Ceratitis capitata* (Diptera: Tephritidae). Journal of Economic Entomology, 79: 392–394.

McDonald, R.E., Shaw, P.E., Greany, P.D., Hatton, T.T. and Wilson, C.W., 1987. Effects of gibberellic acid on selected physical and chemical properties of grapefruit. Tropical Science, 27: 17–22.

Miyake, T., 1919. Studies on the fruit flies of Japan. I. Japanese orange fly. Bulletin of the Imperial Central Agricultural Expt. Station, 2: 87–165.

Neilson, W.T.A., 1967. Development and mortality of the apple maggot, *Rhagoletis pomonella*, in crab apples. Canadian Entomologist, 99: 217–219.

Ortu, S., 1978. Preliminary observations about the temporary resistance of fruits of some species of citrus fruit to the attacks of *Ceratitis capitata*. Wied. Report to the IOBC meeting on fruit fly biology and control, held in Sassari, Italy, spring of 1978.

Pree, D.J., 1977. Resistance to development of larvae of the apple maggot in crab apples. Journal of Economic Entomology, 70: 611–614.

Prokopy, R.J., 1977. Stimuli influencing trophic relations in Tephritidae. Colloques Internationaux du CNRS, 265: 305–336.

Prokopy, R.J., 1982. Tephritid relationships with plants. In: R. Cavalloro (Editor), Fruit Flies of Economic Importance. Balkema, Rotterdam, The Netherlands, pp. 230–239.

Reissig, W.H., 1979. Survival of apple maggot larvae, *Rhagoletis pomonella* (Diptera: Tephritidae), in picked and unpicked apples. Canadian Entomologist, 111: 181–187.

Rivnay, E., 1950. The Mediterranean fruit fly in Israel. Bulletin of Entomological Research, 41: 321–341.

Rossiter, M.C., Howard, D.J. and Bush, G.L., 1982. Symbiotic bacteria of *Rhagoletis pomonella*. In: R. Cavalloro (Editor), Fruit Flies of Economic Importance. Balkema, Rotterdam, The Netherlands, pp. 77–84.

Schoonhoven, L.M., 1982. Relations of fruit flies and host plants (report on session). In: R. Cavalloro (Editor), Fruit Flies of Economic Importance. Balkema, Rotterdam, The Netherlands, pp. 228–229.

Seo, S.T., Farias, G.J. and Harris, E.J., 1982. Oriental fruit fly: Ripening of fruit and its effect on index of infestation of Hawaiian papayas. Journal of Economic Entomology, 75: 173–178.

Seo, S.T. Tang, C.S., Sanidad, S. and Takenaka, T.H., 1983. Hawaiian fruit flies (Diptera: Tephritidae): variation of index of infestation with benzyl isothiocyanate concentration and color of maturing papaya. Journal of Economic Entomology, 76: 535–538.

Spitler, G.H., Armstrong, J.W. and Couey, H.M., 1984. Mediterranean fruit fly (Diptera: Tephritidae) host status of commercial lemon. Journal of Economic Entomology, 77: 1441–1444.

Styer, S.C. and Greany, P.D., 1983. Increased susceptibility of laboratory-reared vs. wild Caribbean fruit fly, *Anastrepha suspensa* (Loew) (Diptera: Tephritidae), larvae to toxic citrus allelochemics. Environmental Entomology, 12: 1606–1608.

Swanson, R.W. and Baranowski, R.M., 1972. Host range and infestation by the Caribbean fruit fly, *Anastrepha suspensa* (Diptera: Tephritidae), in south Florida. Proceedings of the Florida State Horticultural Society, 85: 271–274.

Szentesi, A., Greany, P.D. and Chambers, D.L., 1979. Oviposition behavior of laboratory-reared and wild Caribbean fruit flies (*Anastrepha suspensa*; Diptera: Tephritidae): I. Selected chemical influences. Entomologia Experimentalis et Applicata, 26: 227–238.

Tingey, W.M. and Singh, S.R., 1980. Environmental factors influencing the magnitude and expression of resistance. In: F.G. Maxwell and P.R. Jennings (Editors), Breeding Plants Resistant to Insects. Wiley-Interscience, New York, pp. 87–113.

Tsiropoulos, G.J., 1981. Effect of antibiotics incorporated into defined adult diets on survival and reproduction of the walnut husk fly, *Rhagoletis completa* Cress. (Dipt., Trypetidae). Zeitschrift für Angewandte Entomologie, 91: 100–106.

Von Windeguth, D.L., Pierce, W.H. and Steiner, L.F., 1973. Infestations of *Anastrepha suspensa* in fruit on Key West, Florida, and adjacent islands. Florida Entomologist, 56: 127–131.

Yokoyama, H., Hsu, W.J., Hayman, E. and Poling, S., 1984. Bioregulation of plant constituents. In: Recent Advances in Phytochemistry–Phytochemical Adaptations to Stress. Plenum Press, New York, 18: 231–250.

General Index

Species Index

Anastrepha

A. antunesi 15
A. atrox 22
A. bicolor 22
A. bistrigata 15
A. dentata 22
A. dissimilis 22
A. distincta 15, 86
A. elegans 22
A. fenestrata 22
A. fraterculus 15, 18, 21, 108, 190, 193, 196, 207, 213, 296, 297, 330, 331, 346
A. grandis 15, 20
A. hamata 18
A. interrupta 20
A. limae 20
A. ludens 15, 22, 76, 85, 104, 145, 190, 193, 196, 202, 203, 225, 315, 330–332, 345, 355, 356
A. manihoti 15
A. montei 15
A. obliqua 15, 21–23, 108, 330
A. pallens 20, 23
A. pickeli 15, 22
A. punctata 22
A. serpentina 15, 21, 86
A. sororcula 15
A. stonei 22
A. striata 15, 21, 86
A. suspensa 15, 22, 23, 77, 104, 145, 190, 192, 193, 195–203, 207, 225, 297, 308, 310, 311, 316, 317, 330, 343, 344, 346–348, 354–359
A. tripunctata 17

Ceratitis

C. capitata 29, 39, 60, 67, 73, 89, 93, 96, 104, 143, 154, 158, 179, 181, 182, 186, 190, 197, 207, 213, 224–226, 242, 261, 275, 330–334, 336, 337, 344, 346, 348, 354, 356, 358
C. catoirii 57
C. cosyra 55
C. discussa 55
C. hispanica 46
C. malgassa 57
C. pedestris 55
C. quinaria 55
C. rosa 55, 108, 158, 225, 226
C. rubivora 55
C. tananarivana 55, 57

Dacus

D. aglaiae 329
D. aquilonis 12, 70, 225, 329
D. armatus 9
D. atrisetosus 68
D. brevistylus 59
D. brunneus 136
D. bryoniae 68, 259, 329
D. cacuminatus 68, 134, 207, 224, 258, 329, 332
D. caudatus 59
D. ciliatus 59, 95, 146, 157, 329, 331
D. correctus 59, 222
D. cucumis 69, 108, 156, 158, 258, 259, 329
D. cucurbitae 29, 59, 60, 63, 67, 73, 84, 95, 104, 156, 157, 165–167, 207, 224, 225, 277, 284, 296, 297, 316, 329, 331, 332, 345, 356
D. curvipennis 71
D. decipiens 68
D. decurtans 329
D. demmerezi 57, 59
D. diversus 222, 329
D. dorsalis 28, 29, 59, 63, 84, 104, 156, 165–167, 224, 225–233, 278, 280, 284, 296, 297, 316, 329, 331–334, 336, 355–357
D. endiandrae 329
D. facialis 28, 71
D. fraterculus 71
D. frauenfeldi 67
D. hageni 59, 329
D. halfordiae 329
D. jarvisi 69, 215, 258, 260, 329
D. kirki 71
D. kraussi 329
D. longistylus 59
D. melanotus 28, 71
D. musae 68, 134, 280, 329
D. neohumeralis 134, 167, 224, 348
D. occipitalis 156
D. ochrosiae 223
D. oleae 39, 95, 104, 124, 131, 157, 169, 176, 186, 190, 207, 223, 242, 243, 256, 258, 278, 280, 284, 285, 297, 308–310, 313, 315–317, 319, 329, 332, 347
D. passiflorae 71, 329
D. psiddi 70
D. scutellaris 329
D. serpentinus 17
D. tau 59
D. tenuifascia 225, 330
D. trivialis 68
D. tryoni 7, 12, 28, 29, 67, 104, 134, 143, 156, 165–167, 222, 224, 232–234, 236, 242–245, 256, 258–261, 280, 284, 285, 296, 297, 309, 310, 313, 315, 325, 326, 329, 331, 332, 334, 335–337, 355
D. tsuneonis 330, 354
D. umbrosus 68, 331
D. vertebratus 223
D. xantodes 71
D. zonatus 59, 60, 103, 157, 221, 222, 329

Rhagoletis

R. alternata 121